Enzyme Synthesis and Degradation in Mammalian Systems

Enzyme Synthesis and Degradation in Mammalian Systems

Edited by: *Miloslav Rechcigl*, Jr., Ph.D.
Chief, Research and Institutional Grants Division, Agency for International Development,
U.S. Department of State, Washington, D.C.

With 73 figures and 23 tables

University Park Press · Baltimore · London · Tokyo   1971

Originally published by S. Karger AG, Basel, Switzerland
Distributed exclusively in the United States of America and Canada by University Park
Press, Baltimore, Maryland

Library of Congress Catalog Card Number LC 73-141818
International Standard Book Number (ISBN) 0-8391-0594-0

S. Karger · Basel · München · Paris · London · New York · Sydney
Arnold-Böcklin-Strasse 25, CH–4000 Basel 11 (Switzerland)

List of Contributors

Alois Čihák, RNDr., Ph.D., Research Worker in Molecular Biology, Institute of Organic Chemistry and Biochemistry, Czechoslovak Academy of Sciences, Prague (Czechoslovakia)

Richard A. Freedland, Ph.D., Professor of Physiological Chemistry, Department of Physiological Sciences, School of Veterinary Medicine, University of California, Davis, Calif. (USA)

Paul J. Fritz, Ph.D., Associate Professor, Department of Pharmacology, The Milton S. Hershey Medical Center, The Pennsylvania State University, College of Medicine, Hershey, Pa. (USA)

Ray W. Fuller, Ph.D., Head, Department of Metabolic Research, The Lilly Research Laboratories, Eli Lilly and Company, Indianapolis, Ind. (USA)

Thomas Gelehrter, M.D., Fellow in Medical Genetics, Division of Medical Genetics, Department of Medicine, University of Washington, Seattle, Wash.; currently: Assistant Professor of Medicine and Pediatrics, Division of Medical Genetics, Yale University School of Medicine, New Haven, Conn. (USA)

Osmo Hänninen, M.D., Ph.D., Associate Professor, Departments of Physiology and Biochemistry, University of Turku, Turku (Finland)

Joel H. Kaplan, Ph.D., Staff Member, General Electric Research and Development Center, Schenectady, N.Y. (USA)

Richard J. Milholland, D.D.S., Senior Cancer Research Scientist, Department of Experimental Therapeutics, Roswell Park Memorial Institute, Buffalo, N.Y. (USA)

Florence Moog, Ph.D., Professor of Biology, Department of Biology, Washington University, St. Louis, Mo. (USA)

Kenneth Paigen, Ph.D., Associate Chief Cancer Research Scientist, Department of Experimental Biology, Roswell Park Memorial Institute, Buffalo, N.Y. (USA)

Henry C. Pitot, M.D., Ph.D., Professor of Oncology and Professor and Chairman, Department of Pathology, McArdle Laboratory, University of Wisconsin Medical School, Madison, Wis. (USA)

Brian Poole, Ph.D., Research Associate, The Rockefeller University, New York, N.Y. (USA)

Miloslav Rechcigl, Jr., Ph.D., Chief, Research and Institutional Grants Division and Nutrition Advisor, Agency for International Development, U.S. Department of State, Washington, D.C. (USA); formerly: Special Assistant for Nutrition and Health, Regional Medical Programs Service, Health Services and Mental Health Administration; and Senior Investigator, Laboratory of Biochemistry, National Cancer Institute, National Institutes of Health, U.S. Department of Health, Education, and Welfare

Fred Rosen, Ph.D., Principal Cancer Research Scientist, Department of Experimental Therapeutics, Roswell Park Memorial Institute, Buffalo, N.Y. (USA)

Bela Szepesi, Ph.D., Research Chemist, Carbohydrate Nutrition Laboratory, Human Nutrition Research Division, Agricultural Research Service, U.S. Department of Agriculture, Beltsville, Md. (USA)

Elliot S. Vesell, M.D., Professor and Chairman, Department of Pharmacology and Professor of Genetics, The Milton S. Hershey Medical Center, The Pennsylvania State University, College of Medicine, Hershey, Pa. (USA)

K. Lemone Yielding, M.D., Chief, Laboratory of Molecular Biology, Professor of Biochemistry and Associate Professor of Medicine, The University of Alabama in Birmingham, Birmingham, Ala. (USA)

Preface

This treatise offers a comprehensive review of the present knowledge of the various factors that control enzyme activity as well as the regulatory mechanisms involved in enzyme synthesis and degradation, with special emphasis on the mammalian systems.

To this end it presents detailed discussions by the leading authorities in the field of such topics as the genetics of enzyme realization, the control of enzyme activity in early development and in old age, the control of enzyme activity by hormonal and nutritional factors, the regulation of enzyme activity through specific modifications in structure, enzyme induction and repression, the translational regulation of enzyme levels, enzyme turnover and the roles of synthesis and degradation in the regulation of enzyme levels, the rhythmic changes in enzyme activity and their control, factors affecting the activity and distribution and synthesis and degradation of isozymes, and synthesis and degradation of enzymes in relation to cellular structure.

The present undertaking represents a pioneer effort in the area which is relatively new but which clearly stands in the forefront of today's biology. As such, the book should be of particular interest and assistance to both students and research workers in a wide range of disciplines including general biologists, biochemists, molecular and cell biologists, enzymologists, comparative and developmental biologists, zoologists, physiologists, pathologists, geneticists, embryologists, endocrinologists, pharmacologists and nutritionists. When reading through it should be kept in mind, however, that our purpose is not just to inform but also/or rather to stimulate the reader, be it a beginning graduate student or an advanced researcher, to explore, to question and, whenever possible, to test various hypotheses advanced herein, always keeping in mind the immortal words of Heraclitus that 'there is nothing permanent except change'.

In order to make the publication more accessible from the point of view of a student or a researcher who are embarking on their first venture into the fascinating field of cellular regulation, the book has been appended with a thorough index and a glossary. For further reading each chapter is accompanied with numerous references.

The Editor wishes to acknowledge with deep appreciation and gratitude the generous cooperation of the individual collaborators which has made this volume a reality. He is also indebted to Dr. *E. Brad Thompson* who has critically read various parts of the manuscript. Mrs. *Eva Rechcigl* and *Jack* and *Karen Rechcigl* rendered indispensable help during the compilation of the author and subject indexes. Acknowledgement is also due to Mrs. *Virginia Waller* for her invaluable assistance in undertaking the many chores incident to publication.

Last but not least, the encouraging and gracious guidance of the Editorial and Production staffs of the Karger Co. is recognized with many thanks.

Miloslav Rechcigl, Jr.

Washington, D.C.
September 1970

Abbreviations and Symbols

ACTH Adrenocorticotrophic hormone
ADH   Alcohol dehydrogenase
AIA   Allylisopropylacetamide
ALA   Aminolevulinic acid
AMD   Actinomycin D
AMP   Adenosine 5′-phosphate
AT   3-Amino-1,2,4-triazole
ATC   Aspartate transcarbamylase
ATP   Adenosine 5′-triphosphate
ATPase Adenosine triphosphatase

cAMP Cyclic adenosine 3′,5′-monophosphate
CoA   Coenzyme A

DEAE O-(diethylaminoethyl) [cellulose]
DNA   Deoxyribonucleic acid

EDTA Ethylenediaminetetraacetic acid

GAT   Glutamic-alanine transaminase
GTT   Glutamic-tyrosine transaminase

HTC   Hepatoma tissue culture

IMP   Inosine 5′-monophosphate

LDH   Lactate dehydrogenase

k_D First order rate constant for degradation
k_S Rate constant for synthesis
k_M Michaelis constant

MAO   Monoamine oxidase
mRNA Messenger RNA

NAD   Nicotinamide-adenine dinucleotide
NADH   Reduced nicotinamide-adenine dinucleotide
NADP   Nicotinamide-adenine dinucleotide phosphate
NADPH   Reduced nicotinamide-adenine dinucleotide phosphate

OMP   Orotidine-5'-monophosphate
OTC   Ornithine transcarbamylase

PEP   Phosphoenolpyruvate
PEPCK   Phosphoenolpyruvate carboxykinase
PPC   Phosphopyruvate carboxylase

RNA   Ribonucleic acid
RNase   Ribonuclease
rRNA   Ribosomal RNA

SDH   Serine dehydratase

$t\frac{1}{2}$ Half-life
TAT   Tyrosine aminotransferase
TCA   Tricarboxylic acid
TPO   Tryptophan oxygenase
tRNA   Transfer RNA

UDP   Uridine 5'-diphosphate

V_{max} Maximal velocity

Contents

I. *Control of Enzyme Activity*

The Genetics of Enzyme Realization
Paigen, Kenneth (Buffalo, N.Y.)

Contents

The Control of Enzyme Activity in Mammals in Early Development and in Old Age
Moog, Florence (St. Louis, Mo.)

Control of Enzyme Activity by Glucocorticoids
Rosen, Fred and Milholland, Richard J. (Buffalo, N.Y.)

Control of Enzyme Activity: Nutritional Factors
Freedland, Richard A. (Davis, Calif.) *and Szepesi, Bela* (Beltsville, Md.)

II.   Regulatory Mechanisms of Enzyme Synthesis and Degradation

Regulation of Protein Activity and Turnover Through Specific Modifications in Structure
Yielding, K. Lemone (Birmingham, Ala.)

Regulatory Mechanisms of Enzyme Synthesis: Enzyme Induction
Gelehrter, Thomas D. (Seattle, Wash.)

Contents

Enzyme Repression
Hänninen, Osmo (Turku)

Translational Regulation of Enzyme Levels in Liver
Pitot, Henry C. (Madison, Wis.); *Kaplan, Joel* (Schenectady, N.Y.) *and Čihák,
Alois* (Prague)

Intracellular Protein Turnover and the Roles of Synthesis and Degradation in
Regulation of Enzyme Levels
Rechcigl, Miloslav, Jr. (Washington, D.C.)

Contents

III. Special Topics

Rhythmic Changes in Enzyme Activity and Their Control
Fuller, Ray W. (Indianapolis, Ind.)

Factors Affecting the Activity, Tissue Distribution, Synthesis and Degradation of Isozymes
Vesell, Elliot S. and Fritz, Paul J. (Hershey, Pa.)

Contents

Synthesis and Degradation of Proteins in Relation to Cellular Structure
Poole, Brian (New York, N.Y.)

I. Control of Enzyme Activity

Enzyme Synthesis and Degradation in Mammalian Systems, pp. 1–46
(Karger, Basel 1971)

The Genetics of Enzyme Realization

Kenneth Paigen

Department of Experimental Biology, Roswell Park Memorial Institute, Buffalo, N.Y.

Contents

I.   Introduction

Enzyme *realization* is the set of processes acting to produce the final phenotype of a protein. Differentiation is largely the summation of many individual realizations, and the study of enzyme realization is, in part, an effort to reduce the study of differentiation to experimentally accessible dimensions.

The important phenotypic characteristics of an enzyme are its properties as a catalyst, its intracellular locations, the regulation of its synthesis and breakdown, and its tissue distribution. Among these characteristics the catalytic properties are largely a consequence of its amino acid sequence, and, with few exceptions, can be changed only by mutation of the structural gene. These include the substrate specificity, catalytic efficiency, physical stability, and susceptibility to regulatory effectors. The other phenotypic characteristics result from an interaction between the enzyme and other macromolecules of the cell. These other macromolecules must recognize the unique structural features which each enzyme possesses. Hence, changes in some phenotypic properties may result from two classes of mutations: those within the enzyme structural gene that change its recognition features and those outside the enzyme structural gene that change the remaining cell machinery.

Mutations affecting the realization of an enzyme that occur outside its structural gene are useful for analyzing molecular mechanisms in enzyme realization. They can also clarify questions about the functional organization of genetic material. For convenience such mutations can be divided into several classes: regulatory, architectural, and temporal (*Paigen and Ganschow, 1965*).

Regulatory genes control enzyme activity, enzyme synthesis, and enzyme breakdown. Mutations affecting the machinery for one of these processes are unlikely to affect the others since they probably do not share common mechanisms. Structural gene mutations which alter the recognition features of a protein may alter some aspects of its regulation.

Architectural genes determine the site of an enzyme within a cell. Subcellular structures are built by a combination of catalytically-mediated steps and self-assembly steps. At the present time genetic analysis provides the only means for identifying the assembly catalysts which integrate enzymes into these structures. Architectural gene mutants can be mimicked by structural gene changes since the amino acid sequence of an enzyme is important both in defining its self-assembly properties and in providing for its recognition by assembly catalysts.

Temporal genes program the system in time, determining the activation of structural, architectural and regulatory genes at different developmental stages in the various cell types. Despite increasing evidence that such genes exist, the mechanism of this temporal programming is unknown.

The regulation and architecture of a protein can vary throughout a multicellular organism. There is a tissue distribution of the accessory proteins which act in the enzyme's realization, as well as a tissue distribution of enzyme molecules themselves, and the final phenotype will be a resultant of both distributions.

A. Historical Development of Concepts

The first report of a genetically determined difference in enzyme phenotype was by *Figge and Strong* (1941). They described a two-fold difference in liver xanthine oxidase activity between the JK and C_3H strains of mice. Soon after *Khanolkar and Chitre* (1942) reported that C_3H and A mouse strains contain twice as much esterase as C57BL. 1941 was also the year in which the initial publication on inherited nutritional requirements in *Neurospora* appeared. In retrospect it is astonishing that neither of the reports of biochemical variants in higher organisms had any effect on the development of biochemical genetics during the next decade. The investigators working with mice were comparing strains with high and low tumor incidences for clues as to the origin of cancer; neither group had set out to show that enzyme differences are heritable, nor did either question the mode of inheritance of an enzyme difference.

During the following years additional strain differences in enzymatic content of various organisms, especially mice, were described. Nevertheless, it was not until 1952 that the first genetic analysis was carried out on a mammalian enzyme variant (*Law et al.*, 1952). Ironically, the mutant studied, low glucuronidase activity (*Morrow et al.*, 1949, 1950), is one of the few enzyme variants that is recessive when heterozygous. Consequently, the results of that analysis fit the genetic theory current at that time and failed to reveal the codominance which characterizes most enzyme variants.

These events can be placed in perspective by recalling that sickle cell anemia was established as a 'molecular' disease in 1949 when *Neel* deduced that this disease was inherited as a Mendelian recessive and that heterozygotes had an intermediate phenotype. Shortly thereafter, *Pauling* and coworkers (1949) found that the hemoglobins of anemic and normal individuals differed in electrophoretic mobility. It was another decade, however, before *Ingram* (1959) succeeded in demonstrating that the abnormality in sickle cell hemoglobin is the substitution of a single amino acid in the β chain of the molecule.

In succeeding years, studies of biochemical genetics in higher organisms established several concepts not included in either earlier genetic theory or the parallel development of microbial genetics.

1. Codominance

Foremost has been the concept of codominance, which states that, in heterozygotes, both alleles of a gene are expressed. If the homozygous parents contain physically distinguishable forms of enzyme, the heterozygote will have both forms; if the parents differ in their levels of enzyme activity, the heterozygote will contain the average of the two parents. This rule has been confirmed by the study of many electrophoretic variants and by compilations of quantitative data in *Childs and Young* (1963) and in *Harris* (1964). Deviations from the rule usually involve genetic factors in enzyme realization outside the structural gene. The major exception to the rule is the *anucleolate* mutation which deletes all the ribosomal RNA genes. Embryos heterozygous for this defect synthesize as much ribosomal RNA as normal embryos with two sets of ribosomal RNA genes (*Gurdon*, 1967).

In general half the usual amount of enzyme is enough to provide for normal function. This accounts for the normal morphological development and appearance of individuals heterozygous for an enzyme defect and for the apparently recessive character of many mutations. The concepts of genetic recessiveness and dominance, which were a significant feature of classical genetic theory, are now seen as a genetically trivial consequence of the workings of intermediary metabolism. True genetic dominance, i.e. the ability of one allele to enhance or suppress the activity of its partner, has yet to be discovered.

The failure of heterozygotes to synthesize a full complement of enzyme activity suggests that higher organisms are not regulated to produce a fixed level of enzyme activity and are unable to sense their own functional capacities (*Paigen and Ganschow*, 1965). This is in marked contrast to the regulatory systems of microorganisms which control the level of enzyme activity in the cell. The low enzyme levels of heterozygotes of higher organisms apparently do not result from random somatic inactivation of one member of a pair of alleles (*Paigen and Ganschow*, 1965; *McClintock*, pers. comm.) analogous to the well known inactivation of one X chromosome in the somatic cells of female mammals. For further discussion of the genetic evidence concerning the nature of mammalian regulatory systems see pp. 165–199 and 216–235.

2. Polymorphism

Natural populations contain variant forms for a surprisingly high percentage of proteins. *Harris* (1966, 1969), surveying human populations for enzyme electrophoretic variants, found that 8 of the 18 enzymes studied were polymorphic. The frequency of heterozygotes for these 8 enzymes ranged between 4 and 50 %. Similar results were reported for wild populations of *Drosophila pseudoobscura* (*Lewontin and Hubby*, 1966). The actual level of heterozygosity is even higher than these estimates since electrophoretic variants represent only a fraction, about one-fourth, of all enzyme variants.

The presence of a large reservoir of genetic variation explains the ease of finding new variants in humans and the variability observed between inbred strains of laboratory animals. For example, a survey of 8,000 Europeans uncovered 10 electrophoretic variants of hemoglobin (*Sick et al.*, 1967), and a survey of 63 inbred mouse lines for variants involving four aspects of β-glucuronidase realization uncovered three mutant phenotypes (*Ganschow and Paigen*, 1968).

High levels of variation in populations subject to natural selection are not accounted for readily by older theories of population genetics (*Lewontin and Hubby*, 1966). Biochemical factors that might contribute to frequent enzyme polymorphism are the selective advantages provided by having two kinds of protein molecules or by having oligomeric proteins with more than one kind of subunit. An example of the advantage that may accrue from having two kinds of protein molecules is the relative resistance to malaria of heterozygotes for sickle cell anemia (*Allison*, 1964) and for glucose-6-phosphate dehydrogenase deficiency (*Luzzatto et al.*, 1969). An example of the advantage of having a protein with two kinds of subunits is maize alcohol dehydrogenase, a dimeric enzyme. Two types of subunits occur. If the dimer has both monomers of one type, the enzyme has high activity but is physically unstable. If both monomers are of the other type, the enzyme has low activity but is physically stable. Enzyme molecules containing one monomer of each type have both high activity and physical stability (*Schwartz and Laughner*, 1969).

3. Gene Duplication

A randomly breeding population cannot, of course, contain more than 50 % heterozygotes. Gene duplication, followed by fixation of different alleles into adjacent positions on the chromosome, would provide the evolutionary solution to a selective advantage of heterozygosity. All members of the population could become functionally 'heterozygous'. Gene duplication, followed by further mutation and genetic drift, probably accounts for those cases in which closely related polypeptide sequences are produced by linked genes, as for example the β and δ chains of hemoglobin.

More surprising has been the discovery of gene duplications involving *like* alleles. Such duplications do not provide any of the selective advantages that functional heterozygosity does. Instead they may provide a means for increasing rates of protein synthesis. The most thoroughly documented case for a protein is the structural gene for hemoglobin γ chains, which probably occurs as four adjacent copies (*Schroeder et al.*, 1968; *Huisman et al.*, 1969). The genes for ribosomal RNA are the extreme example of genetic redundancy (*Ritossa and Spiegelman*, 1965). In some higher organisms, several thousands of copies are present in each haploid genome (for reviews of the data see *Perry*, 1967; *Ritossa et al.*, 1968).

Duplications of either like or unlike alleles may arise by unequal crossing over. When such crossing over occurs in the middle of a gene, it produces a new structural gene altogether as in the *a* peptide of haptoglobin 2FS, an almost end to end fusion of the *a* peptides of haptoglobins lF and lS (*Smithies et al.*, 1962; *Black and Dixon*, 1968).

II. Structural Genes

A. Identification

Mutations affecting realization of an enzyme which lie outside its structural gene are especially interesting in the analysis of the cellular components required for cell assembly and for regulation of protein synthesis and breakdown. However, because so many steps in enzyme realization require interaction between the enzyme and other cellular components, structural gene mutations which alter the recognition features of the enzyme may mimic genetic alterations in the realization machinery. For this reason, it is very important to identify the structural gene of the protein under study. For example, mutation at the *ce* locus in mice changes the level of liver catalase. Is the *ce* locus the structural gene for this enzyme? Such questions are answered by determining whether the mutation has also changed the physical or catalytic properties of the catalase molecule. In this case, the *ce* locus does not determine catalase structure since the properties of the enzyme are not changed by mutation at this locus and since the *ce* gene segregates independently of another presumed structural gene mutation affecting the catalytic efficiency of catalase (*Ganschow and Schimke*, 1969).

Sometimes efforts to identify structural genes are confused by the presence of multiple components catalyzing the same reaction. For example, liver contains both soluble and mitochondrial NADP-dependent malate dehydrogenases. Are these two activities coded for by the same structural gene? Such questions are answered by testing whether a mutation which alters the physical properties of one component simultaneously alters the other. In the case of the malate dehydrogenases, the two enzymes are thought to be coded for by separate structural genes since a mutation changing the electrophoretic mobility of the soluble component had no effect on the other (*Henderson*, 1966). Liver also contains two β-glucuronidase components, one lysosomal and the other microsomal. These are coded for by the same gene since a mutation increasing the thermolability of the lysosomal enzyme has an identical effect on the thermolability of the microsomal component (*Paigen*, 1961a).

Identifying a mutation as having occurred in a structural gene requires detecting a change in some property of a protein that depends upon its amino acid sequence. The properties most often examined have been susceptibility to physical denaturation, electrophoretic mobility, kinetic constants, immunological specificity, and the peptide sequence. Applying the first three of these techniques requires that the mutant retain a measurable level of activity. A difference in properties between the mutant and wild type activities suggests a structural gene change provided the residual activity is not another enzyme altogether. The residual activity is unlikely to come from another enzyme if the mutant and wild type activities have at least some properties in common (antigenicity, electrophoretic mobility, or substrate specificity). However, this assumption should not be made without supporting evidence since even relatively specific enzymes may present problems. For example, guinea pig strains differ in their ability to catalyze the hydroxylation of cortisol. Adrenal homogenates of high activity strains are about eight times more active. A kinetic analysis of the reaction showed that the high activity strain contains two enzymatic activities, one with a slow maximum velocity but a low Km and hence able to react at low substrate concentrations, and another with a fast maximum velocity but a 30 times higher Km and hence able to react only at much higher substrate concentrations. In the other strain, the faster enzyme with a high Km is missing, and only the slower component is left (*Burstein et al.*, 1967). Without this information, comparing these two strains for another physical parameter of enzyme activity would probably suggest a structural gene mutation. With this information we understand that the faster enzyme is completely missing. However, it is not possible to decide whether this loss resulted from a mutated structural gene producing a totally inactive gene product, or a mutation in another gene affecting the realization of this enzyme.

For some reactions, as the hydrolysis of esters, many enzymatic forms are present with overlapping substrate specificities, each coded for by a different structural gene. In addition many enzymes have a complex subunit interaction with multiple isozymes present. These factors further complicate the application of physical techniques to the identification of structural gene variants. (The problems of isozymes will not be considered in this review since a recent symposium [*Vesell*, 1968], a review [*Markert and Whitt*, 1968], and a chapter in this volume [p. 339] all provide excellent summaries of the subject.)

In evaluating methods for detecting structural changes in enzymes, the most appropriate criterion is the probability that a random substitution in the amino acid sequence of the protein will change the parameter measured (provided of course that the change does not destroy enzyme activity altogether).

1. Physical Denaturation

The most sensitive and accurate test for changes in protein structure is relative thermolability. The probability that a random amino acid substitution produces a protein with altered thermolability is more than 50 %. This probability is higher than for any other test, and the necessary measurements can generally be made with considerable accuracy in dilute solutions of crude enzyme.

This estimate is for the probability that a *random* amino acid change will affect protein stability. However, the raw material for studies of biochemical genetics are mutants with an obvious change in enzyme realization. Chemical reasoning suggests that if minor changes in structures are likely to affect thermolability, then structural changes sufficient to alter the catalytic properties or recognition features of a protein are even more likely to do so.

The sensitivity of thermal inactivation as a test for structural changes is illustrated by findings on carboxypeptidase A. Two forms of this enzyme differ only in the antepenultimate C-terminal amino acid residue. The form with leucine in this position is twice as labile as the form with valine (*Walsh et al.,* 1966).

The data of *Langridge* (1968a) provide one estimate of the probability that a random amino acid substitution would change the thermolability of a protein. A set of mutant β-galactosidases differing by a serine residue in place of the original amino acid at 49 different sites was constructed by combining a serine suppressor with each of a series of β-galactosidase amber mutants. The suppressor permits the insertion of serine at the position of the amber nonsense mutation. Among the resulting protein sequences, approximately 70 % had increased thermolability. The relative increase in thermolability effected by a serine replacement showed periodic oscillation along the peptide chain (fig. 1). This result is understandable if we imagine a protein structure similar to hemoglobin with successive regions of α-helical configuration folded across each other. Non-polar bonding at the 'crossover' points is quite important in holding the structure together. In these points the replacement of a non-polar by a polar amino acid would be expected to have a drastic effect. Since points of helical contact occur periodically along the chain, so will the effect of serine replacement show periodic oscillation.

A similar probability was obtained for hemoglobin. *Lehmann and Carrell* (1969), who surveyed known hemoglobin mutants, estimated that mutation to a polar amino acid would have a drastic effect on protein conformation for about 50 % of the hemoglobin residues.

In addition to changes in hydrophobic bonding, changes in ionic bonding, hydrogen bonding, solvation, or disulfide linkages also affect the resistance of a

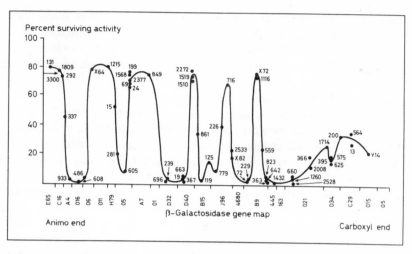

Fig. 1. The per cent enzyme activity remaining after heating β-galactosidase from serine-suppressed amber mutants for 10 min at 57 °C. The mutants are arranged according to their map position in the gene for β-galactosidase. The heat stability of the wild-type enzyme (3300) is marked by an arrow at the beginning of the gene. Adapted from *Langridge* (1968a).

protein to thermal denaturation. (The last change is probably too drastic to permit survival of activity.)

Thermal lability provides a sensitive indicator of protein structural changes because even slight changes in bonding energy affect appreciably the rate of thermal inactivation. Thermal inactivation is often a first order reaction, and the rate of inactivation typically doubles with a one degree rise in temperature. This sharp temperature dependence reflects the large energy of activation for protein denaturation (200–300 Kcal) compared with much lower energies of activation of other biochemical reactions (10–20 Kcal). Despite such large activation energies, doubling the rate of inactivation would require only a decrease of 500 cal in the activation energy. Such a change could easily be produced by an amino acid substitution affecting helix stability or non-polar bonding.

The temperature required to inactivate 50 % of enzyme molecules in a fixed period of time provides a simple relative measure of thermal lability. For most proteins, these temperatures fall into the range 40–80 °C. This temperature is quite sensitive to even small changes in the activation energy of the denaturation reaction so that even the same enzyme from different species may vary widely in thermal lability. (For an interesting survey of mammalian catalases see *Feinstein et al.,* 1967b.)

The application of thermal lability as a test for structural changes is limited to circumstances where other, unrelated, enzymes do not interfere with the assay. Denaturation by urea or extremes of pH and relative susceptibility to proteolytic enzymes have also been used as indicators of protein structure. However, I am not aware of any case in which a difference in sensitivity to one of these agents cannot also be detected as a difference in sensitivity to thermal inactivation.

2. Electrophoretic Mobility

The electrophoretic mobility of an enzyme activity is determined readily by acrylamide or starch gel electrophoresis if the enzymatic reaction can be converted into a suitable staining procedure. The presence of other enzymes with similar catalytic activity generally does not interfere. However, electrophoresis is less sensitive than thermal inactivation in detecting structural changes. The extraordinary variety of electrophoretic variants described (over 100 are known for hemoglobin) reflects the ease of screening rather than the frequency of occurrence for this type of variant. Of the 2,217 possible amino acid substitutions that could occur in hemoglobin by changing one base in the DNA, only 700 would result in a charge alteration (*Sick et al.,* 1967). However, changes in charge are forbidden for about one-quarter of the hemoglobin molecule if a recognizable protein is to be formed (*Lehmann and Carrell,* 1969). These figures suggest that about 25 % of the possible amino acid substitutions can be detected as electrophoretic variants.

3. Kinetic Constants

A study of mutant forms of β-galactosidase is the only source of which I am aware of comparative data on the effect of mutation on the kinetic constants of an enzyme (*Langridge,* 1968b). Among several hundred mutants selected for inability to grow on lactose, most had some small but measurable level of activity. However, only 4–6 % showed any change in substrate affinity. This suggests that most amino acid substitutions are likely to affect the maximum velocity of an enzyme but only rarely affect its affinity for substrates or inhibitors. In practice, this means that a kinetic analysis alone is unlikely to identify structural gene mutants. To determine whether a mutant with less activity has a structural gene mutation reducing the maximum velocity of the enzyme or a regulatory gene mutation affecting the number of enzyme molecules present, it is necessary to determine the specific activity of enzyme molecules. Specific activity is determined either by isolating the enzyme or by immunological titration. Both procedures were used to demonstrate a catalase mutant with reduced catalytic efficiency (*Ganschow and Schimke,* 1968, 1969).

4. Immunological Variation

Despite the legendary ability of antisera to make subtle structural distinctions, this method has not been widely used in the identification of structural gene variants, partly because purification of the protein is required for antibody preparation. Three techniques have been used. Antibody titration curves estimate the catalytic efficiency of individual enzyme molecules but are a poor indicator of changes in antigenic specificity. The Ouchterlony immunodiffusion technique, a qualitative indicator of antigenic identity, only signals the complete loss of an antigenic determinant. Complement fixation may provide some measure of the relative affinity of mutant and wild-type proteins for antibody molecules although the mechanism is unknown.

The relative insensitivity of immunodiffusion procedures in detecting structural changes is illustrated by carbonic anhydrase II proteins from a number of species (*Tashian et al.,* 1968). Despite appreciable differences in amino acid composition, enzymes from humans, pig-tailed macaques, and green monkeys show complete immunological identity in the Ouchterlony technique if tested against antiserum to human enzyme. *Sengbusch* (1965) and *Van Regenmortel* (1967), studying tobacco mosaic virus coat protein, found that amino acid substitutions at 7 out of 19 positions resulted in immunologically distinguishable variants.

Several comparative studies indicate that complement fixation is the most sensitive immunological technique available for detecting structural differences. *Arnheim and Wilson* (1967) were able to demonstrate differences in the lysozymes of birds by microcomplement fixation but not by immunodiffusion. For example the lysozymes of chickens and pheasants are easily distinguished by complement fixation although they show complete identity in the Ouchterlony technique. There is conflicting data on the ability of microcomplement procedures to detect single amino acid changes. *Reichlin et al.* (1964, 1966) could distinguish between hemoglobins A, S, and C, which differ by single amino acid substitutions, if microcomplement fixation was used but not if a quantitative precipitin test was used. Similarly, *Cocks and Wilson* (1969) were able to distinguish by complement fixation 7 out of 8 amino acid substitutions which altered the net charge of bacterial alkaline phosphatase. However, *Murphy and Mills* (1968) failed to distinguish any of 26 mutant forms of *E. Coli* tryptophan synthetase *a* subunit despite the fact that many carried an altered charge.

In summary, microcomplement fixation is the most sensitive immunological technique available for detecting structural changes and may be as sensitive as thermal inactivation. However, the comparative data on this technique were obtained using electrophoretic variants with altered net charge. When variants without charge differences were compared in the tryptophan synthetase series, no differences in structure were detected.

5. Peptide Mapping

The direct demonstration of an amino acid change is, of course, the ultimate proof of genetic alteration. In practice, this achievement comes long after the structural gene has been identified by other means since the technique requires appreciable amounts of purified material and is laborious. The less laborious technique of fingerprinting is no more sensitive than thermal denaturation or electrophoresis. A change in peptide fingerprint requires a peptide with different charge or polarity, and this type of change is likely to be detected by the simpler techniques of thermal denaturation and electrophoresis. Fingerprinting has, however, the important advantage that its result is unambiguous.

6. Secondary Changes in Protein Structure

As the preceding data indicate, a change in amino acid sequence usually results in a detectable change in physical properties of the enzyme. However, changes in the properties of a protein can result from other than structural gene mutations. Once a polypeptide chain has been assembled, it is still subject to secondary alterations by factors which themselves are presumably under genetic control. Hemoglobin A, for example, is subject *in vivo* to mixed disulfide formation with glutathione (*Muller*, 1961), acetylation (*Schroeder et al.*, 1962), Schiff's base formation of the N-terminal amino group (*Holmquist and Schroeder*, 1966), and removal of C-terminal arginine by carboxypeptidase (*Marti et al.*, 1967). Alterations reported for other proteins include variable loss of amide groups, phosphorylation, methylation, conjugation with AMP or carbohydrate moieties, polymerization, and establishment of stable alternate conformational states.

With this array of possibilities, the same polypeptide sequence may have somewhat different properties if enzyme preparations from different tissues are compared. For example, the N form of the E_l esterase of maize differs in urea sensitivity depending upon whether it is extracted from seedlings or kernels (*Schwartz*, 1964; *Endo and Schwartz*, 1966). The effect is specific to the N form since the S and F forms, coded for by other alleles at this locus, do not show a similar difference. Differences can also be discerned in the same enzyme obtained from different sites within the same cell. For example, hepatic β-glucuronidase activities present in lysosomes and microsomes are distinguishable electrophoretically even though they are coded for by the same structural gene (*Ganschow*, pers. comm.).

The properties of an enzyme even may depend upon the sex of the animal synthesizing it (*Komma*, 1968). Male and female *Drosophila*, carrying the same structural allele for glucose-6-phosphate dehydrogenase, produce enzymes with different thermolability, electrophoretic mobility, and substrate affinity. In stocks carrying the autosomal gene *transformer*, XX flies, which are genetic females, become morphological males. These pseudo males produce a glucose-6-phosphate dehydrogenase with properties intermediate between those of normal males and females.

From these examples, it is obvious that some caution should be exercised in assigning a mutation changing the physical properties of an enzyme to the structural gene for that enzyme.

B. Properties of Structural Gene Mutants

The following examples illustrate some of the complexities that may arise in the expression of structural gene mutations.

1. Catalase

Catalase is a tetramer with a molecular weight of 250,000 whose four sub-units are probably identical (*Schroeder et al.,* 1964). Human acatalasemic variants were reported in Japan by *Takahara* (1952), in Switzerland by *Aebi et al.* (1961), and in Israel by *Szeinberg et al.* (1963). The Oriental and Swiss forms are probably not identical mutants since Oriental cases have oral gangrene and Swiss ones are asymptomatic. In mice a series of 5 allelic mutations causing low levels of blood catalase have been reported (*Feinstein et al.,* 1966). In addition there are reports of acatalasemic mutants in guinea pigs (*Rader,* 1960), dogs (*Allison et al.,* 1957) and chickens (*Bather et al.,* 1963). Ducks as a species have only traces of serum catalase (*Nakamura et al.,* 1952; *Feinstein et al.,* 1968). The human and guinea pig mutants exhibit very low tissue catalase levels while mouse and dog mutants, as well as ducks, have appreciable amounts of catalase in solid tissues.

The mouse mutants are almost certainly structural gene variants. The resi-dual enzyme present in homozygous mutants is more readily inactivated by exposure to heat, acid, alkali, trypsin, urea, and guanidine than is wild type enzyme (*Feinstein et al.,* 1967a). The catalase gene has been located in linkage group V of the mouse, and the normal allele designated Cs^a (*Dickerman et al.,* 1968). Blood catalase levels were 2.4 % of normal for the Cs^b mutant and ranged between 11 and 20 % of normal for the Cs^c, Cs^d, Cs^e, and Cs^f variants. For each mutant, residual levels of enzyme varied widely from one tissue to another with a range of 35–80 % for the Cs^{e-f} variants and 6–30 % for Cs^b. Thus a change in

Table I. Glucose-6-phosphate dehydrogenase content as a function of erythrocyte age in normal and A- individuals

| | Erythrocyte content | | | |
| | Glucose-6-phosphate dehydrogenase | | 6-Phosphogluconate dehydrogenase | |
	Normal	A-	Normal	A-
All erythrocytes	6.6	0.93	3.8	4.0
Young cells	17	13	8.7	8.1
Old cells	4.0	0.60	3.1	3.7

From *Yoshida et al.* (1967).

the catalase structural gene has a different quantitative effect in each tissue. Although some of the reduction in tissue catalase levels may be due to a change in the catalytic efficiency of the catalase molecules produced, it also must reflect tissue specific differences in rates of catalase synthesis and breakdown.

The low levels of catalase in blood as compared to other tissues probably reflect the operation of a unique physiological factor in the realization of erythrocyte proteins. Whereas all other cell types undergo turnover and continually replace their protein complement, red blood cells, being incapable of protein synthesis, do not. Furthermore, individual erythrocytes survive for many weeks, a relatively long time compared with rates of protein turnover in other cell types. For these reasons physical aging or denaturation of protein molecules becomes especially significant in red cells, and the enzyme content of older cells may be quite different from that of young ones. Thus, mutations which slightly increase the physical fragility of proteins may be mild in their effects on enzyme levels in solid tissues but may be severe in their effects in red cells.

Aebi and co-workers have concluded that this is precisely what happens in human catalase mutants. Although earlier work had suggested that Swiss acatalasemics do not make a structurally altered catalase (*Aebi et al.,* 1964), recent evidence from the same laboratory has shown that the small amount of residual enzyme in mutants is distinguishable from wild type and is primarily in reticulocytes and juvenile erythrocytes (*Aebi and Cantz,* 1966; *Matsubara et al.,* 1967). In acatalasemic mutants, erythrocyte catalase content declines rapidly as the cells age, in contrast to normal individuals whose older erythrocytes retain enzyme.

An excellent confirmation of the importance of the red cell aging effect comes from work on the A— variant of glucose-6-phosphate dehydrogenase (*Yoshida et al.,* 1967). Homozygous carriers of this defect have about 10 % of normal enzyme levels in erythrocytes. In both normal and A— individuals older cells contain less enzyme, and this loss of enzyme is much more rapid in A— than in A+ individuals (table I).

Thus the complex expression of acatalasemic and hypocatalasemic mutants in both mouse and man can be explained in terms of the secondary effects produced by structural gene changes, and it is no longer necessary to postulate that human acatalasemics carry an altered regulatory system for catalase biosynthesis. However, although changes in catalase realization can be related to structural gene mutations, the manner in which rates of catalase synthesis and degradation depend upon its protein structure is still not understood.

2. *Malate Dehydrogenase*

The malate dehydrogenase of *Neurospora* illustrates the manner in which structural gene changes can affect the catalytic properties of an enzyme by altering its association within a cell organelle (*Munkres and Woodward,* 1966). A

set of mutants mapping in the structural gene for mitochondrial malate dehydrogenase required malate for growth suggesting that they are deficient in this
enzyme. The malate dehydrogenase activity in mitochondria obtained from
mutants had an abnormally high Km for malate; however, if the mutant enzyme
was isolated from mitochondria, its Km was similar to that of wild type. If
mutant malate dehydrogenase was complexed with purified mitochondrial structural protein, its Km was increased. The Km for malate of wild type enzyme was
substantially the same whether the enzyme was assayed in intact mitochondria,
complexed to mitochondrial structural protein, or free in solution. The increased
Km for malate in mutant mitochondria effectively reduced the activity of
mutant enzyme at physiological malate concentrations to a level below that
required for normal growth.

3. β-Glucuronidase

The importance of enzyme structure in determining intracellular location is
illustrated by β-glucuronidase (*Paigen,* 1961a; *Paigen and Ganschow,* 1965). The
livers of wild type mice have enzyme in both lysosomes and endoplasmic reticulum (microsomes). The enzyme at both sites is coded for by the same structural gene. A mutant carrying a structural alteration has glucuronidase at both
sites but in an altered proportion (fig. 2). In mutants the level of lysosomal
enzyme is always greater than the level of microsomal enzyme by a fixed
amount (algebraically, lysosome − constant = microsome). In wild type animals
equal levels of activity are present at the two sites at the lowest concentrations
of enzyme encountered; additional enzyme is partitioned 3/4 to lysosomes and
1/4 to microsomes (algebraically, lysosome + constant = 3 x microsome). Thus,
in both mutant and wild type, the ratio of activity at the two sites is a function
of the total amount of enzyme made; however, the nature of this function
differs for the two structural forms of the enzyme. Heterozygotes behave as
though each structural form of the enzyme located itself independently of the
other, and in crosses the distribution patterns segregated with the structural alleles.

III. Regulatory Genes

A. Criteria for Identification

The criteria for identifying a regulatory gene mutation controlling enzyme
concentration are (1) the mutation must alter the rate of enzyme synthesis or
breakdown, not merely the measurable level of activity, and (2) the mutation
must be at a genetic locus separate from the structural gene for the enzyme.
Among regulatory mutants affecting enzyme synthesis, none completely meets
these criteria. Among regulatory mutants affecting enzyme breakdown, a mu-

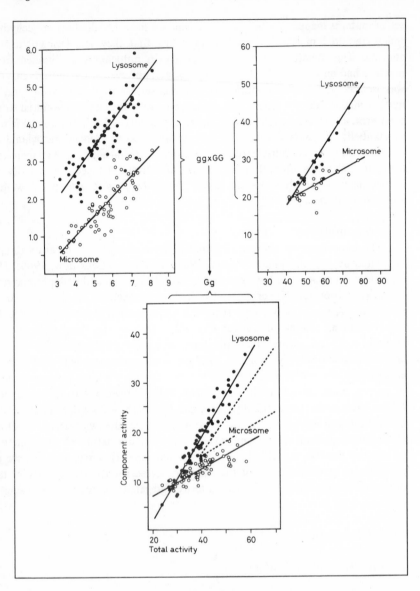

Fig. 2. The lysosomal and microsomal β-glucuronidase activities of homozygous wild type *(GG)*, homozygous mutant *(gg)* and F_1 *(Gg)* animals. The total enzyme present in each component derived from 1 g of liver is plotted as a function of the total amount of enzyme present. The two component activities were separated by an osmotic shock technique. Curves were fitted by least squares. In the *Gg* figure, the curves obtained for GG animals are shown as dashed lines to indicate the nature of the locational shift in heterozygotes. After *Paigen and Ganschow* (1965).

tation specifically affecting hepatic catalase degradation has been identified which is distinct from the catalase structural gene.

Similar criteria apply to the identification of a regulatory gene controlling enzyme activity. The gene must determine the concentration or properties of an agent, separable from the enzyme, capable of controlling enzyme activity *in vitro*. The only mutation meeting these criteria is the *dilute* mutation in mice which determines the presence of a phenylalanine hydroxylase inhibitor.

Because structural gene mutations often produce phenotypic effects similar to those expected for regulatory gene mutants, the problem of distinguishing between regulatory and structural gene changes is very difficult. The 'ideal' structural gene mutant is inherited as a single factor, is codominant so that enzyme content in heterozygotes is the arithmetic mean of the two homozygotes, and is expressed equally in all tissues at all times. There has been a regrettable tendency to label as regulatory any mutant which deviates from this norm. However many structural gene mutations have secondary effects on enzyme concentration and activity, and these are expressed to varying degrees in different tissues.

B.   Regulation of Enzyme Synthesis

None of the postulated regulatory mutants has a negative control analogous to the repressor system of bacteria. Mutations that inactivate repressor function are expected to be recessive if heterozygous and to produce high enzyme levels if homozygous (whether the repressor limits the synthesis, stability, or efficiency of the messenger). The *ce* mutation affecting catalase is the only mutation with increased enzyme levels in homozygotes, and this mutant determines the rate of catalase breakdown, not the rate of catalase synthesis.

It is possible to invent special reasons why such mutants have not been observed; they are lethal for unknown reasons, their phenotype has been predicted incorrectly, etc. However, contemporary genetic experience suggests that negative control of the repressor type is not the common mode of regulation in eukaryotic cells. Rather, the properties of most postulated regulatory mutants suggest that regulatory control is positive and that regulatory gene products are required for enzyme synthesis to proceed. Even in bacteria, positive control systems are known for arabinose (*Englesberg et al.*, 1965) and possibly rhamnose (*Power*, 1967).

1.   δ-Aminolevulinate Dehydratase

Strains of mice differ in their δ-aminolevulinate dehydratase activity. The difference is determined by a single gene. Originally two alleles were described, with Lv^b mice having one-third as much enzyme as Lv^a animals. Heterozygotes have intermediate activities (*Russell and Coleman,* 1963). Recently a third allele, Lv^c, with intermediate levels in homozygotes has been reported (*Hutton and Coleman,* 1969). *Coleman* (1966) isolated the enzyme from homozygous Lv^a and Lv^b mice. The two enzyme preparations showed identical specific activities, sedimentation and diffusion coefficients, electrophoretic mobilities, pH-activity curves, affinities for δ-aminolevulinate, activation energies as derived from Arrhenius plots, and rates of thermal inactivation at 65, 75, and 80°C. Since the two proteins were indistinguishable by all criteria, *Coleman* concluded that the Lv locus might be a regulatory gene although he was puzzled by the intermediate levels of activity in heterozygous animals. *Doyle and Schimke* (1969) confirmed *Coleman's* results and also showed that the two enzyme preparations were indistinguishable with respect to sensitivity to trypsin, inhibition by hematin, and antigenic specificity as determined by immunodiffusion. The difference in enzyme activity between Lv^a and Lv^b mice was shown by immunoprecipitation to be associated with an equivalent difference in the amount of enzyme present. The two strains were compared for their rates of enzyme synthesis and breakdown to determine which had changed. Rates of enzyme synthesis were estimated from the radioactivity of immunologically precipitated enzyme immediately after a brief exposure to labelled amino acid, and rates of breakdown were estimated by following the disappearance of previously incorporated radioactivity. Both strains proved to have similar rates of enzyme breakdown, but Lv^b mice had a slower rate of enzyme synthesis. Heterozygotes had intermediate rates of synthesis along with intermediate levels of activity.

The relationship between fetal and adult δ-aminolevulinate dehydratase is confusing. δ-Aminolevulinate dehydratase activity decreases around the time of birth and increases again later. Both fetal and adult tissues show the difference between Lv^a and Lv^b mice (*Russell and Coleman,* 1963). However, fetal enzyme is more heat labile, less susceptible to trypsin, and catalytically twice as active per unit weight of enzyme antigen as adult enzyme. Nevertheless fetal enzyme does have the same sedimentation coefficient, electrophoretic mobility, and *Km* as adult enzyme.

As *Doyle and Schimke* have pointed out, three possibilities can account for the data on δ-aminolevulinate dehydratase. First, Lv^b may represent a structural gene change, too subtle to be detected, which imposes a limitation on the rate of enzyme synthesis. Second, the Lv locus may specify another protein which functions in the regulation of δ-aminolevulinate dehydratase synthesis. Third, the Lv^a allele may contain multiple copies of the structural gene for this enzyme.

The last hypothesis most completely accounts for the available information by predicting the identity of Lv^a and Lv^b enzyme, the intermediate levels of heterozygotes, and the constant ratio of activities in Lv^a and Lv^b mice in all tissue sites at all ages. It also provides a reasonable explanation for the recent finding of the Lv^c allele with an activity intermediate between Lv^a and Lv^b. The activity of the three alleles is in the ratio 3:2:1, which could be the number of structural gene copies that each carries. However, this hypothesis requires an additional assumption to explain the variant properties of fetal δ-aminolevulinate dehydratase, either that fetal enzyme undergoes a secondary structural change or that adult enzyme is complexed with an inhibitor. An independent identification of the structural gene for this enzyme will help decide among the various possibilities.

2. Orotic Aciduria

Orotic aciduria is a rare recessive autosomal defect characterized by a requirement for pyrimidine nucleosides in the diet and by excretion of large amounts of orotic acid in the urine. Affected individuals have very low levels of the last two enzymes in the pyrimidine biosynthetic pathway, orotidine-5'-monophosphate (OMP) pyrophosphorylase and OMP decarboxylase, and normal levels of the first two enzymes (*Smith et al.*, 1961) (see fig. 3). Heterozygotes have only 20–25 % of normal activity and also excrete orotic acid (*Smith et al.*, 1961; *Fallon et al.*, 1964). These findings have been confirmed in tissue cultures derived from normal, heterozygous and homozygous mutant individuals

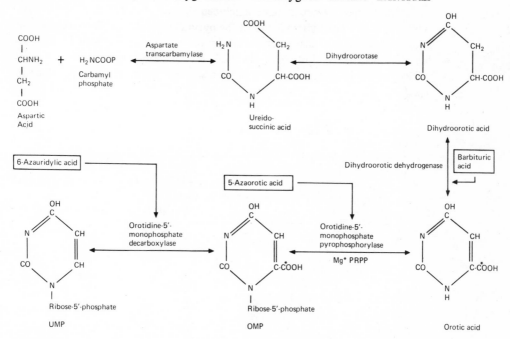

Fig. 3. The *de novo* pathway of uridine-5'-monophosphate biosynthesis, including the presumed sites of action of various enzyme inhibitors. Adapted from *Pinsky and Krooth* (1967b).

(*Krooth,* 1964; *Howell et al.,* 1967). Mutant cultures contain normal amounts of the third enzyme of this five enzyme pathway (*Wuu and Krooth,* 1968).

The pleiotropic effect of one mutation on two metabolically-related enzymes and the unusually low activity of these enzymes in heterozygotes suggest a regulatory defect. *Krooth* and co-workers (*Krooth,* 1964, 1969; *Pinsky and Krooth,* 1967a, 1967b; *Wuu and Krooth,* 1968) extensively analyzed the regulation of these enzymes in tissue cultures derived from mutant *(rr),* heterozygous *(Rr)* and normal *(RR)* individuals and observed the following: (a) Levels of OMP decarboxylase and OMP pyrophosphorylase, but not the first three enzymes of pyrimidine biosynthesis, were markedly elevated in mutant cells growing in the presence of inhibitors of the third, fourth and fifth enzymes of the pathway. The respective inhibitors were barbituric acid, 6-aza-OMP and 6-aza-UMP. Some elevation also occurred in heterozygous and normal cells. (b) The elevation occurred in the presence of cytidine, an end product of the pathway. (c) The levels of OMP decarboxylase and OMP pyrophosphorylase were depressed if inhibitors of the first two reactions were added, such as adenosine or 3-methyl-aspartate. (d) Adding the product of the second reaction, dihydroorotate, elevated enzyme synthesis. (See tables II and III for data.)

Krooth and co-workers have concluded that dihydroorotate stimulates the synthesis of the last two enzymes of the pyrimidine pathway. They suggest that inhibitors of the first two reactions act by preventing the synthesis of dihydroorotate, and that inhibitors of the last three reactions act by preventing its utilization. Cytidine acts independently to suppress the synthesis of the last two enzymes.

Table II. Effect of culture in 5-azaorotate (5-AzOr) on enzyme activities of various cell genotypes

Genotype	Enzyme activity		
	5-AzOr present	OMP decarboxylase	OMP pyrophosphorylase
RR	–	10.6	3.65
	+	12.8	12.3
Rr	–	3.9	1.40
	+	5.7	3.88
rr	–	0.2	0.013
	+	3.7	3.57

From *Pinsky and Krooth* (1967b), enzyme activities are in μmoles/h/mg protein. When present the concentration of 5-AzOr was $6 \times 10^{-5} M$.

Table III. Effect of medium supplements on OMP decarboxylase activity of mutant (rr) cells

Medium supplement	OMP decarboxylase
None	0.032
Cytidine	0.013
3-Methylaspartate	0.011
Cytidine and 3-methylaspartate	0.001
Cytidine, 3-methylaspartate and dihydroorotate	0.023

From *Krooth* (1969), enzyme activities are in μmoles/h/mg protein. Concentrations were $1.2 \times 10^{-4} M$ cytidine, $10^{-2} M$ 3-methylaspartate, and $1.2 \times 10^{-3} M$ dihydroorotate.

The key question which remains is the nature of the lesion in orotic aciduria. Is the regulatory apparatus which responds to dihydroorotate altered in its sensitivity, or do mutant cells have a low concentration of dihydroorotate because some other protein is defective? Usually heterozygotes for other enzyme defects have normal concentrations of intermediary metabolites. On this basis, if the defect in orotic aciduria affected dihydroorotate metabolism, heterozygotes would be expected to have normal enzyme levels. Instead they have only 20–25 % of normal, suggesting that a defect in the regulatory machinery is more likely. A firm decision between the two possibilities could be reached by simultaneous measurements of enzyme synthesis and dihydroorotate concentrations in mutant and wild-type cells growing in different media.

It would also be of interest to know whether the enzyme activities induced in mutant cells are structurally identical (i.e. have similar physical properties) to those present in wild-type cells. An answer to this would help to decide whether the mutation in orotic aciduria occurs within or overlaps any of the structural genes for the enzymes in this pathway.

3. Lactate Dehydrogenase

Lactate dehydrogenase contains two types of subunits which associate in various combinations into a tetrameric enzyme. Evidence for a regulatory function in controlling the B subunit has been presented for a gene, *Ldr-1*, in the mouse (*Shows and Ruddle, 1968*). Animals homozygous for one allele of the *Ldr-1* gene have B subunits in their erythrocyte lactate dehydrogenase and in the lactate dehydrogenases of other tissues. Animals homozygous for the other allele of the *Ldr-1* gene have B subunits in other tissues but lack B subunits and have only A subunits in their erythrocyte lactate dehydrogenase. Heterozygotes have intermediate levels of B subunits in erythrocytes. *Shows and Ruddle* suggest a regulatory function for the *Ldr-1* gene because of the tissue specificity of expression. An alternative explanation, analogous to that already discussed for the erythrocyte deficiencies of glucose-6-phosphate dehydrogenase and catalase, is that one *Ldr-1* allele codes for a less stable form of the B subunit that cannot survive in erythrocytes. Since the structural gene of the B subunit has not been identified, tests for linkage between it and the *Ldr-1* gene are not possible. The physical properties of the B subunit in the two genetic types have not been examined.

Differences in stability of variant forms of the lactate dehydrogenase B subunit also would help to account for species variation in erythrocyte content of the B subunit (*Shows et al.,* 1969).

4. *Other Mutations*

Several other mutations may involve regulatory genes but information is fragmentary. One of these is the recessive glucose-6-phosphatase deficiency which accompanies small deletions covering the region of the albino *(c)* locus in mice (*Erickson et al.,* 1968). If these represented loss of the glucose-6-phosphatase structural gene, heterozygotes would have half normal activity. The fact that homozygous mutants have no activity suggests that the deleted gene performs a positive function in glucose-6-phosphatase synthesis; the fact that heterozygotes have full activity suggests that one gene copy is sufficient to provide for maximum enzyme synthesis.

The alkaline phosphatase activity of mouse intestine is determined by a polygenic system (*Nayudu and Moog,* 1966, 1967). There are four enzyme components with varying ratios of activity against phenylphosphate and β-glycerophosphate (*Moog et al.,* 1966). The ratio of activities against the two substrates varies along the length of the intestine from duodenum to jejunum, and in any one region, from the base to the tip of the villi (*Moog and Grey,* 1967). The system has been extensively analyzed for its developmental features reviewed elsewhere in this volume (pp. 47–76).

The genetic control of nitrate reductase levels in corn seedlings has been studied by *Warner et al.* (1969). If two inbred lines with equal enzyme activities were crossed, the F_1 progeny had nearly double the activity of either parent. *Warner* and co-workers suggest the existence of two loci controlling enzyme levels with high activity dominant over low for both. If each parent strain carried alleles for high activity at one locus and alleles for low activity at the other in a complementary arrangement, the F_1 progeny would be heterozygous at both loci. However, interpretation of the results is confused by several additional factors. The two parent lines differ both in the structural gene for nitrate reductase (one form of enzyme is more thermolabile than the other) and in the rate of enzyme turnover. Moreover, some form of subunit interaction occurs in the F_1 hybrid since the thermolability and turnover properties of this protein appear to be intermediate to those of the parents.

C. Regulation of Enzyme Breakdown

1. Catalase

While studying the decline of liver catalase in tumor-bearing animals, *Greenstein and Andervont* (1945) noted that catalase activity of C57BL mice is only half that of other strains (C3H, A, C, I, Y). Strain C57 originated in 1921 from the Lathrop breeding colony at Rockefeller. A single male (No. 52) was mated with two of his female littermates, numbers 57 and 58. The progeny of the first mating, bred by *Little*, gave rise to the various C57 strains now extant. The progeny of the second mating, bred by *MacDowell*, gave rise to the C58 family. All mice descended from these two matings have low liver catalase, except for

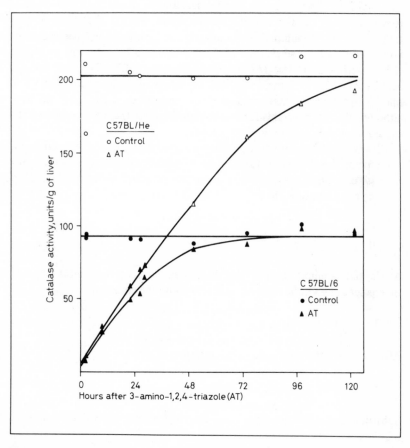

Fig. 4. Liver catalase activity in two substrains of C57BL mice after intraperitoneal injection of 3-amino-1,2,4-triazole (AT). Adapted from *Rechcigl and Heston* (1967).

one subline which was separated from the other C57 lines in the early 1940's (*Rechcigl and Heston,* 1963). C57BL/An and stocks derived from it, such as C57BL/He and C57BL/Ha, show normal levels of liver catalase. *Rechcigl and Heston* assumed, reasonably enough, that in the C57BL/An line the mutated locus had reverted to wild type. However, further work has shown that in reality mice of the C57BL/An line are double mutants, and their superficially normal phenotype results from the interaction of two mutant loci. The first mutation presumably was present in the Lathrop stock from which C57 and C58 arose, and the second mutation probably occurred shortly after the separation of the C57BL/An line. By chance the second mutation was fixed in that small breeding population and now appears in all descendents of this stock.

The second mutation, responsible for the difference in activity of various C57 sublines, affects the rate of catalase turnover (*Rechcigl and Heston,* 1967; *Rechcigl,* 1968; *Ganschow and Schimke,* 1968, 1969). The effect of the mutation was first established by examining the kinetics of catalase reappearance after feeding mice a single dose of the drug 3-amino-1,2,4-triazole which irreversibly inactivates catalase (*Rechcigl and Heston,* 1967). The reappearance of activity with time is shown in figure 4. C57BL/He mice are slower than C57BL/6 mice in returning to their original steady-state level of liver catalase, because, although the rate constants for enzyme synthesis are the same in the two strains, those for enzyme degradation are not.

For catalase, as for other proteins, enzyme synthesis is a zero order process, and enzyme breakdown is a first order one. Under these circumstances, if k_S is the rate of synthesis, k_D the first order rate constant of degradation, and E the enzyme concentration, then

$$\frac{dE}{dt} = k_S - k_D (E)$$

At the time amino-triazole is fed, the catalase concentration falls to zero. If this time is set as $t = 0$, the recovery is described by the equation

$$E_t = \frac{k_S}{k_D} (1 - e^{-k_D t})$$

After recovery the steady state concentration of enzyme is

$$E_\infty = k_S/k_D$$

Substituting this last expression and rearranging terms provides the more useful relationship

$$\ln (E_\infty - E_t) = \ln E_\infty - k_D t$$

The rate constant can be estimated using this equation. Plotting the log of the difference between the enzyme concentration after recovery and the concentration present at some time during the recovery period as a function of time yields a straight line whose slope is $-k_D$. The rate of synthesis is $k_D \cdot E_\infty$.

The data for catalase recovery, transformed in this way, shows that the two mouse strains differ in the rate of catalase degradation and not in the rate of catalase synthesis. In C57BL/He mice liver catalase is degraded at the rate of 1.9 % per hour, rather than 4.5 % per hour as in C57BL/6 (fig. 5). This difference explains why C57BL/He animals have twice as much enzyme as C57BL/6. This is the first example of a genetically determined alteration in the rate of enzyme degradation.

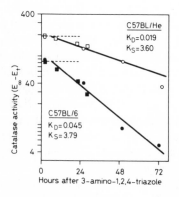

Fig. 5. The data of figure 4 replotted as log E_∞-E_t v. time. K_D is calculated from the slope of each line, and K_s from the relationship $E_\infty = K_s/K_D$. Adapted from *Rechcigl* (1968).

These results have been confirmed and extended by *Ganschow and Schimke* (1968, 1969), who estimated the rate of catalase synthesis by *in vivo* incorporation of labelled amino acids into subsequently purified catalase protein. The rate of degradation was determined using both the kinetic method of following recovery from amino-triazole and a double isotope technique which estimates the half-life of a protein relative to the average half-life of all liver proteins.

A single gene with two alleles, *Ce* and *ce*, determines the difference in catalase turnover between C57BL/He and C57BL/6 (*Heston et al.*, 1965). The *Ce* gene is not sex-linked, but its location among autosomes is unknown. Preliminary data suggest that it is not closely linked to the catalase structural gene (*Ganschow and Schimke*, 1968). The allele for rapid turnover, producing low enzyme, is dominant over that for slow turnover, and intermediate levels of catalase are not seen in *ce/Ce* heterozygotes. Therefore the *Ce* allele product is an essential, but not a rate limiting, factor in the pathway of catalase degradation.

Expression of the *Ce* gene is tissue specific. The kidney catalase levels of *ce/ce* mice are identical to those of *Ce/Ce* mice of the same sex (*Heston et al.,* 1965). A sex difference does exist with male kidneys having approximately twice as much enzyme as female kidneys, but this is unrelated to the strain difference.

Liver catalase levels of the related strains C57BL/He and C57BL/Ha are twice those of C57BL/6 because they degrade enzyme more slowly. Liver catalase activities of DBA/2 and other strains, which do not belong to the C57 and C58 families, also are twice those of C57BL/6, but not because they degrade enzyme more slowly. Instead it is because the enzyme molecules they contain are twice as active (*Ganschow and Schimke,* 1968, 1969). This was deduced from the following facts concerning liver catalase summarized in table IV. (a) All three types of strains, C57BL/Ha, C57BL/6 and DBA/2, synthesize the same number of enzyme molecules per minute, as judged by the incorporation of labelled amino acids into catalase. (b) DBA/2 mice degrade enzyme rapidly like C57BL/6 and not slowly like C57BL/Ha. (c) DBA/2 and C57BL/6 mice have the same number of enzyme molecules in liver, namely half that of C57BL/6, as measured by immunological titration. This fact confirms the equal rate of synthesis in the three strains and the slower rate of degradation in C57BL/6. (d) Purified enzyme from DBA/2 is twice as active per molecule as enzyme from either C57BL/6 or C57BL/Ha. Thus DBA/2 livers have twice the catalase activity of C57BL/6 livers because, although they contain the same number of molecules, each molecule is twice as active; C57BL/Ha livers have twice the catalase activity of C57BL/6 livers because, although their molecules are no more active, they contain twice as many of them.

No other feature of the enzyme isolated from DBA/2 was distinguishably different, including its antigenic properties, electrophoretic mobility, pH optimum, sedimentation coefficient, hematin content, or resistance to heat or proteolytic inactivation.

Thus, the original Lathrop stock which gave rise to C57 and C58 had a low enzyme activity in both liver and kidney because of a mutation which reduced the catalytic efficiency of catalase molecules. This mutation is expressed as a codominant allele in heterozygotes and presumably occurred in the structural gene for catalase (*Ganschow and Schimke,* 1968). Subsequently, a second mutation, slowing the rate of catalase degradation in liver and increasing the half-life of the enzyme, occurred in the branch of the C57 family tree containing C57BL/An, C57BL/He and C57BL/Ha. Mice homozygous for this mutation now had more enzyme molecules per liver, but the same number per kidney. This mutant allele is recessive, and heterozygotes have wild-type rates of catalase degradation.

Table IV. Catalase in strains of inbred mice

	DBA / 2	C57BL / 6	C57BL / An, He, Ha
Genotype			
Presumed catalase structural gene	Cs^a/Cs^a	Cs^g/Cs^g	Cs^g/Cs^g
Degradation gene	*Ce/Ce*	*Ce/Ce*	*ce/ce*
Liver			
Rate of synthesis	1.0	1.0	1.0
Rate of degradation	2.0	2.0	1.0
No. of molecules per cell	0.5	0.5	1.0
Activity per molecule	2.0	1.0	1.0
Observed activity	1.0	0.5	1.0
Kidney			
Rate of synthesis	1.0	1.0	1.0
Rate of degradation	1.0	1.0	1.0
No. of molecules per cell	1.0	1.0	1.0
Activity per molecule	2.0	1.0	1.0
Observed activity	2.0	1.0	1.0

All values are expressed as relative to the same quantity in the C57BL / An, He, Ha phenotype.

The molecular basis for the reduced rate of catalase turnover in *ce/ce* mice is unknown. However, as *Ganschow and Schimke* (1968, 1969) have pointed out, the properties of this mutant imply that turnover systems act on only a limited family of proteins, that the rates at which different proteins turnover reflect the presence of specific degrading systems, and that such systems can be tissue specific in their action. The existence of tissue specific factors has been confirmed for the breakdown, as well as the synthesis, of lactate dehydrogenase (*Fritz et al.,* 1969).

D.   Regulation of Enzyme Activity

1.   Phenylalanine Hydroxylase

The *dilute* mutation in mice may be a case of mutationally altered control of enzyme activity rather than enzyme concentration. Dilute animals are deficient in their ability to oxidize phenylalanine to tyrosine and have a low phenylalanine hydroxylase activity in liver (*Coleman,* 1960). However, the low enzyme activity is due to an excess of a particulate enzyme inhibitor rather than a lack of enzyme protein. Animals homozygous for the *dilute* allele are extremely nervous and very susceptible to audiogenic seizures. Since phenylalanine metabolism is diverted towards phenylacetate production rather than towards tyrosine, these neurologic effects of the dilute mutation can be explained in terms of the pharmacologic action of phenylacetate on catecholamine metabolism.

E.   Multiple Gene Effects

Although the realization of any enzyme is determined by many loci, the effects of mutations at more than one locus have been studied in only a few cases. The most extensive effort has been the study of murine tyrosinase by *Coleman* (1962). Several loci, many with multiple alleles, are known to affect pigmentation in the mouse. Their effect on tyrosinase activity, as measured by the incorporation of C^{14} tyrosine into melanin, has been compared. The *C* locus is probably the structural gene for tyrosinase itself since tyrosine incorporation is proportionately reduced among a series of alleles with varying intensities of pigmentation, since alleles show codominance in heterozygous combinations, and since the ability to incorporate tyrosine is thermolabile in the Himalayan mutant ($c^h c^h$) which is more heavily pigmented at low temperatures.

The black locus *(B)* controls the shade of brown-black pigmentation. The two extreme alleles of this gene are black *(B)* and brown *(b)*. If a normal *C* gene is present, homozygous *bb* mice have twice as much tyrosinase activity as *BB* mice despite the fact that they are less intensely pigmented. *Bb* and *BB* mice have the same color and tyrosinase activity. *B* is therefore dominant to *b* both with respect to coat color and its effect in regulating tyrosinase concentration. In a series of tyrosinase structural gene mutants with various levels of tyrosinase activity, the presence of *BB* or *bb* alleles only affected tyrosinase activity if the animals could make at least one-third of the wild-type level of enzyme. This suggests that the regulatory action of the *B* gene product may require the simultaneous presence of a sufficient concentration of a metabolic product of tyrosinase action.

The *agouti* locus *(A)* also controls pigmentation. Homozygous *aa* animals make only brown-black eumelanin, but *Aa* and *AA* animals make both eumelanin and some yellow phaeomelanin. The extreme allele in this series A^y is lethal in homozygotes, permits only phaeomelanin to be formed in heterozygotes, and reduces tyrosinase activity to 1/3 normal in heterozygotes. This latter effect of the A^y allele is expressed at all enzyme levels set by the *B* and *C* alleles present.

Among other pigmentation mutants, *ruby,* which affects the rate of pigment granule development, and *maltese dilution* and *leaden,* which control the fine clumping of pigment granules, had little effect on tyrosinase activity. However the mutation *pink-eyed dilution,* which drastically changes the structure of the pigment granule, permitted only half normal tyrosinase activity to be present.

In summary, at least three loci, *black*, *agouti* and *pink-eyed dilution*, affect the level of tyrosine incorporating activity, a reaction probably catalyzed by tyrosinase, a product of the *C* locus.

F.   Coordinancy

The problem of enzyme regulation extends beyond that of single proteins. The relative levels of different enzymes must be correlated with each other in order to effectively integrate the regulation of intermediary metabolism and the assembly of intracellular structures. The available facts are worth citing since they raise important questions of mechanism.

The cytochromes of mitochondria are present in stoichiometric ratios indicating a coordinated control over these proteins either at the level of synthesis or in the process of organelle assembly (*Chance and Hess,* 1959). In contrast, the hepatic content of at least two lysosomal enzymes, acid phosphatase and β-glucuronidase, is not controlled coordinately suggesting that lysosomes do not have a fixed enzymatic content (*Paigen and Ganschow,* 1965). Rather, lysosomes seem capable of receiving proteins in any proportion.

Table V. An equimolar constant-proportion group of glycolytic enzymes

Enzyme	Molecular weight	Specific activity (μM reaction/ h/mg protein)	Calculated activity ratio for an equimolar mixture	Observed ratios from *Pette et al.* (1962)
Triosephosphate isomerase	$\sim 5 \times 10^4$	2.95×10^5	~ 19	6−12
Glyceraldehyde-3-phosphate dehydrogenase	1.4×10^5	5.5×10^3	1	1
Phosphoglycerate kinase	3.4×10^4	1.85×10^4	0.82	0.2−1.2
Phosphoglycerate mutase	5.7×10^4	7.8×10^3	0.58	0.6−1.2
Enolase	8.5×10^4	2.72×10^3	0.30	0.15−0.3

From Mier and Cotton (1966). Activity ratios are calculated relative to glyceraldehyde-3-phosphate dehydrogenase.

Within a metabolic sequence, different cells in the same organism may not contain the component enzymes in constant proportion. This is true for the pyrimidine (*Kretschmer et al.,* 1966; *Krooth,* 1969) and cysteine (*Mudd et al.,* 1965) biosynthetic pathways. However, a constant proportionality does exist among the group of five soluble enzymes that function in the conversion of triosephosphate to phosphoenolpyruvate. This was first shown for insect flight muscles by *Vogell et al.* (1959) and extended to a variety of avian tissues, mammalian tissues, and baker's yeast by *Pette et al.* (1962a, 1962b). The five enzymes involved (triosephosphate isomerase, glyceraldehyde-3-phosphate dehydrogenase, phosphoglycerate kinase, phosphoglycerate mutase, and enolase) were present in a constant ratio of activities in all tissues examined despite wide variations in the absolute amounts of activity present. As pointed out by *Pette* and co-workers, the enzymes of this 'constant-proportion group' catalyze a metabolic sequence that fits between two junction points in the network of carbohydrate metabolism.

The constant proportionality of enzyme activities in the carbohydrate group must have a strong selective advantage if it has been preserved throughout evolution. The proportion selected probably reflects some biochemical requirement(s) for catalyzing this metabolic sequence. It is suggestive that the order of enzyme activities is the same as the sequence of reactions they catalyze. The first enzyme is the most active and so on down to the last which is the least active. It would be most interesting to know whether this same constant-proportion group occurs in higher plants, especially in their photosynthetic tissues.

The five members of the constant-proportion group apparently occur in equimolar amounts (*Mier and Cotton,* 1966). The activity ratios expected for an equimolar mixture of enzymes agree rather well with the findings of *Pette* and co-workers (table V). *Mier and Cotton* have suggested that equimolarity may result from synthesis of all five enzymes on one polycistronic messenger. If this is so, then at least two further conditions must apply. All enzymes in the group must have the same rate of turnover in all cells, and the entire group of structural genes must be closely linked genetically. While the first of these predictions may be difficult to test, the second would not.

IV. Architectural Genes

The process of building a cell is probably a combination of self-assembly steps and catalytically mediated reactions. In this process, the structure of an enzyme determines its own intracellular site both by defining its self-assembly properties and by providing for its recognition by assembly catalysts.

Several major features of the assembly process are vague at the present time. These concern the types of chemical linkages holding molecules in structures, the chemical nature of the reactions mediated by assembly catalysts, the sub-strate specificities of these catalysts, and the mechanisms for regulating the synthesis of different sub-cellular elements to produce an integrated cell struc-ture. The identification and characterization of assembly mutants should help to clarify these features.

A. Enzyme Integration Mutants

1. β-Glucuronidase

Endoplasmic glucuronidase omission (eg^o) is the only mutant known lacking the ability to integrate a single protein into an otherwise normal cell organelle (Paigen and Ganschow, 1965; Ganschow and Paigen, 1967). Normal mice contain approximately equal amounts of β-glucuronidase in the lysosomes and endoplasmic reticulum of liver. The two components are probably coded for by the same gene since they exhibit nearly identical physical and catalytic properties and since a single mutation increases the thermolability of both com-ponents in an identical fashion (Paigen, 1961a).

The livers of mice homozygous for the eg^o allele do not contain glucuroni-dase in endoplasmic reticulum although they have normal amounts of lysosomal enzyme. This deficiency was demonstrated by conventional fractionation of sub-cellular particles (where the endoplasmic reticulum enters the microsome fraction) (table VI) and by sucrose density gradient analysis (fig. 6). The defect in eg^o animals was specific for glucuronidase since no change in other micro-somal proteins was detected. Thus glucuronidase must possess recognition

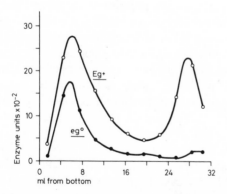

Fig. 6. Sucrose gradient centrifugation of cytoplasmic particles from Eg^+ and eg^o mouse liver. β-Glucuronidase activity of each fraction is plotted as a function of distance along the gradient. The peak to the left corresponds to lysosomes and the one to the right to microsomes. Adapted from Ganschow and Paigen (1967).

Table VI. Intracellular distribution of β-glucuronidase in Eg^+ and eg^o mouse liver

		Nuc.	Mit.	Lys.	Mic.	Super	Total homogenate
Percent total dry wt.	Eg^+	14	22	6	20	37	100
	eg^o	14	18	7	20	37	100
Total enzyme activity	Eg^+	9.1	19.7	11.7	13.4	9.1	64
	eg^o	1.9	8.7	4.9	0.8	6.0	24
Enzyme specific activity	Eg^+	180	249	541	186	68	177
	eg^o	36	129	189	10	44	64
Ratio specific activity	eg^o/Eg^+	0.20	0.32	0.35	0.05	0.65	0.36

Recalculated from *Ganschow and Paigen* (1967). Enzyme units are mmoles product/h at 56 °C. Total activity is the enzyme present in the fraction derived from 1 g of fresh tissue. Specific activity is enzyme units per mg dry wt.

features which are not shared by other proteins of the endoplasmic reticulum. These recognition features occupy only a portion of the glucuronidase molecule since Eg^+, but not eg^o animals can insert different structural forms of glucuronidase into endoplasmic reticulum. Although the products of the *Eg* and glucuronidase structural genes interact, the two loci are not linked genetically.

The *Eg* gene product might function in one of several ways: modification of the structure of β-glucuronidase to permit insertion into endoplasmic reticulum, synthesis of a membrane receptor site for the enzyme, or catalysis of the attachment process. Since a single dose of the Eg^+ allele is sufficient for normal integration of β-glucuronidase, the function of the *Eg* gene product is not rate limiting in enzyme attachment.

2. β-D-N-Acetylhexosaminidase

I suggest that the enzymatic lesion in Tay-Sachs disease may originate from a similar assembly defect for β-D-N-acetylhexosaminidase. Tay-Sachs disease, which occurs almost exclusively among Ashkenazic Jews, is inherited as a Mendelian recessive. Affected individuals, who invariably die within the first few years of life, are unable to break down ganglioside GM_2 which accumulates in massive amounts in the brain. Tissues from normal individuals can liberate the terminal N-acetyl galactosamine residue from labelled GM_2 ganglioside, but tissues from Tay-Sachs individuals cannot (*Kolodny et al.*, 1969).

The enzyme responsible for this reaction, β-D-N-acetylhexosaminidase, normally occurs in human cells as two electrophoretically separable forms which are thought to possess the same polypeptide sequence but to differ in the number of conjugated sialic acid residues (*Robinson and Stirling*, 1968). The intracellular distribution of total enzyme activity paralleled that of β-glucuronidase. *Okada and O'Brien* (1969) have shown that one of the electrophoretic components is released from particulate form by osmotic shock, and that it is this component that is completely absent from the tissues of Tay-Sachs individuals.

The simplest explanation of Tay-Sachs disease is that, like β-glucuronidase, a single species of the β-D-N-acetylhexosaminidase is normally present in both lysosomes and endoplasmic reticulum and that Tay-Sachs individuals are unable to incorporate enzyme into lysosomes. In the eg^0 defect, loss of the endoplasmic reticulum component leaves the remaining enzyme completely sensitive to osmotic shock; in the Tay-Sachs defect, loss of the lysosomal component leaves the remaining enzyme completely resistant to osmotic shock. The Tay-Sachs defect, then, like the eg^0 defect, is an assembly deficiency, but one involving enzyme incorporation into lysosomes rather than endoplasmic reticulum.

Two aspects of this hypothesis are testable: if the two electrophoretic forms of β-D-N-acetylhexosaminidase are located in lysosomes and microsomes, this should be revealed by additional studies of the intracellular distribution. If the two forms are coded by the same structural gene, this should become apparent from studies of enzyme structural variants.

A number of other human diseases may also represent assembly defects, including many sphingolipidoses and defects of mucopolysaccharide metabolism. In each case, normal tissues contain at least two enzymatic components that carry out the same acid hydrolytic reaction. Although the data is fragmentary, one component seems to be lysosomal and the other microsomal. In these disease states, one enzyme component, presumably the lysosomal one, is missing. This hypothesis would account for the findings that, for example, the A form of arylsulfatase is missing in metachromatic leukodystrophy (*Austin et al.*, 1963, 1965; *Mehl and Jatzkewitz*, 1965, 1968) and that only one of the β-galactosidase components is lost in Hurler's syndrome (*Ho and O'Brien*, 1969).

3. Mitochondrial Structural Protein

Several mutations derange the organization of mitochondria or chloroplasts, causing complex phenotypic changes in the realization of many enzymes. Such mutations may show either nuclear or cytoplasmic inheritance. For most such mutants, it has not been possible to relate the phenotypic changes to a specific protein lesion. However, two of the cytoplasmically inherited *Neurospora* mitochondrial mutants produce an altered mitochondrial structural protein (*Wood-ward and Munkres,* 1966). The normal mitochondrial structural protein is composed of subunits of approximately 23,000 molecular weight. In the *mi-3* mutant, this subunit contains one less tryptophan residue, and in the *mi-1* mutant it contains one less tryptophan residue and one additional cysteine residue. These changes in structure are accompanied by changes in function. If wild type mitochondrial structural protein is complexed with malate dehydrogenase extracted from mitochondria, it does not change the affinity of the enzyme for malate. However, if mitochondrial structural protein from either the *mi-1* or *mi-3* mutant is complexed with wild type malate dehydrogenase, the enzyme has a greatly decreased affinity for substrate and the resulting complex is virtually inactive (*Munkres and Woodward,* 1966).

V.   Temporal Genes

Cellular differentiation proceeds by a process of selective gene activation reflected in such phenomena during embryogenesis as chromosome puffing, heterochromatization and changes in the species of RNA synthesized. Encoded in the genome must be mechanisms for this selective activation. Genetic experiments provide evidence that a set of genetic elements exists capable of 'mutating' or undergoing changes in state or function in somatic cells at fixed times during development. Reviews of some of the major systems studied are in papers by *Allen* (1965), *Baker* (1967), *Becker* (1966), *Brink* (1964), *Harrison and Fincham* (1964), *McClintock* (1965, 1967) and *Nanney* (1968). Biochemical evidence for possible mechanisms of selective gene activation comes from mutants with altered developmental timing of enzyme synthesis and from studies in interspecific hybrids of the time during development at which different alleles of the same gene are expressed.

A.   Protein Realization Mutants

1.   β-Glucuronidase
Changes in the concentration of β-glucuronidase during development are determined by a genetic element either closely linked to or identical with the structural gene for this enzyme (*Paigen,* 1961b). The livers of normal (DBA/2) mice maintain a constant enzyme concentration during post-natal growth, but the livers of mutant (C_3H) mice are low in enzyme activity at birth and decline even further starting at 12 days of age (fig. 7). The factor for this developmental abnormality showed no recombination (less than 0.5 %) with the β-glucuronidase structural mutation in the C_3H strain. Whether the two phenotypic effects result

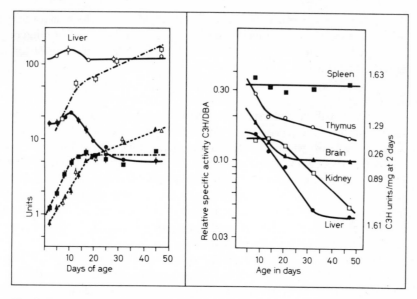

Fig. 7. The growth and glucuronidase content of the liver as a function of age. Open symbols are wild-type (DBA) animals; closed symbols are mutant (C3H) animals. The vertical bars indicate the standard errors of each group of measurements. The values of the arbitrary units are: liver weight (△,▲) 100 mg net weight: total enzyme activity (□,■) 100 enzyme units; and specific activity (○,●) 0.1 enzyme units/mg wet weight. Adapted from *Paigen* (1961b).

Fig. 8. The relative β-glucuronidase activities of various organs in mutant (C3H) and wild-type (DBA) mice at various ages. In each case the specific activity of the mutant tissue has been compared with the equivalent specific activity of wild-type animals of the same age. Adapted from *Paigen* (1961b).

from a single genetic lesion or two closely linked ones is unknown. An apparently unrelated gene *(rd),* causing retinal degeneration which also starts at day 12, is close to *glucuronidase (Paigen and Noell,* 1961), and both have been located on linkage group XVII of the mouse *(Sidman and Green,* 1965).

Thymus, brain and kidney also showed genetically determined changes in glucuronidase concentration with time although the spleen did not (fig. 8). The developmental changes are probably not in response to any generalized physiological change, such as the appearance of a factor circulating in blood, since they occur at a different time in each tissue. There is no suggestion of a feed-back mechanism involving the substrate or products of β-glucuronidase since the developmental changes are not correlated with levels of enzyme activity. The most plausible explanation is that mutation within or adjacent to the *glucuronidase* gene has changed its response to a stimulus appearing at a different time during the development of each tissue. This stimulus could be either a somatic genetic change of the type mentioned in the introduction to this section or the appearance of a regulatory system to which only the mutant allele for this enzyme is sensitive.

2. Esterase

Mutants with altered expression during development have been reported in maize for the E_1 esterase. Seven alleles have been described for this dimeric enzyme, each of which specifies a monomer unit with a different charge (*Schwartz, 1960; Schwartz et al., 1965*). In homozygotes, only one dimer with identical subunits is formed. In heterozygotes, both alleles are expressed codominantly and three dimers are formed, two with identical subunits and one with mixed subunits. If enzyme, extracted from heterozygotes containing the $E_1{}^N$ and $E_1{}^S$ alleles, is subjected to starch gel electrophoresis, three bands are found corresponding to the SS, SN, and NN dimers.

Altered forms of some of the E_1 structural alleles differ in the timing of gene action (*Schwartz, 1962, 1963*). For example, the allele F' produces an esterase monomer with the same mobility as that of the F allele. However, the F allele functions throughout the development of the corn kernel endosperm while the F' allele functions only for the first 14 days after pollination. By 19 days after pollination, corn kernels homozygous for F' have very little esterase in their endosperm while those homozygous for F have normal amounts. The F' allele is codominant. Extracts of immature kernels from F'/S heterozygotes show three enzyme bands in electrophoresis ($F'F'$, $F'S$, and SS); extracts of older kernels show only the SS band. Analogous forms of the N, S, and T alleles also produce the corresponding enzyme monomers early in kerly development but not later. The developmental abnormality is tissue specific, and the expression of the 'prime' allele is indistinguishable from that of the normal allele in developing seedlings. The timing effect is closely linked to the structural gene and is not separable by recombination (less than 0.04 %).

3. Thalassemia

All normal hemoglobins contain two a subunits and two other subunits. In fetal hemoglobin (HbF) these subunits are γ chains. In the predominant form of adult hemoglobin (HbA) these are β chains. A few percent of hemoglobin molecules in adults have two δ chains (HbA$_2$). Although the switch from γ chain synthesis to β and δ chain synthesis occurs around birth, it is determined by fetal age and not by birth itself. Individuals homozygous for β-thalassemia mutations are unable to initiate β chain synthesis and produce very little or no HbA. They survive by continuing to produce HbF and elevated levels of HbA$_2$. The β-thalassemia mutations may be temporal gene mutations similar to those seen for β-glucuronidase and E_1 esterase.

The β_{thal} gene is closely linked to the β structural gene, and recombination has not been observed between them (*Weatherall,* 1967). As in the case of the E_1 esterase and probably glucuronidase, the β_{thal} defect is codominant and only affects the expression of the chromosome it is on. For example, β_{thal} β^A/β^+_{thal} β^S heterozygotes (with a β_{thal} mutation and a normal β^A structural allele on one chromosome and a normal β^+_{thal} gene and a β^S structural gene variant on the other chromosome) make predominantly β^S chains. The mutant β_{thal} allele has prevented expression of its adjacent β^A allele but has not suppressed expression of the β^S allele on the other chromosome.

The β structural gene is also closely linked to the δ structural gene. This was originally established with the *Lepore* hemoglobin variant in which a deletion has excised the carboxyl end of the δ gene and the amino end of the β gene and left the remaining fragments joined as a single gene (*Baglioni,* 1962). The β_{thal} gene is distinct from the β and δ structural genes as shown by the production of δ chains with normal sequence in all β_{thal} mutants and by the production of β chains with normal sequence in some β_{thal} mutants.

The order of these three genes can be deduced from the rate of hemoglobin *Lepore* synthesis by assuming the existence of an analogous δ_{thal} gene controlling δ structural gene expression and occupying a corresponding position adjacent to the δ gene. The wild type gene order must then be either $\delta^+_{thal}-\delta-\beta^+_{thal}-\beta$ or $\delta-\delta^+_{thal}-\beta-\beta^+_{thal}$. In normal adults δ chains are synthesized at a rate only a few percent that of β chains suggesting that δ^+_{thal} determines a much slower rate of gene expression than β^+_{thal}. The rate of synthesis of δ chains is increased somewhat in individuals producing structurally defective β chains, but this rate never approaches the rate of β chain synthesis. In hemoglobin *Lepore* the region between the β and δ genes is deleted, deleting either the β^+_{thal} gene, if the first sequence is correct, or the δ^+_{thal} gene, if the second sequence is correct. Individuals heterozygous for hemoglobin *Lepore* synthesize this protein at a rate corresponding to a δ^+_{thal} gene suggesting that it is the β^+_{thal} gene which has been deleted. Thus the first sequence appears to be correct and β^+_{thal} lies between the δ and β structural genes. The peptide sequence of hemoglobin *Lepore* establishes the amino to carboxyl orientation of the hemoglobin structural genes. If β^+_{thal} lies between the two structural genes, it must be at the amino end of the β structural gene.

4. Pyrimidine Degradation

A single locus with two alleles, Pd^a and Pd^b, has been reported to determine the levels of three successive enzymes in pyrimidine degradation, dihydrouracil dehydrogenase, dihydropyrimidinase, and 3-ureidopropionase (*Dagg et al.,* 1964). Heterozygotes had intermediate levels of activity. Neither mutant nor wild type animals showed appreciable enzyme activity at birth, and both developed enzyme at the same rate during the first 6 days of life. However, wild type strains continued to develop enzyme after 6 days while mutant strains did not. The possibility was suggested that the *Pd* gene determines a repressor controlling synthesis of the three enzymes. However, this interpretation is not easily reconciled with the intermediate activity of heterozygotes. A more likely possibility is that the *Pd* mutation has occurred in a temporal gene and that either the three enzyme structural genes are linked or the three enzymes share a common subunit.

B. Interspecific Hybrids

Isozyme variants provide an ingenious tool for determining when gene loci are activated during development. This was first carried out for lactate dehydrogenase by crossing related frog species carrying different forms of LDH-1 (*Wright and Moyer*, 1966). Lactate dehydrogenase is a tetramer made up of A and B type subunits (or M and H depending on the system of notation) in various combinations. Three closely related species, *Rana pipiens pipiens, Rana p. sphenocephala,* and *Rana palustris,* carry electrophoretically distinguishable forms of the LDH-1 (B_4) isozyme. Progeny of reciprocal interspecific crosses had only the maternal type of enzyme until stage 19 (heart beat).

As *Wright and Moyer* suggested, this result could reflect (a) the carryover of LDH synthesized during oogenesis prior to fertilization; (b) the carryover of LDH messenger synthesized before fertilization; or (c) the preferential transcription of maternal genes. A decision between these possibilities could be made by examining the progeny of a cross between a female heterozygous for two B subunit structural alleles and a male homozygous for a third B subunit structural allele. If possibility (a) is correct, early embryos should contain LDH with equal amounts of both maternal subunit types; if possibility (b) is correct, early embryos should make LDH molecules in which both maternal subunit types eventually become associated with the paternally determined type; and if possibility (c) is correct, each offspring should make new LDH containing only one of the maternal subunit types, and lacking the paternal subunit type. Such a cross was made between a female *Rana p. sphenocephala* heterozygous at the B locus and a male *Rana p. p.* homozygous for a different structural allele (*Wright and Moyer*, 1968). Early embryos were found to contain LDH with both maternal subunits, and later embryos made LDH containing the paternal subunit and one of the maternal types. These results indicate that LDH synthesized during oogenesis accounts for the persistence of maternal type enzyme, possibility (a). The same explanation probably accounts for the persistence of LDH of maternal type in interspecific trout hybrids (*Hitzeroth et al.,* 1968; *Goldberg et al.,* 1969).

Although preferential activation of maternal alleles does not occur in the cases above, it does with two other proteins in hybrids between brown trout *(Salmo trutta)* and rainbow trout *(Salmo irideus)* (*Hitzeroth et al.,* 1968). Only hybrids from a rainbow mother could be studied; the reciprocal cross hybrid was not viable. Alcohol dehydrogenase, absent in early embryos, is synthesized in liver only after this organ begins to differentiate. In hybrids the expression of the maternal allele begins 150 days after fertilization, but expression of the paternal allele is delayed and, even at 192 days, is not expressed to the same extent. A similar delay occurs with the unusual C class of LDH subunit synthesized only in fish retina. In hybrids the maternal type appears at 95 days, and the paternal type appears only at 135 days.

A similar delay in activation of a paternal allele also occurs for alcohol dehydrogenase of fowl hybrids (*Castro-Sierra and Ohno,* 1968). No alcohol dehydrogenase is present during early embryonic development; it appears in liver in the later stages of embryogenesis. No other tissue contains the enzyme until after hatching when some appears in kidney. Japanese quail are polymorphic for this enzyme with several alleles known at the *alcohol dehydrogenase* structural gene. The inheritance pattern suggests a dimeric enzyme with random association of subunits. Domestic chickens contain an alcohol dehydrogenase distinct from any of the quail forms. Offspring of matings between chicken fathers and Japanese quail mothers form quail type enzyme beginning 19 days after fertilization. This is the same time at which normal chickens and quail begin enzyme synthesis. If this enzyme is of the A or C quail type, the paternal chicken allele is not expressed until 3 days later. If this is of the B quail type, the paternal chicken allele is not expressed at all.

The results obtained with alcohol dehydrogenase in fowl hybrids and both alcohol and retinal lactate dehydrogenases of fish suggest that alleles of the same gene can respond differently to the temporal programming of differentiation even if in a common cytoplasm. Linked to each structural gene appears to be a genetic element which determines its responsiveness during development. This linked genetic element is revealed in interspecific hybrids because of the unusual regulatory systems that occur when chromosome sets, which have been separated for a long period of evolution, are combined. There is, for example, no difference in the time of expression of maternal and paternal alcohol dehydrogenase alleles in quail x quail crosses (*Ohno et al.,* 1969).

C. *t* Genes as a Special Class of Genetic Elements

The parallel behavior of the genetic systems affecting β-glucuronidase, E_1 esterase, and hemoglobin β chains is striking; especially since they involve three such diverse proteins and organisms. In each case a genetic region next to the structural gene determines the time in development at which its adjacent structural gene can be expressed without affecting the allele present on the homologous chromosome. The behavior of the interspecific hybrids fits these findings.

Extrapolation from these cases suggests that temporal controlling elements may be attached to many, if not all, structural genes in eukaryotic organisms. For brevity I shall call them *t* (temporal) as distinct from *s* (structural) genes. The properties of the glucuronidase system suggest that *t* genes respond to a timed signal generated within the cell. There is adequate precedent for providing such signals by genetic events that occur at fixed stages of development. The properties of the hemoglobin system suggest that *t* genes are adjacent to the amino end of the structural genes they activate.

Both the codominance and position of *t* genes are reminiscent of bacterial operators. Nevertheless, there is little justification for equating the two. Bacterial operators are components of a regulatory apparatus controlling enzyme synthesis by responding to concentration changes of low molecular weight effectors. The *t* genes of multicellular organisms are components in a regulatory system which carries an inherent timing program for enzyme synthesis during differentiation. So far, no evidence suggests that this system responds to low molecular weight effectors. More importantly, operators serve as sites of inhibition to prevent the progression of RNA polymerase along the DNA. Operator function is negative, and the loss of operator function is followed by an unrestricted expression of adjacent structural genes. However, *t* gene mutations result in a relative loss in the ability to express adjacent genes implying that *t* gene function is positive. Some event, perhaps initiating RNA synthesis, must occur at the *t* gene for the adjacent structural gene to be expressed.

VI. References

Aebi, H.; Baggiolini, M.; Dewald, B.; Lauber, E.; Suter, H.; Micheli, A., and Frei, J.: Observations in two Swiss families with acatalasia. Enzym. biol. clin. *4:* 121–151 (1964).

Aebi, H. und Cantz, M.: Über die zelluläre Verteilung der Katalase im Blut homozygoter und heterozygoter Defektträger (Akatalasie). Humangenetik *3:* 50–63 (1966).

Aebi, H.; Heiniger, J.P.; Bütler, R., and Hässig, A.: Two cases of acatalasia in Switzerland. Experientia *17:* 466 (1961).

Allen, S.L.: Genetic control of enzymes in *Tetrahymena*. Brookhaven Symp. Biol. *18:* 27–54 (1965).

Allison, A.C.: Polymorphism and natural selection in human populations. Cold Spr. Harb. Symp. quant. Biol. *29:* 137–149 (1964).

Allison, A.C.; Rees, W.A.P., and Burn, G.P.: Genetically controlled differences in catalase activity of dog erythrocytes. Nature, Lond. *180:* 649–650 (1957).

Arnheim, N., Jr. and Wilson, A.C.: Quantitative immunological comparison of bird lysozymes. J. biol. Chem. *242:* 3951–3956 (1967).

Austin, J.; Armstrong, D., and Shearer, L.: Metachromatic form of diffuse cerebral sclerosis. V. The nature and significance of low sulfatase activity: a controlled study of brain, liver, and kidney in four patients with metachromatic leukodystrophy (MLD) urine. Arch. Neural. *13:* 593–614 (1965).

Austin, J.; Balasubramanian, A.S.; Pattabiraman, T.N.; Saraswathi, S.; Basu, D.K., and Bachhawat, B.K.: A controlled study of enzyme activities in three human disorders of glyco-lipid metabolism. J. Neurochem. *10:* 805–816 (1963).

Baglioni, C.: The fusion of two peptide chains in hemoglobin Lepore and its interpretation as a genetic deletion. Proc. nat. Acad. Sci., Wash. *48:* 1880–1886 (1962).

Baker, W.K.: A clonal system of differential gene activity in *Drosophila.* Develop. Biol. *16:* 1–17 (1967).

Bather, R.; Dziubalo, S., and Darcel, C.L.Q.: Catalase levels in Line 15 East Lansing White Leghorns. Poult. Sci. *42:* 1023–1024 (1963).

Becker, H.J.: Genetic and variation mosaics in the eye of *Drosophila.* Curr. Topics develop. Biol. *1:* 155–171 (1966).

Black, J.A. and Dixon, G.H.: Amino acid sequence of a-chains of human haptoglobins. Nature, Lond. *218:* 736–741 (1968).

Brink, R.A.: Genetic repression in multicellular organisms. Amer. Naturalist. *98:* 193–211 (1964).

Burstein, S.; Bhavnani, B.R., and Bauer, C.W.: Genetic aspects of cortisol hydroxylation in guinea pigs: kinetics of enzymatic 2-a- and 6-β-hydroxylation by adrenal homogenates. Endocrinology *80:* 663–678 (1967).

Castro-Sierra, E. and Ohno, S.: Allelic inhibition at the autosomally inherited gene locus for liver alcohol dehydrogenase in chicken-quail hybrids. Biochem. Genet. *1:* 323–335 (1968).

Chance, B. and Hess, B.: Metabolic control mechanisms. I. Electron transfer in the mammalian cell. J. biol. Chem. *234:* 2404–2412 (1959).

Childs, B. and Young, W.J.: Genetic variations in man. Amer. J. Med. *34:* 663–673 (1963).

Cocks, G.T. and Wilson, A.T.: Immunological detection of single amino acid substitutions in alkaline phosphatase. Science *164:* 188–189 (1969).

Coleman, D.L.: Phenylalanine hydroxylase activity in dilute and nondilute strains of mice. Arch. Biochem. *91:* 300–306 (1960).

Coleman, D.L.: Effect of genic substitution on the incorporation of tyrosine into the melanin of mouse skin. Arch. Biochem. *96:* 562–568 (1962).

Coleman, D.L.: Purification and properties of δ-aminolevulinate dehydratase from tissues of two strains of mice. J. biol. Chem. *241:* 5511–5517 (1966).

Dagg, C.P.; Coleman, D.L., and Fraser, G.M.: A gene affecting the rate of pyrimidine degradation in mice. Genetics *49:* 979–989 (1964).

Dickerman, R.C.; Feinstein, R.N., and Grahn, D.: Position of the acatalasemia gene in linkage groupe V of the mouse. J. Hered. *59:* 177–178 (1968).

Doyle, D. and Schimke, R.T.: The genetic and developmental regulation of hepatic δ-aminolevulinate dehydratase in mice. J. biol. Chem. *244:* 5449–5459 (1969).

Endo, T. and Schwartz, D.: Tissue specific variations in the urea sensitivity of the E_1 esterase in maize. Genetics *54:* 233–239 (1966).

Englesberg, E.; Irr, J.; Power, J., and Lee, N.: Positive control of enzyme synthesis by gene C in the L-arabinose system. J. Bacteriol. *90:* 946–957 (1965).

Erickson, R.P.; Gluecksohn-Waelsch, S., and Cori, C.F.: Glucose-6-phosphate deficiency caused by radiation-induced alleles at the albino locus in the mouse. Proc. nat. Acad. Sci., Wash. *59:* 437–444 (1968).

Fallon, H.J.; Smith, L.H.; Graham, J.B., and Burnett, C.H.: A genetic study of hereditary orotic aciduria, New Engl. J. Med. *270:* 878–881 (1964).

Feinstein, R.N.; Braun, J.T., and Howard, J.B.: Acatalasemic and hypocatalasemic mouse mutants. II. Mutational variations in blood and solid tissue catalases. Arch. Biochem. *120:* 165–169 (1967).

Feinstein, R.N.; Faulhaber, J.T., and Howard, J.B.: Acatalasemia and hypocatalasemia in the dog and the duck. Proc. Soc. exp. Biol., N.Y. *127:* 1051–1054 (1968).

Feinstein, R.N.; Howard, J.B.; Braun, J.T., and Seaholm, J.E.: Acatalasemic and hypocata-lasemic mouse mutants. Genetics *53:* 923–933 (1966).

Feinstein, R.N.; Sacher, G.A.; Howard, J.B., and Braun, J.T.: Comparative heat stability of blood catalase. Arch. Biochem. *122:* 338–343 (1967).

Figge, F.H.J. and Strong, L.C.: Xanthine oxidase (dehydrogenase) activity in livers of mice of cancer-susceptible and cancer-resistant strains. Canc. Res. *1:* 779–784 (1941).

Fritz, P.J.; Vesell, E.S.; White, E.L., and Pruitt, K.M.: The roles of synthesis and degradation in determining tissue concentrations of lactate dehydrogenase-5. Proc. nat. Acad. Sci., Wash. *62:* 558–565 (1969).

Ganschow, R. and Paigen, K.: Separate genes determining the structure and intracellular location of hepatic glucuronidase. Proc. nat. Acad. Sci., Wash. *58:* 938–945 (1967).

Ganschow, R. and Paigen, K.: Glucuronidase phenotypes of inbred mouse strains. Genetics *59:* 335–349 (1968).

Ganschow, R. and Schimke, R.T.: Genetic control of catalase in inbred mice; in *San Pietro, Lamborg and Kenney* Regulatory mechanisms for protein synthesis in mammalian cells, pp. 377–398 (Acad. Press, New York 1968).

Ganschow, R. and Schimke, R.T.: Independent genetic control of the catalytic activity and the rate of degradation of catalase in mice. J. biol. Chem. *244:* 4649–4658 (1969).

Goldberg, E.; Cuerrier, J.P., and Ward, J.C.: Lactate dehydrogenase ontogeny, paternal gene activation, and tetramer assembly in embryos of brook trout, lake trout and their hybrids. Biochem. Genet. *2:* 335–350 (1969).

Greenstein, J.P. and Andervont, H.B.: The liver catalase activity of tumor-bearing mice and the effect of spontaneous regression and of removal of certain tumors. J. nat. Canc. Inst. *2:* 345–355 (1945).

Gurdon, J.B.: Control of gene activity during the early development of *Xenopus laevis;* in *Brink* Heritage from Mendel, pp. 203–241 (Univ. Wisconsin Press, Madison 1967).

Harris, H.: Genetic control of enzyme formation in man; in Second Int. Conf. on Congenital Malformations, pp. 135–144 (International Medical Congress, Ltd., New York 1964).

Harris, H.: Enzyme polymorphisms in man. Proc. roy. Soc. (Biol.) *164:* 298–310 (1966).

Harris, H.: Enzyme and protein polymorphism in human populations. Brit. med. Bull. *25:* 5–13 (1969).

Harrison, B.J. and Fincham, J.R.S.: Instability at the *Pal* locus in *Antirrhinum majus.* I. Effects of environment on frequencies of somatic and germinal mutation. Heredity *19:* 237–258 (1964).

Henderson, N.S.: Isozymes and genetic control of NADP-malate dehydrogenase in mice. Arch. Biochem. *117:* 28–33 (1966).

Heston, W.E.; Hoffman, H.A., and Rechcigl, M., Jr.: Genetic analysis of liver catalase acti-vity in two substrains of C57 BL mice. Genet. Res. *6:* 387–397 (1965).

Hitzeroth, H.; Klose, J.; Ohno, S., and Wolf, V.: Asynchronous activation of parental alleles at the tissue-specific gene loci observed on hybrid trout during early development. Biochem. Genet. *1:* 287–300 (1968).

Ho, M.W. and O'Brien, J.S.: Hurler's syndrome: deficiency of a specific β-galactosidase isoenzyme. Science *165:* 611–613 (1969).

Holmquist, W.R. and Schroeder, W.A.: A new N-terminal blocking group involving a Schiff base in hemoglobin. Biochemistry, Wash. *5:* 2489–2503 (1966).

Howell, R.R.; Klingenberg, J.R., and Krooth, R.S.: Enzyme studies on diploid cell strains developed from patients with hereditary orotic aciduria. Johns Hopk. med. J. *120:* 81–88 (1967).

Huisman, T.H.J.; Schroeder, W.A.; Dozy, A.M.; Shelton, J.R.; Shelton, J.B.; Boyd, E.M., and Apell, G.: Evidence for multiple structural genes for the γ-chain of human fetal hemoglobin in hereditary persistence of fetal hemoglobin. Ann. N.Y. Acad. Sci. *165:* 320–331 (1969).

Hutton, J.J. and Coleman, D.L.: Linkage analyses using biochemical variants in mice. II. Levulinate dehydratase and autosomal glucose 6-phosphate dehydrogenase. Biochem. Genet. *3:* 517–523 (1969).

Ingram, V.M.: Abnormal human hemoglobins. III. The chemical difference between normal and sickle cell hemoglobins. Biochem. biophys. Acta *36:* 402–411 (1959).

Khanolkar, V.R. and Chitre, R.G.: Studies in esterase (butyric) activity. I. Esterase content of serum of mice from certain cancer-resistant and cancer-susceptible strains. Canc. Res. *2:* 567–570 (1942).

Kolodny, E.H.; Brady, R.O., and Volk, B.W.: Demonstration of an alteration of ganglioside metabolism in Tay-Sachs disease. Biochem. biophys. Res. Comm. *37:* 526–531 (1969).

Komma, D.J.: Glucose-6-phosphate dehydrogenase in *Drosophila:* a sex-influenced electrophoretic variant. Biochem. Genet. *1:* 229–237 (1968).

Kretschmer, N.; Hurwitz, R., and Raiha, N.: Some aspects of urea and pyrimidine metabolism during development. Biol. Neonat. *9:* 187–196 (1966).

Krooth, R.S.: Properties of diploid cell strains developed from patients with an inherited abnormality of uridine biosynthesis. Cold Spr. Harb. Symp. quant. Biol. *29:* 189–212 (1964).

Krooth, R.S.: Studies on the regulation of UMP synthesis in human diploid cells; in *Padykula* Control mechanisms in expression of cellular phenotypes (Academic Press, New York 1970).

Langridge, J.: Genetic and enzymatic experiments relating to the tertiary structure of β-galactosidase. J. Bacteriol. *96:* 1711–1717 (1968a).

Langridge, J.: Genetic evidence for the disposition of the substrate binding site of β-galactosidase. Proc. nat. Acad. Sci., Wash. *60:* 1260–1267 (1968b).

Law, L.W.; Morrow, A.G., and Greenspan, E.M.: Inheritance of low liver glucuronidase activity in the mouse. J. nat. Canc. Inst. *12:* 909–916 (1952).

Lehmann, H. and Carrell, R.W.: Variations in the structure of human hemoglobin. With particular reference to the unstable hemoglobins. Brit. med. Bull. *25:* 14–23 (1969).

Lewontin, R.C. and Hubby, J.L.: A molecular approach to the study of genic heterozygosity in natural populations. II. Amount of variation and degree of heterozygosity in natural populations of *Drosophila pseudoobscura.* Genetics *54:* 595–609 (1966).

Luzzatto, L.; Usanga, E.A., and Reddy, S.: Glucose-6-phosphate dehydrogenase deficient red cells; resistant to infection by malarial parasites. Science *164:* 839–841 (1969).

Markert, C.L. and Whitt, G.S.: Molecular varieties of isozymes. Experientia *24:* 977–991 (1968).

Marti, H.R.; Beale, D., and Lehmann, H.: Hemoglobin Koelliker: a new hemoglobin appearing after severe haemolysis: a-2 minus 141 Arg β-2. Acta haemat., Basel *37:* 174–180 (1967).

Matsubara, S.; Suter, H., and Aebi, H.: Fractionation of erythrocyte catalase from normal, hypocatalatic and acatalatic humans. Humangenetik *4:* 29–41 (1967).

McClintock, B.: The control of gene action in maize. Brookhaven Symp. Biol. *18:* 162–184 (1965).

McClintock, B.: II. The role of the nucleus. Genetic systems regulating gene expression during development. Develop. Biol. Suppl. *1:* 84–112 (1967).

Mehl, E. and Jatzkewitz, H.: Evidence for the genetic block in metachromatic leucodystrophy (ML). Biochem. biophys. Res. Comm. *19:* 407–411 (1965).

Mehl, E. and Jatzkewitz, H.: Cerebroside 3-sulfate as a physiological substrate of arylsulfatase A. Biochem. biol. phys. Acta *151:* 619–627 (1968).

Mier, P.D. and Cotton, D.W.K.: Operon hypothesis: new evidence from the 'constant-proportion' group of the Embden-Meyerhog pathway. Nature, Lond. *209:* 1022–1023 (1966).

Moog, F. and Grey, R.D.: Spatial and temporal differentiation of alkaline phosphatase on the intestinal villi of the mouse. J. Cell Biol. *32:* C1–5 (1967).

Moog, F.; Vire, H.R., and Grey, R.D.: The multiple forms of alkaline phosphatase in the small intestine of the young mouse. Biochim. biophys. Acta *113:* 336–349 (1966).

Morrow, A.G.; Greenspan, E.M., and Carroll, D.M.: Liver-glucuronidase activity of inbred mouse strains. J. nat. Cancer Inst. *10:* 657–661 (1949).

Morrow, A.G.; Greenspan, E.M., and Carroll, D.M.: Comparative studies of liver-glucuronidase activity in inbred mice. J. nat. Cancer Inst. *10:* 1199–1203 (1950).

Mudd, S.H.; Finkelstein, J.F.; Irreverre, F., and Laster, L.: Transsulfuration in mammals. J. biol. Chem. *240:* 4382–4392 (1965).

Muller, C.J.: in Molecular evolution; a comparative study on the structure of mammalian and avian hemoglobins (Van Gorcum, Assen, Netherlands 1961).

Munkres, K.D. and Woodward, D.O.: On the genetics of enzyme locational specificity. Proc. nat. Acad. Sci., Wash. *55:* 1217–1224 (1966a).

Murphy, T.M. and Mills, S.E.: Immunochemical comparisons of mutant and wild-type a-subunits of tryptophan synthetase. Arch. Biochem. *127:* 7–16 (1968).

Nakamura, H.; Yoshiya, M.; Kajiro, K., and Kikuchi, G.: Anenzymia catalasea, a new type of constitutional abnormality. Physiological significance of catalase in the animal organism. Proc. Jap. Acad. *28:* 59–64 (1962).

Nanney, D.L.: Ciliate genetics: patterns and programs of gene action. Ann. Rev. Genet. *2:* 121–140 (1968).

Nayudu, P.R.V. and Moog, F.: Intestinal alkaline phosphatase: regulation by a strain-specific factor in mouse milk. Science *152:* 656–657 (1966).

Nayudu, P.R.V. and Moog, F.: The genetic control of alkaline phosphatase activity in the duodenum of the mouse. Biochem. Genet. *1:* 155–170 (1967).

Neel, J.V.: The inheritance of sickle cell anemia. Science *110:* 64–66 (1949).

Ohno, S.; Christian, L.; Stenius, C.; Castro-Sierra, E., and Muramoto, J.: Developmental genetics of the alcohol dehydrogenase locus of the Japanese quail. Biochem. Genet. *2:* 361–369 (1969).

Okada, S. and O'Brien, J.S.: Tay-Sachs disease, generalized absence of a β-D-N-acetylhexosaminidase component. Science *165:* 698–700 (1969).

Paigen, K.: The effect of mutation on the intracellular location of β-glucuronidase. Exp. Cell Res. *25:* 286–301 (1961a).

Paigen, K.: The genetic control of enzyme activity during differentiation. Proc. nat. Acad. Sci., Wash. *47:* 1641–1649 (1961b).

Paigen, K. and Ganschow, R.: Genetic factors in enzyme realization. Brookhaven Symp. Biol. *18:* 99–114 (1965).

Paigen, K. and Noell, W.K.: Two linked genes showing a similar timing of expression in mice. Nature *190:* 148–150 (1961).

Pauling, L.; Itano, H.A.; Singer, S.J., and Wells, I.C.: Sickle cell anemia, a molecular disease. Science *110:* 543–548 (1949).

Perry, R.P.: The nucleolus and the synthesis of ribosomes. Progr. Nucl. Acid Res. *6:* 219–257 (1967).

Pette, D.; Luth, W., and Bücher, T.: A constant-proportion group in the enzyme activity pattern of the Embden-Meyerhof chain. Biochem. biophys. Res. Comm. *7:* 419–424 (1962a).

Pette, D.; Klingenberg, M., and Bücher, T.H.: Comparable and specific proportions in the mitochondrial enzyme activity pattern. Biochem. biophys. Res. Comm. *7:* 425–429 (1962b).

Pinsky, L. and Krooth, R.S.: Studies on the control of pyrimidine biosynthesis in human diploid cell strains. I. Effect of 6-azauridine on cellular phenotype. Proc. nat. Acad. Sci., Wash. *57:* 925–932 (1967a).

Pinsky, L. and Krooth, R.S.: Studies on the control of pyrimidine biosynthesis in human diploid cell strains. II. Effects of 5-azaorotic acid, barbituric acid, and pyrimidine precursors on cellular phenotype. Proc. nat. Acad. Sci., Wash. *57:* 1267–1274 (1967b).

Power, J.: The L-rhamnose genetic system in *Escherichia coli* K-12. Genetics *55:* 557–568 (1967).

Rader, T.: Inheritance of hypocatalasemia in guinea-pigs. J. Genet. *57:* 169–172 (1960).

Rechcigl, M., Jr.: Relative role of synthesis and degradation in regulation of catalase activity; in *San Pietro, Lamborg and Kenney* Regulatory mechanisms for protein synthesis in mammalian cells, pp. 399–416 (Acad. Press, New York 1968).

Rechcigl, M., Jr. and Heston, W.E.: Tissue catalase activity in several C57BL substrains and in other strains of inbred mice. J. nat. Cancer Inst. *30:* 855–864 (1963).

Rechcigl, M., Jr. and Heston, W.E.: Genetic regulation of enzyme activity in mammalian system by the alteration of the rate of enzyme degradation. Biochem. biophys. Res. Comm. *27:* 119–124 (1967).

Reichlin, M.; Bucci, E.; Fronticelli, C.; Wyman, J.; Antonini, E.; Ioppolo, C., and Rossi-Fanelli, A.: The properties and interactions of the isolated α- and β-chains of human hemoglobin. IV. Immunological studies involving antibodies against the isolated chains. J. molec. Biol. *17:* 18–28 (1966).

Reichlin, M.; Hay, M., and Levine, L.: Antibodies to human A1 hemoglobin and their reaction with A2, S, C, and H hemoglobins. Immunochemistry *1:* 21–30 (1964).

Ritossa, F.M.; Atwood, K.C., and Spiegelman, S.: Internat. Symp. Biochem. of Ribosomes and Messenger RNA, Reinhardsbrunn (1968).

Ritossa, F.M. and Spiegelman, S.: Localization of DNA complementary to ribosomal RNA in the nucleolus organizer region of *Drosophila melanogaster.* Proc. nat. Acad. Sci., Wash. *53:* 737–745 (1965).

Robinson, D. and Stirling, J.L.: N-acetyl-β-glucosaminidases in human spleen. Biochem. J. *107:* 321–327 (1968).

Russell, R.L. and Coleman, D.L.: Genetic control of hepatic δ-aminolevulinate dehydratase in mice. Genetics *48:* 1033–1039 (1963).

Schroeder, W.A.; Cua, J.T.; Matsuda, G., and Fenninger, W.D.: Hemoglobin F1, an acetyl-containing hemoglobin. Biochem. biophys. Acta *63:* 532–534 (1962).

Schroeder, W.A.; Huisman, T.H.J.; Shelton, J.R.; Shelton, J.B.; Kleihauer, E.F.; Dozy, A.M., and Robberson, B.: Evidence for multiple structural genes for the γ chain of human fetal hemoglobin. Proc. nat. Acad. Sci., Wash. *60:* 537–544 (1968).

Schroeder, W.A.; Shelton, J.R.; Shelton, J.B., and Olson, B.M.: Some amino acid sequences in bovine-liver catalase. Biochim. biophys. Acta *89:* 47–65 (1964).

Schwartz, D.: Genetic studies on mutant enzymes in maize: synthesis of hybrid enzymes by heterozygotes. Proc. nat. Acad. Sci., Wash. *46:* 1210–1215 (1960).

Schwartz, D.: Genetic studies on mutant enzymes in maize. III. Control of gene action in the synthesis of pH 7.5 esterase. Genetics *47:* 1609–1615 (1962).

Schwartz, D.: Tissue specificity and the control of gene action in structure and function of the genetic material, p. 201 Erwin-Baur-Gedächtnisvorlesungen III (Akademie Verlag, Berlin 1963).

Schwartz, D.: Genetic studies on mutant enzymes in maize. V. *In vitro* interconversion of allelic isozymes. Proc. nat. Acad. Sci., Wash. *52:* 222–226 (1964).

Schwartz, D.; Fuchsman, L., and McGrath, K.H.: Allelic isozymes of the pH 7.5 esterase in maize. Genetics *52:* 1265–1268 (1965).

Schwartz, D. and Laughner, W.J.: A molecular basis for heterosis. Science *166:* 626–627 (1969).

Sengbusch, P. von: Aminosäureaustausche und Tertiärstruktur eines Proteins. Vergleich von Mutanten des Tabakmosaikvirus mit serologischen und physikochemischen Methoden. Z. Vererbungslehre *96:* 364–386 (1965).

Shows, T.B.; Massaro, E.J., and Ruddle, F.H.: Evolutionary evidence for a regulator gene controlling the lactate dehydrogenase *B* gene in rodent erythrocytes. Biochem. Genet. *3:* 525–536 (1969).

Shows, T.B. and Ruddle, F.H.: Function of the lactate dehydrogenase *B* gene in mouse erythrocytes: evidence for control by a regulatory gene. Proc. nat. Acad. Sci., Wash. *61:* 574–581 (1968).

Sick, K.; Beale, D.; Irvine, D.; Lehmann, H.; Goodall, P.T., and MacDougall, S.: Hemoglobin G Copenhagen and hemoglobin J Cambridge. Two new β-chain variants of hemoglobin A. Biochim. biophys. Acta *140:* 231–242 (1967).

Sidman, R.L. and Green, M.C.: Retinal degeneration in the mouse: location of the *rd* locus in linkage group XVII. J. Hered. *56:* 23–29 (1965).

Smith, L.H., Jr.; Sullivan, M., and Huguley, C.M., Jr.: Pyrimidine metabolism in man. IV. The enzymatic defect of orotic aciduria. J. clin. Invest. *40:* 656–664 (1961).

Smithies, O.; Connell, G.E., and Dixon, G.H.: Chromosomal rearrangements and the evolution of haptoglobin genes. Nature, Lond. *196:* 232–236 (1962).

Szeinberg, A.; De Vries, A.; Pinkhas, J.; Djaldetti, M., and Ezra, R.: A dual hereditary red blood cell defect in one family: hypocatalasemia and glucose-6-phosphate dehydrogenase deficiency. Acta genet. med. *12:* 247–255 (1963).

Takahara, S.: Progressive oral gangrene due to lack of catalase in the blood. (Acatalasemia): report of nine cases. Lancet *2:* 1101 (1952).

Tashian, R.E.; Shreffler, D.C., and Shows, T.B.: Genetic and phylogenetic variation in the different molecular forms of mammalian erythrocyte carbonic anhydrases. Ann. N.Y. Acad. Sci. *151:* 64–77 (1968).

Van Regenmortel, M.H.V.: Serological studies on naturally occurring strains and chemically induced mutants of tobacco mosaic virus. Virology *31:* 467–480 (1967).

Vesell, E.S.: Symposium: multiple molecular forms of enzymes. Ann. N.Y. Acad. Sci. *151:* 1–689 (1968).

Vogell, W.; Bishai, F.R.; Bücher, T.; Klingenberg, M.; Pette, D., and Zebe, E.: On the structural and enzymatic types of muscles in the migratory locust. Biochem. Z. *332:* 81–117 (1959).

Walsh, K.A.; Ericsson, L.H., and Neurath, H.: Bovine carboxypeptidase A variants resulting from allelomorphism. Proc. nat. Acad. Sci., Wash. *56:* 1339–1344 (1966).

Warner, R.L.; Hageman, R.H.; Dudley, J.W., and Lambert, R.J.: Inheritance of nitrate reductase activity in *Zea mays* L. Proc. nat. Acad. Sci., Wash. *62:* 785–792 (1969).

Weatherall, D.J.: The thalassemias. Progr. med. Genet. *5:* 8–57 (1967).

Woodward, D.O. and Munkres, K.D.: Alterations of a maternally inherited mitochondrial structural protein in respiratory deficient strains of *Neurospora*. Proc. nat. Acad. Sci., Wash. *55:* 872–880 (1966).

Wright, D.A. and Moyer, F.H.: Parental influences on lactate dehydrogenase in the early development of hybrid frogs in the genus *Rana*. J. exp. Zool. *163:* 215 (1966).

Wright, D.A. and Moyer, F.H.: Inheritance of frog lactate dehydrogenase patterns and the persistence of maternal isozymes during development. J. exp. Zool. *167:* 197 (1968).

Wuu, K.D. and Krooth, R.S.: Dihydroorotic acid dehydrogenase activity of human diploid cell strains. Science *160:* 539–541 (1968).

Yoshida, A.; Stamatoyannopoulos, G., and Motulsky, A.G.: Negro variant of glucose-6-phosphate dehydrogenase deficiency (A-) in man. Science *155:* 97–99 (1967).

Author's address: Dr. *Kenneth Paigen*, Department of Experimental Biology, Roswell Park Memorial Institute, *Buffalo, NY 14203* (USA)

Enzyme Synthesis and Degradation in Mammalian Systems, pp. 47–76
(Karger, Basel 1971)

The Control of Enzyme Activity
in Mammals in Early Development and in Old Age

Florence Moog

Department of Biology, Washington University, St. Louis, Mo.

Contents

I.   Introduction

The life of a mammal is a continuum that begins with the primordial germ cell and ends with the death of the individual formed by the union of 2 mature germ cells. Within this continuum it is sometimes useful, and feasible, to delineate a series of phases characterized by distinctive properties; but these phases are by no mean discontinuous. Thus fertilization, the most narrowly circumscribed event in the life cycle, essentially allows the egg to express the potentialities invested in it during the course of oogenesis. And even birth, despite its drama, in a sense merely permits the fetus to enter upon a life-style for which many of the molecular preparations have been made well in advance.

Between fertilization and birth, the mammal passes through the most radical and fast-moving of the series of events that constitute development. The egg cleaves, and subsequently the ectoderm, endoderm, and mesoderm are formed. Although it has not yet been possible to explore the potentialities of parts of the mammalian embryo at this stage, in other vertebrates it is well established that at a comparable level of development, the future fate of each part of the embryo has become irreversibly fixed, or 'determined'. This state is followed by the period of primary differentiation, in which the invisible determination is translated into visible differences that are the first evidences of future tissues and organs. Some time later each organ rudiment passes through a more or less prolonged period of definitive differentiation, in which it acquires the chemo-structural characteristics that enable it to assume its proper function. The identification of the factors that control these developmental steps remains a major issue in biochemical embryology.

In the prenatal life of the mammal, the shifting relations between embryo or fetus and mother afford opportunities for interaction that have no counterpart in other vertebrates. The cleaving mammalian egg is exposed first to the secretions of the oviduct, subsequently to those of the uterus. Beginning with the blastocyst stage, it establishes a closer relation to the maternal system, either through the apposition of chorionic membrane and endometrial surface, or through partial or complete implantation in the endometrium. In either case it receives from the maternal system not only basic foodstuffs, but also vitamins, cofactors of various sorts, hormones, normal and abnormal metabolites, any of which may participate in the regulation of enzyme activity in the embryonic system. Conversely the fetus and placenta profoundly affect the physiology of the mother (75). At birth the separation from maternal and placental factors abruptly imposes on the newborn the need to activate new regulatory mechanisms; at the same time, the colostrum and milk provide a renewed source of maternal influences.

The control of enzyme activity in mammals thus involves a dimension of complexity not encountered in work on the self-sufficient embryonic systems of birds and lower vertebrates. Because of the practical difficulties in experimenting with the mammalian embryo and fetus, studies on the control of their enzymology, with the exception of the liver in the perinatal period, are not abundant. In the sections that follow, reference will be made to the chick embryo, and other embryos, particularly where the results may be applicable to the mammal. Nevertheless, the special conditions of intra-uterine life do require that the regulation of enzyme activity in mammalian development be elucidated in mammals.

The subject of enzyme development has been reviewed recently in terms of the differentiation of metabolic pathways (79, 139), in relation to functional differentiation (116), and with special reference to the late fetal perinatal periods (119).

II. The Preimplantation Phase

The vertebrate egg is nurtured in the ovary, where it acquires the structural and metabolic characteristics that are required for its post-fertilization development (176). That the pattern of energy metabolism of the mouse zygote is carried over from the oocyte stage has been demonstrated by *in vitro* experiments in which potential substrates are added to the medium individually. Oocytes will undergo maturation, and zygotes will cleave once, in the presence of pyruvate or oxalacetate, but neither event will occur if lactate, phosphoenolpyruvate, or glucose are used (12), although the egg is permeable to ^{14}C-DL-lactate (169), and lactic dehydrogenase is abundant (17). From the 2-cell stage on, lactate and phosphoenolpyruvate will support development (18). Other intermediates in the Embden-Meyerhof pathway, as well as metabolites in the citric acid cycle, are ineffective at the 2-cell stage, but do permit development to proceed from the 8-cell stage to the blastocyst, as glucose and alanine do also (21). The rabbit embryo can use pyruvate, lactate and phosphoenolpyruvate to support development for 24 h from the 2-cell stage, but apparently cannot utilize glucose or citric acid cycle intermediates until the beginning of blastulation, or later (39, 58).

The inference that the citric acid cycle becomes increasingly important in preimplantation development is supported by the fact that malic dehydrogenase activity rises steadily in the mouse in this period (19), as succinic dehydrogenase does in the hamster and rabbit (87). Lactic dehydrogenase on the other hand decreases sharply after the 8-cell stage (17), as glucose-6-phosphate dehydrogenase does also (20), this result possibly indicating a shift from the pentose shunt to the Embden-Meyerhof pathway in late cleavage.

The mouse egg cannot develop even through the first cleavage without an exogenous energy source (12), in contrast to the equally small and microlecithal sea urchin egg, which requires only inorganic ions to support development to the larval stage (78). This fact suggests that the mammalian egg became adapted in evolution to depend from the outset on nutrients supplied by the maternal environment. During maturation and first cleavage the follicle cells can support the ovum by metabolizing lactate or glucose, which the ovum cannot utilize for itself (12). Subsequently the cleaving egg passes slowly through the Fallopian tube, bathed in secretions that are probably rich in substrates and may differ in composition between the isthmus and ampulla of the tube. The availability of substrates in the environment may thus constitute a basic mechanism for regulating enzymic activity, with permeability also playing a critical role. ^{14}C-L-malate cannot enter mouse eggs until the 8-cell stage, at which time it can be utilized (168).

Whether the shifting pattern of metabolism reflects activation of pre-existing enzyme molecules, or synthesis of new enzymes, or changes in degradation rates, is unknown. *Manes and Daniel* (105) detected no qualitative change in isotopically labelled proteins synthesized in rabbit ova during cleavage stages, but the data do not preclude the appearance of a new enzyme in physiologically effective amount. It may be significant that in contrast to sea urchin and amphibian eggs, in which inhibition of synthesis of new heterogeneous RNA does not prevent development as far as gastrulation (74, 92), the mammalian egg appears to be influenced by its own genome at an earlier stage. Uptake of ^{3}H-uridine goes on at a low but significant rate in the mouse ovum during cleavage stages (111), as might be expected from the increased cytoplasmic basophilia at this time, when ^{3}H-uridine appears to pass from nucleolus to cytoplasm (109). The t^{12} gene, when homozygous, causes death as early as the morula stage (158), and in the lethal condition the normal increase of basophilia either fails to occur (109, 110), or appears irregular (25). The tubal and uterine fluids may contain novel regulators responsible for controlling early gene-dependent events: 'blastokinin', a protein in the uterine fluid that appears to be needed for the cavitation of the rabbit morula (98), may be the first example of such a protein. So far there is evidence suggesting the *de novo* origin of only one enzyme protein in preimplantation ova. This is the A subunit of lactic dehydrogenase, which *Auerbach and Brinster* (4) and *Rapola and Koskimies* (142) have reported to be absent from the mouse ovum until the blastocyst implants; according to the former authors, the event is dependent on the action of estrogen in providing the essential conditions for implantation.

III. Determination and Early Differentiation

Within the blastocyst, gastrulation is completed, the primitive streak appears, and the ground plan of the future embryo is determined. Although the steps that immediately follow have not been examined, there is little reason to doubt that the mammalian embryo, like that of the amphibian (82) or bird (148), undergoes a series of stepwise determinations that gradually specify organ, tissue, cell type. Probably each specific differentiation is preceded by a specific inductive event. Defining differentiation by morphological criteria leaves a dark period between the time when a cell group becomes committed to develop in a given direction, and the time when visible difference emerges. Defining differentiation in terms of enzymes (or other molecules with readily recognizable characteristics) is however making it possible to penetrate into the dark period and to work back toward the inductive event itself. The results of such investigations may reveal that embryonic induction and enzyme induction, which in their limited senses do not seem to have much in common, are in reality analogous events occurring at different stages in an epigenetic continuum. Despite the difficulties imposed by the small size of embryos and the paucity of ready definable end points in the period when many inductive events are occurring, some progess has been made. Work on the pancreas, on cartilage and muscle, and on blood-forming tissues, among others, is beginning to suggest answers to fundamental questions about the control of differentiation.

A.   The Exocrine Pancreas

The pancreas has been studied with particular reference to the formation of the specific enzymes that it produces for secretion into the intestine. In the mouse embryo the pancreas first appears as an outpouching from the midgut endoderm on day 11 of gestation. Provided that the epithelium remains in the presence of mesoderm (64), the rudiment will differentiate in culture, and much that has been learned about its biochemical development is based on the fact that under appropriate conditions, the explanted rudiment rather precisely keeps pace with its counterpart *in vivo* (90, 150). In both situations cell division proceeds actively at 11 days, but 2 days later the central cells fail to incorporate ^{3}H-thymidine, and mitoses are soon confined to the periphery; after 4 days the zymogen granules that are characteristic of exocrine secretion appear in the non-dividing central cells (171).

Although these results indicate that in this tissue replication of DNA may be incompatible with a high level of synthesis of specific protein product, *Rutter* and his coworkers (149, 150) report that lipase, trypsin, carboxypeptidase, and other enzymes that are components of the pancreatic juice, are detectable in the pancreatic rudiment as early as the 12th day. At the same time, these enzymes are present in much smaller quantities, or are undetectable, in other tissues. On the basis of these results, which have unfortunately never been published in detail, *Rutter et al.* (150), have suggested that the pancreatic acinar cells pass through a series of states of differentiation: (1) a 'predifferentiated' state in which the pancreas, though committed, has neither the morphological nor the biochemical characteristics of differentiation; (2) a 'protodifferentiated' state, extending from the 12th to the 17th day, in which traces of enzymes products may be detected; and (3) a fully differentiated state reached shortly before birth, after an intensive burst of synthetic activity that raises the concentration of the enzyme products by several orders of magnitude in a period of approximately 3 days. Incorporation of ^{3}H-leucine in the period of rapid increase indicates that the enzymes are being synthesized, not merely activated (150).

The shift from the protodifferentiated to the fully differentiated state is dependent on the presence of mesoderm (64, 171), or of a particulate fraction, derived from mesodermal tissues, that is heat labile, sensitive to trypsin, but unaffected by RNAase or DNAase (150). In the presence of mesoderm, or mesodermal factor, the epithelium proliferates and at the same time produces

stable, actinomycin-inhibitable RNA that is essential to subsequent differentiation (172, 173). If 13-day pancreatic rudiments are cultured in the presence of
actinomycin D for 3 days, with ^{3}H-leucine being added 3 days later, the labelling
of total protein is reduced by about 40 %, but that of chymotrypsin and amylase
are reduced by more than 90 %; by contrast the addition of actinomycin D on
days 5 and 6 in culture has a relatively slight effect on the labelling of the
2 enzymes (150). In this case, apparently, the transcription of the genome
occurs several days in advance of the massive translational activity that marks
the shift from the protodifferentiated to the fully differentiated state. The early
presence of largely idle messenger RNA may account for the occurrence of
traces of specific products in the protodifferentiated state.

The work of *Rutter* and his colleagues (149, 150) implies that when enzyme
activities level off at the end of the intensive burst of synthesis that occurs in
late fetal life, the fully functional state has been reached, with enzyme output
being subsequently subject only to limited fluctuations ('modulation'). In the
newborn rat, however, the activities of a-amylase (141) and lipase (146) are constant in the pancreas to the age of 10 days. Thereafter a new upsurge, culmi-
nating at 20 days or a little later, raises both activities about 10-fold, to the adult
level. The overall course of activity from embryo to adult thus resembles that of
alkaline phosphatase in the mouse duodenum; in late fetal life this activity surges
up from a low level only to pause before a second, corticoid-dependent rise
occurs in the third postnatal week (112, 113). The postnatal increases of both
a-amylase and lipolytic activity can be elicited precociously in the infant rat by
the administration of cortisone (141, 146). Whether these increases occurring
some time after birth should be correctly interpreted as modulatory changes
depends in part on whether the level of activity would return to that of the
newborn following adrenalectomy. Duodenal phosphatase falls when the adrenals are removed, but not as low as the infant level (120, 167).

B. Cartilage and Muscle,
    and the Relationship between Proliferation and Differentiation

Cartilage and muscle forming cells have been the subjects of extensive
investigation from the period of their initial divergence. Both types of cells are
contained in the mesodermal somites that are laid down in antero-posterior
progression alongside the neural tube when the embryonic body is first taking
shape. If small groups of somites from 3-day chick or 9-day mouse embryos are
explanted, they produce a few myoblasts and myotubes. No cartilage appears,
however, unless spinal cord or notochord (but not other tissues) are added (83).
About a day later, the isolated somites alone form cartilage, indicating that an
inductive influence has acted on them during the intervening time. Experiments

in which somites are exposed to spinal cord across a membrane filter for short periods show that the inductive event occurs in the first few hours (99).

Recognizable cartilage does not appear until about 3 days after the initial induction, either *in vivo* or *in vitro*. Cartilage is characterized by a unique glycosaminoglycan, chondroitin sulfate, and it was once thought that this substance could be used as an early indicator of cartilage induction (83). What has been found, however, is that chondroitin sulfate is present in somite cells at the stage at which they are inducible (57, 107). The action of the inducer therefore is to reinforce an existing pathway, not to evoke a new one. The condition of the inducible somite cells thus appears to be analogous to that of the pancreas: in both, specific products are present in minute quantities for several days before an accelerated rate of synthesis comes into play.

What may go on in the period of protodifferentiation is suggested by *Medoff*'s (108) study of the enzymic pathway leading to chondroitin sulfate. Using dissociated limb mesenchyme from the 3-day old chick embryo, under conditions that permitted the differentiation of cartilage exclusively, *Medoff* found that sulfate-activating enzymes, uridine diphosphate-D-glucose dehydrogenase, uridine diphosphate-N-acetylglucosamine-4-epimerase, and a polymerase, all increased in specific activity during a 7-day culture period; but the activity of glucose-6-phosphate dehydrogenase and ATPase, which are not directly involved in cartilage synthesis, did not rise. As the tissue mass increased, the rates of synthesis of the cartilage-related enzymes increased sharply. This shift in rates, which occurs after 2–3 days, suggests the accumulation in the culture medium of a factor that may stimulate the production of specific enzymes, and thus make possible the rapid deposition of chondroitin sulfate. The enzymes under study were all readily detectable in the undifferentiated limb mesenchyme (108), and also seem to be present at the same stage in non-cartilage-forming tissues (107). This fact again indicates that the initial inducting event is the selection and stabilization of a set of pre-existing conditions.

The embryonic somites give rise to muscle as well as cartilage, spatial segregation of the presumptive cell types beginning in the chick embryo in the anterior somites late on the second day of incubation and spreading posteriorly. As noted before, explanted somites alone produce a few recognizable muscle cells, but do not produce large amounts of muscle (myosin) unless acted on by spinal cord (83). By use of fluorescein-labelled antimyosin it can be shown that a cross-reacting material is present in somites, but not elsewhere, in the stage 16 (52 h) chick embryo, about half a day earlier than striations can be demonstrated by conventional histological methods (83, 85).

In the differentiation of skeletal muscle, proliferation is restricted to mononucleated mesenchyme cells which differentiate into myoblasts and subsequently fuse to form multinucleate muscle fibers. The massive accumulation of myosin that is essential to the functional differentiation of muscle occurs only in the multinucleate structures (7, 32). On the other hand, myosin has not been demonstrated in myoblasts in which DNA synthesis is taking place (83, 133); nor does it accumulate in cultured myogenic cells in which cell division and growth are abnormally prolonged by 5-bromodeoxyuridine (33, 161). These facts have been interpreted to mean that cells must withdraw from the mitotic cycle before beginning to translate for a specialized product (134). The possibility cannot be ruled out, however, that a small amount of myosin, below the resolving power of the fluorescent antibody technique, is part of the mechanism that maintains a mesenchyme cell in the protodifferentiated state that is the prelude to myogenesis. Creatine kinase, which is quite specific to muscle, is present in dividing cells (32), and Ca^{++}-ATPase activity is present in both somites and unsegmented mesoderm of stage 11 (about 40 h) chick embryos (41). The disparity between these results and those obtained with fluorescent-labelled antibodies probably reflects the intrinsically greater sensitivity of enzymatic methods.

The apparent antagonism between DNA synthesis and specific protein synthesis that seems to exist in muscle has sometimes been considered to be a general mechanism by which the onset of the differentiated state is regulated (83, 84, 134). The situation in the endocrine pancreas seems to fit this hypothesis (173), as does the differentiation of lens fibres (160); and there are similar cases. That cessation or even diminution of DNA synthesis is not a prerequisite for the emergence or maintenance of a differentiated state is shown however by the fact that both pigmented retinal cells (23) and chondrocytes (24) can synthesize their special products while growing rapidly in culture. In the young mouse embryo, hemoglobin synthesis, which is the subject of the following section, proceeds at a high rate in yolk sac cells that are actively dividing (28).

C.   Erythropoietic Tissues

The differentiation of red blood cells has lent itself to the study of control mechanisms in development in a variety of ways. In mammals the first erythroid cells arise in blood islands on the yolk sac, to be followed by populations originating in the liver, spleen, and bone marrow; in fetal mice the process begins on the yolk sacs on the 8th day, in the liver on the 12th, in the spleen on the 16th, and in bone marrow at birth (106). In the chick embryo, in which red blood cells also develop first from blood islands in the yolk sac, the readiness with which the sac can be explanted has made it a rewarding subject for studies of the initial steps in erythroid differentiation. Though the blood islands are mesodermal, the presence of the endoderm is necessary for attainment of full hemoglobin-forming capacity (177), possibly because it transmits essential precursors from yolk to mesoderm.

Hemoglobin normally appears in the chick blastoderm at about 34 hours of incubation, or at a comparable stage in culture. The system necessary for the synthesis of hemoglobin is however laid down at an earlier stage, for actinomycin D does not block the appearance of hemoglobin unless applied more than 12 h before the protein appears (100, 178). If δ-aminolevulinic acid (ALA) is added to young colorless blastoderms, heme synthesis begins within 2 h; a day later half-blastoderms treated with ALA have produced 2.7 times as much hemoglobin as the untreated halves (100, 101). Since this effect is prevented by puromycin, it is probably not due to interaction of heme with preformed globin. Rather it appears that synthesis of globin depends on the availability of heme (100). A similar control of translation seems to be exerted in the yolk sac of the mouse, in which differentiation of erythroid cells is initiated on day 8. After day 10, hemoglobin becomes labelled with ^{14}C-ALA in the presence of sufficient actinomycin D to inhibit RNA synthesis by 95 % (54). In the chick blastoderm, glycine and succinyl-CoA, which are condensed to yield ALA, have no stimulatory effect on Hb synthesis, indicating that the critical step preventing translation in the earliest stages is a lack of ALA synthetase (101).

The hypothesis that the availability of ALA synthetase regulates hemoglobin synthesis is supported by the demonstration that 11-ketopregnanolone, and 2 closely related steroids, strongly elevate the activity of ALA synthetase in explanted blastoderms, at the same time increasing hemoglobin output in treated halves about 2.3-fold above the level in control halves (102). Actinomycin D prevents the stimulatory effect of the steroids. *Levere et al.* (102) suggest that certain 5β-HC$_{19}$ and C$_{21}$ steroids may serve as physiological derepressors, permitting the ALA synthetase gene to be transcribed and thus activating the processes that culminate in the appearance of hemoglobin. The critical role of ALA in controlling heme synthesis may be a general one, having first been described in chick embryo liver (68, 91). In reticulocytes of iron-deficient rabbits, heme promotes hemoglobin production, the effect being halted by puromycin or cycloheximide (69).

Fetal blood contains hemoglobins not found in postnatal stages. In the blood of the 15-day mouse *in utero* there are 4 Hb, 3 of which contain components somewhat more acidic than the β chain of the adult type (61). Although the source of the embryonic hemoglobins was for some time in doubt, it is now clear that they are associated with the nucleated erythrocytes from the yolk sac, whereas the non-nucleated erythrocytes, which in the 15-day mouse are of liver origin, produce adult hemoglobin (36, 97). That liver erythroid may in fact have no capacity to produce embryonic hemoglobin is indicated by the finding that the fetal mouse liver, even when taken as early as 11 days and cultured under a variety of conditions, produces only adult hemoglobins (140). By working with yolk sacs from 10–14 day embryos, *Fantoni et al.* (54) have been able to show that the 3 embryonic hemoglobins change in relative proportions, form II (containing a and y chains) being synthesized much more rapidly than form III (a and z chains). Since the production of these hemoglobins goes on in the presence of actinomycin D, the authors suggest that the rates of synthesis of the different globin chains are controlled by factors not involving supply of RNA's (54). Whether such sequential controls are exerted in a single cell is not known, though it is clear that yolk sac erythrocytes, in contrast to those from other sources, gradually lose ribosomes during their life in the blood stream (97).

In the adult mammal, the red cell supply is maintained by recruitment from a stem cell population in the bone marrow, although the liver and spleen may also produce erythrocytes under conditions of stress. The production of erythrocytes is regulated by erythropoietin, a hormone apparently made in the kidney (63). This substance is believed to act on stem cells themselves, causing some of their progeny to enter on the series of differentiative steps that lead to the enucleate, hemoglobin-laden erythrocyte (63). Fetal liver in organ culture, even as early as 10 3/4 days, responds to erythropoietin by incorporating ^{59}Fe into heme, and up to 15 days erythropoiesis *in vitro* depends on the presence of the hormone (31). Yolk sac, however, which may be induced by an endodermal factor (177), is insensitive to erythropoietin (31). Whether there is any connection between receptivity to a given signal, and the type of hemoglobin synthesized, can only be answered by fuller investigation of the potentialities and limitations of the precursor cells, and of the molecular alterations of the induced state.

IV. Control Mechanisms in Functional Differentiation

A. The Liver

1. Enzymes Involved in Gluconeogenesis

The capacity for function represents the culmination of the differentiative processes that have gone forward in a cell or tissue. Functional capacity may be attained early, as with hemopoietic tissues, or late, as with the estrogen-induced maturation of the uterine endometrium. The event may be regularly repeated, as in the intestinal epithelium, or it may occur sporadically, as with case in synthesis in the mammary gland. The phenomenon of birth in mammals entails a series of regulatory problems that the newborn organism must solve promptly if it is to survive. The liver, which must abruptly assume functions previously managed by the maternal system, has proved a particularly useful source of information about control mechanisms in mammalian development.

That birth itself affects enzyme activity was demonstrated more than a decade ago, by experiments showing that tryptophan pyrrolase activity in the rabbit liver rises 10-fold in the 24 h following birth, whether this event occurs at normal term or is experimentally premature or postmature (126). Neither tryptophan nor adrenal corticoids, which enhance tryptophane pyrrolase activity in the adult rat (55), alter activity in the unborn rabbit fetus (126). This situation has not been further elucidated. *Spiegel and Spiegel* (159) have however shown that in the rat liver, in which tryptophan pyrrolase activity begins to rise only at the end of the 2nd postnatal week, substantial activity can be elicited by assaying in the presence of EDTA; the authors attribute this effect to the high content of copper, an inhibitor of tryptophan pyrrolase, in the liver of the infant rat.

The rat liver does however contain 2 enzymes that do rise abruptly at birth, and appear to be controlled by glucagon secreted in response to hypoglycemia resulting in the newborn from the loss of glucose previously derived from the mother's blood. One of these enzymes is phosphopyruvate carboxylase (PPC), which catalyzes an essential step in the formation of glucose from pyruvate:

$$\text{---> pyruvate ---> oxalacetate ---> phosphoenolpyruvate -> -> -> glucose}$$

$$\begin{array}{cc} \uparrow & \uparrow \\ \text{pyruvate} & \text{phosphopyruvate} \\ \text{carboxylase} & \text{carboxylase} \end{array}$$

A trace of PPC activity is present in the particulate fraction of the 17-day fetal liver (6), but the enzyme remains barely detectable until birth occurs (166, 182). Pyruvate carboxylase activity however rises in late fetal life, reaching the adult level by the time birth occurs, and hexosediphosphatase and glucose-6-phosphatase, which participate in later steps in the pathway leading to glucose, are also rising rapidly before the animal is born (116, 182). If fetuses are delivered a day and a half before term, PPC activity immediately begins rising at the normal rate (180). The effect is inhibited by the administration of glucose to the newly delivered fetus, but glucagon accelerates the rise; the influence of glucagon is not reduced by glucose, indicating that the sugar exerts its inhibitory effect in the pancreas, not on PPC-activity directly (180). The inference that postnatal hypoglycemia elicits secretion of glucagon is supported by the fact that blood glucose levels decline about 80 % in the first 2 h after birth (180). This drop is prevented in prematurely delivered fetuses by administration of $3'5'$-cyclic AMP, which concomitantly elevates PPC activity (181).

These results imply that glucose deficiency, by evoking the appearance of one essential enzyme, opens the whole pathway of gluconeogenesis. This is evidently the case, for the ability of neonatal liver slices to incorporate ^{14}C-pyruvate into glucose rises in parallel with the increase of PPC activity (179). The pathway is definitely *not* regulated by coordinated synthesis of the enzymes involved in it.

2. Tyrosine Aminotransferase and Related Enzymes

Tyrosine aminotransferase (TAT) which participates in the oxidative degradation of phenylalanine by converting tyrosine to 2-oxoglutarate, rises sharply just after birth in the livers of rat, rabbit, guinea pig, and probably man (5, 104, 154). Starting from a barely detectable level, TAT activity increases to about twice that of the adult within 12 h after birth; the increase is inhibited by adrenalectomy, but activity cannot be elicited in the fetus *in utero* by the administration of hydrocortisone (72, 154). Glucagon, however, does elicit a more than 15-fold increase of TAT activity within 5 h if injected into fetuses a day or two before birth (70, 72). The same treatment also elevates the activity of serine dehydratase and glucose-6-phosphatase (72), but not of NADP dehydrogenase (70). Dibutyryl-3'5'-cyclic AMP has a similar selective effect; and *Yeung and Oliver* (181) have reported that this compound, which also induces phosphopyruvate carboxylase, does not affect fructose diphosphatase or pyruvate kinase in fetal rat liver.

Although hepatic TAT is inducible from late fetal life to old age (56), its responsiveness changes with time. Glucagon brings about precocious increase in rat liver during the 2 days before term, but not earlier; 3'5' cyclic AMP (or its dibutyryl derivative) is however effective 4 days before term (71). Thus the maturation of the adenyl cyclase system may be an essential prerequisite to the induction of TAT. Once birth occurs, the system becomes responsive to, and indeed dependent on, adrenal corticoids (72, 154). For a few days after birth glucagon and hydrocortisone are equally effective, but at 50 days the responsiveness to glucagon is lost (70). The prenatal resistance to induction by corticoids *in vivo* is apparently due to repression, for fetal rat liver cultured in defined medium does respond to added hydrocortisone: liver from near-term fetuses begins increasing in TAT activity after about 6 h in the presence of the hormone, and even 17-day liver (i.e., 4—5 days before term) responds after being maintained in culture for 42 h (174). In the liver of the 5-day chick embryo UDP-glucuronyltransferase, which normally appears only after hatching, rises to adult levels within 3 days in culture in defined media without the addition of any hormones or other inducing substances (157). The implication that this result indicates the presence of a repressor in the embryonic system is supported by the fact that the enzyme remained undetectable when livers were grafted onto the chorio-allantoic membrane.

Actinomycin D prevents the rapid increase of PPC (180) and TAT (81) activity in the liver of the prematurely delivered rat, and also blocks the induction of TAT or glucose-6-phosphatase in fetal liver by glucagon or dibutyryl 3'5'-cyclic AMP (71). In liver maintained in culture, actinomycin D reduces TAT induction in parallel with its inhibitory effect on incorporation of ^{3}H-orotate into RNA, and cycloheximide stops protein synthesis and enzyme induction (174). These results are consistent with the finding that hydrocortisone elicits an increase of an enzyme protein immunochemically identical with adult tyrosine aminotransferase; but $10^{-7}M$ hydrocortisone, which is maximally effective in elevating TAT activity *in vitro,* does not affect the rate of incorporation of ^{14}C-orotate or ^{3}H-leucine (174), as it does in adult liver (93). Although this paradox has not been explained, the effect of actinomycin D does implicate *de novo* RNA synthesis in TAT induction in the fetal liver. At term there normally appears to be a slight increase in the diversity of the RNA molecules being produced (29), and hydrocortisone may bring about a qualitative change in this population; the hormone has been reported to alter the distribution of messenger RNA's being produced in normal and regenerating adult liver (46).

The non-coordinate development of functionally related enzymes is again illustrated by hepatic phenylalanine hydroxylase, which oxidizes phenylalanine to tyrosine, and thus provides substrate for TAT. It was previously thought that this enzyme is lacking at birth in several animals, so that the newborn would suffer from a transient phenylketonuria (94, 116). *Brenneman and Kaufman* (16) have however shown that in the livers of fetal rats 2 days before birth the activity of phenylalanine hydroxylase is about one-third of the adult level, which is attained 12 h after birth. If the enzyme from fetal or newborn liver is assayed with added tetrahydropteridine, moreover, the activity rises close to that of the adult under the same conditions, and the further addition of dihydropteridine reductase brings both to the same level. The implication that the activity of the apoenzyme is regulated by the supply of a pteridine cofactor is supported by the fact that newborn liver contains only half as much cofactor activity as livers from rats 20 h old, or older (16).

B. The Small Intestine: Enzymes of the Surface Membrane

The dynamic architecture of the small intestine makes it an attractive system for studying the control of enzyme activities in spatial as well as temporal terms. The intestinal surface is continually replaced, the turnover time of the whole population of epithelial cells being about 2 days. New cells are proliferated in the crypts, and assume their functional characteristics as they emerge from the crypts to glide up the sides of the villi toward the extrusion zones at the tips. Since the unique functions of the small intestine are properties of the surface (37, 163, 164), it is significant that the surface expands as the cells migrate, According to *Brown* (22), the microvilli of the human jejunum become longer, narrower, and more densely packed during migration, with the consequence that their total surface area increases from $225\ \mu^2$/crypt cell to $1717\ \mu^2$/cell at the villus crest. Similar form change and surface enlargement occur in the rat jejunum (138) as well as in the mouse duodenum (27). These considerations have served to focus attention on disaccharidase, dipeptidases, and alkaline phosphatases, which are components of the outer membrane of the microvilli (49, 86, 89, 128, 132, 137).

Alkaline phosphatase, which probably participates in the breakdown and absorption of phosphorylated compounds in the ingesta, is known from histochemical studies to be highly active on the villi sides and tips in vertebrates of all kinds. By quantitative assay of sections cut at right angles to the length of the villi, it has been shown that specific activity rises steadily from base to tip in rat (130, 170), mouse (122) and chicken (73). In the mouse duodenum the increase is not simply quantitative, but involves also a change in the characteristics of the phosphatase, the activity at the tip being higher with phenylphosphate as substrate than with beta-glycerophosphate; this change is conveniently expressed as PhP/bGP ratio, which is about 2 at the villus bases, about 6 at the tips (122, 123). The duodenal phosphatase of the juvenile or adult mouse is a series of multiple forms that differ in chromatographic and electrophoretic properties, and have PhP/bGP ratios ranging from 2 to 10 (51, 52). These isozymes are not distributed at random, but appear to succeed each other in a systematic way as the cells travel toward the villi tips (121).

That intestinal phosphatase activity falls in adrenalectomized animals and is restored by adrenocortical hormones has been known for many years (120, 167). In the intact mouse, cortisone elevates duodenal phosphatase activity and also increases the PhP/bGP ratio (114). What is more unexpected is that actinomycin D, and puromycin, also elevate phosphatase activity and PhP/bGP ratio (115, 118), and analysis of cross sections of villi shows that the PhP/bGP ratio may rise as high as 5 at the villi bases and 9 at the tips (123). More than 24 h is required for these effects to reach significant levels. Cortisone and antibiotics have not been tested in combination in the adult mouse, but in the nursling (in which cycloheximide, colchicine, and ethionine also elicit an increase of phosphatase activity and PhP/bGP ratio), the effect of cortisone is augmented by actinomycin D, the results being apparently additive (117, 118). The amount of actinomycin D administered in these experiments is approximately half that required for 95 % inhibition of protein synthesis in the mouse (56). In the adult rat, actinomycin D enhances an increase of duodenal phosphatase induced by the stress of surgery (15).

These results indicate that in the mouse the increase of phosphatase activity and elevation of PhP/bGP ratio that go on as the cells migrate toward the villi tips are processes capable of being enhanced by agents that have in common the ability to interfere with protein synthesis. This suggestion is interesting in the light of *Overton*'s (136) conclusion that the expansion of the microvillus surface, a process that goes on concurrently with the alteration of the characteristics of phosphatase, may not involve *de novo* synthesis. Putting these facts together raises the possibility that differentiation of phosphatase and alteration of microvillus morphology may not involve protein synthesis directly; on the contrary, they seem to be restrained by a mechanism that does require protein synthesis. This hypothesis is in accord with the fact that new material is added to the surface as the cells travel upward (80, 127, 145). In the swift process of regeneration in the intestine of the metamorphosing bullfrog, microvilli appear to grow by the incorporation of vesicles that are abundant in the epithelial cells (14). Such vesicles are not seen in cells that are not under the peculiar stress of metamorphosis, but it is nevertheless possible that preformed elements sufficient to permit some enlargement of the surface are present at any time. Antibiotics, by lowering the rate of synthesis that does go on in cells that have moved out onto the villi (40, 155), might divert energy into microvillus enlargement. A critical question is whether microvilli reach exceptional lengths when antibiotics elevate the PhP/bGP ratio of phosphatase to abnormally high levels, and work in progress indicates that these results may occur in parallel; microvilli in intestines of laboratory mammals generally have length/width ratios of 5–10, but in the duodena of adult mice treated with 1.0 μg actinomycin D/gram of body weight, length/width ratios of 14–22 have been observed (135).

The hypothesis that the differentiation of phosphatase is intimately related to the differentiation of microvilli is strengthened by the fact that these two entities also develop in parallel in fetal and infant intestines. A regular array of microvilli appears in the rat duodenum during the last 3 days before birth (47), and phosphatase increases at the same time (30). In the mouse phosphatase activity and microvillus length increase sharply during the 3 days preceding birth, and both increase again midway through the 3rd postnatal week (112, 114, 136). Up to the end of the 2nd week in the mouse the phosphatase present has a PhP/bGP ratio of about 0.7 (51, 52, 114), and is uniformly distributed along the villi (123). The family of isozymes characterized by PhP/bGP ratios of 2 and higher may be present in inactive form in the infant intestine (50), but they only manifest their high activity in the 3rd week. Then the tendency for the PhP/bGP ratio to rise as cells migrate up the villus appears (123), and concurrently microvillus morphology shifts from the short, broad form of the infant intestine to the long, narrow, mature form (136). The appearance of high-activity, high-ratio alkaline phosphatase may be elicited more than a week ahead of the normal time by the administration of cortisone (113, 114), and the same treatment brings about premature lengthening of microvilli (136). As pointed out before, the effect of cortisone on phosphatase in the infant intestine is augmented by actinomycin D, cycloheximide, and puromycin (117, 118).

If the differentiation of phosphatase is affected by qualitative and quantitative changes in the intestinal surface, it might be expected that other surface-related enzymes would show similar patterns. In the adult rat, not only phosphatase but also lactase, sucrase, maltase, leucyl naphthylamidase, and 1-alanyl-1-proline dipeptidase are weak or absent in the crypts, but increase several fold, in parallel with phosphatase, as cells migrate up the villi (130, 170). On the other hand, acid phosphatase, cytochrome oxidase, and glucose-6-phosphate dehydrogenase, which are intracellular enzymes, show little change from crypt to tip (131, 170). Striking parallels are also found in early life. Lactase, which is necessary for the utilization of milk sugar, rises about 15-fold in the 4 days before term in the rat (1, 42), and several dipeptidases show a similar pattern (103). In the fetal mouse intestinal leucyl naphthylamidase climbs steadily from the 16th day *in utero* until birth on the 21st day (13). Sucrase is present in only trace amounts in the infant rat intestine, but it surges up in the latter part of the 3rd week (43, 147), when alkaline phosphatase activity also rises in the rat duodenum (77), and leucyl naphthylamidase activity approximately doubles in rat jejunum (129). Sucrase and leucyl naphthylamidase activity show similar patterns of increase in the mouse intestine (118).

What is known of the prenatal control of enzyme development in the mammalian intestine clearly implicates the pituitary-adrenal system (10, 165). In the infant rat, adrenalectomy on the 15th day prevents the usual rise of sucrase activity (95), just as it prevents the increase of phosphatase in the mouse (113). Conversely, glucocorticoids elicit sucrase activity-prematurely in the rat intestine *in vivo* and *in vitro* (43), and repair the effect of adrenalectomy (95). Cortisone elicits sucrase activity prematurely in mouse duodenum and jejunum, and also elevates leucyl naphthylamidase activity, though the latter effect occurs in the jejunum only (118). Again more surprising is the fact that both actinomycin D and cycloheximide elicit sucrase activity and increase leucyl naphthylamidase activity; these agents are effective singly, and if administered together with cortisone, they augment the influence of the hormone (118).

To interpret properly the mechanisms by which corticoids and antibiotics influence the development of surface-related enzymes, whether in early life or in maturity, it is necessary to know what is involved in the commitment of some crypt cells to migrate while others remain as stem cells, and it is also necessary to learn in more detail what structural and biochemical events follow from this commitment, and in what sequence such events occur. By immunofluorescent labelling, *Doell, Rosen and Kretchmer* (44) have shown that sucrase appears first in the crypts and then spreads up the sides of the villi at the rate at which cell migration proceeds; this pattern holds both in normal development and under the influence of exogenous corticoid. Whether other enzymes react according to the same pattern is not known. An important but unresolved question is whether the expression of enzyme activity, and of microvillus morphology, are fixed at the time of commitment, or are subject to unprogrammed modification by influences that can act after the cell emerges from the crypt.

If a general mechanism for control of enzyme-active surface does exist, it must also be affected by special factors that produce local or transient patterns. It is well known that phosphatases, disaccharidases, and dipeptidases are not uniformly distributed along the length of the intestine (95). As pointed out before, these three types of enzymes all increase in activity as cells move up to the distal thirds of the villi; near the tips, however, phosphatase activity remains high but the activity of disaccharidases and dipeptidases falls (73, 130, 170). This situation may reflect the different relationships of the enzymes to the microvillus membrane, alkaline phosphatase being incorporated in the continuous phase, whereas sucrase and leucyl naphthylamidase are localized in bulblets that can be stripped from the membrane by papain (49, 89, 132). The loss of these enzymes from the villi tips is probably due to the wear and tear of intestinal activity, rather than to conventional degradative processes. Thus the evidence again suggests that the enzymes localized on the intestinal surface are regulated by specialized mechanisms that may have evolved in conjunction with the peculiar architecture, cell dynamics, and functional activity of the epithelium.

C.   The Overshoot Phenomenon

Enzymes that undergo a large increase in activity during a short period of time frequently rise above their adult level and subsequently fall back to a relatively stable level. The liver provides numerous examples. In the rat liver within a day after birth glucose-6-phosphatase activity rises to more than twice the adult level, whether expressed in terms of whole liver protein (182) or microsomal protein (38). The specific activities of pyruvate carboxylase (182) and tyrosine aminotransferase (154) in the infant rat liver are also more than double the adult levels, as is the activity of tryptophane pyrrolase, per gram of liver, in the newborn guinea pig (126). The intestine shows similar patterns, both at birth and later. In the rat intestine glycyl-glycine dipeptidase rises from 4.2 U at 10 days to 15.8 U on the last (22nd) day *in utero*, then drops to 2.4 U, which is about half the adult value, after suckling has begun; *L*-alanyl-*L*-glutamic acid dipeptidase falls directly to the adult level after suckling, and *L*-alanyl-*L*-proline diptidase falls in two steps, the first at suckling and the second in the third postnatal week (103). Alkaline phosphatase overshoots the adult level in the duodenum, jejunum, and ileum of the 16-day mouse (114). In the duodenum the rise continues to 20 days, when the activity is more than 3 times that of the adult, the difference being more pronounced in a strain characterized by high intestinal phosphatase (125); the subsequent decline affects all levels of the villi (123). Among other tissues in which this tendency to exceed adult activities has been observed is the chicken retina, in which glutamine synthetase activity rises steadily from 17 days *in ovo* to 5 weeks after hatching, after which the activity is reduced by about half (124).

Whether overshooting is of physiological significance is a question that can be answered with some confidence only for intestinal lactase (neutral β-galactosidase). This enzyme is maximally active in early infancy in rats (42) and other mammals (96), and is quite clearly involved in the breakdown and absorption of milk sugar (95, 131). In most cases, however, there is no obvious relation between elevated enzyme activity and special needs of infancy.

The factors that cause elevated activities to fall back to lower levels are equally obscure. In rat liver, phosphoserine phosphatase is maximal at 19 days *in utero*, and then declines in the period when other enzymes are surging up; hydrocortisone accelerates the decrease in both prenatal and neonatal periods, but glucagon, thyroxine, and serine have no effect (88). Intestinal lactase, which is normally declining as sucrase activity is rising, is however not affected by doses of hydrocortisone that elicit a precocious increase of sucrase activity (43). The abrupt drop of intestinal dipeptidase activity that occurs after suckling has been attributed to an inhibitor in the colostrum (103); the effect however is not temporary. Neither the tendency to rise above adult levels, nor the patterns of decrease, are necessarily coordinated in functionally related enzymes: for example, phosphopyruvate carboxylase rises only slightly above the adult level in the liver of the newborn rat, but pyruvate carboxylase surges up to 3 times its adult level (182). A structural relationship does not impose a similar developmental pattern either. In rat liver microsomes, only glucose-6-phosphatase is known to overshoot, reaching more than 200 % of its adult activity on the third day after birth; NADPH-diaphorase and NADPH-cytochrome c reductase also surge up suddenly at birth, but only to the adult value, while the corresponding NADH-enzymes, as well as inosine diphosphatase, increase at a slow rate (38).

Excessively high activities in developmental stages may be simply the epiphenomenal consequence of the characteristics of the regulator mechanisms. If the rising titer of a hormone or other inducer turns on or enhances the activity of a responsive enzyme-forming system for a limited period, it might be expected that an enzyme with a long half-life would tend to accumulate to a greater extent than one with a short half-life. This would in a sense be the converse of the situation in adult stages, in which a long half-life may mask the response to a hormone because the newly synthesized molecules make only a small addition to those in stock (11). It is however also possible that at the time of its initial appearance in quantity, an enzyme may be transiently long-lived because the degradative system affecting it may develop more slowly, or slightly later, than the synthesizing system. The fact that catalytic activity and degradation rate may be under the control of independent genes (59, 144) makes this possibility seem quite likely. None of these hypotheses has however proved correct in the case of δ-aminolevulinate dehydratase, which is very active in mouse liver before birth and then plummets down to about a tenth of the maximum; subsequently it rises to about half the starting level. *Doyle and Schimke* (45) have shown that the fetal enzyme, though immunochemically indistinguishable from that of the adult, is twice as active catalytically, and differs in heat stability and in resistance to digestion by trypsin. Although this may be a special situation, with the loss of fetal enzyme actually reflecting the loss of erythropoietic cells in the liver, *Doyle and Schimke*'s work does provide a model for the kind of studies that may clarify the overshoot phenomenon.

V. Levels of Enzyme Activity and Their Control in Old Age

The senesence of metazoa is among the least understood of biological phenomena. Although the life span of a species may be programmed (34), that of its members is subject to great variability, and partly for this reason the time-dependent changes that precede the termination of an individual life are more complex than those that follow its initiation. In the context of this chapter, several questions are pertinent: (1) Do enzyme activities tend to decline in aging animals? (2) If the activity of an enzyme does decline, is this due to some failure in the intrinsic regulatory system, or does it result from changes in the supply of extrinsic controlling agents? (3) Whether basal activity decreases or not, does the enzyme nevertheless remain capable of responding to regulatory agents?

The evidence bearing on the first question indicates that real losses in enzyme activity in old age are not common, and are seldom more than trivial. Comparing brain, heart, skeletal muscle and kidney, which do not carry on mitosis in full-grown animals, and liver, which remains capable of mitosis, *Schmuckler and Barrows* (152) found no changes in malic dehydrogenase activity in these tissues in female Wistar rats between 12 and 24 months; LDH activity also remained unchanged, except in skeletal muscle, in which a 15 % decrease occurred. In 2-year-old C57B1/6J mice, *Zorzoli* (184) found no change in fumarase activity in liver or intestine, and a decrease of only 10 % in diaphragm and kidney. Glucose-6-phosphatase also remains at a constant level in the livers of old mice (183), as does tyrosine aminotransferase (56). In wild rats, levels of liver succinoxidase, *D*-amino acid oxidase, and alkaline phosphatase are the same at 14 and 35 months, although succinoxidase activity declines about 20 % in the kidney (8). The succinoxidase activity of mitochondria from aged rat liver is not different from that in young animals, and histochemical observations indicate that the same is true in biopsy specimens of human liver, despite marked changes in the form and number of the mitochondria (162). Hepatic mitochondria do not decline in phosphorylative efficiency in old rats (62).

The only report of a serious loss of enzyme activity with age seems to be that of *Singh and Kanungo* (156), who examined lactic dehydrogenase activity in the 12,000 g supernatant fractions of organs of Wistar rats. Specific activity of heart extract fell from 712 U at 30 weeks to 272 U at 96 weeks, with smaller but significant losses in brain, skeletal muscle, and liver. The difference between these findings and those of *Schmuckler and Barrows* (152), who determined activity in whole homogenates, may indicate a rising tendency for lactic dehydrogenase to bind to membranes in old age. *Singh and Kanungo* (156) claim that the losses they observed are due to decrease of muscle-type subunits, without increase of heart-type. *Schmuckler and Barrows* (153), however, failed to find any difference in isozymic composition of lactic dehydrogenase (or other dehydrogenases) extracted from the same organs as *Singh and Kanungo* used, or from kidney.

It is perhaps not surprising if enzymes that participate in cellular respiration are not free to change greatly if the tissues containing them are to remain viable. But there is not much evidence that enzymes involved in specific functions are substantially affected by age either. In the kidney alkaline phosphatase, a constituent of the brush border, maintains a constant level in wild rats to extreme age (9). It has been reported that this enzyme decreases in the intestinal mucosa of 34-month-old mice (151), but this result most probably reflects the dilution of the epithelium with the coarse fibrous tissue that becomes abundant in the mucosa in old age (2). The capacity of the intestinal surface to absorb 6-deoxy-D-glucose does not diminish with age (26).

Since alterations of tissue structure are by no means limited to intestinal mucosa (3), quantitative assays of enzyme activity may frequently be to some extent misleading. An interesting illustration of the value of histochemical observation is provided by a study of non-specific esterases in rats. In livers of animals 199-day-old staining was uniform, but in 855-day livers some parenchymal cells stained less intensely than those in young animals, others much more intensely (53). A quantitative measurement of activity in the older livers would be a statistical average of questionable significance. Also using a histochemical technique, *Ray and Pinkerton* (143) have shown that the alkaline phosphatase activity of leukocytes drops about 50 % during the 7th decade in both men and women.

Although lysosomal enzymes are of special interest in senescence, not much is known of their activity at different stages. Cathepsin rises significantly in livers of old rats, but is unchanged in the kidneys of the same animals (9). Hepatic acid ribonuclease maintains a constant level of activity in mice and hamsters to 20 months and in guinea pigs to 50 months; the proportion of activity in the soluble fraction seems to decrease after late adolescence, but shows no clear relation to aging (65).

For such real decreases of enzyme activity as may occur, the likeliest explanation is that they result from declining supply of hormones, and perhaps of other factors, in the blood. Leukocyte alkaline phosphatase is strongly affected by sex steroids (48, 143), and its low level in the elderly may reflect the lowered titer of both male and female hormones. The supply of other hormones also decreases with age. Although *Finch et al.* (56) failed to find any difference in serum corticosterone in fasted C57B1/6J mice at 6 and 26 months, *Grad and Khalid* (67), in a carefully controlled study, demonstrated that the corticosterone level in serum of males falls from 3.2 μg/100 ml at 8 months to 1.6 at 26 months, the values in females at the same ages being 11.1 μg and 2.6 μg/100 ml. In old rats, the rate of thyroxine secretion may be as much as a third lower than in young (175). The well-known propensity of living systems to evade the obvious, however, makes it unsafe to conclude that changes like those cited will necessarily be reflected in the activities of hormone-dependent enzymes. *Grad* has recently shown that old rats are not hypothyroid, despite their low thyroid secretion rate (66).

To the question whether intrinsic regulatory capacity is subject to alteration by age, no answer can be given now. In 4-month-old mice, hepatic tyrosine aminotransferase activity rises within half an hour in response to cold stress, but in 26-month-old animals the response seems to be delayed by 2 h (56). The enzyme in the older animals does respond promptly to injected insulin or corticosterone, however; and since the capacity of the adrenal cortex itself to react to cold is unimpaired, the delay in tyrosine aminotransferase increase in the old mice under cold stress must be ascribed to unknown factors (56). To gain deeper insight into the problem of aging of the regulatory mechanisms affecting enzymes will require studies of rates of both synthesis and degradation, the level of enzyme activity at any time being a function of relative rates of these 2 processes (11). Such an investigation has been attempted with tryptophan pyrrolase, which is known to be subject to transient inductive increase that is cut short by the appearance of a repressor (60). In 1-month-old rats the induced rise proceeds for only 4 h, and at 12 months for more than 6 h (35). This result might suggest that the repressor system is losing responsiveness. In a subsequent study extended to 24 months, however, *Haining and Correll* (76) demonstrated that rates of synthesis and degradation fluctuate with time: both rates are higher at 12 months than at either 6 or 24 months. As the authors point out, alterations in feeding habits, endocrine balance, free amino acid pools, and demands of other synthetic processes could account for changes which can certainly not be ascribed simply to the passage of time.

Parts of this manuscript were reviewed by my colleagues Dr. *Oscar Chilson*, Dr. *David Kirk*, and Dr. *Patricia Olds*. I am most grateful for their help. Original work reported in this paper was supported by research grant HD-03490 and training grant HD-00012 from the United States Public Health Service.

VI. References

1   *Alvarez, A. and Sas, J.:* β-galactosidase changes in the developing intestinal tract of the rat. Nature, Lond. *190:* 826–827 (1961).

2 *Andrew, W. and Andrew, N.:* An age involution in the small intestine of the mouse. J. Geront. *12:* 136–149 (1957).

3 *Andrew, W.; Shock, N.W.; Barrow, C.F., and Yiengst, J.J.:* Correlation of age changes in histological and chemical characteristics in some tissues of the rat. J. Geront. *14:* 405–414 (1959).

4 *Auerbach, S. and Brinster, R.L.:* Lactate dehydrogenase isozymes in the early mouse embryo. Exp. Cell Res. *46:* 89–92 (1967).

5 *Auerbach, V.H. and Waisman, H.A.:* Tryptophan peroxidase-oxidase, histidase and transaminase activity in the liver of the developing rat. J. biol. Chem. *234:* 304–306 (1959).

6 *Ballard, F.J. and Hanson, R.W.:* Phosphoenolpyruvate carboxykinase and pyruvate carboxylase in developing rat liver. Biochem. J. *104:* 866–871 (1967).

7 *Baril, E.F. and Herrmann, H.:* Immunological and enzymatic properties and accumulation of chromatographically homogenous myosin by the leg musculature of the developing chick. Develop. Biol. *15:* 318–333 (1967).

8 *Barrows, C.H.; Falzone, J.A., and Shock, N.W.:* Age differences in the succinoxidase activity of homogenates and mitochondria from the livers and kidneys of rats. J. Geront. *15:* 130–133 (1960).

9 *Barrows, C.H.; Roeder, M.L., and Falzone, J.A.:* Effect of age on the activities of enzymes and the concentration of nucleic acids in the tissues of female wild rats. J. Geront. *17:* 144–147 (1962).

10 *Bearn, J.G.:* The influence of the foetal pituitary-adrenal axis on the accumulation of alkaline phosphatase in the duodenal epithelium of the foetal rabbit; in *Hamburgh and Barrington* Hormones and Development (Appleton-Century-Crofts, New York, in press).

11 *Berlin, C.M. and Schimke, R.T.:* Influence of turnover rates on the responses of enzymes to cortisone. Molec. Pharmacol. *1:* 149–156 (1965).

12 *Biggers, J.D.; Whittingham, D.G., and Donahue, R.P.:* The pattern of energy metabolism in the mouse oocyte and zygote. Proc. nat. Acad. Sci., Wash. *58:* 560–567 (1967).

13 *Birkenmeier, E.H.; Glazier, H.S., and Moog, F.* (unpublished).

14 *Bonneville, M.A. and Weinstock, M.:* Brush border development in the intestinal absorptive cells of Xenopus during metamorphosis. J. Cell Biol. *44:* 151–171 (1970).

15 *Börnig, H.; Horn, A. und Muller, W.:* Zum Mechanismus der Aktivitätsänderung der alkalischen Phosphatase in Leber und Darm der Ratte nach Ligatur der Ductus choledechus. Acta biol. med. germ. *22:* 537–549 (1969).

16 *Brenneman, A.R. and Kaufman, S.:* Characteristics of the hepatic phenylalanine hydroxylating system in newborn rats. J. biol. Chem. *240:* 3617–3622 (1965).

17 *Brinster, R.L.:* Lactate dehydrogenase activity in the preimplantation mouse embryo. Biochim. biophys. Acta *110:* 439–441 (1965).

18 *Brinster, R.L.:* Studies on the development of mouse embryos *in vitro.* II. The effect of energy source. J. exp. Zool. *158:* 59–68 (1965).

19 *Brinster, R.L.:* Malic dehydrogenase activity in the preimplantation mouse embryo. Exp. Cell Res. *43:* 131–135 (1966).

20 *Brinster, R.L.:* Glucose-6-phosphatase dehydrogenase activity in the preimplantation mouse embryo. Biochem. J. *101:* 161–163 (1966).

21 *Brinster, R.L. and Thomson, J.L.:* Development of 8-cell mouse embryos *in vitro.* Exp. Cell Res. *42:* 308–315 (1966).

22 *Brown, A.L.:* Microvilli of the human jejunal epithelial cell. J. Cell Biol. *12:* 623–627 (1962).

23 *Cahn, R.D. and Cahn, M.B.:* Heritability of cellular differentiation: clonal growth and expression of differentiation in retinal pigment cells *in vitro* Proc. nat. Acad. Sci., Wash. *55:* 106–113 (1966).

24 *Cahn, R.D. and Lasher, R.:* Simultaneous synthesis of DNA and specialized cellular products by differentiation cartilage cells *in vitro.* Proc. nat. Acad. Sci., Wash. *58:* 1131–1138 (1967).

25 *Calarco, P.G. and Brown, E.H.:* Cytological and ultrastructural comparison of t^{12}/t^{12} and normal mouse morulae. J. exp. Zool. *168:* 169–186 (1968).

26 *Calingaert, A. and Zorzoli, A.:* The influence of age on 6-deoxy-D-glucose accumulation by mouse intestine. J. Geront. *20:* 211–214 (1965).

27 *Caramia, F. and Moog, F.* (unpublished).

28 *de la Chapelle, A.; Fantoni, A., and Marks, P.A.:* Differentiation of mammalian somatic cells: DNA and hemoglobin synthesis in fetal mouse yolk sac erythroid cells. Proc. nat. Acad. Sci., Wash. *63:* 812–819 (1969).

29 *Church, R.B. and McCarthy, B.J.:* Ribonucleic acid synthesis in regenerating embryonic liver. J. molec. Biol. *23:* 477–486 (1967).

30 *Cohen, A.:* Développement de l'activité phosphatasique alcaline au niveau de l'intestin d'embryons. Compt. rend. Soc. Biol. *151:* 918–921 (1957).

31 *Cole, R.J. and Paul, J.:* The effect of erythropoietin on haem synthesis in mouse yolk sac and cultured foetal liver cells. J. Embryol. exp. Morph. *15:* 245–260 (1966).

32 *Coleman, J.R. and Coleman, A.W.:* Muscle differentiation and macromolecular synthesis. J. Cell Physiol. *72:* suppl. 1: 19–24 (1968).

33 *Coleman, J.R.; Coleman, A.W., and Hartline, E.J.:* A clonal study of the reversible inhibition of muscle differentiation by the halogenated thymidine analog 5-bromodeoxyuridine. Develop. Biol. *19:* 527–549 (1969).

34 *Comfort, A.:* Ageing, the biology of senescene (Holt, Rinehart, and Winston, New York 1964).

35 *Correll, W.W.; Turner, M.D., and Haining, J.L.:* Changes in tryptophan pyrrolase induction with age. J. Geront. *20:* 507–510 (1965).

36 *Craig, M.L. and Southard, J.L.:* Some factors associated with a developmental change in hemoglobin components in early fetal mice. Develop. Biol. *16:* 331–340 (1967).

37 *Crane, R.K.:* Digestive-absorptive surface of the small bowel mucosa. Ann. Rev. Med. *19:* 57–68 (1968).

38 *Dallner, G.; Siekevitz, P., and Palade, G.:* Biogenesis of endoplasmic reticulum membranes. J. Cell Biol. *30:* 97–117 (1966).

39 *Daniel, J.C.:* The pattern of utilization of respiratory metabolic intermediates by preimplantation rabbit embryos. Exp. Cell Res. *47:* 619–623 (1967).

40 *Das, B.C. and Gray, G.M.:* Protein synthesis in small intestine: localization and correlation with DNA synthesis and sucrase activity. Biochim. biophys. Acta *195:* 255–256 (1969).

41 *Deuchar, E.M.:* ATPase activity in early somite tissue of the chick embryo. J. Embryol. exp. Morph. *8:* 251–258 (1960).

42 *Doell, R.G. and Kretchmer, N.:* Studies of small intestine during development. I. Distribution and activity of β-galactosidase. Biochim. biophys. Acta *62:* 353–362 (1962).

43 *Doell, R.G. and Kretchmer, N.:* Intestinal invertase: precocious development after injection of hydrocortisone. Science *143:* 42–44 (1964).

44 *Doell, R.G.; Rosen, G. and Kretchmer, N.:* Immunochemical studies of intestinal dissacchardiases during normal and precocious development. Proc. nat. Acad. Sci., Wash. *54:* 1268–1273 (1965).

45 *Doyle, D. and Schimke, R.T.:* The genetic and developmental regulation of hepatic δ-aminolevulinate dehydratase in mice. J. biol. Chem. *244:* 5449–5576 (1969).

46 *Drewes, J. and Brawerman, G.:* Alteration in the nature of ribonucleic acid synthesized in rat liver during regeneration and after cortisol administration. J. biol. Chem. *242:* 801–808 (1967).

47 *Dunn, J.S.:* The fine structure of the absorptive epithelial cells of the developing small intestine of the rat. J. Anat. *101:* 57–69 (1967).

48 *Ebadi, M. and McCoy, E.:* Progesterone-mediated increases of leucocyte alkaline phosphatase in rabbits. Biochim. biophys. Acta *130:* 502–510 (1966).

49 *Eichholz, A.:* Studies on the organization of the brush border in intestinal epithelial cells. Biochim. biophys. Acta *163:* 101–107 (1968).

50 *Etzler, M. and Moog, F.:* Inactive alkaline phosphatase in the nursling mouse: immunological evidence. Science *154:* 1037–1038 (1966).

51 *Etzler, M. and Moog, F.:* Immunochemical characterization of alkaline phosphatase isozymes of the mouse duodenum. Biochim. biophys. Acta *154:* 151–161 (1968).

52 *Etzler, M. and Moog, F.:* Biochemical identification and characterization of the multiple forms of alkaline phosphatase in the developing duodenum of the mouse. Develop. Biol. *18:* 515–535 (1968).

53 *Falzone, J.A.; Samis, H.V., and Wulff, V.J.:* Cellular compensation and controls in the aging process. J. Geront. *22:* suppl. 42–52 (1967).

54 *Fantoni, A.; de la Chapelle, A., and Marks, P.A.:* Synthesis of embryonic hemoglobins during erythroid cell development in fetal mice. J. biol. Chem. *244:* 675–681 (1969).

55 *Feigelson, P.; Feigelson, M., and Greengard, O.:* Comparison of the mechanisms of hormonal and substrate induction of rat liver tryptophan pyrrolase. Recent Progr. Hormone Res. *18:* 491–512 (1962).

56 *Finch, C.E.; Foster, J., and Mirsky, A.E.:* Ageing and the regulation of cell activities during exposure to cold. J. gen. Physiol. *54:* 690–712 (1969).

57 *Franco-Browder, S.; DeRydt, J., and Dorfman, A.:* The identification of a sulfated mucopolysaccharide in chick embryos, stages 11–23. Proc. nat. Acad. Sci., Wash. *49:* 643–647 (1963).

58 *Fridhandler, L.:* Pathways of glucose metabolism in fertilized rabbit ova at various implantation stages. Exp. Cell Res. *22:* 303–316 (1961).

59 *Ganschow, R.E. and Schimke, R.T.:* Independent genetic control of the catalytic activity and the rate of degradation of catalase in mice. J. biol. Chem. *244:* 4649–4658 (1969).

60 *Garren, L.; Howell, R.; Tomkins, G., and Crocco, R.:* A paradoxical effect of actinomycin D: the mechanism of regulation of enzyme synthesis by hydrocortisone. Proc. nat. Acad. Sci., Wash. *52:* 1121–1129 (1964).

61 *Gilman, J.G. and Smithies, O.:* Fetal hemoglobin variants in mice. Science *160:* 885–886 (1968).

62 *Gold, P.H.; Gee, M.V., and Strehler, B.:* Effect of age on oxidative phosphorylation in the rat. J. Geront. *23:* 509–512 (1968).

63 *Goldwasser, E.:* Biochemical control of erythroid cell development. Current topics in develop. biol., vol. 1, pp. 173–211 (Academic Press, New York 1966).

64 *Golosow, N. and Grobstein, C.:* Epitheliomesenchymal interaction in pancreatic morphogenesis. Develop. Biol. *4:* 242–255 (1962).

65 *Goto, S.; Takano, T.; Mizuno, D.; Nakano, T., and Imaizumi, K.:* Aging and location of acid ribonuclease in liver. J. Geront. *24:* 305–308 (1969).

66 *Grad, B.:* The metabolic responsiveness of young and old female rats to thyroxine. J. Geront. *24:* 5–11 (1969).

67 *Grad, B. and Khalid, R.:* Circulating corticosterone levels of young and old, male and female mice. J. Geront. *23:* 522–528 (1968).

68 *Granick, S.:* The induction *in vitro* of the synthesis of δ-aminolevulinic acid synthetase in chemical porphyria. J. biol. Chem. *241:* 1359–1375 (1966).

69 *Grayzel, A.; Hörchner, P., and London, I.:* The stimulation of globin synthesis by heme. Proc. nat. Acad. Sci., Wash. *55:* 650–655 (1966).

70 *Greengard, O.:* Enzymic differentiation in mammalian liver. Science *163:* 891–895 (1969).

71 *Greengard, O.:* The hormonal regulation of enzymes in prenatal and postnatal rat liver. Biochem. J. *115:* 19–25 (1969).

72 *Greengard, O. and Dewey, H.K.:* Initiation by glucagon of the premature development of tyrosine aminotransferase, serine dehydratase, and glucose-6-phosphatase in fetal rat liver. J. biol. Chem. *242:* 2986–2991 (1967).

73 *Grey, R.D. and LeCount, T.S.:* Distribution of leucyl naphthylamidase and alkaline phosphatase on the villi of the chick duodenum. J. Histochem. Cytochem. (in press).

74 *Gross, P.R.:* Biochemistry of differentiation. Ann. Rev. Biochem. *38:* 631–660 (1968).

75 *Grumbach, M.M.:* The endocrine control of fetal growth; in *Waisman and Kerr* Fetal growth and development (McGraw-Hill, New York, in press).

76 *Haining, J.L. and Correl, W.W.:* Turnover of tryptophan-induced tryptophan pyrrolase as a function of age. J. Geront. *24:* 143–148 (1969).

77 *Halliday, R.:* The effect of steroid hormones on the absorption of antibody by the young rat. J. Endocrin. *18:* 56–66 (1959).

78 *Herbst, C.:* Über die zur Entwicklung der Seeigellarven notwendigen anorganischen Stoffe. Roux Arch. Entwmech. *5:* 649–793 (1897).

79 *Herrmann, H. and Tootle, M.L.:* Specific and general aspects of the development of enzymes and metabolic pathways. Physiol. Rev. *44:* 289–371 (1964).

80 *Holmes, R. and Crane, R.K.:* Protein turnover in the digestive-absorptive surface of the rat small intestine. Gut *8:* 630 (1967).

81 *Holt, P.G. and Oliver, I.T.:* Factors affecting the premature induction of tyrosine aminotransferase in foetal rat liver. Biochem. J. *108:* 333–339 (1968).

82 *Holtfreter, J. and Hamburger, V.:* Amphibians; in *Willier, Weiss and Hamburger* Analysis of development, pp. 230–296 (W.B. Saunders, Philadelphia 1955).

83 *Holtzer, H.:* Aspects of chondrogenesis and myogenesis; in *Rudnick* Molecular and cellular structure, pp. 35–87 (Ronald Press, New York 1961).

84 *Holtzer, H.:* Mitosis and cell transformation; in *Mazia and Tyler* General physiology of cell specialization, pp. 80–90 (McGraw-Hill, New York 1963).

85 *Holtzer, H.; Marshall, J., and Finck, H.:* An analysis of myogenesis by the use of fluorescent antimyosin. J. biochem. biophys. Cytol. *3:* 705–724 (1957).

86 *Hugon, J. and Borgers, J.M.:* Ultrastructural localization of alkaline phosphatase activity in the absorbing cells of the duodenum of mouse. J. Histochem. Cytochem. *14:* 629–640 (1966).

87 *Ishida, K. and Chang, M.C.:* Histochemical demonstration of succinate dehydrogenase in hamster and rabbit eggs. J. Histochem. Cytochem. *13:* 470–475 (1965).

88 *Jamdar, S. and Greengard, O.:* Phosphoserine phosphatase: development, formation, and hormonal regulation in rat tissues. Arch. Biochem. *134:* 228–232 (1969).

89 *Johnson, C.F.:* Disaccharidase localization in hamster intestine brush borders. Science *155:* 1670–1672 (1967).

90 *Kallman, F. and Grobstein, C.:* Fine structure of differentiating mouse pancreatic exocrine cells in transfilter culture. J. Cell Biol. *20:* 399–413 (1964).

91 *Kappas, A.; Song, C.; Levere, R.; Sachson, R., and Granick, S.:* The induction of
    δ-aminolevulinic acid synthetase *in vivo* in chick embryo by natural steroids. Proc. nat.
    Acad. Sci., Wash. *61:* 509–513 (1968).
92 *Kedes, L.H. and Gross, P.R.:* Synthesis and function of messenger RNA during early
    embryonic development. J. molec. Biol. *42:* 559–575 (1969).
93 *Kenney, F.T.:* Induction of tyrosine a-ketoglutarate transaminase in rat liver. J. biol.
    Chem. *237:* 3495–3498 (1962).
94 *Kenney, F.T.; Reem, H., and Kretchmer, N.:* Development of phenylalanine hydro-
    xylase in mammalian liver. Science *127:* 86–87 (1958).
95 *Koldovský, O.:* Development of the functions of the small intestine in mammals and
    man (Karger, Basel/New York 1969).
96 *Koldovský, O.; Heringova, A.; Jirsová, V.; Chytil, F. and Hosková, J.:* Postnatal
    changes in β-galactosidase activity in the jejunum and ileum of mice, rabbits and guinea
    pigs. Canad. J. Biochem. *44:* 523–527 (1966).
97 *Kovach, J.; Marks, P.; Russel, E., and Epler, H.:* Erythroid cell development in fetal
    mice: ultrastructural characteristics and hemoglobin synthesis. J. molec. Biol. *25:*
    131–142 (1967).
98 *Krishnan, R.S. and Daniel, J.C.:* 'Blastokinin': inducer and regulator of blastocyst
    development in the rabbit uterus. Science *158:* 490–492 (1967).
99 *Lash, J.W.:* Tissue interaction and specific metabolic responses; in *Locke* Cytodiffer-
    entiation and macromolecular synthesis, pp. 235–260 (Academic Press, New York
    1963).
100 *Levere, R.D. and Granick, S.:* Control of hemoglobin synthesis in the cultured chick
    blastoderm by δ-aminolevulinic acid synthetase. Proc. nat. Acad. Sci., Wash. *54:*
    134–137 (1965).
101 *Levere, R.D. and Granick, S.:* Control of hemoglobin synthesis in the cultured chick
    blastoderm. J. biol. Chem. *242:* 1903–1911 (1967).
102 *Levere, R.D.; Kappas, A., and Granick, S.:* Stimulation of hemoglobin synthesis in
    chick blastoderms by certain 5-androstane and 5-pregnane steroids. Proc. nat. Acad.
    Sci., Wash. *58:* 985–990 (1967).
103 *Lindberg, T. and Owman, C.:* Development of dipeptidase activity in the small intes-
    tine of the rat as related to the development of the intestinal mucosa. Acta physiol.
    scand. *68:* 141–151 (1966).
104 *Litwack, G. and Nemeth, A.M.:* Development of liver tyrosine aminotransferase acti-
    vity in the rabbit, guinea pig and chicken. Arch. Biochem. *109:* 316–321 (1965).
105 *Manes, C. and Daniel, J.C.:* Quantitative and qualitative aspects of protein synthesis in
    the preimplantation rabbit embryo. Exp. Cell Res. *55:* 261–268 (1969).
106 *Marks, P. and Kovach, S.:* Development of mammalian erythroid cells. Current topics
    in develp. biol. vol. 1, pp. 213–252 (Academic Press, New York 1966).
107 *Marzullo, G. and Lash, J.W.:* Acquisition of the chondrocytic phenotype; in *Hagen,
    Wechsler and Zilliken* Morphological and biochemical aspects of cytodifferentiation, pp.
    213–219 (Karger, Basel/New York 1967).
108 *Medoff, J.:* Enzymatic events during cartilage differentiation in the chick embryo limb
    bud. Develop. Biol. *16:* 118–143 (1967).
109 *Mintz, B.:* Synthetic processes and early development in the mammalian egg. J. exp.
    Zool. *157:* 85–100 (1964).
110 *Mintz, B.:* Gene expression in the morula stage of mouse embryos, as observed during
    development of t^{12}/t^{12} lethal mutants *in vitro*. J. exp. Zool. *157:* 267–272 (1964).
111 *Monesi, V. and Salfi, V.:* Macromolecular syntheses during early development in the
    mouse embryo. Exp. Cell Res. *46:* 632–635 (1967).

112 *Moog, F.:* Differentiation of alkaline phosphomonoesterase in the duodenum of the mouse. J. exp. Zool. *118:* 187–208 (1951).

113 *Moog, F.:* Influence of the pituitary-adrenal system on the differentiation of phosphatase in the duodenum of the mouse. J. exp. Zool. *124:* 329–346 (1953).

114 *Moog, F.:* Functional differences in the alkaline phosphatases in the small intestine of the mouse from birth to one year. Develop. Biol. *3:* 153–174 (1961).

115 *Moog, F.:* Intestinal phosphatase activity: acceleration of increase by puromycin and antinomycin. Science *144:* 414–416 (1964).

116 *Moog, F.:* Enzyme development in relation to functional differentiation; in *Weber* Biochemistry of animal development, vol. 1, pp. 307–365 (Academic Press, New York 1965).

117 *Moog, F.:* The regulation of alkaline phosphatase activity in the duodenum of the mouse from birth to maturity. J. exp. Zool. *161:* 353–368 (1966).

118 *Moog, F.:* Corticoids and the enzymic maturation of the intestinal surface: alkaline phosphatase, leucyl naphthylamidase and sucrase; in *Hamburgh and Barrington* Hormones and development (Appleton-Century-Crofts, New York, in press).

119 *Moog, F.:* Enzyme development and functional differentiation in the fetus; in *Waisman and Kerr* Fetal growth and development (McGraw-Hill, New York, in press).

120 *Moog, F.; Bennett, C., and Thomas, E.R.:* Differential effect of adrenalectomy and adrenocortical hormones on the alkaline phosphatase of the duodenum, jejunum and ileum of the adult mouse. Anat. Rec. *120:* 777 (1954).

121 *Moog, F.; Etzler, M., and Grey, R.D.:* Differentiation of alkaline phosphatase in the small intestine. Ann. NY Acad. Sci. *166:* 447–461 (1969).

122 *Moog, F. and Grey, R.D.:* Spatial and temporal differentiation of alkaline phosphatase on the intestinal villi of the mouse. J. Cell Biol. *32:* No. 2, C1–C6 (1967).

123 *Moog, F. and Grey, R.D.:* Alkaline phosphatase isozymes in the duodenum of the mouse: attainment of pattern of spatial distribution. Develop. Biol. *18:* 481–500 (1968).

124 *Moscona, A. and Hubby, J.:* Experimentally induced changes in glutamotransferase activity in embryonic tissue. Develop. Biol. *7:* 192–206 (1963).

125 *Nayudu, P.R.V. and Moog, F.:* The genetic control of alkaline phosphatase activity in the duodenum of the mouse. Biochem. Genet. *1:* 155–170 (1967).

126 *Nemeth, A.M.:* Mechanisms controlling changes in tryptophan perioxidase activity in developing mammalian liver. J. biol. Chem. *234:* 2921–2924 (1959).

127 *Neutra, M. and Leblond, C.P.:* Radioautographic comparison of the uptake of galactose-H^3 and glucose-H^3 in the Golgi zone of various cells secreting glycoproteins. J. Cell Biol. *30:* 137–150 (1966).

128 *Nishi, Y.; Yoshida, T., and Takesue, Y.:* Electron microscope studies on structure of rabbit intestinal sucrase. J. molec. biol. *37:* 441–444 (1968).

129 *Noack, R.; Koldovský, O.; Friedrich, M.; Heringova, A.; Jirsová, V., and Schenk, G.:* Proteolytic and peptidase activities of the jejunum and ileum of the rat during postnatal development. Biochem. J. *100:* 775–778 (1966).

130 *Nordstrom, C.; Dahlqvist, A. and Josefsson, L.:* Quantitative determination of enzymes in different part of the villi and crypts of rat small intestine. J. Histochem. Cytochem. *15:* 713–721 (1967).

131 *Nordstrom, C.; Koldovský, O., and Dahlqvist, A.:* Localization of β-galactosidases and acid phosphatase in the small intestinal wall. J. Histochem. Cytochem. *17:* 341–347 (1969).

132 *Oda, T. and Seki, S.:* Molecular basis of structure and function of the plasma membrane of the microvilli of intestinal epithelial cells. 6th Internat. Congr. Electron Micros., Tokyo, 1966, p. 387.

133 *Okazaki, K. and Holtzer, H.:* An analysis of myogenesis *in vitro* using fluorescein labelled antimyosin. J. Histochem. Cytochem. *13:* 726–739 (1965).

134 *Okazaki, K. and Holtzer, H.:* Myogenesis: fusion, myosin synthesis and the mitotic cycle. Proc. nat. Acad. Sci., Wash. *56:* 1484–1488 (1966).

135 *Olson, G. and Moog, F.* (unpublished).

136 *Overton, J.:* Fine structure of the free surface in developing mouse intestinal mucosa. J. exp. Zool. *159:* 195–201 (1965).

137 *Overton, J.; Eichholz, A., and Crane, R.K.:* Studies on the organization of the brush border in intestinal epithelial cells, J. Cell Biol. *26:* 693–706 (1965).

138 *Palay, S.L. and Karlin, L.J.:* An electron microscopic study of the intestinal villus. J. biochem. biophys. Cytol. *5:* 363–372 (1959).

139 *Papaconstantiou, J.:* Metabolic control of growth and differentiation in vertebrate embryos; in *Weber* Biochemistry of animal development, vol. 2, pp. 58–114 (Academic Press, New York 1967).

140 *Patton, D.E.; Kirk, D.L., and Moscona, A.:* Hemopoiesis in embryonic mouse liver tissue and cells *in vitro.* Exp. Cell Res. *54:* 181–186 (1969).

141 *Procházka, P.; Hahn, P.; Koldovský, O.; Nohynek, M., and Rokos, J.:* The activity of a-amylase in homogenates of the pancreas of rats during early postnatal development. Physiol. bohemoslov. *13:* 288–291 (1964).

142 *Rapola, J. and Koskimies, O.:* Embryonic enzyme patterns: characterization of the single lactate dehydrogenase isozyme in preimplanted mouse ova. Science *157:* 1311–1312 (1967).

143 *Ray, P.K. and Pinkerton, P.H.:* Leucocyte alkaline phosphatase. The effect of age and sex. Acta haemat. jap. *42:* 18–22 (1969).

144 *Rechcigl, M. and Heston, W.E.:* Genetic regulation of enzyme activity in a mammalian system by the alteration of the rate of enzyme degradation. Biochem. biophys. res. Commun. *27:* 119–124 (1967).

145 *Revel, J-P. and Ito, S.:* The surface components of cells: in *Davis and Warren* The specificity of cell surfaces, pp. 211–234 (Prentice-Hall, Englewood 1967).

146 *Rokos, J.; Hahn, P.; Koldovský, O., and Procházka, P.:* The postnatal development of lipolytic activity in the pancreas and small intestine of the rat. Physiol. bohemoslov. *12:* 213–218 (1963).

147 *Rubino, A.; Zimbalatti, F., and Auricchio, S.:* Intestinal disaccharidase activities in adult and suckling rats. Biochim. biophys. Acta *92:* 305–312 (1964).

148 *Rudnick, D.:* Teleosts and birds; in *Willier, Weiss and Hamburger* Analysis of development, pp. 297–314 (W.B. Saunders, Philadelphia 1955).

149 *Rutter, W.; Ball, W.; Bradshaw, W.; Clark, W., and Saunders, T.:* Levels of regulation in cytodifferentiation; in *Hagen, Wechsler and Zilliken* Morphological and biochemical aspects of cytodifferentiation, pp. 110–124 (Karger, Basel/New York 1967).

150 *Rutter, W.; Kemp, J.; Bradshaw, W.; Clark, W.; Ronzio, R., and Saunders, T.:* Regulation of specific protein synthesis in cytodifferentiation. J. Cell Physiol. *72:* suppl. 1: 1–18 (1968).

151 *Sayeed, M. and Blumenthal, H.T.:* Age difference in the intestinal phosphomonoesterase activity of mice. Proc. Soc. exp. Biol., NY *129:* 1–3 (1968).

152 *Schmuckler, M. and Barrows, C.H.:* Age difference in lactic and malic dehydrogenases in the rat. J. Geront. *21:* 109–111 (1966).

153 *Schmuckler, M. and Barrows, C.H.:* Effect of age on dehydrogenase heterogeneity in the rat. J. Geront. *22:* 8–13 (1967).

154 *Sereni, F.; Kenney, F.T., and Kretchmer, N.:* Factors influencing the development of tyrosine transaminase activity in rat liver. J. biol. Chem. *234:* 609–612 (1959).

155 *Shorter, R.G. and Creamer, B.:* Ribonucleic acid and protein metabolism in the gut. Gut *3:* 118–128 (1962).

156 *Singh, S.N. and Kanungo, M.S.:* Alterations in lactate dehydrogenase of the brain, heart, skeletal muscle and liver of rats of various ages. J. biol. Chem. *243:* 4526–4529 (1968).

157 *Skea, B.R. and Nemeth, A.M.:* Factors influencing premature induction of UDP-glucuronyltransferase activity in cultured chick embryo cells. Proc. nat. Acad. Sci., Wash. *64:* 795–802 (1969).

158 *Smith, L.J.:* A morphological and histochemical investigation of a preimplantation lethal in the house mouse. J. exp. Zool. *132:* 51–79 (1956).

159 *Spiegel, E. and Spiegel, M.:* Regulation of mammalian tryptophan pyrrolase activity during development. Exp. Cell Res. *36:* 427–429 (1964).

160 *Stewart, J. and Papaconstantinou, J.:* Lactate dehydrogenase isozymes and their relationship to lens cell differentiation. Biochim. biophys. Acta *121:* 69–79 (1966).

161 *Stockdale, F.; Okozaki, O.; Nameroff, M., and Holtzer, H.:* 5-bromodeoxyuridine effect on myogenesis *in vitro.* Science *146:* 533–535 (1964).

162 *Tauchi, H. and Sato, T.:* Age changes in size and number of mitochondria of human hepatic cells. J. Geront. *23:* 454–461 (1968).

163 *Ugolev, A.M.:* Membrane (contact) digestion. Physiol. Rev. *44:* 555–595 (1965).

164 *Ugolev, A.M.:* Physiology and pathology of membrane digestion (Plenum Press, New York 1968).

165 *Verne, J. et Hebert, S.:* L'apparition de l'activité phosphomonoesterasique de l'intestin au cours du développement et ses rapports avec le fonctionnement cortico-surrénal. Ann. Endocrin., Paris *10:* 456–460 (1949).

166 *Vernon, R.G. and Walker, D.G.:* Changes in activity of some enzymes involved in glucose utilization and formation in developing rat liver. Biochem. J. *106:* 321–330 (1968).

167 *Verzar, F.; Sailer, E., und Richterich, R.:* Einfluss der Nebennierenrinde auf die alkalische Phosphatase der Dünndarmschleimhaut. Helv. physiol. Acta *10:* 231–246 (1952).

168 *Wales, R.G. and Biggers, J.D.:* The permeability of two- and eight-cell mouse embryos to l-malic acid. J. Repro. Fert. *15:* 103–111 (1968).

169 *Wales, R.G. and Whittingham, D.G.:* A comparison of the uptake and utilization of lactate and pyruvate by one- and two-cell mouse embryos. Biochem. biophys. Acta *148:* 703–712 (1967).

170 *Webster, H.L. and Harrison, D.D.:* Enzymic activities during the transformation of crypt to columnar epithelial cells. Exp. Cell Res. *56:* 245–253 (1969).

171 *Wessels, N.K.:* Tissue interactions and cytodifferentiation. J. exp. Zool. *157:* 139–152 (1964).

172 *Wessels, N.K.:* Acquisition of actinomycin D insensitivity during differentiation of pancreas exocrine cells. Develop. Biol. *9:* 92–114 (1964).

173 *Wessels, N.K. and Wilt, F.H.:* Action of actinomycin D on pancreas cell differentiation. J. molec. Biol. *13:* 767–779 (1965).

174 *Wicks, W.D.:* Inductions of tyrosine transaminase in fetal rat liver. J. biol. Chem. *243:* 900–906 (1968).

175 *Wilansky, D.; Newsham, L., and Hoffman, M.:* The influence of senescence on thyroid function: functional changes evaluated with I^{131}. Endocrinology *61:* 327–336 (1957).

176 *Williams, J.D.:* Chemical constitution and metabolic activities of animal cells; in *Weber* Biochemistry of animal development, vol. 1, pp. 14–72 (Academic Press, New York 1965).

177 *Wilt, F.H.:* Erythropoiesis in the chick embryo: the role of endoderm. Science *147:* 1588–1590 (1965).
178 *Wilt, F.H.:* Regulation of the initiation of chick embryo hemoglobin synthesis. J. molec. Biol. *12:* 331–340 (1965).
179 *Young, D. and Oliver, I.T.:* Development of gluconeogenesis in neonatal rat liver. Biochem. J. *105:* 1229–1233 (1967).
180 *Yeung, D. and Oliver, I.T.:* Factors affecting the premature induction of carboxylase in neonatal rat liver. Biochem. J. *108:* 325–333 (1968).
181 *Yeung, D. and Oliver, I.T.:* Induction of phosphopyruvate carboxylase in neonatal rat liver by 3'5'-cyclic AMP. Biochemistry *7:* 3231–3239 (1968).
182 *Yeung, D.; Stanley, R., and Oliver, I.T.:* Development of gluconeogenesis in neonatal rat liver. Biochem. J. *105:* 1219–1228 (1967).
183 *Zorzoli, A.:* The influence of age on mouse liver glucose-6-phosphatase activity. J. Geront. *17:* 359–362 (1962).
184 *Zorzoli, A.:* Fumarase activity in mouse tissues during development and aging. J. Geront. *23:* 506–508 (1968).

Author's address: Dr. *Florence Moog,* Department of Biology, Washington University, *St. Louis, MO 63130* (USA)

Enzyme Synthesis and Degradation in Mammalian Systems, pp. 77–102
(Karger, Basel 1971)

Control of Enzyme Activity by Glucocorticoids

Fred Rosen and Richard J. Milholland

Department of Experimental Therapeutics, Roswell Park Memorial Institute,
Buffalo, N.Y.

Contents

I. Introduction

During the past 20 years many reports have appeared indicating that glucocorticoids are capable of selectively stimulating and depressing the *de novo* synthesis of specific enzymes in liver and other tissues. Examples of enzymes which are increased in activity following treatment with glucocorticoids include tryptophan oxygenase (*L*-tryptophan:oxygen oxiodoreductase, EC 1.11.1.4), tyrosine aminotransferase (*L*-tyrosine:2-oxolglutarate amino transferase, EC 2.6.1.5.), serine dehydratase (*L*-serine hydro-lyase (deaminating), EC 4.2.1.13), and alanine aminotransferase (*L*-alanine:2-oxolglutarate amino transferase, EC 2.6.1.2); those observed to undergo depression include 3-phosphoglycerate dehydrogenase and *D*-glycerate dehydrogenase (*D*-glycerate:NAD oxidoreductase, EC 1.1.1.29). Among the adaptive enzymes which respond to the adrenal corticoids as well as to various physiological conditions, e.g. alloxan diabetes, starvation and variation in protein intake, differences have been recognized in their half-lives, rate of response to the steroid, effects of adrenalectomy or hypophysectomy, and in their inducibility by protein hormones and substrates. The basis for most of these differences remains to be determined. With the use of antibodies specific for these adaptive enzymes, and puromycin and cycloheximide which inhibit protein synthesis, it is now evident that the hormonal response of these enzymes involves an enhanced rate of synthesis, rather than activation or stabilization against degradation of the endogenous enzyme. Also, studies with actinomycin D have provided substantial evidence that enzyme induction by corticosteroids and other inducing agents requires new RNA synthesis.

Knox and associates (55, 57), in studies reported in the early 1950's, demonstrated that the hepatic enzyme tryptophan oxygenase could be markedly increased in activity following treatment of rats with tryptophan or cortisone. Two important conclusions that emerged from these studies were: (1) a mechanism existed for altering enzyme activity in animals which appeared to be analogous to the substrate-induced enzyme adaptation that occurs in microorganisms and (2) treatments or stress situations that stimulate the pituitary-adrenal axis and thus the secretion of corticosterone by the adrenals can cause an increase in the activity of a cortisone-responsive enzyme. These early observations were of considerable interest and many investigators have since used tryptophan oxygenase as a model system for studying various aspects of the control of enzyme activity in mammalian tissues.

Some of the problems of enzyme regulation in mammalian tissues which have been investigated using the cortisol-responsive enzymes as test systems include: (1) factors involved in the control of enzyme activity *in vivo*, (2) the manner in which glucocorticoids, other hormones, and amino acids stimulate a rise in the activity of specific enzymes, (3) relationship between altered enzyme activity and the physiological action of the hormone, (4) protein turnover, and (5) control of enzyme activity in malignant tissues. These and other topics related to the glucocorticoid-responsive enzymes have been considered in many comprehensive reviews (45, 87, 104, 112). With the intent to be selective and critical, rather than exhaustive, this presentation will be limited mainly to tryptophan oxygenase, tyrosine and alanine aminotransferases and serine dehydratase. These adaptive enzymes, which show many differences with respect to the regulation of enzyme synthesis, have been studied extensively and will serve to highlight some of the unsolved problems in this area of research.

II. Enzymes Responsive to Glucocorticoids

That treatment with hormones, with few exceptions can stimulate new protein synthesis in responsive tissues is now generally accepted. The glucocorticoids have been studied extensively with regard to their influence on the induction of specific enzymes in liver (104) and lymphoid tissues (68, 97). Other steroid hormones have also been demonstrated to selectively increase the activity of certain enzymes in their target tissues (24, 81, 135). Alterations in the levels of specific proteins also appear to be the basis for the physiological effects of the protein hormones and cyclic AMP (47, 134). There are a number of reports (125, 129) in which cyclic AMP appears to be acting as an allosteric effector in the regulation of enzyme activity. Although it was suggested many years ago that the hormonal control of cellular functions could be mediated by changes in the activity of certain key enzymes (25), there is at present no well-documented example of a physiological effect of a hormone that is controlled by a change in enzyme activity.

Various enzymes concerned with the metabolism of amino acids and carbohydrates have been shown to be responsive to the glucocorticoids (104). Much of the information that is available in this area has been obtained in the rat on liver enzymes. Differences in the magnitude of response of these enzymes to corticoids is dependent on a number of factors. These include the dosages used, duration of the injection period, route of administration, the age and sex of the animal, and the species used (104). Moreover, most of these enzymes show an adaptive response to the endocrine, nutritional and pathological status of the animal, which in turn influences their response to cortisol (33, 34, 109).

It is reasonable to assume that enzymes which increase in activity as a consequence of glucocorticoid treatment will show a decrease in activity after adrenalectomy. For many enzymes this is the case; however, for some enzymes such as tyrosine aminotransferase normal activity is maintained in adrenalectomized rats (64, 98). Since this enzyme is induced by glucagon (9, 10, 42) and insulin (42), and is responsive to alterations in the protein content of the diet (102), it is possible that these factors are involved in this effect. General cachexia, rather than a depletion of corticosterone, may explain the depression in activity of certain enzymes following adrenalectomy; the activity of such enzymes is usually restored to values normal for the intact animal when cortisol is given, but otherwise is not responsive to corticoid treatment.

A. Differences in the Rate of Response of Enzymes to Glucocorticoids

In view of the various physiological effects exerted by the glucocorticoids, it is reasonable to expect that the underlying biochemical changes would occur at different times. In this regard it is of interest that marked differences are observed in the rate at which the adaptive enzymes respond to cortisol administration. The route and period of cortisol administration are important considerations in eliciting an increase in enzyme activity of glucocorticoids. Maximal responses of tyrosine aminotransferase, tryptophan oxygenase, and the branched chain amino acid aminotransferases (1) have been obtained within a few hours following intraperitoneal injection of the glucocorticoid. On the other hand subcutaneous injections of the hormone for several days were required to obtain a significant induction of alanine aminotransferase (107). The urea cycle enzymes (111), glucose-6-phosphatase (130) and serine dehydratase (17) also show a slow adaptive response to cortisol administration. Thus, to determine whether an enzyme is inducible by cortisol it is necessary to measure changes in activity within several hours as well as after several days of corticoid treatment.

The induction of alanine transaminase occurs over a period of a few days. Following a single subcutaneous injection of cortisol, the activity of this enzyme reaches a maximum between the second and fourth day after treatment (105). When daily doses of cortisol are given, this enzyme responds maximally and reaches its highest level after the fifth day of treatment. In rats receiving subcutaneous injections of the steroid, the rate of enzyme loss following cessation of treatment is markedly slower than the rate at which the enzyme responds during induction with cortisol. This effect can probably be attributed to the accumulation and gradual loss of the steroid from subcutaneous depots.

While the time required for several glucocorticoid-responsive enzymes to show a maximal response to the steroid varies considerably, this is largely related to differences in the half-life of the enzymes rather than lack of sensitivity to the steroid. Thus tryptophan oxygenase (13) and tyrosine aminotransferase (48, 65), with half-lives of less than 3 h, show a maximal response to corticoid treatment within 4 to 6 h, while alanine aminotransferase, with a half-life calculated to be from 40 to 80 h (32, 118), does not reach maximal activity until 2 to 4 days following a single dose of cortisol.

After studying the increase in the rate of synthesis following steroid treatment, *Berlin and Schimke* (4) concluded that the ratio of the increased rate of synthesis to the basal rate is comparable in each of 4 enzymes studied (alanine- and tyrosine aminotransferase, tryptophan oxygenase, and arginase). They also suggested that initiation of this increased rate of synthesis probably occurs simultaneously for all the glucocorticoid inducible enzymes. However, more recently *Kim* (54), using the pulse-label antibody technique, observed that the rate of alanine aminotransferase synthesis was unchanged up to 6.5 h after cortisol administration and concluded 'that the apparent effects of glucocorticoid hormone on the synthesis of liver proteins are sequential rather than simultaneous'.

B.   Tissue and Species Specificity

A previous review (104) considered the earlier literature in this area and indicated that most studies on the response of the adaptive enzymes to glucocorticoids have been carried out in rat liver. Although their activity is often at the minimal level of detection, most of the adaptive enzymes are present in other tissues as well. However, with few exceptions, they have not been found to be responsive to either cortisol or substrate induction (65, 108). There is still no adequate explanation as to why an enzyme may be responsive in one tissue but not in another in the same animal.

Even more puzzling is why one enzyme such as alanine aminotransferase can be induced in thymus gland and Walker carcinoma 256, whereas in the same tissues tyrosine aminotransferase activity is not appreciably elevated either by cortisol or tyrosine administration (97, 108).

In the rat, the response of tyrosine aminotransferase and tryptophan oxygenase to cortisol administration is remarkably similar. However, within certain strains of mice the differential response of these two enzymes to either cortisol or tryptophan is quite noteworthy (79, 104). Also the alanine aminotransferase level in several strains of mice, in the rabbit, the hamster, and in the guinea pig was relatively unresponsive to daily treatment with cortisol for 1 week. In each of these species, the endogenous level of this transaminase in liver was noted to be several-fold higher than in the rat. It is of interest that in man, *Altman and Greengard* (2) have observed that treatment with glucocorticoids can also induce a significant rise in hepatic tryptophan oxygenase within a period of 5 h. The properties of tissues which enable them to undergo specific changes in enzyme activity and the possibility that enzymic adaptations are characteristic mainly of tissues known to be physiologically responsive to glucocorticoids remain to be determined.

Studies by *Holt and Oliver* (41) as well as *Miller and Litwack* (76) offer a possible explanation for many of these enigmatic observations. These authors have presented data which indicate the existence of multiple forms of tyrosine aminotransferase. The differences in the electrophoretic profile of this enzyme which are seen following induction by hydrocortisone, glucagon, insulin, cyclic AMP and, pyridoxine suggest that perhaps different forms of the enzyme are present in different tissues. The failure to observe the expected response to cortisol may merely reflect an atypical distribution of these multiple forms of this enzyme. The existence of two forms of serine dehydratase reported by *Inoue and Pitot* (44) suggests that additional adaptive enzymes may also exist as isozymes.

C. Diurnal Variation in the Activity of Certain Adaptive Enzymes

It is now well-established that tyrosine aminotransferase activity in rat liver varies over a 3-fold range each day, with peak activity occurring between 8 and 11 p.m. (92). This rhythm persists even in the absence of the pituitary or adrenal glands (120, 136). Tryptophan oxygenase also demonstrates a daily rhythmicity in activity in rat liver but this is abolished following adrenalectomy (93). Some of the factors which appear to be involved in maintaining the circadian rhythm of tyrosine aminotransferase include: (1) alterations in the lighting schedule (6), availability of food and particularly access to protein (139), level of tryptophan in the diet (137), treatments which alter norepinephrine levels (5), and adrenal function. It is important to recognize that many of the earlier investigations designed to study the effects of potential inducing agents or treatments (e.g. hypophysectomy, adrenalectomy) on hepatic tyrosine aminotransferase activity did not adequately provide controls for the changes produced by the diurnal variation in activity of this enzyme. A detailed review on this important aspect of enzyme regulation by R.W. Fuller is included in this volume (viz. p. 311).

D. Hormone Specificity

Among the steroids which have been tested to date, only those with gluco-corticoid activity have produced significant increases in the activity of the adaptive enzymes. In some cases treatment of intact rats with relatively large doses of other hormones has been observed to cause a rise in enzyme activity. In these instances, comparable studies are needed in adrenalectomized animals to rule out the effects produced in intact animals as a reaction to stress. There are only a few studies in which comparisons have been made of a series of different gluco-corticoids with respect to their relative activity as inducers of adaptive enzymes. These are summarized in an earlier review (104).

The data presented in figure 1 indicate that the response of 3 adaptive enzymes to cortisol in both intact and adrenalectomized rats is dose-dependent. The unique responsiveness of tyrosine aminotransferase to small doses of cortisol in the adrenalectomized rat indicates some function of the adrenals in regulating the activity of this enzyme. It is of interest that daily administration of deoxy-corticosterone to rats without adrenals reduces the sensitivity of this enzyme to cortisol (98). The increased responsiveness of tyrosine aminotransferase to cortisol in adrenalectomized rats could be associated with impaired capacity to maintain the level of a repressor.

It is now established that in addition to the glucocorticoids, the pancreatic hormones, glucagon and insulin, can initiate an increase in the activity of tyrosine aminotransferase in rat liver (9, 10, 26). In 1958, *Schorr and Frieden* (117) reported that large doses of insulin induced 3- to 4-fold rises in tryptophan oxygenase in livers of intact, fasted, and adrenalectomized rats. Thus both of these adaptive enzymes are responsive to glucocorticoids and insulin, whereas only the aminotransferase can be induced by glucagon. There is evidence that the induction of tyrosine aminotransferase by the pancreatic hormones differs in several respects from the manner whereby this enzyme responds to the gluco-

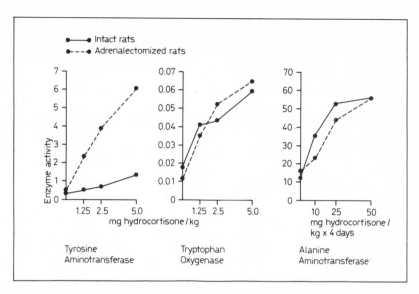

Fig. 1. Comparison of the response of tyrosine and alanine aminotransferase and tryptophan oxygenase to cortisol in intact and adrenalectomized rats. Enzyme activity is expressed as millimoles of product formed per g of protein per h. Each *point* represents the average of individual assays on four animals per group.

corticoids. The rate and magnitude of the response of this enzyme is dependent on which of these hormones is given. When either of the pancreatic hormones is given, tyrosine aminotransferase activity reaches a peak value in about 2 h, while cortisol administration is followed by a lag period of 60 to 90 min before the enzyme begins to rise in activity (26, 42). The maximal stimulation of this enzyme seen after glucagon or insulin is administered is 3- to 5-fold in 2 to 3 h; whereas for cortisol, increases as large as 7- to 10-fold are elicited within 4 to 6 h after treatment. Other studies indicated that when optimal doses of glucagon and insulin were combined, this treatment was no more effective in stimulating a rise in enzyme activity than either hormone given alone. Cortisol, however, acted additively in combination with either or both of these hormones. Although different mechanisms appear to be involved in the induction of aminotransferase by the pancreatic hormones and cortisol, each of these hormones acts via a process that is dependent on RNA synthesis since their action is inhibited by actinomycin (10, 26, 42). Other enzymes in liver have also been reported to be increased in activity following glucagon treatment (11, 74, 121). Whether the induction of these enzymes, similar to tyrosine aminotransferase, also involves new protein and RNA synthesis is not known. Of considerable interest are the recent reports indicating that the induction of tyrosine aminotransferase (134) and serine dehydratase (46, 47) in rat liver by the pancreatic hormones may be mediated by an increase in the level of cyclic AMP.

E.   Endocrine Gland Ablation

The individual differences that have been observed in the response of several adaptive enzymes to endocrine ablation serve to illustrate the complex hormonal relationships which are involved in the regulation of enzyme activity in mammals. For example, in addition to altering the basal activity of certain corticoid-responsive enzymes in a selective manner, adrenalectomy and hypophysectomy also appear to influence the response of some of these enzymes to hydrocortisone treatment (98, 108), and a high-protein diet (109). Conceivably, hormonal imbalances will also be found to play an important role in mediating the specific action of various physiological and pathological conditions on the activity of the adaptive enzymes. This section considers the effect of adrenalectomy on only a few of the enzymes that are known to be glucocorticoid-responsive.

Following adrenalectomy of the adult rat, the hepatic levels of alanine aminotransferase (32) and tryptophan oxygenase (56, 127) are significantly lowered, while the activity of tyrosine aminotransferase remains essentially unchanged (64). In general, adaptive enzymes which are involved in amino acid metabolism show a depression in activity when the adrenals are removed. By contrast, the cortisol inducible enzymes concerned with carbohydrate metabolism do not undergo any significant change in activity after adrenalectomy (131). Although the factors involved in maintaining enzyme activity in adrenalectomized rats have not been identified, it is likely that the pancreatic hormones, glucagon and insulin, which induce tyrosine aminotransferase, are responsible for the apparent lack of effect of adrenalectomy on this enzyme.

Age was found to be an important factor in obtaining a fall in alanine aminotransferase activity in adrenalectomized rats. Adrenalectomy of immature rats less than 7 weeks of age did not result in a loss of enzyme activity, whereas in older rats, with an elevated basal level of the enzyme, removal of the adrenals was followed by a marked reduction in enzyme activity within 48 h (32).

The selective effect of adrenalectomy on the developmental pattern of ornithine aminotransferase in the liver and kidney of rats has been reported (37, 38). The developmental formation of this enzyme in liver which begins during the third week of postnatal life could be stimulated by hydrocortisone treatment and delayed by adrenalectomy. Estrogen treatment alone retarded the development of this transferase, and inhibited the action of hydrocortisone in stimulating this process. In contrast, during development, the kidney enzyme was unresponsive to glucocorticoids and adrenalectomy, but showed an elevation in activity when estrogen was administered. Glucocorticoids, which were effective in raising hepatic ornithine aminotransferase during the developmental period, were ineffective when given to postweanling rats; similarly, adrenalectomy performed in rats 7 weeks old did not lower the activity of this enzyme in liver. Estrogen, on the other hand, stimulated increases in kidney ornithine aminotransferase in both preweanling and adult animals. It can be concluded that an enzyme in different tissues which catalyzes the same reaction, is not necessarily under the same hormonal regulatory control. Also, these studies reveal the temporal and sexual differences that are involved in the hormonal control of ornithine aminotransferase.

F.   Response to Amino Acids and Pyridoxine

There are many reports indicating that the adaptive enzymes, tryptophan oxygenase and tyrosine aminotransferase, can be increased in activity in the liver of rats treated with the substrates of these enzymes, tryptophan and tyrosine (55, 57, 65, 99). It is important to note, however, that not all of the inducible mammalian enzymes show an adaptive response to substrate treatment. For example the administration of substrates of alanine aminotransferase (98), serine dehydratase (22) and the urea cycle enzymes (111) did not lead to a rise in the activity of these enzymes. Thus while each of these enzymes is responsive to glucocorticoids, their induction following substrate treatment or ingestion of casein hydrolysate (60, 102) is selective and the basis for such effects remains to be explained.

Various amino acids (65, 103, 105), indoles (99) and non-specific compounds have been found to be capable of producing increases in the hepatic level of tryptophan oxygenase and tyrosine aminotransferase. Many of these treatments are effective only in intact rats, suggesting that their effects are due to a stress-mediated release of corticosterone from the adrenals. Of these treatments, tryptophan has been clearly demonstrated to act as an inducing agent of tryptophan oxygenase (55) and tyrosine aminotransferase (99) in the livers of intact or adrenalectomized rats. In contrast, tyrosine is an effective inducer of tyrosine aminotransferase in intact rats, but ineffective in adrenalectomized rats (65). Substrate stabilization of tryptophan oxygenase thus prolonging its rate of degradation, has been proposed to explain the manner in which tryptophan administration produces a rise in the activity of this enzyme (113, 114, 115). More recently, *Labrie and Korner* (60) have obtained data indicating that the induction of this oxygenase by relatively small doses of tryptophan can be inhibited by concomitant treatment with actinomycin D. A requirement for new RNA synthesis as determined by the blocking action of actinomycin D, also appears to be involved in the mechanism by which α-methyl tryptophan induces tyrosine aminotransferase (27). In view of these results, it is not unlikely that tryptophan induction of tryptophan oxygenase requires both substrate stabilization of the enzyme and new RNA synthesis; additional evidence concerning this possibility is needed.

A several-fold increase in the activity of tyrosine aminotransferase is observed in intact or adrenalectomized rats following treatment with an enzymatic hydrolysate of casein, whereas tryptophan oxygenase is unresponsive to this treatment (102). The effect of casein hydrolysate is inhibited by actinomycin D or cycloheximide, indicating that the synthesis of new RNA and protein is associated with the response of tyrosine aminotransferase to this amino acid digest. A dose of tryptophan, equivalent to that contained in 1 g of casein hydrolysate, did not produce a rise in tyrosine aminotransferase. In a subsequent study (103), proline and glutamic acid, independently, were found to induce tyrosine aminotransferase in the livers of adrenalectomized rats, and in the isolated perfused rat liver. Proline, glutamate and tryptophan, when administered together in amounts equal to their content in 1 g of casein was equally as effective (3-fold response) in inducing tyrosine aminotransferase, as 1 g of the casein hydrolysate. The effects of proline and glutamate are dependent on new RNA and protein synthesis and specific for tyrosine aminotransferase.

It is evident that there are many glucocorticoid-responsive enzymes that can respond independently to protein administration, and that in some instances the effects of these treatments are additive with glucocorticoids (88). Tryptophan oxygenase is unique in this regard, since it responds to tryptophan and glucocorticoids but not to variation in protein intake (102). *Peraino* (88, 89) has reported that several adaptive enzymes differ with respect to their responses to protein, carbohydrate and cortisone; protein, but not cortisone, was found to be necessary for the induction of serine dehydratase and ornithine transaminase, while glucose-6-phosphate dehydrogenase was responsive to both of these agents and to carbohydrate. These results suggest that probably many metabolites of proteins and carbohydrates, either independently or with various hormones, are involved in the regulation of enzyme activity in mammalian tissues.

Two adaptive enzymes in liver which are stimulated (serine dehydratase) and unaffected (ornithine aminotransferase) following glucocorticoid treatment, undergo marked increases in activity in rats fed a diet containing 60 % protein (88). Treatment with glucocorticoids inhibits the increase in the level of ornithine aminotransferase produced by high dietary protein, whereas the response of serine dehydratase to dietary protein was stimulated by glucocorticoids. Further studies are needed to elucidate the mechanism by which glucocorticoids exert different regulatory effects on specific enzymes.

In addition to the regulation of tyrosine aminotransferase by hormones and certain amino acids, another mechanism was proposed by *Greengard and Gordon* (28) who found that administration of large doses of pyridoxine increased the level of this enzyme in the livers of adrenalectomized rats. The increase in activity of this transaminase was reported to be preceded by elevated levels of pyridoxal phosphate in liver and thus was termed 'cofactor induction'. Induction of this enzyme by pyridoxine was inhibited by puromycin but not by actinomycin D and thus was analogous to the substrate-induced induction of tryptophan oxygenase by tryptophan (29). Subsequently, *Holten et al.* (43) observed that the pyridoxine-induced response of tyrosine aminotransferase is prevented by actinomycin D and that pyridoxine did not effect a stabilization of the enzyme *in vivo*. These authors concluded that the induction of this transaminase by pyridoxine was similar to the hormonal induction of this enzyme in that it appears to require the synthesis of RNA. Because of the very large dose (1 g/kg) of pyridoxine required to produce induction of tyrosine aminotransferase in adrenalectomized rats, the suggestion was made that pyridoxine may be acting indirectly to affect the hormonal balance which is involved in regulating the activity of this enzyme (43). Norepinephrine which competitively inhibits aminotransferase *in vitro* by forming a complex with the cofactor pyridoxal phosphate, abolished the pyridoxine-induced rise of this enzyme in the fasted, adrenalectomized rat (7).

G.   Response to Cyclic AMP

Cyclic adenosine 3′,5′-monophosphate (cyclic AMP), which mediates the action of many protein hormones, also appears to play a role in the induction of tyrosine aminotransferase by glucagon and insulin (134), and of serine dehydratase by glucagon and a mixture of amino acids (46, 47). Studies both *in vivo* and in organ culture have demonstrated that cyclic AMP increases the activity of tyrosine aminotransferase (132), serine dehydratase (46, 47) and P-enolpyruvate carboxykinase (138). A single injection of glucagon or the intubation of a mixture of amino acids stimulated the formation of a very high concentration of cyclic AMP in rat liver (47); glucocorticoids to our knowledge have not been tested in this regard. That glucocorticoids are acting to induce the synthesis of specific enzymes by a mechanism which does not involve cyclic AMP as a 'second messenger' is suggested by the additive and sometimes synergistic response of tyrosine aminotransferase elicited by the combined treatment of rats with corticosteroids and glucagon (42) or casein hydrolysate (102).

It is tempting to speculate that the induction of tyrosine aminotransferase by tryptophan, proline, glutamate, or pyridoxine and, perhaps in part, the substrate-induced increases of tryptophan oxygenase may be cyclic AMP-mediated. It would also be important to determine whether the glycogenolytic action in liver exerted by such compounds as actinomycin D (94), puromycin (39), cycloheximide (15) and various amino acids (83) including tryptophan can be attributed to an elevation in cyclic AMP levels. The results of studies to test these possibilities will be awaited with interest.

III. Possible Mechanisms by which Glucocorticoids Stimulate Enzyme Induction

A. Nucleic Acids

Many investigators have asked the question, 'How do glucocorticoids stimulate enzyme synthesis?'. Earlier reports indicated that treatment with moderate doses of glucocorticoids elicited slight increases in the RNA content of rat liver (23, 122). In 1962, *Feigelson et al.* (12) noted that the incorporation of P^{32} and glycine-2-C^{14} *in vivo* into the RNA of all the subcellular fractions of normal and regenerating liver was significantly stimulated by treatment with a single dose of cortisone. A requirement for new RNA synthesis in enzyme induction was implied by the inhibitory action of actinomycin D on the response of tyrosine aminotransferase and tryptophan oxygenase to hydrocortisone (29).

These observations were followed by more definitive experiments indicating that hydrocortisone treatment stimulated a 2- to 3-fold increase in the incorporation of radioactive precursors into liver nuclear RNA (19, 51). Evidence was also obtained in these studies that the earliest effect of the hormone was obtained in the nucleus, and that nuclear RNA with increased template activity in an *E. coli* cell-free system could be demonstrated in the hydrocortisone-treated animals. The relatively rapid changes in RNA metabolism elicited by hydrocortisone *in vivo* correlate well in time with the earliest detectable adrenal corticoid-induced responses of tyrosine aminotransferase and tryptophan oxygenase. *Sekeris* and co-workers (116) have found a direct effect of hydrocortisone on RNA template activity in isolated rat liver nuclei. Of considerable interest in later studies was the observation that a messenger RNA fraction coding for tyrosine aminotransferase could be obtained from nuclei exposed to the steroid (61).

There is ample evidence that the three major species of RNA are stimulated in rat liver following injection of glucocorticoids. The major fraction, ribosomal RNA or its precursor in nuclei, undergoes a 2- to 3-fold increase in its rate of synthesis within 2 h after hydrocortisone treatment (31). A fraction characterized operationally as DNA – like RNA, with a rapid turnover time and comprising only about 1.5 % of total liver RNA, was also stimulated in its rate of synthesis by hydrocortisone. Transfer RNA and a heterogenous RNA that sediments in a sucrose gradient like transfer RNA showed 2- to 3-fold increases in their *de novo* synthesis when the adrenal corticoid was given (133).

In view of the compelling evidence that glucocorticoids elicit a stimulatory effect on hepatic RNA synthesis it seems reasonable to propose that the effects of these steroids on protein and enzyme biosynthesis are secondary to changes in RNA metabolism. However, it also seems clear that various amino acids (14, 102), ammonium sulfate (14), growth hormone (52) and the pancreatic hormones (42) are also capable of affecting a rise in RNA synthesis in liver. In some instances, these different treatments lead to an increase in the activity of tyrosine aminotransferase, but are without effect on tryptophan oxygenase. Whether these various compounds which elicit a stimulation in RNA synthesis act specifically with regard to a particular species of RNA has not yet been determined. At any rate this lack of specificity cautions against interpreting some of the present data too precisely in terms of a relationship between changes in RNA synthesis and the effects of glucocorticoids on enzyme induction.

Much of the current work to determine the biochemical basis for the actions of the glucocorticoids on lymphoid cells has been carried out on rat thymus gland. In this tissue, glucocorticoids have been observed to produce inhibitory effects on the incorporation of radioactive precursors into nucleic acids (36, 40) and protein (69, 70) as well as depress RNA polymerase activity (84) and glucose utilization (82). Many of these effects have also been seen in spleen (124), the cortisol-sensitive line of lymphosarcoma P1798 (110) and in malignant cells grown in culture (18). It is of considerable interest that in these tissues the glucocorticoids exert effects on the synthesis of RNA and protein which are opposite to those seen in liver.

Several enzymes in normal and neoplastic lymphoid tissues undergo a rise in activity following treatment with the adrenal corticoids (85). Such changes are difficult to explain in view of the concomitant depression in RNA synthesis that occurs in these tissues during their exposure to a glucocorticoid.

B.   Amino Acid Pools

It is becoming increasingly evident that alterations in amino acid levels may play a central role in explaining the action of the glucocorticoids on liver. Specific amino acids can induce adaptive enzymes in adrenalectomized rats, stimulate new RNA synthesis in liver, and act synergistically with small doses of hydrocortisone to induce tyrosine aminotransferase (50) and tryptophan oxygenase (102). Furthermore, treatment with glucocorticoids exerts a rapid effect on the uptake and concentration of amino acids in liver (96) and such effects have been suggested to be the basis for the gluconeogenic action of this steroid (62).

Earlier studies in our laboratory (104) indicated a correlation between changes in the activity of alanine aminotransferase and the metabolic pool of amino acids. Conditions associated with accelerated gluconeogenesis, such as diabetes or starvation, involve mobilization of proteins from peripheral tissues and the transfer of amino acids to the liver. Other conditions, such as hypophysectomy, aging, and feeding a high protein diet also would be expected to enlarge the pool of amino acids in liver. Each of these treatments or conditions has been observed to stimulate a significant rise in hepatic alanine aminotransferase. In contrast, tumor growth, (21, 33), liver regeneration (34), and the growth of fetal tissue (35) require amino acids for the synthesis of new proteins, and as a consequence would lower the level of free amino acids in liver. In each of these conditions, the hepatic activity of alanine aminotransferase is depressed.

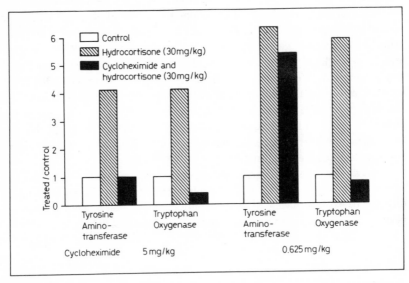

Fig. 2. Selective action of cycloheximide on the induction of tyrosine aminotransferase and tryptophan oxygenase by cortisol. Both compounds were given intraperitoneally. Cortisol was given at 0 time, cycloheximide 30 min later and the animals killed 4 h after the steroid was given. Each value represents the mean value for 5 animals in each group.

The induction of tyrosine aminotransferase and tryptophan oxygenase by hydrocortisone is prevented when this steroid is given together with actinomycin D (29), puromycin (29), or cycloheximide (100) in doses which are sufficiently large to block new RNA and protein synthesis. Typical results obtained when an adequate (5 mg/kg) amount of cycloheximide is given together with hydrocortisone are presented in figure 2. However, it is also evident that cycloheximide can act selectively to impair the induction of tryptophan oxygenase, but not tyrosine aminotransferase, when a dose (0.63 mg/kg) of this antibiotic is administered which impairs protein synthesis by only 60 %. Furthermore, tyrosine aminotransferase in particular shows a rise in activity in the livers of intact or adrenalectomized rats treated only with small doses of actinomycin D (106) or cycloheximide (16). *Kenney* (49) noted that endogenous levels of tyrosine aminotransferase were maintained in adrenalectomized rats treated with cycloheximide; evidence that this effect might be due to an impairment in the normal rate of degradation of this enzyme was obtained with the use of the antibody of this enzyme. On the basis of these results, it was suggested that cycloheximide might be acting to inhibit the synthesis of a protein that was required for the turnover and inactivation of this transaminase. It is difficult to relate this explanation to the fact that tryptophan oxygenase and tyrosine aminotransferase have approximately the same turnover time, 2 to 3 h, and presumably their rates of synthesis and degradation are regulated by similar control mechanisms. Thus, if the synthesis of new protein is involved in the turnover of tyrosine aminotransferase, one might expect the same conditions to apply for tryptophan oxygenase. However, the latter enzyme is remarkably sensitive to inhibition by cycloheximide, under the same conditions in which tyrosine aminotransferase is either resistant or induced by treatment with this antibiotic.

Since tyrosine aminotransferase is responsive to casein hydrolysate and the magnitude of its response to hydrocortisone can be regulated by the protein content of the diet (102), it was of interest to determine whether an increase in the free amino acid content of liver might be associated with the induction of this enzyme by cycloheximide (101). When cycloheximide was given alone or when it was administered together with cortisol there was an appreciable rise in the free amino acid levels in liver. This increase occurred maximally within 15 to 30 min after injection of the antibiotic. The results of this work are consistent with the concept that for the optimal translation of the RNA template for tyrosine aminotransferase an increased supply of amino acid is required. In contrast, several lines of evidence suggest that the adaptive response of tryptophan oxygenase is mainly dependent on new RNA synthesis and that increased levels of amino acids play but a minor role in the induction of this enzyme.

The foregoing suggests that alterations in the amino acid pool in liver may mediate, at least in part, the induction of adaptive enzymes by glucocorticoids. Thus, treatment with glucocorticoids is known to produce a marked catabolic effect on protein metabolism in peripheral tissues (67) and to decrease the rate of incorporation of amino acids into the proteins of such tissues as skeletal muscle, thymus, and cartilage (71, 72). Such effects in extrahepatic tissues would be expected to elevate the level of amino acids in plasma (123) and to favor the accumulation of amino acids in liver. Furthermore, glucagon and insulin, as well as hydrocortisone, stimulated the uptake of a-aminoisobutyric acid-1-^{14}C into the isolated, perfused rat liver (8). These observations support the concept that the induction of tyrosine aminotransferase, and probably other adaptive enzymes with the exception of tryptophan oxygenase, by certain hormones and other treatments may be due to increased levels of amino acids in liver.

C. Repressors

Treatment of intact rats with small doses of actinomycin, which permitted survival of the animals for several days, resulted in marked increases in the levels of four hepatic enzymes: tyrosine aminotransferase, alanine aminotransferase, serine dehydratase and tryptophan oxygenase (106). Actinomycin also induced responses of tyrosine aminotransferase and serine dehydratase in the livers of adrenalectomized rats. To explain these findings it was suggested that the negative nitrogen balance and cachexia which occurred after administration of toxic doses of actinomycin might enlarge the pool of free amino acids in liver and thus selectively promote the induction of hepatic enzymes responsive to increased protein intake or fasting (106).

Other observations of the stimulatory action of actinomycin on various enzymes have been reported with quite different experimental systems (79, 91). On the basis of studies indicating that delayed treatment with actinomycin after the injection of hydrocortisone further stimulated the levels of tyrosine aminotransferase and tryptophan oxygenase, *Garren et al.* (20) postulated that the plateau as well as the fall in the activities of these enzymes could be due to the synthesis of a repressor(s). Reports from other laboratories (10, 77), however, have failed to confirm the initial observations made by Garren and co-workers.

Rat hepatoma cells (HTC) and Reuber H-35 hepatoma grown in continuous culture have been shown to undergo an increased rate of synthesis of tyrosine aminotransferase when exposed to adrenal corticoids (95, 126). In both cell lines, preinduction of tyrosine aminotransferase by hydrocortisone, followed by exposure to actinomycin with or without the corticosteroid, results in the 'superinduction' of this enzyme. Different mechanisms have been proposed to explain this effect of actinomycin in hepatoma cells. *Reel and Kenney* (95), on the basis of studies employing the antibody of tyrosine aminotransferase to precipitate the isotopically labelled enzyme, concluded that the further increase in enzyme activity following exposure to actinomycin was due to a differential effect on its rate of synthesis and degradation. Thus the elevation in transaminase activity produced by actinomycin was attributed to a marked depression in its rate of degradation which occurred together with a progressive decrease in the rate of synthesis of the enzyme.

A different mechanism to explain the 'superinduction' of tyrosine aminotransferase by actinomycin in HTC cells was advanced by *Tompkins et al.* (128). These workers reported that actinomycin did not affect the rate of degradation of this enzyme (90). To explain the paradoxical effects of actinomycin on the level of this enzyme, as well as the glucocorticoid-mediated induction of this enzyme, it was postulated that a protein repressor of this transaminase existed which functioned both to inhibit messenger translation and to stimulate messenger degradation (128). Induction of the enzyme by glucocorticoids was suggested to involve antagonism of the posttranscriptional repressor by the hormone, whereas the action of actinomycin was attributed to inhibition of RNA synthesis which prevented the formation of the repressor. The manner by which the steroid hormone inactivates the repressor, leading to an increase in the concentration of the messenger and thus a stimulation in the rate of enzyme synthesis, is unknown. Although this hypothesis offers an attractive explanation of enzyme induction by glucocorticoids and of the untoward action of actinomycin on tyrosine aminotransferase in HTC cells, substantially more evidence is required before it can be viewed as other than a hypothesis.

The results of other studies also suggest that posttranscriptional events may be concerned with the hydrocortisone-mediated response of tyrosine amino-transferase. The purine analog 8-azaguanine inhibits protein synthesis by being incorporated into template RNA (59). When given with hydrocortisone, this anti-metabolite acts differentially to block the induction of tryptophan oxygenase, but not that of tyrosine aminotransferase (63). However, in rats treated at 3-h intervals with the steroid and 8-azaguanine the activity of the transaminase was blocked almost completely by 9 to 10 h, in contrast to the progressive rise seen when hydrocortisone was given alone. These results were interpreted to mean that template RNA for the transaminase is relatively stable and undergoes activation and renewal during the induction process. Further, it was postulated that glucocorticoids may affect 'the nucleocytoplasmic barrier, allowing for the transport to the cytoplasm of certain RNA species which are normally restricted to the nucleus'. Evidence that cortisone treatment markedly reduces the degradation of hepatic microsomal RNA has also been reported (86). Thus, it appears that both the synthesis and breakdown of RNA may be involved in the gluco-corticoid-induced response of certain adaptive enzymes.

Although there is some evidence for the presence of receptors in HTC cells which may function in the steroid induction of tyrosine aminotransferase (3), these macromolecules do not appear to be the labile repressor predicted in the model that has been proposed by *Tomkins et al.* (128).

Only a limited amount of information is available on the accumulation of glucocorticoids in subcellular fractions, and in particular their binding to specific receptors which may give a clue as to mechanism of action. *Litwack et al.* (66) studied the uptake and subcellular distribution of radioactivity from ^{14}C-cortisol in rat liver. The counts were highest at 45–60 min, and were concentrated mainly in the supernatant and microsomal fractions. No differences in the retention of the radioactivity were noted in these fractions. In more recent studies by *Litwack* and co-workers evidence was obtained that labelled hydrocortisone binds mainly to the smooth membrane of the microsomal fraction (73) and that two negatively charged metabolites of this steroid bind to two proteins in rat liver (80).

IV. Biological Significance

The biological significance of increased enzyme activity in liver as a consequence of hydrocortisone administration is not yet clear. Changing the amounts of particular enzymes could by altering the tissue's potential for the turnover of substrates affect the functional and growth capacity of a tissue. Although changes in the level of cortisol-responsive enzymes in liver can be correlated with changes in the capacity for gluconeogenesis (107) there is no conclusive evidence that changes in the amount of particular enzymes may regulate this important function. Studies (30) relating the induction of tyrosine aminotransferase and tryptophan oxygenase to the gluconeogenic response of liver to hydrocortisone are difficult to interpret since the inhibitors (puromycin, actinomycin D) which were used to block enzyme induction also stimulated glycogenolysis (105). Several workers (58, 119) have concluded that the gluconeogenic function of glucocorticoids is not dependent on new enzyme formation. It has been suggested that glucocorticoids may function in gluconeogenesis by impairing protein synthesis in peripheral tissues, thus favoring the accumulation and use of amino acids in liver for carbohydrate synthesis. The enhancement of gluconeogenic enzymes by adrenal corticoids may further facilitate carbohydrate synthesis. Recently, the glucocorticoids were reported to exert a controlling effect on the glycogen synthetase-activating system in liver and it was proposed that changes in active synthetase, in response to glucocorticoids, were related to alterations in hepatic glycogen (75).

Whether increased enzyme activity is reflected in the enhanced metabolism of its substrate *in vivo* is a question of fundamental importance. The relevance of increases in tyrosine aminotransferase and tryptophan oxygenase has recently been examined by *Kim and Miller* (53). Using the isolated perfused rat liver, as well as the intact and adrenalectomized rat, they were unable to detect increases in the metabolism of radioactive substrates commensurate with the increases in enzyme activity as determined *in vitro* and thus concluded that such increases in activity may be of limited functional importance in the animal. It was proposed that any increase in the metabolic products of these two enzymes was largely due to elevated substrate levels *per se* rather than to the increase in enzyme activity.

Acknowledgement

Research cited in this review which originated in the authors' laboratory was supported in part by Public Health Service Grants AM-04389 and CA-05671.

V. References

1 *Aki, K.; Ogawa, K., and Ichihara, A.:* Transaminases of branched chain amino acids. IV. Purification and properties of two enzymes from rat liver. Biochim. biophys. Acta *159:* 276–284 (1968).

2 *Altman, K. and Greengard, O.:* Tryptophan pyrrolase induced in human liver by hydrocortisone: effect on the excretion of kynurenine. Science *151:* 332–333 (1966).

3 *Baxter, J.D. and Tomkins, G.M.:* The relationship between glucocorticoid binding and tyrosine aminotransferase induction in hepatoma tissue culture cells. Proc. nat. Acad. Sci., Wash. *65:* 709–715 (1970).

4 *Berlin, C.M. and Schimke, R.T.:* Influence of turnover rates on the responses of enzymes to cortisone. Molec. Pharmacol. *1:* 149–156 (1965).

5 *Black, I.B. and Axelrod, J.:* Elevation and depression of hepatic tyrosine transaminase activity by depletion and repletion of norepinephrine. Proc. nat. Acad. Sci., Wash. *59:* 1231–1234 (1968a).

6 *Black, I.B. and Axelrod, J.:* Regulation of the daily rhythm in tyrosine transaminase activity by environmental factors. Proc. nat. Acad. Sci., Wash. *61:* 1287–1291 (1968b).

7 *Black, I.B. and Axelrod, J.:* Inhibition of tyrosine transaminase activity by norepinephrine. J. biol. Chem. *244:* 6124–6129 (1969).

8 *Chambers, J.W.; Georg, R.H., and Bass, A.D.:* Effect of hydrocortisone and insulin on uptake of α-aminoisobutyric acid by isolated perfused rat liver. Molec. Pharmacol. *1:* 66–76 (1965).

9 *Civen, M.; Trimmer, M., and Brown, C.B.:* The induction of hepatic tyrosine-α-ketoglutarate and phenylalanine pyruvate transaminases by glucagon. Life Sci. *6:* 1331–1338 (1967).

10 *Csanyi, V.; Greengard, O., and Knox, W.E.:* The inductions of tyrosine aminotransferase by glucagon and hydrocortisone. J. biol. Chem. *242:* 2688–2692 (1967).

11 *Curry, D.M. and Beaton, G.H.:* Glucagon administration in pregnant rats. Endocrinology, Springfield *63:* 252–254 (1958).

12 *Feigelson, M.; Gross, P.R., and Feigelson, P.:* Early effects of cortisone on nucleic acid and protein metabolism of rat liver. Biochim. biophys. Acta *55:* 495–504 (1962).

13 *Feigelson, P.; Dashman, T., and Margolis, F.:* The half-lifetime of induced tryptophan peroxidase *in vivo.* Arch. Biochem. *85:* 478–482 (1959).

14 *Feigelson, P. and Feigelson, M.:* Studies on the mechanism of regulation by cortisone of the metabolism of liver purine and ribonucleic acid. J. biol. Chem. *238:* 1073–1077 (1963).

15 *Fiala, S. and Fiala, E.:* Hormonal dependence of actidione (cycloheximide) action. Biochim. biophys. Acta *103:* 699–701 (1965).

16 *Fiala, S. and Fiala, E.:* Induction of tyrosine transaminase in rat liver by actidione. Nature *210:* 530–531 (1966).

17 *Freedland, R.A. and Avery, E.H.:* Studies on threonine and serine dehydrase. J. biol. Chem. *239:* 3357–3360 (1964).

18 *Gabourel, J.D. and Aronow, L.:* Growth inhibitory effects of hydrocortisone on mouse lymphoma ML-388 *in vitro.* J. Pharmacol. exp. Ther. *136:* 213–221 (1962).

19 *Garren, L.D.; Howell, R.R., and Tomkins, G.M.:* Mammalian enzyme induction by hydrocortisone: the possible role of RNA. J. molec. Biol. *9:* 100–108 (1964).

20 *Garren, L.D.; Howell, R.R.; Tomkins, G.M., and Crocco, R.M.:* A paradoxical effect of actinomycin D: The mechanism of regulation of enzyme synthesis by hydrocortisone. Proc. nat. Acad. Sci., Wash. *52:* 1121–1129 (1964).

21   *Girkin, G. and Kampschmidt, R.F.:* Changes in liver tyrosine-a-ketoglutarate trans-
     aminase activity during growth of Walker carcinosarcoma 256. Proc. Soc. exp. Biol.,
     N.Y. *105:* 221–223 (1960).

22   *Goldstein, L.; Knox, W.E., and Behrman, E.J.:* Studies on the nature, inducibility, and
     assay of the threonine and serine dehydrase activities of rat liver. J. biol. Chem. *237:*
     2855–2860 (1962).

23   *Goodlad, G.A.J. and Munro, H.N.:* Diet and the action of cortisone on protein meta-
     bolism. Biochem. J. *73:* 343–348 (1959).

24   *Gorski, J.:* Early estrogen effects on the activity of uterine RNA polymerase. J. biol.
     chem. *239:* 889–892 (1964).

25   *Green, D.E.:* Enzymes and trace substances. Adv. Enzymol. *1:* 177–198 (1941).

26   *Greengard, O. and Baker, G.T.:* Glucagon, starvation, and the induction of liver en-
     zymes by hydrocortisone. Science *154:* 1461–1462 (1966).

27   *Greengard, O.; Baker, G.T.; Horowitz, M.L., and Knox, W.E.:* Effects of irradiation and
     starvation on the regulation of rat liver enzymes. Proc. nat. Acad. Sci., Wash. *56:*
     1303–1309 (1966).

28   *Greengard, O. and Gordon, M.:* The cofactor-mediated regulation of apoenzyme levels
     in animal tissues. I. The pyridoxine-induced rise of rat liver tyrosine transaminase level
     in vivo. J. biol. Chem. *238:* 3708–3710 (1963).

29   *Greengard, O.; Smith, M.A., and Acs, G.:* Relation of cortisone and synthesis of ribo-
     nucleic acid to induced and developmental enzyme formation. J. biol. Chem. *238:*
     1548–1551 (1963).

30   *Greengard, O.; Weber, G., and Singhal, R.L.:* Glycogen deposition in the liver induced
     by cortisone: dependence upon enzyme synthesis. Science *141:* 160–161 (1963).

31   *Greenman, D.L.; Wicks, W.D., and Kenney, F.T.:* Stimulation of ribonucleic acid
     synthesis by steroid hormones. II. High molecular weight components. J. biol. Chem.
     240: 4420–4426 (1965).

32   *Harding, H.R.; Rosen, F., and Nichol, C.A.:* Influence of age, adrenalectomy, and
     corticosteroids on hepatic transaminase activity. Amer. J. Physiol. *201:* 271–275
     (1961).

33   *Harding, H.R.; Rosen, F., and Nichol, C.A.:* Depression of alanine transaminase activity
     in the liver of rats bearing Walker Carcinoma 256. Cancer Res. *24:* 1318–1323 (1964).

34   *Harding, H.R.; Rosen, F., and Nichol, C.A.:* Effect of partial hepatectomy and preg-
     nancy on tumor growth and alanine-a-ketoglutarate transaminase activity. Proc. Soc.
     exp. Biol., N.Y. *122:* 561–565 (1966a).

35   *Harding, H.R.; Rosen, F., and Nichol, C.A.:* Effects of pregnancy on several cortisol-
     responsive enzymes in rat liver. Amer. J. Physiol. *211:* 1361–1365 (1966b).

36   *Haynes, R.C., Jr. and Sutherland, E.W., III:* Altered metabolism of DNA in rat thymus,
     an early response to cortisol. Endocrinology, Springfield *80:* 297–301 (1967).

37   *Herzfeld, A. and Greengard, O.:* Endocrine modification of the developmental forma-
     tion of ornithine aminotransferase in rat tissues. J. biol. Chem. *244:* 4894–4898
     (1969).

38   *Herzfeld, A. and Knox, W.E.:* The properties, developmental formation, and estrogen
     induction of ornithine aminotransferase in rat tissues. J. biol. Chem. *243:* 3327–3332
     (1968).

39   *Hofert, J.F.; Gorski, J.; Mueller, G., and Boutwell, R.K.:* The depletion of liver glyco-
     gen in puromycin-treated animals. Arch. Biochem. *97:* 134–137 (1962).

40   *Hofert, J.F. and White, A.:* Effect of a single injection of cortisol on the incorporation
     of ^{3}H-thymidine and ^{3}H-deoxycytidine into lymphatic tissue DNA of adrenalectomiz-
     ed rats. Endocrinology, Springfield *82:* 767–776 (1968).

41 *Holt, P.G. and Oliver, I.T.:* Multiple forms of tyrosine aminotransferase in rat liver and their hormonal induction in the neonate. FEBS Letters *5:* 89–91 (1969).

42 *Holten, D. and Kenney, F.T.:* Regulation of tyrosine-a-ketoglutarate transaminase in rat liver. VI. Induction by pancreatic hormones. J. biol. Chem. *242:* 4372–4377 (1967).

43 *Holten, D.; Wicks, W.D., and Kenney, F.T.:* Studies on the role of vitamin B₆ derivatives in regulating tyrosine-a-ketoglutarate transaminase activity *in vitro* and *in vivo.* J. biol. Chem. *242:* 1053–1059 (1967).

44 *Inoue, H. and Pitot, H.:* Regulation of the synthesis of serine dehydratase isozymes. Advances in Enzyme Regulation, vol. 8, in press (1969).

45 *Jervell, K.F.:* Early effects of glucocorticoids on ribonucleic acid and protein metabolism in rat liver. Acta endocrin., Kbh., Suppl. 88, pp. 1–36 (1963).

46 *Jost, J.-P.; Hsie, A.W.; Hughes, S.D., and Ryan, L.:* Role of cyclic adenosine 3′,5′-monophosphate in the induction of hepatic enzymes. I. Kinetics of the induction of rat liver serine dehydratase by cyclic adenosine 3′,5′-monophosphate. J. biol. Chem. *245:* 351–357 (1970).

47 *Jost, J.-P.; Hsie, A.W., and Rickenberg, H.V.:* Regulation of the synthesis of rat liver serine dehydratase by adenosine 3′,5′-cyclic monophosphate. Biochem. biophys. Res. Comm. *34:* 748–754 (1969).

48 *Kenney, F.T.:* Induction of tyrosine-a-ketoglutarate transaminase in rat liver. IV. Evidence for an increase in the rate of enzyme synthesis. J. biol. Chem. *237:* 3495–3498 (1962).

49 *Kenney, F.T.:* Turnover of rat liver tyrosine transaminase: stabilization after inhibition of protein synthesis. Science *156:* 525–528 (1967).

50 *Kenney, F.T. and Flora, R.M.:* Induction of tyrosine-a-ketoglutarate transaminase in rat liver. I. Hormonal nature. J. biol. Chem. *236:* 2699–2702 (1961).

51 *Kenney, F.T. and Kull, F.J.:* Hydrocortisone-stimulated synthesis of nuclear RNA in enzyme induction. Proc. nat. Acad. Sci., Wash. *50:* 493–499 (1963).

52 *Kenney, F.T.; Wicks, W.D., and Greenman, D.L.:* Hydrocortisone stimulation of RNA synthesis in induction of hepatic enzymes. J. cell. comp. Physiol. *66:* 125–136 (1965).

53 *Kim, J.H. and Miller, L.L.:* The functional significance of changes in activity of the enzymes, tryptophan pyrrolase and tyrosine transaminase, after induction in intact rats and in the isolated, perfused rat liver. J. biol. Chem. *244:* 1410–1416 (1969).

54 *Kim, Y.S.:* The sequential increase in the rate of synthesis of enzymes in rat liver after glucocorticoid administration. Molec. Pharmacol. *4:* 168–172 (1968).

55 *Knox, W.E.:* Two mechanisms which increase *in vivo* the liver tryptophan peroxidase activity: specific enzyme adaptation and stimulation of the pituitary-adrenal system. Brit. J. exp. Path. *32:* 462–469 (1951).

56 *Knox, W.E. and Auerbach, V.H.:* The hormonal control of tryptophan peroxidase in the rat. J. biol. Chem. *214:* 307–313 (1955).

57 *Knox, W.E. and Mehler, A.H.:* The adaptive increase of the tryptophan peroxidase-oxidase system of liver. Science *113:* 237–238 (1951).

58 *Kvam, D.C. and Parks, R.E., Jr.:* Hydrocortisone-induced changes in hepatic glucose-6-phosphatase and fructose diphosphatase activities. Amer. J. Physiol. *198:* 21–24 (1960).

59 *Kwan, S.-W. and Webb, T.E.:* A study of the mechanism of polyribosome breakdown induced in regenerating liver by 8-azaguanine. J. biol. Chem. *242:* 5542–5548 (1967).

60 *Labrie, F. and Korner, A.:* Actinomycin-sensitive induction of tyrosine transaminase and tryptophan pyrrolase by amino acids and tryptophan. J. biol. Chem. *243:* 1116–1119 (1968).

61 Lang, N.; Herrlich, P., and Sekeris, C.E.: On the mechanism of hormone action. VII. Induction by cortisol of a messenger RNA coding for tyrosine-α-ketoglutarate transaminase in an *in vitro* system. Acta endocrin., Kbh. *57:* 33–44 (1968).

62 Lardy, H.A.; Foster, D.O.; Young, J.W.; Shrago, E., and Ray, P.D.: Hormonal control of enzymes participating in gluconeogenesis and lipogenesis. J. cell. comp. Physiol. *66:* 39–53 (1965).

63 Levitan, I.B. and Webb, T.E.: Posttranscriptional control in the steroid-mediated induction of hepatic tyrosine transaminase. Science *167:* 283–285 (1970).

64 Lin, E.C.C. and Knox, W.E.: Adaptation of the rat liver tyrosine-α-ketoglutarate transaminase. Biochem. biophys. Acta *26:* 85–88 (1957).

65 Lin, E.C.C. and Knox, W.E.: Specificity of the adaptive response of tyrosine-α-ketoglutarate transaminase in the rat. J. biol. Chem. *233:* 1186–1189 (1958).

66 Litwack, G.; Sears, M.L., and Diamondstone, T.I.: Intracellular distribution of tyrosine-α-ketoglutarate transaminase and 4-C^{14}-hydrocortisone activities during induction. J. biol. Chem. *238:* 302–305 (1963).

67 Long, C.N.H.; Katzin, B., and Fry, E.G.: The adrenal cortex and carbohydrate metabolism. Endocrinology, Springfield *26:* 309–344 (1940).

68 Macleod, R.M.; King, C.E., and Hollander, V.P.: Effect of corticosteroids on ribonuclease and nucleic acid content in lymphosarcoma P1798. Cancer Res. *23:* 1045–1050 (1963).

69 Makman, M.H.; Dvorkin, B., and White, A.: Alterations in protein and nucleic acid metabolism of thymocytes produced by adrenal steroids *in vitro.* J. biol. Chem. *241:* 1646–1648 (1966).

70 Makman, M.H.; Dvorkin, B., and White, A.: Influence of cortisol on the utilization of precursors of nucleic acids and protein by lymphoid cells *in vitro.* J. biol. Chem. *243:* 1485–1497 (1968).

71 Manchester, K.L.; Randle, P.J., and Young, F.G.: The effect of growth hormone and of cortisol on the response of isolated rat diaphragm to the stimulating effect of insulin on glucose uptake and on incorporation of amino acids into protein. J. Endocrin. *18:* 395–408 (1959).

72 Mankin, H.J. and Conger, K.A.: The effect of cortisol on articular cartilage of rabbits. I. Effect of a single dose of cortisol on glycine-C-14 incorporation. Lab. Invest. *15:* 794–800 (1966).

73 Mayewski, R.J. and Litwack, G.: ^{3}H-Cortisol radioactivity in hepatic smooth endoplasmic reticulum. Biochem. biophys. Res. Commun. *37:* 729–735 (1969).

74 McLean, P. and Novello, F.: Influence of pancreatic hormones on enzymes concerned with urea synthesis in rat liver. Biochem. J. *94:* 410–422 (1965).

75 Mersmann, H.J. and Segal, H.L.: Glucocorticoid control of the liver glycogen synthase-activating system. J. biol. Chem. *244:* 1701–1704 (1969).

76 Miller, J.E. and Litwack, G.: Studies on soluble and mitochondrial tyrosine aminotransferase: evidence for a physical change in the cytosol enzyme during induction. Biochem. biophys. Res. Comm. *36:* 35–41 (1969).

77 Mishkin, E.P. and Shore, M.L.: Inhibition by actinomycin D of the induction of tryptophan pyrrolase by hydrocortisone. Biochim. biophys. Acta *138:* 169–174 (1967).

78 Monroe, C.B.: Induction of tryptophan oxygenase and tyrosine aminotransferase in mice. Amer. J. Physiol. *214:* 1410–1414 (1968).

79 Moog, F.: Intestinal phosphatase activity: acceleration of increase by puromycin and actinomycin. Science *114:* 414–416 (1964).

80 Morey, K.S. and Litwack, G.: Isolation and properties of cortisol metabolite binding proteins of rat liver cytosol. Biochemistry *8:* 4813–4821 (1969).

81 *Mueller, G.C.:* The role of RNA and protein synthesis in estrogen action; in *Karlson* Mechanisms of hormone action, pp. 228–245 (Academic Press, New York 1965).

82 *Munck, A.:* Metabolic site and time course of cortisol action on glucose uptake, lactic acid output, and glucose-6-phosphate levels of rat thymus cells *in vitro.* J. biol. Chem. *243:* 1039–1042 (1968).

83 *Munro, H.N.; Clark, C.M., and Goodlad, G.A.J.:* Loss of liver glycogen after administration of protein or amino acids. Biochem. J. *80:* 453–458 (1961).

84 *Nakagawa, S. and White, A.:* An acute decrease in RNA polymerase activity of rat thymus in response to cortisol injection. Proc. nat. Acad. Sci., Wash. *55:* 900–904 (1966).

85 *Nichol, C.A. and Rosen, F.:* Changes in alanine transaminase activity related to corticosteroid treatment or capacity for growth. Advances in Enzyme Regulation, vol. 1, pp. 341–361 (Pergamon Press, New York 1963).

86 *Ottolenghi, C.; Romano, B., and Barnabei, O.:* Reduced breakdown of rat liver microsomal ribonucleic acid after administration of cortisone. Nature, Lond. *212:* 1267–1268 (1966).

87 *Pardee, A.B. and Wilson, A.C.:* Control of enzyme activity in higher animals. Cancer Res. *23:* 1483–1490 (1963).

88 *Peraino, C.:* Interactions of diet and cortisone in the regulation of adaptive enzymes in rat liver. J. biol. Chem. *242:* 3860–3867 (1967).

89 *Peraino, C.:* Regulatory effects of glucocorticoids on ornithine aminotransferase and serine dehydratase in rat liver. Biochim. biophys. Acta *165:* 108–112 (1968).

90 *Peterkofsky, B. and Tomkins, G.M.:* Effect of inhibitors of nucleic acid synthesis on steroid-mediated induction of tyrosine aminotransferase in hepatoma cell cultures. J. molec. Biol. *30:* 49–61 (1967).

91 *Pollock, M.R.:* The differential effect of actinomycin D on the biosynthesis of enzymes in *Baccillus subtilis* and *Bacillus cereus.* Biochim. biophys. Acta *76:* 80–93 (1963).

92 *Potter, V.R.; Gebert, R.A., and Pitot, H.C.:* Enzyme levels in rats adapted to 36-h fasting. Advances in Enzyme Regulation, vol. 4, pp. 341–361 (Pergamon Press, New York 1963).

93 *Rappaport, M.R.; Feigin, R.D.; Bruton, J., and Beisel, W.R.:* Circadian rhythm for tryptophan pyrrolase activity and its circulating substrate. Science *153:* 1642–1644 (1966).

94 *Ray, P.D.; Foster, D.O., and Lardy, H.A.:* Mode of action of glucocorticoids. I. Stimulation of gluconeogenesis independent of synthesis *de novo* of enzymes. J. biol. Chem. *239:* 3396–3400 (1964).

95 *Reel, J.R. and Kenney, F.T.:* 'Superinduction' of tyrosine transaminase in hepatoma cell cultures: differential inhibition of synthesis and turnover by Actinomycin D. Proc. nat. Acad. Sci., Wash. *61:* 200–206 (1968).

96 *Riggs, T.H.:* Hormones and the transport of nutrients across cell membranes; in Actions of Hormones on Molecular Processes, pp. 1–57 (Wiley and Sons, New York 1964).

97 *Rosen, F.:* Enzymes in tissues responsive to corticosteroids. Cancer Res. *23:* 1181–1497 (1963).

98 *Rosen, F.; Harding, H.R.; Milholland, R.J., and Nichol, C.A.:* Corticosteroids and transaminase activity. VI. Comparison of the adaptive increases of alanine- and tyrosine-*a*-ketoglutarate transaminase. J. biol. Chem. *238:* 3725–3729 (1963).

99 *Rosen, F. and Milholland, R.J.:* Glucocorticoids and transaminase activity. VII. Studies on the nature and specificity of substrate induction of tyrosine-*a*-ketoglutarate transaminase and tryptophan pyrrolase. J. biol. Chem. *238:* 3730–3735 (1963).

100 *Rosen, F. and Milholland, R.J.:* Selective effects of cycloheximide on the adaptive enzymes tyrosine-α-ketoglutarate transaminase (TT) and tryptophan pyrrolase (TP). Fed. Proc. *25:* 285 (1966).

101 *Rosen, F. and Milholland, R.J.:* Induction of tyrosine-α-ketoglutarate transaminase (TT) in relation to amino acid levels as altered by hydrocortisone (HC) and cycloheximide (CH). Fed. Proc. *26:* 392 (1967).

102 *Rosen, F. and Milholland, R.J.:* Effects of casein hydrolysate on the induction and regulation of tyrosine-α-ketoglutarate transaminase in rat liver. J. biol. Chem. *243:* 1900–1907 (1968).

103 *Rosen, F.; Milholland, R.J.; Hoffman, W.W., and Raina, P.N.:* Induction of tyrosine-α-ketoglutarate transaminase by glutamate, proline, and tryptophan. Abstracts, 156th National meeting of the American Chemical Society, Atlantic City, N.J.: Biol. 074 (1968).

104 *Rosen, F. and Nichol, C.A.:* Corticosteroids and enzyme activity. Vitamins Hormones, N.Y., vol. 21, pp. 135–214 (Academic Press, Inc., New York 1963).

105 *Rosen, F. and Nichol, C.A.:* Studies on the nature and specificity of the induction of several adaptive enzymes responsive to cortisol. Advances in Enzyme Regulation, vol. 2, pp. 115–135 (Pergamon Press, New York 1964).

106 *Rosen, F.; Raina, P.N.; Milholland, R.J., and Nichol, C.A.:* Induction of several adaptive enzymes by Actinomycin D. Science *146:* 661–663 (1964).

107 *Rosen, F.; Roberts, N.R.; Budnick, L.E., and Nichol, C.A.:* An enzymatic basis for the gluconeogenic action of hydrocortisone. Science *127:* 287–288 (1958).

108 *Rosen, F.; Roberts, N.R.; Budnick, L.E., and Nichol, C.A.:* Corticosteroids and transaminase activity: the specificity of the glutamic-pyruvic transaminase response. Endocrinology, Springfield *65:* 256–264 (1959).

109 *Rosen, F.; Roberts, N.R. and Nichol, C.A.:* Glucocorticoids and transaminase activity. I. Increased activity of glutamic-pyruvic transaminase in four conditions associated with gluconeogenesis. J. biol. Chem. *234:* 476–480 (1959).

110 *Rosen, J.M.; Rosen, F.; Milholland, R.J., and Nichol, C.A.:* Effects of cortisol on DNA metabolism in the sensitive and resistant lines of mouse lymphoma P1798. Cancer Res. *30:* (1970).

111 *Schimke, R.T.:* Studies on factors affecting the levels of urea cycle enzymes in rat liver. J. biol. Chem. *238:* 1012–1018 (1963).

112 *Schimke, R.T.:* Studies on the roles of synthesis and degradation in the control of enzyme levels in animal tissues. Bull. Soc. Chim. biol. *48:* 1009–1030 (1966).

113 *Schimke, R.T.; Sweeney, E.W., and Berlin, C.M.:* An analysis of the kinetics of rat liver tryptophan pyrrolase induction: the significance of both enzyme induction and degradation. Biochem. biophys. Res. Comm. *15:* 214–219 (1964).

114 *Schimke, R.T.; Sweeney, E.W., and Berlin, C.M.:* The roles of synthesis and degradation in the control of rat liver tryptophan pyrrolase. J. biol. Chem. *240:* 322–331 (1965a).

115 *Schimke, R.T.; Sweeney, E.W., and Berlin, C.M.:* Studies on the stability *in vivo* and *in vitro* of rat liver tryptophan pyrrolase. J. biol. Chem. *240:* 4609–4620 (1965b).

116 *Schmid, W.; Gallwitz, D., and Sekeris, C.E.:* On the mechanism of hormone action. VIII. Further studies on the effect of cortisol on isolated rat-liver nuclei. Biochim. biophys. Acta *134:* 80–84 (1967).

117 *Schor, J.M. and Frieden, E.:* Induction of tryptophan peroxidase of rat liver by insulin and alloxan. J. biol. Chem. *233:* 612–618 (1958).

118 *Segal, H.L. and Kim, Y.S.:* Environmental control of enzyme synthesis and degradation. J. cell. comp. Physiol. *66:* 11–22 (1965).

119 *Segal, H.L. and Lopez, C.G.:* Early effects of glucocorticoids on precursor incorporation into glycogen. Nature, Lond. *200:* 143–144 (1963).

120 *Shambaugh, G.E., III; Warner, D.A., and Beisel, W.R.:* Hormonal factors altering the rhythmicity of tyrosine-a-ketoglutarate transaminase in rat liver. Endocrinology, Springfield *81:* 811–818 (1967).

121 *Shrago, E.; Hardy, H.A., Nordlie, R.C., and Foster, D.O.:* Metabolic and hormonal control of phosphoenolpyruvate carboxykinase and malic enzyme in rat liver. J. biol. Chem. *238:* 3188–3192 (1963).

122 *Silber, R.H. and Porter, C.C.:* Nitrogen balance, liver protein repletion and body composition of cortisone treated rats. Endocrinology, Springfield *52:* 518–525 (1953).

123 *Smith, O.K. and Long, C.N.H.:* Effect of cortisol on the plasma amino nitrogen of eviscerated adrenalectomized diabetic rats. Endocrinology, Springfield *80:* 561–566 (1967).

124 *Stevens, W. and Dougherty, T.F.:* Effect of continued treatment with cortisol on thymidine incorporation into mouse lymphatic tissue nucleic acid. Proc. Soc. exp. Biol., N.Y. *124:* 542–545 (1967).

125 *Sutherland, E.W. and Robison, G.A.:* The role of cyclic AMP in the control of carbohydrate metabolism. Diabetes, N.Y. *18:* 797–819 (1969).

126 *Thompson, E.B.; Tompkins, G.M., and Curran, J.F.:* Induction of tyrosine-a-ketoglutarate transaminase by steroid hormones in a newly established tissue culture cell line. Proc. nat. Acad. Sci., Wash. *56:* 296–303 (1966).

127 *Thomson, J.F. and Mikuta, E.T.:* Effect of total body X-irradiation on the tryptophan peroxidase activity of rat liver. Proc. Soc. exp. Biol., N.Y. *85:* 29–32 (1954).

128 *Tomkins, G.M.; Gelehrter, T.D.; Granner, D.; Martin, D., Jr.; Samuels, H.H., and Thompson, E.B.:* Control of specific gene expression in higher organisms. Science *166:* 1474–1480 (1969).

129 *Walsh, D.A.; Perkins, J.P., and Krebs, E.G.:* An adenosine 3′,5′-monophosphate-dependent protein kinase from rabbit skeletal muscle. J. biol. Chem. *243:* 3765–3767 (1968).

130 *Weber, G.; Allard, C.; DeLamirande, G., and Cantero, A.:* Increased liver glucose-6-phosphatase activity after cortisone administration. Biochim. biophys. Acta *16:* 618–619 (1955).

131 *Weber, G.; Banerjee, G., and Bronstein, S.P.:* Role of enzymes in homeostasis. III. Selective induction of increases in liver enzymes involved in carbohydrate metabolism. J. biol. Chem. *236:* 3106–3111 (1961).

132 *Wicks, W.D.:* Induction of hepatic enzymes by adenosine 3′,5′-monophosphate in organ culture. J. biol. Chem. *244:* 3941–3950 (1969).

133 *Wicks, W.D.; Greenman, D.L., and Kenney, F.T.:* Stimulation of ribonucleic acid synthesis by steroid hormones. I. Transfer ribonucleic acid. J. biol. Chem. *240:* 4414–4419 (1965).

134 *Wicks, W.D.; Kenney, F.T., and Lee, K.-L.:* Induction of hepatic enzyme synthesis *in vivo* by adenosine 3′,5′-monophosphate. J. biol. Chem. *244:* 6008–6013 (1969).

135 *Williams-Ashman, H.G.:* Androgenic control of nucleic acid and protein synthesis in male accessory genital organs. J. cell. comp. Physiol. *66:* 111–124 (1965).

136 *Wurtman, R.J. and Axelrod, J.:* Daily rhythmic changes in tyrosine transaminase activity of rat liver. Proc. nat. Acad. Sci., Wash. *57:* 1594–1597 (1967).

137 *Wurtman, R.J.; Shoemaker, W.J., and Larin, F.:* Mechanism of daily rhythm in hepatic tyrosine transaminase activity: role of dietary tryptophan. Proc. nat. Acad. Sci., Wash. *59:* 800–807 (1968).

138 *Yeung, D. and Oliver, I.T.:* Induction of phosphopyruvate carboxylase in neonatal rat liver by adenosine 3′,5′-cyclic monophosphate. Biochemistry *7:* 3231–3239 (1968).

139 *Zigmond, M.J.; Shoemaker, W.J.; Larin, F., and Wurtman, R.J.:* Hepatic tyrosine transaminase rhythm: interaction of environmental lighting, food consumption and dietary protein. J. Nutr. *98:* 71–75 (1969).

Authors' address: *Fred Rosen,* Ph.D. and *Richard J. Milholland,* D.D.S., Department of Experimental Therapeutics, Roswell Park Memorial Institute, *Buffalo, NY 14203* (USA)

Enzyme Synthesis and Degradation in Mammalian Systems, pp. 103–140
(Karger, Basel 1971)

Control of Enzyme Activity: Nutritional Factors[1]

R.A. Freedland and B. Szepesi[2]

Department of Physiological Sciences, School of Veterinary Medicine, University of California, Davis, Calif.

Contents

I. Introduction

A. Current Trends in Biochemical Nutrition

Biochemical nutrition is an approach of using both, the tools and techniques of biochemistry and those of nutrition. For a simple definition of biochemical nutrition we may consider the following: a discipline which deals with the effects of nutrients, regardless of the mode of administration, on biochemical processes in the animal; these processes include the metabolism and utilization of the aforementioned nutrients as well as more generalized biochemical

1 Supported in part by USPHS Grant AM-04732.
2 US Department of Agriculture, Agricultural Res. Service, Human Nutrition Res. Division, Beltsville, Md.

changes. These changes can involve alterations in enzymic activity, concentration of intermediates within tissues and changes in metabolic flow patterns which may be unique to various tissues or general throughout the animal being studied.

During the last 20 years, considerable progress has occurred in the field of biochemical nutrition. A significant contribution to this progress has resulted from studying the adaptations of rat liver enzymes to diets and hormones. Although the rat has been the predominant experimental animal in these studies, the reader should be reminded that many metabolic phenomena and control mechanisms found in the rat may differ in other species.

Through the combination of biochemical and nutritional approaches, many basic as well as practical observations have been made. A classic example of using this discipline led to the understanding and treatment of certain metabolic diseases such as phenyl-ketoneuria and galactosemia to mention only 2. Many other basically important observations have been made via the biochemical-nutritional approach. For example, to date the best example in the animal system of any control mechanism approaching that known as repression in bacterial system has been observed via feeding creatine to both rats and chicks and observing the effect of creatine on the glycine-arginine transaminidinase activity (1, 2). The basic actions and generalized mechanisms of various hormones have been examined using nutritional approaches. These range from the specificity of insulin as an enzymic supressor (3) to the effects that various hormones have on the synthesis and degradation of certain enzyme systems (4, 5).

In studies on nutritional and hormonal interrelationships it has been noted that the nutritional regimen of the animal can profoundly influence the results observed after hormonal administration (6). These general approaches have also allowed the estimation of the requirement for certain exogenous factors which would not normally be considered as dietary requirements, such as the amount of exogenous glucose required by an animal to prevent an increased response in gluconeogenesis (7).

In this present chapter we hope to cover considerably more areas and hopefully will consider possible mechanisms as well as future directions for biochemical-nutritional study.

II. Dietary Effects on Enzyme Levels

A. Starvation, Starvation-refeeding

Starvation is an extreme nutritional condition during which there is an absence of external sources of amino acids or external sources of energy. It can be deduced a priori that under such circumstances the animal's metabolism would undergo severe changes. The direction of the changes must be such as to catabolize components which are less essential and to preserve components which are essential for the maintenance of life. Since starvation is an easily reproducible dietary regimen, there is a considerable literature about starvation and its effects.

In general, starvation increases the activities of a number of enzymes involved in gluconeogenesis, amino acid catabolism and lipid catabolism. These changes involve an increase in the activity of glucose 6-phosphatase (8—10), PEP-carboxykinase (11), glutamic-pyruvic transaminase and glutamic-oxaloacetic transaminase (9), serine dehydratase and tyrosine transaminase (12). The activity of liver branch-chain amino acid transaminase has been reported to remain unchanged during starvation (13), as has the activity of transglucosylase (14) and phosphofructokinase, pyruvate kinase and aldolase in adipose tissue (15). The activities of a large number of enzymes are decreased and these include phosphorylase (14, 16), hexokinase in adipose tissue (15, 17), glucokinase (18), aldolase, fumarase and alkaline phosphatase (12), phosphoglucomutase, phosphohexose-isomerase, malic enzyme and glucose 6-phosphate dehydrogenase (19), pentose phosphate oxidation (20), citrate cleavage enzyme (21) and β-hydroxy-β methyl glutaryl-CoA reductase (22). It is interesting that in spite of the large changes in many different metabolic activities of the liver, the number of nuclei in the liver have not appreciably changed even after 7 days of starvation (23).

A number of metabolic parameters are drastically altered during starvation. Among these are the reduction of the level of liver RNA and protein (24). In studying the rate of synthesis and breakdown of ribosomes (25), *Hirsch and Hiatt* found that the half-life of these particles in the fed animal is about 5 days and is decreased during starvation. Apparently, the RNA and protein of the ribosome are degraded together and both have the same half-life.

One of the earliest effects of starvation is a drop in blood glucose (26). In man the drop in blood glucose is accompanied by a drop in the level of circulating insulin (27, 28). The decreased level of insulin may reduce the amount of glucose entering the muscles. The usefulness of such a homeostatic mechanism may be of special importance during the later phases of starvation when the extent of gluconeogenesis is decreased (27). However, the circulating levels of free fatty acids, betahydroxybutyrate and acetoacetate are increased (27). At the same time the activity of D(-) — β-hydroxybutyrate dehydrogenase is increased 7-fold in the brain (29). The increase in the amount of ketogenesis in starvation and also in diabetes may be part of an important physiological adaptation; the reduction of glucose inflow to the muscles is partially counteracted by an increase in the inflow of ketone bodies, which can serve as an energy source. This adaptation may be also of great importance to the brain which is dependent on a constant source of glucose from the blood as an energy source. In this case an increased ability of using ketone bodies may compensate the brain for the reduction in blood glucose.

In rat liver mitochondria, the amount of acetoacetate is increased by 50 % (30) while the amount of acetoacetyl CoA is reduced during starvation. This and the increase in the NADP to NADPH ratio (31) would tend to reduce the amount of lipogenesis in liver during starvation.

It appears, therefore, that starvation reduces the activities of liver enzymes which can produce NADPH, such as malic enzyme and glucose 6-phosphate dehydrogenase. Starvation in chicks also reduces the activity of malic enzyme and glucose 6-phosphate dehydrogenase in the liver (32). But in pigeon liver, a 72-hour long starvation reduces the level of liver glycogen, but does not change the activity of glucose 6-phosphatase, phosphoglucomutase or phosphorylase (33). The response of the animals to starvation may also be changed by age; for example in the newborn chick, the incorporation of uniformly C^{14}-labeled glucose and the level of malic enzyme rises after birth only if the chicks are fed (34). The level of citrate cleavage enzyme will increase even if the chick is not fed, although the increase in activity is much faster after birth if the chick is fed (34). In the 20-day-old chick, after the activities of a number of enzymes have increased, uniformly C^{14}-labeled glucose incorporation, the level of malic enzyme and citrate cleavage enzyme are decreased by starvation and increased if the animal is refed while the level of isocitrate dehydrogenase is increased by starvation and decreased by refeeding. However, the activity of glucose 6-phosphate dehydrogenase and 6-phosphogluconate dehydrogenase are not altered very greatly either by starvation or by starvation and refeeding (34). This indicates that the role of glucose 6-phosphate dehydrogenase in lipogenesis may not be as important in the liver of the bird as in rat or mice, but the importance of malic enzyme in the bird may be greater than in the rat.

The difficulty of expressing enzyme activity becomes especially acute with respect to starvation. One example of this is an earlier work in which the activities of 12 enzymes were studied during starvation (24). It was found that 6 enzymes had an increased activity if activity was expressed as per gram of fresh liver weight or per milligram of liver protein. The activities of 4 enzymes were increased in terms of final body weight. All enzymes decreased, however, in terms of initial body weight and 11 enzymes were decreased on the basis of per liver nuclei. It has been suggested that in starvation the activities of enzymes should be expressed in terms of original body weight (35). Since the level of liver protein or body weight may change rapidly during starvation, the expression of enzyme activity on these bases may actually obscure changes in enzyme activity. Expressing enzyme activity on the basis of original body weight, it was found (35) that the activities of enzymes associated with the Embden-Meyerhof pathway, gluconeogenesis and the TCA cycle were maintained preferentially with respect to the activities of other enzymes. From this it was concluded that in starvation the activities of enzymes which are concerned with energy metabolism or the formation of glucose, functions which are vital for the maintenance of

life, are maintained in preference to other enzymes (35). In terms of physiological requirements, therefore, the changes which are observed in the activities of
enzymes during starvation are partially deducible. Such enzymes as glucose
6-phosphate dehydrogenase which are involved in metabolic pathways not essential during starvation are decreased.

Another nutritional phenomenon affecting liver enzyme activities is starvation-refeeding. Starvation causes a reduction in the level of a number of enzymes, as noted, and also in the level of fatty acid synthesis. When starved
animals are refed, a number of metabolic alterations occur, such as hyperglyco-
genesis, the return of a number of enzyme activities to normal and the overshoot
of normal activity by a number of other enzymes. The overshoot of enzyme
activity is particularly noticeable with glucose 6-phosphate dehydrogenase (34,
36, 37, 38). Enzymes which return to normal following refeeding include phosphorylase (14, 39), citrate cleavage enzyme (21), β-methylglutaryl CoA reductase (22) and hexokinase in the rat epididymal fat pad (15). Refeeding is followed by super-normal fatty acid synthesis in the liver calculated to vary between
5 times normal (40) to 90 times normal (41). Supernormal lipogenesis has also
been noted in heart and kidney (41). The starve-refeed phenomenon is altered

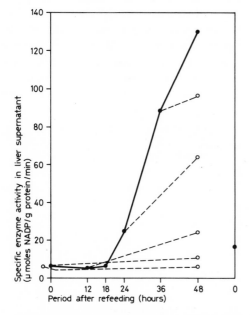

Fig. 1. Effect of 8-azaganine on the starve-refeed response of rat liver glucose 6-phos-
plate dehydrogenase: ─•─ specific activity after refeeding the rats with a high-carbohydrate
diet; ─○─ specific activity after injection of 8-azaguanine to refed rats (43, 46).

by cold; if rats are kept in the cold the overshoot in lipid synthesis after refeeding is reduced from 5-fold to 2-fold (42).

High dietary fat inhibited the overshoot of glucose 6-phosphate dehydrogenase after refeeding (43). The overshoot in glucose 6-phosphate dehydrogenase activity was also abolished by dehydroepiandrosterone (44, 45). Injection of antiinsulin serum after refeeding again reduced the extent of glucose 6-phosphate dehydrogenase overshoot. The overshoot of glucose 6-phosphate dehydrogenase has been found to be abolished by actinomycin D (46) or by 8-azaguanine (43). It was found that the 3 isozymes of glucose 6-phosphate dehydrogenase were equally decreased in starvation and were just about equally increased after refeeding (43). Since these rats were starved acutely, meaning that they were starved for at least 4–6 days (43, 46), the apparent requirement for *de novo* RNA synthesis (figure 1) could be explained by earlier suggestions that messenger RNA may have a life span of 4 days in mammals (25).

Later experiments using 8-azaguanine as an inhibitor of RNA synthesis (47) have indicated that after a 48-hour starvation-refeeding, *de novo* RNA synthesis is required for the overshoot but not for the recovery of normal level of enzyme activity (figure 2). Expansion of the same experiments (48) have indicated that the additional RNA synthesis occurs between 24 and 48 h after refeeding (figure 2). Actual measurements of RNA in the cell have indicated that the amount of RNA present in the cell decreases substantially after 48 h of starvation and remains relatively steady during progressive starvation. After refeeding, the largest amount of increase in RNA occurs between 24 and 48 h after refeeding. The data, therefore, would support previous observations (47) that the additional RNA synthesis required in the overshoot following refeeding of starved rats occurs between 24 and 48 h after refeeding. In animals which were

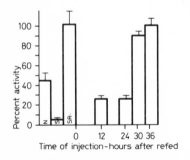

Fig. 2. Effect of 8-azaguanine (8AG) on the starve-refeed response of rat liver glucose 6-phosphate dehydrogenase. N = rats fed *ad libitum;* S = rats starved for 48 h; SR = rats starved for 48 h and refed for 48 h. Vertical bars represent 1 standard error of the mean (52, 55).

starved for much longer periods, the largest increase in RNA synthesis occurred between 48 h and 96 h after refeeding. It may be, therefore, that the actual resynthesis of RNA in the cell following starvation is dependent on the length of previous starvation.

B.   Dietary Protein and Amino Acids

The affects of dietary protein in the rat might be better understood by first examining the effects of protein deprivation. In general, protein deprivation is accompanied by a sharp reduction in the excretion of urea, a general and marked reduction in protein synthesis in the muscle (49, 50) and a smaller reduction in liver size (50).

One would presume that the animal's survival during protein deprivation would be promoted by mechanisms which would tend to conserve amino nitrogen and essential amino acids. To some degree the adaptation to a protein-free diet does indeed consist of changes in enzyme activities in accordance with these expectations. It has been noted that protein depletion decreases the oxidation of L-leucine and D-amino acids (51–53), phenylalanine (52) and tyrosine (54). Protein-depletion reduces the activity of serine dehydratase and threonine dehydratase (55) since the 2 enzymes appear to be the same (55, 56). A protein-free diet was found to decrease the activity of arginase (51), argininosuccinase (57) and histidase (58). These changes would promote conservation of essential amino acids and amino nitrogen. A rise in the activities of amino acid activating enzymes in the liver (57, 59), but not in muscle (57) could serve as an additional mechanism for the preservation of amino acids reaching the liver. Glutamic-pyruvic transaminase, glutamic-oxaloacetic transaminase and tyrosine-a-keto-glutarate transaminase are also lower in rats fed a protein-free diet than in rats fed a diet containing 25 % protein (6). In addition to the enzymes already mentioned, glutamate dehydrogenase (60), ribonuclease (51), succinic dehydrogenase (51, 61, 62) and alkaline phosphatase (63) are increased, xanthine oxidase and uricase are decreased (51), while ATPase (51), cytochrome oxidase (51, 60, 61, 64), glutamate and alanine oxidation (52) and phosphorylase remain unchanged or decreased in some cases (60). In general, the brain is most resistant to protein-depletion, followed by the kidney, skeletal muscle and spleen (65). However, the activities of glutamic-pyruvic transaminase, glutamic-oxaloacetic transaminase, fructose 1,6-diphosphate aldolase, malic dehydrogenase and lactic dehydrogenase have been found to be decreased in the heart of protein-depleted animals (66). It has been noted that the rate of loss of some liver components occur at different rates (60), but if enzyme activity is expressed in terms of original body weight then catalase, arginase, alkaline phosphatase, xanthine oxidase and cathepsin are all decreased (67).

Table I. Toxicity of ammonium acetate and amino acids injected intraperitoneally

Treatment	Injection	Dose mmoles/kg	% Mortality Control	High-protein-fed
Fed[1]	ammonium acetate	11.7	100	47
Unfed	ammonium acetate	9.1	61	61
Fed + arginine[2]	ammonium acetate	11.7	100	7
Fed	glycine	53–67	56	100
Fed	*L* threonine	34	10	50

1 Fed rats were given a 25 % casein diet (control) or an 80 % casein diet (high-protein-fed).
2 Arginine HC1, 1.5 mmoles/kg, 1 h before injection of ammonium acetate. Taken from *Wergedal and Harper* (77).

Rats fed a high-protein diet must undergo adaptations to manufacture glucose from amino acids. Such adaptations require increased catabolism of amino acids and an increased output of urinary nitrogen. It has been observed that urea production is proportional to the amount of dietary protein (68) and amount of liver arginase (69). High-protein diets increase the activity of liver arginase (69–72) and the activities of urea cycle enzymes appear to vary together (68, 71, 73–76).

In order to ascertain the relevance of threonine dehydratase and urea cycle enzyme activities, as measured *in vitro*, to the function of the enzymes *in vivo*, a number of experiments were performed by *Wergedal and Harper* (77). Rats were injected with relatively large doses of ammonium acetate, glycine, or *L*-threonine (table I). In rats fed a 25 % casein diet the amount of ammonium acetate injected caused 100 % mortality, but mortality was only 47 % in rats adapted to an 80 % casein diet. This indicated that rats adapted to the high-protein diet not only had higher activities of the urea cycle enzymes, but were able to handle excess ammonia better. Fasted rats were also more resistant to ammonia poisoning. Exogenous arginine injected intra-peritoneally did not improve the resistance of rats fed the 25 % casein diet, but did improve the resistance of rats fed the 80 % casein diet. Therefore, in rats fed the 25 % casein diet, it is the level of urea cycle enzymes which limits the animal's ability to detoxify excess ammonia, while in high-protein-fed rats, survival is limited by the level of urea cycle intermediates. To test further the relevance of enzyme activity to *in vivo* function, rats were injected with glycine or *L*-threonine. In both cases mortality was higher in rats adapted to the 80 % casein diet. This indicated that the rats which were adapted to the high-protein diet had a greater capacity to catabolize either amino acid resulting in ammonia toxicity. It appears, therefore, that the output of the urea cycle is related to the activities of the urea cycle enzymes as measured *in vitro*, at least in the cases studied.

Table II. Relationship between xanthine oxidase activity and growth rate in rats fed different types of protein[1]

Protein	Xanthine oxidase $\mu 10_2/h/$ % diet protein	Growth rate g gained/week/ % diet protein
Casein	5.9	2.1
Gliadin	1.0	0.49
Gliadin + tryptophan	1.2	0.49
Gliadin + lysine	4.0	2.0
Gliadin + tryptophan + lysine	5.1	1.8

1 The data were taken from *Litwack et al.* (99, 100).

In addition to the urea cycle enzymes, the activities of a number of enzymes involved in amino acid catabolism are also increased by a high-protein diet or the administration of a casein hydrolysate. These include: branched-chain amino acid transaminase (78), *D*-amino acid oxidase (79), glutamic dehydrogenase (80), glutamic-oxaloacetic transaminase (3, 79, 81–84), glutamic pyruvic-transaminase (3, 79, 83–86), histidase (58), *L*-leucine transaminase (78), methionine-activating enzyme (87), cystathionine synthetase and cystathionase (87), ornithine transaminase (82), serine dehydratase (3, 55, 88, 89), tryptophan pyrrolase (90–92), tyrosine-*a*-ketoglutarate transaminase (3, 83, 88, 93, 94), glucose 6-phosphatase (3, 83, 95), fructose 1,6-diphosphatase (3, 83, 96, 97), lactic dehydrogenase (3), glucose 6-phosphate dehydrogenase (3, 83, 98), xanthine oxidase (79, 99–102), succinic dehydrogenase (79), uricase (79), uracil reductase (98), thymine reductase (98) and transketolase (103, 104). Other authors found no change in the activity of cathepsin (79) while a decrease in pyruvate kinase (3, 105) and glutamine synthetase (106) have been reported. The activities of serine dehydratase and the enzymes responsible for its synthesis seem to vary in reciprocal fashion; under conditions which increase serine dehydratase activity the activities of enzymes in the synthetic pathways decrease (107–109). Since, in different species and during development the extent of utilization of the phosphorylated and nonphosphorylated pathways of serine synthesis vary (110, 111), the extent to which dietary serine can lead to the reduction in the activities of enzymes involved in its synthesis remains to be determined.

There have been attempts to use enzyme data from nutritional experiments for the evaluation of protein, the evaluation of metabolic pathways and to study the mechanism of hormone action. Among these should be mentioned the work with xanthine oxidase. The activity of xanthine oxidase was found to be correlated with the ability of a dietary protein to sustain growth (99, 100). The correlation was high among widely differing proteins (table II), but broke down in later experiments (101). Enzyme data have been used to predict the metabolic fate of some dietary components. Also, it had been suggested that glutamic-pyruvic transaminase may be a limiting enzyme in gluconeogenesis in 4 conditions associated with increased gluconeogenesis: starvation, diabetis, high-protein regimen and glucocorticoid treatment (85, 86). In a fifth gluconeogenic condition (fructose feeding) the activity of glutamic-pyruvic transaminase was not increased (3), hence it was suggested that the role of the enzyme is in the catabolism of alanine and, although indirectly involved in gluconeogenesis, it most likely was not limiting during gluconeogenesis (3).

A further indication that glutamic-pyruvic transaminase is associated with protein catabolism rather than gluconeogenesis can be found by comparing the effect of dietary protein on glucose 6-phosphatase and glutamic-pyruvic trans-aminase activities (figure 3). It was observed (7) that glucose 6-phosphatase is induced when the diet has less than 33 caloric percent of glucose. This is true even if part of the dietary protein is substituted by fat; hence, glucose 6-phosphatase activity is increased as a response to increased demand for gluconeogenesis. Glutamic-pyruvic transaminase activity, however, increases as a function of the dietary protein level even at lower levels of dietary protein. It appears, therefore, that glutamic-pyruvic transaminase is associated with protein catabolism.

It is important to remember that in the 4 conditions associated with gluconeogenesis referred to by *Rosen et al.* (85, 86) increased gluconeogenesis was accompanied by increased amino acid catabolism. Under such conditions increased transaminase activity could be viewed as either a requirement for gluconeogenesis, increased amino acid catabolism or both. The involvement of a transaminase in gluconeogenesis can be better evaluated by determining its activity under conditions which increase gluconeogenesis, but not amino acid catabolism. In the case of glutamic-pyruvic transaminase, its activity was not increased in rats fed a 70 % margarine diet and was only slightly increased in glycerol-fed rats

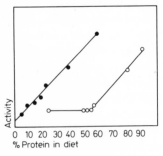

Fig. 3. Comparison of the protein effect on glutamic-pyruvic transaminase (•) and glucose 6-phosphatase (○) with varying content of protein in the diet (7, 85).

(*Freedland and Cunliffe,* unpublished data). Again the data indicate that glutamic-pyruvic transaminase is associated with amino acid catabolism.

Another important aspect of enzyme regulation was deduced from the induction of tryptophan pyrrolase by tryptophan (112). In this case the induction by the glucocorticoid was inhibited by actinomycin D, but not the induction produced by tryptophan. Using antibody-precipitation it was found that the hormone increased the activity of the enzyme by increasing its rate of synthesis, while tryptophan increased tryptophan pyrrolase activity by decreasing its degradation (113). This type of substrate protection has been found so far only with tryptophan pyrrolase.

An interesting dietary phenomenon was noticed in rats adapted to cold temperature. The animals had elevated glucose 6-phosphatase (114, 115), glutamic-pyruvic transaminase, glutamic-oxoloacetic transaminase, arginase (114, 116), tryptophan pyrrolase and tyrosine-a-ketoglutarate transaminase activities (116). This adaptation, which resembles the adaptation to a high-protein diet, was accompanied by a substantial increase in food intake in cold-adapted rats (116). It was found that the increases in glutamic-pyruvic transaminase, glutamic-oxaloacetic transaminase and arginase activities were abolished by restricting the protein intake of cold-adapted rats to that eaten by controls kept at room temperature. However, the increases in tryptophan pyrrolase and tyrosine-a-ketoglutarate transaminase activities were not abolished by protein restriction (116). It was suggested, therefore, that the rise in glutamic-pyruvic transaminase, glutamic-oxaloacetic transaminase and arginase are due to increased protein intake, whereas the rises in tryptophan pyrrolase and tyrosine-a-ketoglutarate transaminase activities are mediated by other factors stimulated by cold-adaptation (116).

In interpreting the possible physiological importance of enzyme changes, cellular localization is often an important factor to be considered. Thus, the separation of glutamic-oxaloacetic transaminase between mitochondrial and extramitochondrial space is the basis for the belief that only one of the isoenzymes participates in amino acid catabolism. Indeed, only the extramitochondrial enzymes is inducible by high-protein diets and cortisol (117, 118).

One important question concerning enzyme adaptation in mammals is whether increased enzyme activity is due to increased enzyme protein from increased *de novo* synthesis or to activation of inactive precursors. Another question is whether such changes in enzyme activity require additional *de novo* RNA synthesis. The evidence available so far indicates that with the exception of phosphorylase, and possibly tryptophan pyrrolase (113) other liver enzymes do not have active and inactive forms. It has been shown using antibody precipitation that an increase in the activity of tryptophan pyrrolase (113, 119, 120), tyrosine-a-ketoglutarate transaminase (121), glutamic-pyruvic transaminase (122), arginase (123) and serine dehydratase (124, 125) is accompanied by a proportional rise in titratable enzyme protein. Similar correlation has been found with catalase (126, 127) using its prosthetic group as a marker.

Finally, the protective effect of tryptophan on tryptophan pyrrolase has given rise to a number of experiments in which induction of various enzymes was attempted by feeding or injecting their substrates. Such experiments include the study of tyrosyluria due to feeding a 5 % tyrosine diet (128), the increase in xanthine oxidase activity in chicks given excess inosine or adenine (129) and the increase in liver ornithine transaminase in rats fed a diet containing 1 % arginine, DL-ornithine or L-proline (130). There are difficulties associated with these types of experiments which render interpretation of the data obtained difficult. For example, injection of large quantities of a nonessential amino acid could lead to the formation of a derivative or derivatives in high enough concentrations to facilitate or decrease the effect of a primary inducer. A large dose of amino acid could also lead to toxity and massive hormonal response. All these factors can combine to render interpretation of the data difficult and the conclusions drawn tentative. A perhaps better approach to the problem is to present the animal with a complete inducing system, then leave out one or a combination of amino acids. Such experiments were performed studying the induction of ornithine transaminase and threonine dehydratase (131). The best induction was obtained by an equimolar mixture of 10 essential amono acids (131). No induction of either enzyme was noted if tryptophan was left out, but leaving out any other amino acid caused only slight reduction (131). No induction could be produced by giving only one amino acid at a time. The effects of leaving out two amino acids at a time were also investigated. Leaving out arginine and lysine decreased the induction of threonine dehydratase whereas leaving out threonine and phenylalanine decreased the induction of ornithine transaminase. Other combinations of omissions were much less effective (131).

It appears very plausible, therefore, that the enzyme adaptations which occur due to feeding a high-protein diet are in some measure mediated by the amino acids ingested. The manner of action of amino acids is not well defined, but may include factors other than simply availability for protein synthesis.

A special case of amino acid effect occurs in rats fed excess phenylalanine. While the feeding of excess threonine (55, 132), histidine (133) and arginine (133) does not appear to affect the enzymes associated with their catabolism, phenylalanine does affect the enzyme associated with its catabolism namely, phenylalanine hydroxylase (134, 135). This amino acid appears to be unique in a further aspect. On the basis of our knowledge of metabolism and the physiological role of enzymes, we would expect that excess phenylalanine should, if it has any effect on phenylalanine hydroxylase, increase the activity of this enzyme. However, this does not appear to be the case and it appears that excess phenylalanine in the diet causes a marked reduction in phenylalanine hydroxylase activity at least in rat liver. Since liver is the only tissue which has phenylalanine hydroxylase activity this decrease would in effect decrease the potential of the animal to degrade phenylalanine by the normal process which is via tyrosine. Since an absence of this enzyme is the primary defect in the disease phenylketoneuria (136), it has been well studied in various experimental animals. Most of the studies have been conducted in the rat, which from an enzymological basis may be an unfortunate choice. This is due to the very high activity of the enzyme in this species, compared to most other species examined (137). Thus, in the presence of excess phenylalanine, even though there is an apparent depression of enzyme activity, there is still sufficient activity to handle even large excesses of phenylalanine.

It has been shown that the phenylalanine hydroxylase system is not a single enzyme, but consists of at least 2 enzymes; and a heat-stable cofactor (138, 139).

In the case of phenylketoneuria, it is apparently enzyme one that is missing, and normal concentrations of enzyme two and heat stable cofactor are observed (140, 141). In the case of feeding excess phenylalanine to rats, it has been observed that it is enzyme one which is decreased with little or no effect on the concentration of enzyme two or the heat stable cofactor (135). Several other closely related aromatic amino acids, particularly tyrosine and tryptophan when fed in excess, do not have as marked an effect on this enzyme as does phenylalanine (135).

C. Dietary Carbohydrate and Fat

The reader should be reminded at this point of an experimental difficulty which conceptually renders the study of effects of dietary composition on enzyme levels very difficult. The difficulty arises because the energy-containing portion of a diet may consist of protein, carbohydrate and fat, singly or in combinations. It is impossible, therefore, to vary one single component of the 3 dietary components without varying at least one other. It is possible, however, to keep one constant, while varying the other 2.

In the mammal the most abundant carbohydrate is glucose. Glucose synthesized in the liver or ingested in the diet is carried through the blood to the various nerve-centers of the brain, the muscles, endocrine organs, the reproductive organs, etc. In turn these organs provide the animal with sensation and direction, motive power and working capacity, hormones and insure the survival of the species. It is important, therefore, that such a continuous flow of glucose to all these organs be maintained. We have already noted that in starvation the animal's biochemical machinery responds to manufacture glucose by catabolizing part of itself, this is the response to an extreme emergency. In rats fed a high-protein diet gluconeogenesis proceeds from dietary amino acids. But in rats fed a high-carbohydrate diet gluconeogenesis becomes of lesser importance. In such condition the animal begins to deposit fat. The extent of lipogenesis and lipid deposition will vary; depending on the total calories ingested and the composition of the diet. It has been suggested that the rate-limiting enzyme in lipogenesis may be the citrate cleavage enzyme. This was suggested, because the level of citrate cleavage enzyme has been found to be correlated with lipogenesis (21, 142); increased lipogenesis was found to be accompanied by higher activities of citrate cleavage enzyme. It has been found, however, that in the hyperthyrotic state, which decreases lipogenesis (143), the level of NADP-linked dehydrogenases and citrate cleavage enzyme activity were increased (144). The correlation, therefore, is not always present. A further lack of correlation has been found between the rates of recovery of lipogenesis and citrate cleavage enzyme activity (145), in this case the recovery of enzyme activity was found to lag behind increased lipogenesis by some 12 h.

As dietary components are absorbed they are carried to the liver by the portal circulation. If the ingested food is high in carbohydrate the activity of glucokinase will be increased (146, 147). Glucokinase is a high Km hexokinase primarily found in the liver (146). Glucose entering the liver by the portal circulation will be released to the blood stream, converted to glycogen or fat. It is well known that lipid synthesis requires NADPH; hence increased lipogenesis in the liver, as in other tissues, requires NADPH formation. There are three primary sources of NADPH in the mammal: the pentose phosphate pathway, NADP-linked malic dehydrogenase known as malic enzyme and NADP-linked isocitrate dehydrogenase. Force-feeding carbohydrate has been shown to elevate glucose 6-phosphate dehydrogenase activity (148). This was accompanied by a fall in the C_6/C_1 ratio (the ratio of $C^{14}O_2$ coming from glucose-6-C^{14} and glucose-1-C^{14}) from 0.312 to 0.142 (174); a strong indication of increased utilization of the shunt pathway. The involvement of shunt activity is also illustrated by elevated shunt activity in obese mice (149). It is very interesting, however, that if a protein-free diet is fed (which is let us say 90 % glucose) the shunt activity drops, food intake decreases and the animal begins to lose weight (150). Increased fat deposition then is dependent on some dietary protein. If the animal is fed a high-protein, carbohydrate-free diet the activity of glucose 6-phosphate dehydrogenase will increase if the antecedent diet was a high-carbohydrate, protein-free diet (83). The requirement of dietary protein then is very stringent with respect to shunt enzyme activity and increased lipid deposition. The nature of this protein effect is unknown at present.

Adaptation to a high-carbohydrate diet is altered by environmental temperature. In rats adapted to the cold, food intake and glucokinase activity are increased while glucose 6-phosphate dehydrogenase activity is decreased (151).

Another type of adaptation takes place when glucose is replaced in the diet by fructose or sucrose. In fructose-fed rats the size of the liver is greatly enlarged and liver glycogen can be as much as 7 % of the liver (152). The liver becomes yellow, a sign of increased lipid content. The activities of a number of enzymes are increased, including glucose 6-phosphatase and fructose 1,6-diphosphatase (97, 152–154), PEP-carboxykinase (152), phosphoglucomutase (153, 154), phosphohexose isomerase (152–154), pyruvate kinase (105, 152), α-glycerophosphate dehydrogenase (152–154), glucose 6-phosphate dehydrogenase (152, 154), 6-phosphogluconate dehydrogenase (154), transketolase (103), aldolase (154), 3-phosphoglycerate dehydrogenase (154), 3-phosphoglycerate kinase (154), lactic dehydrogenase (154) and malic enzyme (154). The activities of aconitase, isocitrate dehydrogenase, malic dehydrogenase (NAD-linked), fumarase and glutamic-oxaloacetic transaminase were unchanged by fructose feeding (154). A slight increase in glutamic-pyruvic transaminase was noted (154) and later disputed (3). It appears, therefore, that the adaptations to fructose-feeding are unique, since these involve an increase in lipogenic enzymes, gluconeogenic enzymes and are accompanied by increased glycogen deposition (152). Again there was no decrease in glucose 6-phosphatase activity in rats fed the 65 % fructose diet for one week (152). This may be an indication of the obligatory conversion of fructose to glucose prior to peripheral utilization.

Adaptations to carbohydrate diets containing different sugars have been used to evaluate metabolic pathways in the rat (97). In the case of fructose it was predicted that this sugar would be broken down to the triose level in the

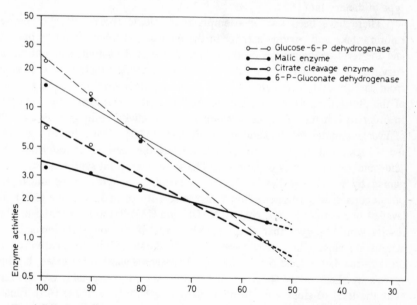

Fig. 4. Effect of ratio of calories of carbohydrate to calories of carbohydrate plus fat in the diet upon the activities of enzymes in rat livers (163).

liver and enter gluconeogenesis at that level. In the case of galactose, phosphorylation and epimerization would convert the sugar to glucose 6-phosphate. Hence, it was suggested that the adaptation to the fructose diet would elicit the induction of both, glucose 6-phosphatase and fructose 1,6-diphosphatase, whereas the adaptation to the galactose diet would induce only glucose 6-phosphatase. The predictions were based on the known or suspected pathways of metabolism in the liver and minimal peripheral utilization. The expected results were indeed observed in rats fed fructose, galactose, mannose and sorbitol (97). The induction of glucose 6-phosphatase independently of fructose 1,6-diphosphatase cast considerable doubt on the probability that these 2 enzymes are in the same operon as suggested by *Weber* (155).

The effects of dietary carbohydrate on enzyme levels are modified not only by dietary protein, but also by dietary fat. This is particularly striking in the case of the starved-refed animal; a high-fat diet prevents completely the overshoot in glucose 6-phosphate dehydrogenase activity (43, 156, 157). High-fat diets have been found to reduce the activities of a number of enzymes in *ad-libitum*-fed rats. These enzymes include: glucose 6-phosphate dehydrogenase and malic enzyme (158), phosphorylase (159), ATP: D-hexose 6-phosphotransferase (160) and transketolase (104). Fatty liver, however, does not develop unless the diet is very high in fat (161). In studying the affects of fructose-feeding on enzyme activities, it was noted that these can be increased or decreased depending on the type of dietary fat (162).

There have been several attempts to elucidate the relationship between caloric intake and enzyme activity in the pentose phosphate shunt on one hand and the level of lipogenesis on the other. In the rat the activities of a number of enzymes have been found to change as a logarithmic function of the calories from carbohydrate over calories from carbohydrate + fat ratio (163). The slope of the plot (figure 4) can be taken as the sensitivity to repression by dietary fat. In order to test the relevance of shunt enzyme activity to lipogenesis a number of liver parameters were examined by *Patterson* (164). It was found that in rats fed a high-fat diet, lipogenesis is reduced but the level of nonesterified fatty acids and ketone bodies is increased. The lower rate of fat synthesis was accompanied by lower activities of glucose 6-phosphate dehydrogenase and 6-phosphogluconate dehydrogenase, but a higher activity of malic enzyme (164). The level of nucleotides (NAD, NADH, NADP, and NADPH) were unchanged. These results would suggest that shunt activity, nucleotide concentration and lipogenesis are closely linked. However, in starved rats refed a high-carbohydrate, protein-free diet glucose 6-phosphate dehydrogenase was not increased, but lipogenesis by slices or the incorporation of hexoses into fatty acids was the same as if the level of glucose 6-phosphate dehydrogenase were higher (45). Finally, there is good evidence to suggest that the type of dietary fat may influence the extent of enzyme adaptation to diet (162) or the type of lipogenesis (165).

D. Food Intake Pattern

The ingestion of food within a relatively short period during a feeding cycle is a special type of dietary and nutritional stress, leading to hyperphagia, gastric hypertrophy (166) and some important metabolic alterations. The role of food intake pattern (167) and the difference between 'meal-eater' and 'nibbler' rats (168) in terms of metabolic adaptations have been noted. Experimenters should be cautioned in view of recent findings (169) that 'pair-fed' rats in some cases may consume all of their food within a relatively short time. In such cases, particularly if the pair-fed amount is much less than the animals would consume *ad libitum,* accidental meal-feeding may be initiated. For example, it was noted that rats fed a high fluoride diet ate less but ate over a longer period (169). The drop in liver glucose 6-phosphate dehydrogenase in rats fed the high-fluoride diet was not duplicated by pair-feeding, but was duplicated when the pair-fed amount was fed in 24 different portions, one portion per hour (169).

Ingestion of the daily food intake within a time-span of 2 h led to a temporary decrease in food intake and loss of weight (170). After 3—4 days the food intake rose to nearly normal and rats began to gain weight (170). Adaptation to this dietary regimen increased the incorporation of acetate-1-C^{14} into lipids of the adipose tissue 25-fold by the end of one week adaptation. Increased lipogenesis in the adipose tissue was accompanied by increased activities of glucose 6-phosphate dehydrogenase and 6-phosphogluconate dehydrogenase. Similar changes were noted in liver, but of lesser magnitude. It was also noted that the level of liver glycogen in meal-fed rats was more resistant to starvation than in nibbler rats (170). Rats maintained on a meal-feeding regimen for 10 weeks ate more and gained more weight. The hyperlipogenic effects of a 10-week adaptation period persisted even if the rats were allowed to nibble for one week, but a 20-day readaptation to nibbling was sufficient to bring the rate of lipogenesis back to normal (170).

Increased lipogenesis in meal-fed rats was later confirmed by others (171—173). It was reported that while lipogenesis is increased in rat adipose tissue 5—7 days after meal-eating began, the activities of glucose 6-phosphate dehydrogenase, 6-phosphogluconate dehydrogenase and malic enzyme did not increase until 9 days after meal-feeding was commenced (171). In chicks increased lipogenesis was noted 7 days after the start of meal-feeding, but no rise in the activities of NADP-linked dehydrogenases was observable (171). It was also noted that in the rat, 2 h after a meal, lipogenesis was increased in adipose tissue without an increase in NADP-linked dehydrogenase activities (172). These findings indicated a lack of correlation between NADP-linked dehydrogenases and lipogenesis, at least in the adipose tissue. It was noted, however, that increased lipogenesis in adipose tissue is accompanied by a fall in free fatty acids (170); increased NADPH formation may follow fatty acid depletion. The adi-

Table III. Tissue enzyme levels in meal-fed and nibbling rats. Activity is expressed as units/mg protein

Diet	Tissue			
	Adipose		Liver	
	G6PD	ME	G6PD	ME
High carbohydrate				
Meal eater	1.543	3.51[1]	0.232[1]	0.288[1]
Nibbler	0.62	1.15	0.113	0.112
High fat				
Meal eater	0.181[2]	0.162[2]	0.032[2]	0.121[2]
Nibbler	0.166	0.176	0.032	0.084

1 Differs significantly ($P < 0.001$) from nibblers.
2 Not significant.
The data were taken from *Leveille* (173).

pose tissue may, however, be limited in lipogenesis by the availability of α-glycerophosphate which is indicated by the increases in the rate of lipogenesis if pyruvate, oxaloacetate, glucose or glucose and insulin were added to the incubation medium (172). These effects disappear, however, after the meal.

An interesting case of meal-feeding is the work in pigs (174). Although these animals adapted to meal-feeding in terms of consuming the daily food-intake during a relatively short period of the day, the adapted animals did not develop hyperlipogenesis, or elevated NADP-linked dehydrogenase activity. It has been suggested that pigs probably do not reach the post-absorptive state; however, the anomaly is open to further explanation (174).

It appears that the gluconeogenic enzymes-glucose 6-phosphatase (175) and PEP-carboxykinase (176) are not changed by meal-feeding per se. Glucose 6-phosphate dehydrogenase and malic enzyme activities were found to be increased by meal-feeding in both liver and adipose tissue (173). These effects of meal-feeding were completely blocked by a high-fat diet (table III). Dietary fat then exerts its effects even in meal-fed rats. In further studies (176) it was noted that enzyme adaptations due to meal-feeding would tend to favor lipogenesis, especially in adipose tissue.

Serine dehydratase, glutamic-pyruvic transaminase and glutamic-oxaloacetic transaminase were found to be increased by a high-protein diet in meal-fed rats (177). Some of the dietary effects which occur in *ad-libitum*-fed rats were found to be less in meal-fed rats. Of the 18 enzymes examined under 8 different dietary conditions the transaminases, serine dehydratase and glucose 6-phosphatase were increased by the high-protein diet, while dietary fructose increased glucose 6-phosphatase, but did not increase the activities of other enzymes more than dietary glucose, 24 h after the change in the composition of the meal (177).

The extent of enzyme adaptations i meal-fed rats came under strong scrutiny following the discovery by *Potter et al.,* of a diurnal rhythm in the level of liver tyrosine transaminase (178). The variation was dependent on the level of dietary protein. The diurnal rhythm in tyrosine transaminase activity and its dependence on dietary protein was later confirmed by *Wurtman and Axelrod* (179). In further studies (180) the tyrosine transaminase rhythm was shown to be absent if a tryptophan-free diet was fed.

The 8-hour fed 40-hour starved feeding cycle is long enough for 2 tyrosine transaminase peaks; one during feeding, and the second during the starvation period (181). The first peak appears to be mediated by an increased intake of protein (182); while the second appears to arise as a result of increased glucocorticoid output (181) and is abolished by adrenalectomy (figure 5). The existence of a feeding rhythm has also been shown with glycogen level (181), transport of amino acids into the liver (183), the activity of serine dehydratase (184) and the incorporation of labeled orotate into RNA (185). Interestingly, the incorporation of oratate fell during feeding and increased during starvation (185). This may be an indication of increased RNA turnover during the starvation phase.

In order to test if possible daily rhythms in enzyme activities of meal-fed rats were due to the time of feeding, or if these rhythms were inherent in the cycle of day, experiments were conducted in which rats meal-fed 6 or 12 h apart were killed at the same time (*Szepesi and Freedland,* unpublished data). Feeding rhythms were found in the activities of liver glycogen (figure 6), tyrosine-α-keto-glutarate transaminase (figure 6), glucose 6-phosphate dehydrogenase (figure 6), pyruvate kinase (figure 6) and phosphohexose isomerase (figure 6). Since the

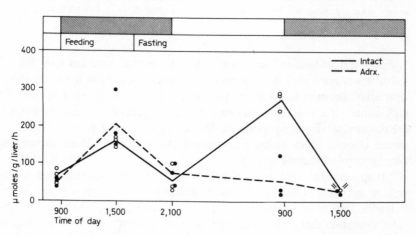

Fig. 5. Effect of adrenalectomy on the activity of tyrosine transaminase in meal fed rats (157).

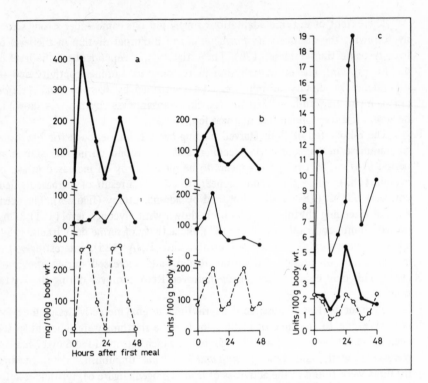

Fig. 6. a, b, c. Fress-urnal variation due to eating. a = level of liver glycogen, b = pyruvate kinase activity (also denotes glucose 6-phosphate dehydrogenase and phosphohexose isomerase) and c = tyrosine transaminase: 0 = rats fed a 65 % glucose, 25 % casein diet; ●—● = after changing to a 90 % casein diet and ●—● after changing to a 65 % fructose, 25 % casein diet.

12-hour- and 24-hour-fed rats were killed at the same time but were fed 12 h apart, it was concluded that the rhythm in enzyme activity is due to the elapsed time after the meal is fed and not truly diurnal (due to the time of day). The differential effects of the fructose and glucose diets on glucose 6-phosphate dehydrogenase (figure 6), pyruvate kinase (figure 6) and phosphokexose isomerase (figure 6) may be of importance in terms of translational or transcriptional control of enzyme activity.

It appears then that the meal-fed rat adapts to store ingested food in terms of liver glycogen, protein and lipids and also in the adipose tissue in a short time following feeding and catabolizes these stores during the post-absorptive state. Since it appears that certain apparent 'diurnal rhythms' are due to feeding, the phenomenon should be renamed. A possibility is to call such adaptation 'fressurnal'; i.e. a variation initiated by the ingestion of metabolizable nutrients.

III. On the Possible Mechanism of Dietary Effects

A. Repression of Enzyme Synthesis by Dietary Factors

Two cases resembling classical types of repression of enzyme synthesis in the animal system have been reported. The first of these is the repression of threonine dehydratase and ornithine transaminase by glucose (124, 186). In rats previously fasted for a short period, force-feeding of casein hydrolysate or the administration of a glucocorticoid hormone will cause a substantial and relatively fast rise in threonine dehydratase and ornithine transaminase activities. The rise in activity was completely prevented by oral intubation of glucose in the case of threonine dehydratase induced by casein hydrolysate (88, 186). In rats treated with the hormone the effect of glucose was much less. The effect of intubated glucose was also much less in the case of ornithine transaminase; the casein hydrolysate-mediated increase was partially reduced, while the hormone-mediated increase was not altered significantly (186). This indicates that casein hydrolysate and cortisol may alter the activities of these enzymes by different mechanisms.

In an attempt to elucidate the mechanism of the glucose repression on threonine dehydratase *Jost, Khairallah and Pitot* have used isotope labeling and antibody precipitation techniques (124). It was found that glucose actually stopped the incorporation of amino acids into threonine dehydratase. This then is the first reported case of catabolite repression in higher animals (124). As such it resembles the bacterial system, however it appears that the repression is at the translational rather than transcriptional level.

Another example of a classical type of repression occurs in the animal system in the creatine synthesis pathway. In higher animals creatine phosphate serves as a limited storage pool of energy in the muscle. It is synthesized from glycine, arginine and methionine, the first unique enzyme in the pathway is arginine-glycine transamidinase. In rats the majority of creatine is made in the kidney and in the liver in birds. It has been found that in rat kidney the activity of the enzyme was decreased to about one third normal after a week of feeding a 2.5 % or 3 % creatine diet (1, 187). The repression of the activity of the first unique enzyme of a synthetic pathway is much like the end product repression of β-galactosidase, except that in the animal the time-scale of events is exaggerated.

This repression is even more dramatic in young chicks (2). In this case maximum repression of enzyme activity occurs in about 80 min after the administration of creatine (figure 7). The extent is clearly dependent on the amount of creatine fed and disappears if a creatine-free diet is fed (1, 2). In 10-day-old chicks the activity of the transamidinase was also decreased by guanidinoacetate, a precursor of creatine, and by starvation (188). Recovery of enzyme activity after

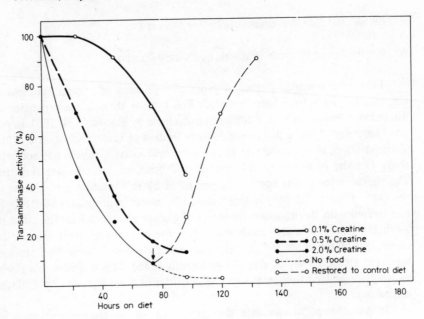

Fig. 7. Arginine-glycine transamidinase in the liver of 6-day-old chicks (2).

withdrawal of creatine from the diet was partially inhibited by ethionine and the ethionine effect was reversible by methionine (188) suggesting *de novo* enzyme synthesis. It was further noted that the recovery of transamidinase activity is tryptophan-dependent. When chicks were fed acid hydrolysed casein there was no recovery of transamidinase activity, in contrast to when enzyme hydrolysed casein was fed in which case the recovery was complete. Since, addition of tryptophan to the diet containing the acid hydrolysed casein restored the potency of the diet with respect to recovery of transamidinase activity, the recovery process was considered to be dependent on dietary tryptophan. The fact that creatinine has no effect on transamidinase levels suggests that whatever the mechanism of the effect it is specific for creatine.

B.   Translational Control

The half-life of messenger RNA in animals is much longer than in bacteria (189); this introduces the possibility of extensive translational control of protein synthesis. Indeed if the message is relatively stable this would necessitate considering the possibility that increased protein synthesis might be due to an increase in the rate of reading the message, that is to an increased rate of translation. There are, in fact, several well-documented instances which suggest that increases in protein synthesis in higher organisms can occur without a prior increase in RNA synthesis. These include the appearance of certain types of protein in the developing sea urchin embryo (190, 191) which occurs even if the synthesis of RNA is blocked (191). The importance of ribosomes in protein synthesis was also suggested from work in the rat (192). In rat diaphragm, increased protein synthesis due to treatment with insulin was not abolished by actinomycin D (193) and the importance of translational factors have also been indicated in the case of hemoglobin (194), triose phosphate dehydrogenase, tryptophan pyrrolase and β-galactosidase synthesis (195).

The importance of translational control of protein synthesis has recently been demonstrated at the level of polysome formation (196). It was shown that preparations made up of a longer fraction of polysomes incorporated more amino acids (196). Disaggregation by a tryptophan-free amino acid mixture was shown to be reversible: the lower rate of amino acid incorporation by the disaggregated polysomes was also reversible (196). The types of polysome patterns obtained under 3 dietary conditions: protein-free diet, tryptophan-free regimen and post-absorptive state, are illustrated in figure 8. It is noteworthy that these treatments change only the extent of aggregation.

The study of polysome distribution has resulted in one very disturbing report on the effect of actinomycin D on polysome profiles (197). If the report is further substantiated, that actinomycin D causes polysomal disaggregation, then previous use of this inhibitor as an indication of requirement for *de novo* RNA synthesis and the conclusions derived therefrom, will be severely questioned.

It may be well to point out some limitations of the use of polysome profiles. Obviously the most severe of these limitations is that the amount of polysomes present includes the RNA systems specific for a large number of proteins synthesized in the liver; hence, the changes represent the average change for all polysomes, whereas individual RNA systems may vary contrary to the general trend. A second limitation concerns the very basic assumption of the use of this technique: is the RNA extracted by mild means representative of the total cellular RNA. Obviously the answer to these questions are not available at present.

An interesting example of translational and transcriptional control has been reported in the case of pyruvate kinase. This was based on the finding that in rats fed a high-glucose, protein-free diet the activity of pyruvate kinase was temporarily increased by a high-protein, carbohydrate-free diet (83), and in the reverse dietary shift the same temporary elevation in pyruvate kinase activity was noted (150). Since, the long term effect of dietary protein is to decrease pyruvate kinase activity (3,105) of the liver, a biochemically suitable answer was sought which would account for the unexpected and seemingly counter-physiological response in terms of a mechanistic approach. The possibility that in the high-carbohydrate-fed rats pyruvate kinase activity was limited at the level of translation was tested using actinomycin D and cycloheximide (198). Increased pyruvate kinase activity due to force-feeding casein hydrolysate was blocked only by cycloheximide, but not by actinomycin D (198). In rats prefed the high-protein, carbohydrate-free diet a transcriptional limitation was postulated and the increase in pyruvate kinase activity following force-feeding carbohydrates was inhibited not only by cycloheximide, but also by actinomycin D (198). More recent work using 8-azaguanine instead of actinomycin D confirmed earlier results (199). It was concluded, therefore, that the synthesis and availability of RNA specific for pyruvate kinase in increased by a high-carbohydrate diet independently of dietary protein, but that the lack of dietary protein introduces a limitation on pyruvate kinase synthesis which is at the level of translation (198). When such rats are subjected to a sudden inflow of amino acids the level of pyruvate kinase is increased, because of increased translation even though such a response may appear to be counter-physiological with respect to the need for increased gluconeogenesis. It was indicated that the liver has two isozymes of pyruvate kinase one, the activity of which is relatively stable, and another, the activity of which is altered under different dietary conditions (200).

C.   Effects of Diets on Rates of Enzyme Synthesis and Degradation

In the animal system enzyme amount can be altered either by a change in the rate of synthesis, a change in the rate of degradation or a combination of both. This relationship will hold true as long as activation of inactive precursors can be excluded, irrespective of the mechanism by which rates of synthesis or degradation are altered.

The basic method of *Segal and Kim* (122) was applied by *Szepesi and Freedland* as a tool in the preliminary estimation of dietary and hormonal effects on enzyme synthesis and degradation (5). In essence, the method consists of changing the diet from one of high-glucose, protein-free to a high-protein, carbohydrate-free regimen. The rate at which maximum activity is achieved is a measure of k_D; k_s then is calculated as previously described. Alternately, the value $\ln(A_{final}\text{-}A_t)$ can be plotted versus t and the value of the half-life ($T\frac{1}{2}$) can be read directly from the graph as that value of t during which the value of log ($A_{final}\text{-}A_t$) is halved (if the change is from minimum to maximum activity). In the reverse dietary shift log ($A_t\text{-}A_o$) is plotted against t; where A_t = enzyme activity on any given day and A_o = minimum activity. Using this technique the effects of protein and cortisol were estimated on glutamic-pyruvic transaminase activity (figure 8). Neither the hormone nor the diet changed the rate of degradation of the enzyme. Interestingly, the rate of degradation of the mito-

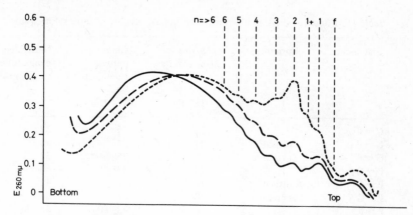

Fig. 8. Polysome patterns on sucrose density gradients. Polysomes were isolated from rat livers 2 h after feeding with the complete amino acid mixture (——) to one group of starved rats and the mixture lacking tryptophan (----) to another group. The polysome patterns are compared with the pattern of polysomes from starved rats (— —) on sucrose gradients (204).

chondrial isozyme ($T^1/_2 = 1$ day) is also uneffected by dietary protein or glucocorticoid (201). This is consistent with findings in which the enzyme is isotopical labeled and the rate of degradation is determined by antibody precipitation (202).

The effects of a number of dietary factors on the rates of synthesis and degradation of glutamic-pyruvic transaminase, serine dehydratase and glutamic-oxaloacetic transaminase are summarized in table IV. In the case of GPT the rate of its degradation is increased by thyroxine and decreased by fructose. The rate of serine dehydratase degradation was decreased by cortisol, uneffected by the diet and increased by thyroxine. The rate of glutamic-oxaloacetic transaminase degradation was increased by cortisol and protein and was uneffected by other factors. All 3 enzymes are inducible by dietary protein and cortisol. The data illustrate that depending on the inducer and the diet, the apparent extent of increase in the rate of synthesis can be obscured or exaggerated, hence, such modifications could obscure coordinate induction of a number of enzymes.

It should be pointed out that this method of estimating rates of synthesis and degradation is only an approximation and should not be used alone as proof of such. However, if the changes of enzyme activities are relatively large then this may be a useful way of obtaining a preliminary estimation, particularly if antibody precipitation is not practicable.

There are a number of strong indications that during induction the half-life of some enzymes is shortened, which would facilitate reaching a steady state in a shorter time. For example, *Schimke* (123) found that changing the diet from one containing 70 % casein to another containing 8 % casein the half-life of

Table IV. Effects of diets and hormones on the degradation of three rat liver enzymes

Diet change	Hormone	Serine dehydratase	Glutamic-pyruvic transaminase	Glutamic-oxaloacetic transaminase
		$T\frac{1}{2}$ h	$T\frac{1}{2}$ h	$T\frac{1}{2}$ h
G→P[1]	–	30	35	37
G→P	Cortisol	44	34	20
G→P	Thyroxine	22	25	32
P→G	–	28	33	60
P→G	Cortisol	40	27	41
P→G	Thyroxine	20	27	86
P→F	–	26–30	72–80	52–76

1 Diet changes G → P = 90 % glucose →90 % casein, P →F = 90 % casein → 90 % fructose. In each case the first diet was prefed for 4 days (90 % glucose) or for 5 days (90 % casein).

arginase was decreased from 5 to 3 days, and eventually back to 5 days again as the new steady state level of activity was reached. In a more extreme dietary change, from 90 % glucose, protein-free to 90 % casein, carbohydrate-free, the half-life of arginase was estimated as approximately one day (203). The half-life of glutamic-pyruvic transaminase was estimated by *Segal and Kim* as 28 h (122) during induction by prednisolone. *Kim* estimated it as between 3 and 5 days during steady state whether or not the rats are treated with the hormone (202). The tabulated values of $T\frac{1}{2}$ for glutamic-pyruvic transaminase agree with the shorter half-life obtained during induction (122). It should be noted that the log $(A_f\text{-}A_t)$ plot for glutamic-pyruvic transaminase deviates considerably from the straight line in changing from the high-glucose to the high-protein diet (figure 9). This could be an indication of continually changing rate of degradation; increasing as induction begins and decreasing again as steady-state values are reached.

In the case of thyroxine-treated rats the decrease due to thyroxine treatment (4) in glutamic-pyruvic transaminase and serine dehydratase activities is parallelled by a similar increase in their degradation, while a lack of thyroxine effect on glutamic-oxaloacetic transaminase activity (4) is consistent with a lack of alteration in the half-life of this enzyme in thyroxine-treated rats (table IV). The thyroxine effect may consist of removing the coenzyme and thus rendering the apoenzyme more susceptible to degradation. This is indicated by findings that the hormone effect is at least partially reversible by increasing the dietary level of vitamin B_6 (4), while stabilization of serine dehydratase by pyridoxine has been also indicated (125).

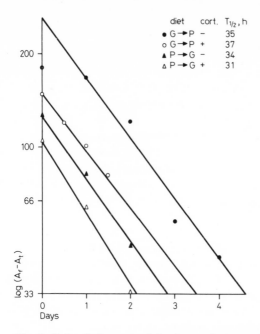

diet	cort.	$T_{1/2}$, h
• G→P	–	35
○ G→P	+	37
▲ P→G	–	34
△ P→G	+	31

Fig. 9. Rate of change in glutamic-pyruvic transaminase activity. Rats were prefed either a 90 % glucose diet for 4 days or a 90 % casein diet for 5 days after which the diet was changed from glucose to casein or from casein to glucose. Cortisol administration was begun with the prefed diet and was continued throughout the experiment (5).

Finally, 'fress-urnal' rhythms in enzyme activity can be expected in the case of those enzymes which have relatively short half-lives under meal-feeding conditions. When a meal is ingested in such animals the rate of translation of a number of enzymes is increased in response to the rapid inflow of nutrients. Since, the rate at which a new steady state level of enzyme activity is achieved is dependent on the half-life, those enzymes with short half-lives will undergo a large increase within a few hours after feeding, while the activities of those enzymes with longer half-lives will appear to remain relatively unchanged. As the post-absorptive state is reached the cessation in nutrient inflow will decrease translation and the activities of enzymes with short half-lives will now decrease quickly while the activities of those enzymes with longer half-lives will appear to decrease more slowly. Thus, one can observe the 'fress-urnal' variation in the activities of some, but not all enzymes of rat liver.

At this point it appears that studying the rates of synthesis and degradation may be helpful and add to our knowledge of the control of enzyme regulation in higher organisms.

D. General Conclusions

Due to the relative infancy of nutritional biochemistry it is safe to say that as this review differs from those preceding it, so will those which will be forthcoming during the 1970's and thereon differ from this one. There are some areas, however, where solid progress has been made and again there are areas where much more diligent work is required.

We have learned for example, that the bulk of enzyme adaptations due to diets and hormones are consistent with deduced physiological requirements. We have learned about the existence of translational control of enzyme synthesis and the type of adaptation which consists of changes in rates of enzymes degradation as well as synthesis. Undoubtedly we shall find more of these and some effects now not known or only suspected.

With respect to future trends it is absolutely vital to ascertain whether enzyme adaptations in higher animals require increased messenger RNA formation and the extent of the transcriptional control. Techniques are needed to differentiate amongst the species of RNA involved in protein synthesis and to ascertain whether any of these may be more important in regulation of enzyme synthesis than the others. Extensive use of tissue and organ cultures, perfusion of organs, synthetic hormones with varying target organs but with similar physiological effects and relatively non-toxic and narrow range inhibitors of protein and RNA synthesis should be looked for and developed. increased attention should be paid to the nutrition of organelles and single cells under varying conditions and finally cell-free systems capable of synthesizing biologically active proteins at rates approaching those *in vivo* are required. Some of the newer methods described in this review may even be dated by the time these words are printed.

IV. References

1   *Walker, J.B.:* Repression of arginine-glycine transamidinase activity by dietary creatine. Biochim. biophys. Acta *36:* 574–575 (1959).

2   *Walker, J.B.:* Metabolic control of creatine biosynthesis. I. Effect of dietary creatine. J. biol. Chem. *235:* 2357–2361 (1960).

3   *Freedland, R.A.; Cunliffe, T.L., and Zinkl, J.G.:* The effect of insulin on enzyme adaptations to diets and hormones. J. biol. Chem. *241:* 5448–5451 (1966).

4   *Szepesi, B. and Freedland, R.A.:* Reversibility of the thyroxine effect on rat liver glutamic-pyruvic transaminase and serine dehydrase by dietary supplementation of vitamin B_6. Life Sci. *8:* 353–355 (1969).

5   *Szepesi, B. and Freedland, R.A.:* A possible method for estimation hormone effects on enzyme synthesis. Arch. Biochem. *133:* 60–69 (1969).

6   *Freedland, R.A.; Murad, S., and Hurvitz, A.I.:* Relationship of nutritional and hormonal influences on liver enzyme activity. Fed. Proc. *27:* 1217–1222 (1968).

7 *Freedland, R.A. and Harper, A.E.:* Metabolic adaptations in higher animals. III. Quantitative study of dietary factors causing response in liver glucose-6-phosphatase. J. biol. Chem. *233:* 1–4 (1958).

8 *Harper, A.E.:* Hormonal factors affecting glucose-6-phosphatase activity. 2. Some effects of diet and of alloxan diabetes in the rat. Biochem. J. *71:* 702–705 (1959).

9 *Fitch, W.M. and Chaikoff, I.L.:* Effect of previous diet on response of hepatic enzyme activities to a 24-hour fast. Biochem. *94:* 380–386 (1961).

10 *Andrassy, K.; Porcsalmy, I., and Vereb, G.:* The effects of feeding, fasting and andrenaline on the glucose-6-phosphatase activity of the liver. Acta Physiol. *26:* 297–304 (1965).

11 *Nordlie, R.C.; Varricchio, F.E., and Holten, D.D.:* Effects of altered hormonal states and fasting on rat-liver mitochondrial phosphoenolpyruvate carboxykinase levels. Biochem. biophys. Acta *97:* 214–221 (1965).

12 *Goswami, M.N.D. and Chatagner, F.:* Starvation-induced adaptation of rat liver tyrosine transaminase and serine dehydrase. Experientia *22:* 1–5 (1966).

13 *Ichihara, A.; Takahashi, H.; Aki, K., and Shirai, A.:* Transaminase of branched chain amino acids. II. Physiological change in enzyme activity in rat liver and kidney. Biochem. Biophys. Res. Com. *26:* 674–678 (1967).

14 *Gutman, A. and Shafrir, E.:* Liver UDP-glucose transglucosylase and phosphorylase in fasted, refed and nephrotic rats. Proc. Soc. exp. Biol., NY *117:* 264–270 (1964).

15 *Pogson, C.I. and Denton, R.M.:* Effect of alloxan diabetes, starvation, and refeeding on glycolytic kinase activities in rat epididymal adipose tissue. Nature, Lond. *216:* 156–157 (1967).

16 *Niemeyer, H.; Gonzales, C., and Rozzi, R.:* The influence of diet on liver phosphorylase. I. Effect of fasting and refeeding. J. biol. Chem. *236:* 610–613 (1961).

17 *Hansen, R.; Pilkis, S.J., and Krahl, M.E.:* Properties of adaptive hexokinase isozymes of the rat. Endocrinology *81:* 1397–1404 (1967).

18 *Vinuela, E.; Salas, M., and Sols, A.:* Glucokinase and hexokinase in liver in relation to glycogen synthesis. J. biol. Chem. *238:* 1175–1177 (1963).

19 *Weber, G. and Cantero, A.:* Effect of fasting on liver enzymes involved in glucose-6-phosphate utilization. Amer. J. Physiol. *190:* 229–234 (1957).

20 *Anderson, J. and Hollifield, G.:* The effects of starvation and refeeding on hexosemonophosphate shunt enzyme activity and DNA, RNA, and nitrogen content of rat adipose tissue. Metab. clin. Exp. *15:* 1098–1103 (1966).

21 *Kornacker, M. and Lowenstein, J.M.:* The relation between the rates of citrate cleavage and fatty acid synthesis in rat liver preparations. Biochem. J. *89:* 272–277 (1963).

22 *Regen, D.; Riepertinger, C.; Hamprecht, B., and Lynen, F.:* The measurement of β-hydroxy-β-methylglutaryl-CoA reductase in rat liver; effects of fasting and refeeding. Biochem. Z. *346:* 78–84 (1966).

23 *Harrison, M.F.:* Effect of starvation on the composition of the liver cells. Biochem. J. *55:* 204–211 (1953).

24 *Allard, C.; De Lamirande, G., and Cantero, A.:* Behavior of enzymes in liver of starved rats. Exp. Cell Res. *13:* 69–77 (1957).

25 *Hirsch, C.A. and Hiatt, H.H.:* Turnover of liver ribosomes in fed and in fasted rats. J. biol. Chem. *241:* 5936–5940 (1966).

26 *Friedmann, B.; Goodman, E.H., Jr., and Weinhouse, S.:* Effects of glucose feeding, cortisol, and insulin on liver glycogen synthesis in the rat. Endocrinology *81:* 486–496 (1967).

27 *Owen, O.E.; Felig, P.; Morgan, A.P.; Wakren, J., and Cahill, G.F., Jr.:* Liver and kidney metabolism during prolonged starvation. J. clin. Invest. *48:* 574–583 (1969).

28 *Felig, P.; Marliss, E.; Owen, O.E., and Cahill, G.F.:* Blood glucose and gluconeogenesis in fasting man. Arch. intern. Med. *123:* 293–298 (1969).

29 *Smith, A.L.; Satterthwaite, H.S., and Sokoloff, L.:* Induction of brain D(-)-β-hydroxybutyrate dehydrogenase activity by fasting. Science *163:* 79–81 (1969).

30 *Burch, R.E. and Triantafillou, D.:* Acetoacetyl coenzyme A deacylase activity liver mitochondria from fed and fasted rats. Biochemistry *7:* 1009–1013 (1968).

31 *Lindall, A.W., Jr. and Lazarow, A.:* A critical study of pyridine nucleotide concentrations in normal fed, normal fasted, and diabetic rat liver. Metabolism *13:* 259–271 (1964).

32 *Leveille, G.A.: In vivo* fatty acid and cholesterol synthesis in fasted and fasted-refed chicks. J. Nutr. *98:* 367–372 (1969).

33 *Bruno, R.; Rinaudo, M.T., and Giunta, C.:* Effect of fasting on pigeon liver glucose-6-phosphatase, phosphoglucomutase, and phosphorylase. Boll. Soc. ital. Biol. sper. *44:* 1640–1643 (1968).

34 *Goodridge, A.G.:* The effect of starvation and starvation followed by feeding on enzyme activity and the metabolism of glucose-U-^{14}C in liver from growing chicks. Biochem. J. *108:* 667–674 (1968).

35 *Freedland, R.A.:* Effect of progressive starvation on rat liver enzyme activities. J. Nutr. *91:* 489–495 (1967).

36 *Weber, G.; Banerjee, G.; Bixler, D., and Ashmore, J.:* Role of enzymes in metabolic homeostasis. II. Depletion and restoration of avian liver carbohydrate-metabolizing enzymes. J. Nutr. *74:* 157–160 (1961).

37 *McDonald, B.E. and Johnson, B.C.:* Metabolic response to realimentation following chronic starvation in the adult male rat. J. Nutr. *87:* 161–167 (1965).

38 *Weber, G.; Banerjee, G., and Bronstein, S.B.:* Selective induction and suppression of liver enzyme synthesis. Amer. J. Physiol. *202:* 137–144 (1962).

39 *Niemeyer, H.; Perez, N.; Garces, E., and Vergara, F.E.:* Enzyme synthesis in mammalian liver as a consequence of refeeding after fasting. Biochim. biophys. Acta *62:* 411–413 (1962).

40 *Jansen, G.R.; Zanetti, M.E., and Hutchinson, C.F.:* Studies in lipogenesis *in vivo.* Fatty acid and cholesterol synthesis during starvation and refeeding. Biochem. J. *101:* 811–818 (1966).

41 *Smith, G.S. and Johnson, B.C.:* Glucose metabolism in the rat during starvation and refeeding following starvation. Proc. Soc. exp. Biol., NY *115:* 438–444 (1964).

42 *Klain, G.J. and Burlington, R.F.:* Effect of cold on glucose metabolism of fasted and fasted-refed rats. Amer. J. Physiol. *213:* 209–215 (1967).

43 *Kagawa, Y.; Ragawa, A., and Shimazono, N.:* Enzymatic studies on metabolic adaptation of hexose monophosphate shunt in rat liver. J. Biochem. *56:* 364–371 (1964).

44 *Willmer, J.S. and Foster, T.S.:* The influence of dehydroepiandrosterone and pregnenolone upon adaptation of hepatic glucose-6-phosphate dehydrogenase to refeeding a high-glucose diet after fasting. Canad. J. Biochem. *43:* 1375–1378 (1965).

45 *Tepperman, H.M.; De La Garza, S.A., and Tepperman, J.:* Effects of dehydroepiandrosterone and diet protein on liver enzymes and lipogenesis. Amer. J. Physiol. *214:* 1126–1132 (1968).

46 *Johnson, B.C. and Sassoon, H.:* Induction of liver glucose-6-phosphate dehydrogenase in the rat. Adv. Enzyme Regul. *5:* 93–106 (1968).

47 *Szepesi, B. and Freedland, R.A.:* Differential requirement for *de novo* RNA synthesis in the starved-refed rat; inhibition of the overshoot by 8-azaguanine after refeeding. J. Nutr. *99:* 449–458 (1969).

48 *Szepesi, B. and Freedland, R.A.:* Effect of administering 8-azaguanine at various times on liver enzyme activities of starved-refed rats (in preparation).

49 Waterlow, J.C.; Alleyne, G.A.O.; Chan, H.; Garrow, J.S.; Hay, A.; Jamas, P.; Picou, D.,
    and Stephen, J.M.L.: The mechanism of adaptation to low protein intakes. Arch.
    latinoamer. Nutr. 16: 175–200 (1966).
50 Waterlow, J.C. and Stephen, J.M.L.: The effect of low-protein diets on the turnover
    rates of serum, liver, and muscle proteins in the rat, measured by continuous infusion
    of L-lysine-14C. Clin. Sci. 35: 287–305 (1968).
51 Wainio, W.W.; Eichel, B.; Eickel, H.J.; Person, P.; Estes, F.L., and Allison, J.B.: Oxida-
    tive enzymes of the liver in protein depletion. J. Nutr. 49: 465–483 (1953).
52 McFarlane, I.G. and Von Holt, C.: Metabolism of amino acids in protein-calorie-defi-
    cient rats. Biochem. J. III: 557–563 (1969).
53 McFarlane, I.G. and Von Holt, C.: Metabolism of leucine in protein-calorie-deficient
    rats. Biochem. J. III: 565–571 (1969).
54 Al-Nejjar, H.Z. and Litwack, G.: Effects of protein availability on resting or induced
    activities of enzymes in the tyrosine oxidase system. Biochim. biophys. Acta 48:
    153–158 (1961).
55 Freedland, R.A. and Avery, E.H.: Studies on threonine and serine dehydrase. J. biol.
    Chem. 239: 3357–3360 (1964).
56 Selim, A.S. and Greenberg, D.M.: Further studies on cystathionine synthetase-serine
    deaminase of rat liver. Biochim. biophys. Acta 42: 211–217 (1960).
57 Stephen, J.M.L.: Adaptive enzyme changes in liver and muscle of rats during protein
    depletion and refeeding. Brit. J. Nutr. 22: 153–163 (1968).
58 Sahib, M.K. and Murti, C.R.K.: Induction of histidine-degrading enzymes in protein-
    starved rats. Biochem. biophys. Acta 149: 615–618 (1967).
59 Gaetani, S.; Paolucci, A.M.; Spadoni, M.A., and Tomassi, G.: Activity of amino acid-
    activating enzymes in tissues from protein-depleted rats. J. Nutr. 84: 173–178 (1964).
60 Pokrovskii, A.A. and Gapparov, M.G.: Changes in mitochondrial enzymes during pro-
    tein depletion. Tsitologiya 10: 1133–1140 (1968).
61 Millman, N.: The effect of protein depletion upon the enzyme activities of organs.
    Proc. Soc. exp. Biol., NY 77: 300–302 (1951).
62 Williams, J.N., Jr.: Response of the liver to prolonged protein depletion. II. The suc-
    cinic oxidase system and its component enzymes. J. Nutr. 73: 210–228 (1961).
63 Rosenthal, O.; Fahl, J.C., and Vars, H.M.: Response of alkaline phosphatase of rat liver
    to protein depletion and in attrition. J. biol. Chem. 194: 299–309 (1952).
64 Stirpe, F. and Schwarz, K.: Effect of dietary protein on the DPNH cytochrome c
    reductase activity of rat liver. Arch. Biochem. 96: 672–673 (1962).
65 Wainio, W.W.; Allison, J.B.; Kremzner, L.T.; Bernstein, E., and Aronoff, M.: Enzymes
    in protein depletion. J. Nutr. 67: 197–204 (1959).
66 Warter, J.; Metais, P., and Keckhut, M.: Transaminases aldolase et déshydrogénases du
    sérum, du foie et du cœur au cours du jeûne protidique chez le rat. Compt. Rend. Soc.
    Biol., Paris 55: 108–111 (1961).
67 Miller, L.L.: The loss and regeneration of rat liver enzymes related to diet protein. J.
    biol. Chem. 186: 253–260 (1950).
68 Schimke, R.T.: Adaptive characteristics of urea cycle enzymes in the rat. J. biol. Chem.
    237: 459–468 (1962).
69 Ashida, K. and Harper, A.E.: Metabolic adaptations in higher animals. IV. Liver argi-
    nase activity during adaptation to high protein diet. Proc. Soc. exp. Biol., NY 107:
    151–156 (1961).
70 Lightbody, H.D. and Kleinman, A.: Variations produced by food differences in the
    concentration of arginase in the livers of white rats. J. biol. Chem. 129: 71–78
    (1939).

71 *Mandelstam, J. and Yudkin, J.:* The effect of variation of dietary protein upon the hepatic arginase of the rat. Biochem. J. *51:* 681–686 (1952).

72 *Block, W.D. and Johnson, D.V.:* Factors influencing xanthine oxidase activity in rat skin. Arch. Biochem. *56:* 137–142 (1955).

73 *Freedland, R.A. and Sodikoff, C.H.:* Effects of diets and hormones on two urea cycle enzymes. Proc. Soc. exp. Biol., NY *109:* 394–396 (1962).

74 *Schimke, R.T.:* Studies on factors affecting the levels of urea cycle enzymes in rat liver. J. biol. Chem. *238:* 1012–1018 (1963)

75 *Freedland, R.A.:* Urea cycle adaptations in intact and adrenalectomized rats. Proc. Soc. exp. Biol., NY *116;* 692–696 (1964).

76 *Freedland, R.A.:* Effect of hypophysectomy on urea cycle enzyme adaptations. Life Sci. *9:* 899–904 (1966).

77 *Wergedal, J.E. and Harper, A.E.:* Metabolic adaptations in higher animals. IX. Effect of high protein intake on amino nitrogen catabolism *in vivo.* J. biol. Chem. *239:* 1156–1163 (1964).

78 *Ichihara, A.; Aki, K.; Ogawa, K.; Shirai, A., and Hisatoshi, T.:* Induction of branched-chain amino acid transaminase in rats; in *Yamada, Kogo* Symp. Pyridoxal Emzymes, 1967, 3rd ed., p. 165 (Maruzen Co. Ltd., Tokyo 1968).

79 *Muramatsu, K. and Ashida, K.:* Effect of dietary protein level on growth and liver enzyme activities of rats. J. Nutr. *76:* 143–150 (1962).

80 *Wergedal, J.E. and Harper, A.E.:* Metabolic adaptations in higher animals. X. Glutamic dehydrogenase activity of rats consuming high-protein diets. Proc. Soc. exp. Biol., NY *166:* 600–607 (1964).

81 *Waldorf, M.A.; Kirk, M.C.; Linksweiler, H., and Harper, A.E.:* Metabolic adaptations in higher animals. VII. Responses of glutamic-oxalacetate and glutamate-pyruvate transaminases to diet. Proc. Soc. exp. Biol., NY *112:* 764–768 (1963).

82 *Waldorf, M.A. and Harper, A.E.:* Metabolic adaptations in higher animals. VIII. Effect of diet on ornithine-α-ketoglutarate transaminase. Proc. Soc. exp. Biol., NY *112:* 955–958 (1963).

83 *Szepesi, B. and Freedland, R.A.:* Alterations in the activities of several rat liver enzymes at various times after initiation of a high protein regimen. J. Nutr. *93:* 301–306 (1967).

84 *Szepesi, B. and Freedland, R.A.:* Time-course of changes in rat liver enzyme activities after initiation of a high protein regimen. J. Nutr. *94:* 463–468 (1968).

85 *Rosen, F.; Roberts, N.R., and Nichol, C.A.:* Glucocorticosteroids and transaminase activity. I. Increased activity of glutamic-pyruvic transaminase in four conditions associated with gluconeogenesis. J. biol. Chem. *234:* 476–480 (1959).

86 *Rosen, F.; Harding, H.R.; Milholland, R.J., and Nichol, C.A.:* Glucocorticoids and transaminase activity. VI. Comparison of the adaptive increases of alanine and thyrosine-α-ketoglutarate transaminase. J. biol. Chem. *238:* 3725–3729 (1963).

87 *Finkelstein, J.D.:* Methionine metabolism in mammals. Effect of age, diet, and hormones of three enzymes of the pathway in rat tissues. ABB *122:* 583–590 (1962).

88 *Pitot, H.C.; Cho, Y.S.; Lamar, C., Jr., and Peraino, C.:* The interaction of external and internal controls on enzyme levels in liver and hepatoma. J. cell comp. Physiol. *66:* 163–174 (1965).

89 *Hoshino, J. and Kroeger, H.:* Properties of L-serine dehydratase purified from rat liver after induction by fasting or feeding casein hydrolysate. Hoppe-Seylers Z. Physiol. *350:* 595–602 (1969).

90 *Knox, W.E.:* The mechanisms which increase *in vivo* the liver tryptophan peroxidase activity: Specific enzyme adaptation and stimulation of the pituitary-adrenal. Brit. J. exp. Path. *32:* 462–469 (1956).

91 *Monti, G.; Luchetti, L., and Sereno, L.:* Comportamento della triptofano-perossi-ossi-dasi nel ratto dopo somministrazione di ormone somatotropo ed a vari regimi dietetici. Folia endocrin. *13:* 872–877 (1960).

92 *Rosen, F. and Milholland, R.J.:* Effects of casein hydrolysate on the induction and regulation of tyrosine-alpha-ketoglutarate transaminase in rat liver. J. biol. Chem. *243:* 1900–1907 (1968).

93 *Lin, E.C.C. and Knox, W.E.:* Adaptation of the rat liver tyrosine-α-ketoglutarate trans-aminase. Biochim. biophys. Acta *26:* 85–88 (1957).

94 *Pitot, H.C.; Potter, V.R., and Morris, H.P.:* Metabolic adaptations in rat hepatomas. I. The effect of dietary protein on some inducible enzymes in liver and hepatoma 5123. Cancer Res. *21:* 1001–1008 (1961).

95 *Freedland, R.A. and Harper, A.E.:* Metabolic adaptations in higher animals. I. Dietary effects on liver glucose-6-phosphatase. J. biol. Chem. *228:* 743–751 (1957).

96 *Mokrasch, L.C.; Davidson, W.D., and McGilvery, R.W.:* The response to glucogenic stress of fructose-1,6-diphosphatase in rabbit liver. J. biol. Chem. *222:* 179–184 (1956).

97 *Freedland, R.A. and Harper, A.E.:* Metabolic adaptations in higher animals. V. The study of metabolic pathways by means of metabolic adaptations. J. biol. Chem. *234:* 1350–1354 (1959).

98 *Ono, T.; Potter, V.R.; Pitot, H.C., and Morris, H.P.:* Metabolic adaptations in rat hepatomas. III. Glucose-6-phosphate dehydrogenase and pyrimidine reductases. Cancer Res. *23:* 385–391 (1963).

99 *Litwack, G.; Williams, J.N., Jr.; Chen, L., and Elvehjem, C.A.:* A study of the relation-ship of liver xanthine oxidase to quality of protein. J. Nutr. *47:* 299–306 (1952).

100 *Litwack, G.; Williams, J. N., Jr., Fatterpaker, P.; Chen, L., and Elvehjem, C.A.:* Further studies relating liver xanthine oxidase to quality of dietary protein. J. Nutr. *49:* 579–588 (1953).

101 *Dju, Mei Yu; Baur, L.S., and Filer, L.J., Jr.:* Assay of biologic value of mille proteins by liver xanthine oxidase determination. J. Nutr. *63:* 437–448 (1957).

102 *Harper, A.E.:* Effect of variations in protein intake on enzymes of amino acid meta-bolism. Canad. J. Biochem. *43:* 1589–1603 (1965).

103 *Benevenga, N.J.; Stileau, W.J., and Freedland, R.A.:* Factors affecting the activity of pentose phosphatemetabolizing enzymes in rat liver. J. Nutr. *84:* 345–350 (1964).

104 *Cheng, C.H.; Koch, M., and Shank, R.E.:* Dietary regulation of transketolase activity of liver. J. Nutr. *98:* 64–70 (1969).

105 *Krebs, H.A. and Eggleston, L.V.:* The role of pyruvate kinase in the regulation of gluconeogenesis. Biochem. J. *94:* 3c–4c (1964).

106 *Raina, P.N. and Rosen, F.:* Induction of glutamine synthetase by cortisol. Biochim. biophys. Acta *165:* 470–475 (1968).

107 *Fallon, H.J.; Hackney, E.J., and Byrne, W.L.:* Serine biosynthesis in rat liver. Regula-tion of enzyme concentration by dietary factors. J. biol. Chem. *241:* 4157–4167 (1966).

108 *Fallon, H.J. and Byrne, W.L.:* Depression of enzyme activity by cortisone: An effect of serine metabolism. Endocrinology *80:* 847–850 (1967).

109 *Fallon, H.J.; Davis, J.L., and Goyer, R.A.:* Effect of protein intake on tissue amino acid levels and the enzymes of serine biosynthesis in the rat. J. Nutr. *96:* 220–226 (1968).

110 *Cheung, G.P.; Cotropia, J.P., and Sallach, H.J.:* Comparative studies of enzymes related to serine metabolism in fetal and adult liver. Biochim. biophys. Acta *170:* 334–340 (1968).

111 *Cheung, G.P.; Cotropia, J.P., and Sallah, H.J.:* Effects of dietary protein on the hepatic enzymes of serine metabolism in the rabbit. Arch. Biochem. *129:* 672–682 (1969).

112 *Greengard, O.; Smith, M.A., and Acs, G.:* Relation of cortisone and synthesis of ribonucleic acid to induced developmental enzyme formation. J. biol. Chem. *238:* 1548–1551 (1963).

113 *Feigelson, P. and Greengard, O.:* Immunochemical evidence for increased titers of liver tryptophan pyrrolase during substrate and hormonal enzyme induction. J. biol. Chem. *237:* 3714–3717 (1962).

114 *Vaughan, D.A.; Hannon, J.P., and Vaughan, L.N.:* Interrelationships of diet and cold exposure on selected liver glycolytic enzymes. Amer. J. Physiol. *201:* 33–36 (1961).

115 *Beaton, J.R.:* The relation of dietary protein level to liver enzyme activities in cold-exposed rats. Canad. J. Biochem. *41:* 1871–1877 (1963).

116 *Klain, G.J.; Vaughan, D.A., and Vaughan, L.N.:* Effect of protein intake and cold exposure on selected liver enzymes associated with amino acid metabolism. J. Nutr. *80:* 107–110 (1963).

117 *Otsuka, Y.:* 細胞内局在性を異にする Transaminase Isozyme の誘導的生成の差異 Vitamin *32:* 499–502 (1965).

118 *Hurvitz, A.I.:* The interrelationship of hydrocortisone, dietary protein and drugs on adaptive rat liver enzymes-a biochemical and morphological study. Doctoral dissertation, Univ. of Calif., Davis Campus 1967.

119 *Schimke, R.T.; Sweeney, E.W., and Berlin, C.M.:* An analysis of the kinetics of rat liver tryptophan pyrrolase induction: The significance of both enzyme synthesis and degradation. Biochem. Biophys. Res. Com. *15:* 214–219 (1964).

120 *Schimke, R.J.; Sweeney, E.W., and Berlin, C.M.:* The roles of synthesis and degradation in the control of rat liver tryptophan pyrrolase. J. biol. Chem. *240:* 322–331 (1965).

121 *Kenney, F.T.:* Induction of tyrosine-α-ketoglutarate transaminase in rat liver. IV. Evidence for an increase in the rate of enzyme synthesis. J. biol. Chem. *237:* 3495–3498 (1962).

122 *Segal, H.L. and Kim, Y.S.:* Glucocorticoid stimulation of the biosynthesis of glutamic-alanine transaminase. Proc. nat. Acad. Sci., Wash. *50:* 912–918 (1963).

123 *Schimke, R.T.:* Importance of both synthesis and degradation in the control of arginase levels in rat liver. J. biol. Chem. *239:* 3808–3817 (1964).

124 *Jost, J.P.; Khairallah, E.H., and Pitot, H.C.:* Studies on the induction and repression of enzymes in rat liver. J. biol. Chem. *243:* 3057–3066 (1968).

125 *Khairallah, E.A. and Pitot, H.C.:* Turnover of serine dehydrase. Amino acid induction, glucose repression, and pyridoxine stabilization; in *Yamada, Kozo* Symp. Pyridoxal Enzymes, Japan 1967, 3rd ed., pp. 159–164 (Maruzen Co. Ltd., Tokyo 1968).

126 *Price, V.E.; Sterling, W.R.; Tarantola, V.A.; Hartley, R.W., Jr., and Rechcigl, M., Jr.:* The kinetics of catalase synthesis and destruction *in vivo.* J. biol. Chem. *237:* 3468–3475 (1962).

127 *Rechcigl, M., Jr. and Heston, W.E.:* Genetic regulation of enzyme activity in mammalian system by the alteration on the rates of enzyme degradation. Biochem. Biophys. Res. Com. *27:* 119–124 (1967).

128 *Knox, W.E.; Linder, M.C.; Lynch, R.D., and Moore, C.L.:* The enzymatic basis of tyrosyluria in rats fed tyrosine. J. biol. Chem. *239:* 3821–3825 (1964).

129 *Stirpe, F. and Corte, E.D.:* Regulation of xanthine dehydrogenase in chick liver. Effect of starvation and of administration of purines and purine nucleosides. Biochem. J. *95:* 309–313 (1964).

130 *Civen, M.; Brown, C.B., and Trimmer, B.M.:* Regulation of arginine and ornithine metabolism at the enzymic level in rat liver and kidney. ABB *120:* 352–358 (1967).

131 *Peraino, C.; Blake, R.L., and Pitot, H.C.:* Studies on the induction and repression of enzymes in rat liver. III. Induction of ornithine a-transaminase and threonine dehydrase by oral intubation of free amino acids. J. biol. Chem. *240:* 3039–3043 (1965).

132 *Goldstein, L.; Knox, W.E., and Behrman, E.J.:* Studies on the nature, inducibility, and assay on the threonine and serine dehydrase activities of rat liver. J. biol. Chem. *237:* 2855–2888 (1962).

133 *Auerbach, V.H. and Waisman, H.A.:* Liver arginase, histidase and phenylalinine transaminase following administration of their respective substrates. Proc. Soc. exp. Biol., NY *98:* 123–124 (1958).

134 *Auerbach, V.H.; Waisman, H.A., ans Wyckoff, L.B.:* Phenylketonuria in the rat associated with decreased temporal discriniation learning. Nature, Lond. *182:* 871–872 (1958).

135 *Freedland, R.A.; Krakowski, M.C., and Waisman, H.A.:* Influence of amino acids on rat liver phenylalanine hydroxylase activity. Amer. J. Physiol. *206:* 341–344 (1964).

136 *Jervis, G.A.:* Phenylpyruvic oligophrenia deficiency of phenylalanine-oxidizing system. Proc. Soc. exp. Biol., NY *82:* 514–515 (1953).

137 *Voss, J.C. and Waisman, H.A.:* The phenylalanine hydroxylase content of livers of various vertebrates. Comp. Biochem. Physiol. *17:* 49–58 (1966).

138 *Mitoma, C.:* Studies on partially purified hydroxylase. Arch. Biochem. *60:* 476–484 (1956).

139 *Kaufman, S.:* The structure of the phenylalanine-hydroxylation cofactor. Proc. nat. Acad. Sci., Wash. *50:* 1085–1093 (1963).

140 *Wallace, H.W.; Kisue, M., and Meister, A.:* Studies on the conversion of phenylalanine to tyrosine in phenylpyruvic oligophrenia. Proc. Soc. exp. Biol., NY *94:* 632–633 (1957).

141 *Mitoma, C.; Auld, R.M., and Udenfriend, S.:* On the nature of the enzymatic defect in phenylpyruvic oligophrenia. Proc. Soc. exp. Biol., NY *94:* 634–635 (1957).

142 *Kornacker, M. and Lowenstein, J.M.:* Citrate cleavage and acetate activation in livers of normal and diabetic rats. Biochim. biophys. Acta *84:* 490–492 (1964).

143 *Fletcher, K. and Myant, N.B.:* Influence of the thyroid on the synthesis of cholesterol by liver and skin *in vitro.* J. Physiol. *144:* 361–372 (1958).

144 *Murad, S. and Freedland, R.A.:* The *in vivo* effect of thyroxine on citrate cleavage enzyme and NADP linked dehydrogenase activities. Life Sci. *4:* 527–534 (196 5).

145 *Foster, D.E. and Srere, P.A.:* Citrate cleavage enzyme and fatty acid synthesi . J. biol. Chem. *243:* 1926–1930 (1968).

146 *Abraham, S.; Borrebaek, B., and Chaikoff, I.L.:* Effect of dietary carbohydrate on glucokinase and mannokinase activities of various rat tissues. J. Nutr. *8.:* 273–288 (1964).

147 *Borrebaek, B.; Abraham, S., and Chaikoff, I.L.:* Glucokinase activities ar d glycogen contents of livers of normal and hypophysectomized, X-irradiated rats subjected to different nutritional treatments. Biochim. biophys. Acta *90:* 451–463 (1964).

148 *Cohn, C. and Joseph, D.:* Effect of rate of ingestion of diet on hexosemonophosphate shunt activity. Amer. J. Physiol. *197:* 1347–1349 (1959).

149 *Fried, G.H. and Antopol, W.:* Enzymatic activities in tissues of obesehyperglycemic mice. Amer. J. Physiol. *211:* 1321–1324 (1966).

150 *Szepesi, B. and Freedland, R.A.:* Alterations in the activities of several rat liver enzymes at various times after the feeding of high carbohydrate diets to rats previously adapted to a high protein regimen. J. Nutr. *94:* 37–46 (1968).

151 *Hannon, J.P. and Vaughan, D.A.:* Effect of prolonged cold exposure on the glycolytic enzymes of liver and muscle. Amer. J. Physiol. *198:* 375–380 (1960).

152 *Szepesi, B. and Freedland, R.A.:* Time-course of enzyme adaptation. I. Effect of sub-stituting dietary glucose and fructose at constant concentrations of dietary protein. Canad. J. Biochem. *46:* 1459–1470 (1968).

153 *Fitch, W.M.; Hill, R., and Chaikoff, I.L.:* The effect of fructose feeding on glycolytic enzyme activities of the normal rat liver. J. biol. Chem. *234:* 1048–1051 (1959).

154 *Fitch, W.M. and Chaikoff, I.L.:* Extent and patterns of adaptation of enzyme activities in livers of normal rats fed diets high in glucose and fructose. J. biol. Chem. *235:* 554–557 (1960).

155 *Weber, G.:* Study and evaluation of regulation of enzyme activity and synthesis in mammalian liver. Adv. Enzyme Regulat. *1:* 1–35 (1963).

156 *Tepperman, H.M. and Tepperman, J.:* Role of hormones in glucose-6-phosphate de-hydrogenase adaptation of rat liver. Amer. J. Physiol. *202:* 401–406 (1962).

157 *Niemeyer, H.; Clark-Turri, L.; Garces, E., and Vergara, F.E.:* Selective response of liver enzymes to the administration of different diets after fasting. ABB *98:* 77–85 (1962).

158 *Vaughan, D.A. and Winders, R.L.:* Effects of diet on HMP dehydrogenase and malic (TPN) dehydrogenase in the rat. Amer. J. Physiol. *206:* 1081–1084 (1964).

159 *Niemeyer, H.; Perez, N.; Radojkovic, J., and Ureta, T.:* The influence of diet on liver phosphorylase. II. Effect of different proportions of carbohydrates, proteins and fats. ABB *96:* 662–669 (1962).

160 *Niemeyer, H.; Clark-Turri, L.; Perez, N., and Rabajille, E.:* Studies on factors affecting the induction of ATP D-hexose 6-phosphotransferase in rat liver. ABB *109:* 634–645 (1965).

161 *Rikans, L.L.; Arata, D., and Cederquist, D.C.:* Fatty livers produced in albino rats by ex-cess niacin in high fat diets. II. Effect of choline supplements. J. Nutr. *85:* 107–112 (1965).

162 *Carroll, C.:* Influences of dietary carbohydrate-fat combinations on various functions associated with glycolysis and lipogenesis in rats. II. Glucose vs. sucrose with corn oil and two hydrogenated oils. J. Nutr. *82:* 163–172 (1964).

163 *Baldwin, R.L.; Ronning, M.; Radanovics, C., and Plange, G.:* Effect of carbohydrate and fat intakes upon the activities of several liver enzymes in rats, guinea piglets, piglets and calves. J. Nutr. *90:* 47–55 (1966).

164 *Patterson, D.S.P.:* The formation of reduced nicotinamide-adenine dinucleotide phos-phate in relation to the impaired hepatic lipogenesis of the rat adapted to a high fat diet. BBA *84:* 198–200 (1964).

165 *Tepperman, H.M. and Tepperman, J.:* Effect of saturated fat diets on rat liver NADP-linked enzymes. Amer. J. Physiol. *209:* 773–780 (1965).

166 *Holeckova, E. and Fabry, P.:* Hyperphagia and gastric hypertrophy in rats adapted to intermittent starvation. Brit. J. Nutr. *13:* 260–266 (1959).

167 *Tepperman, J. and Tepperman, H.M.:* Effects of antecedent food intake pattern on hepatic lipogenesis. Amer. J. Physiol. *193:* 55–64 (1958).

168 *Cohn, C. and Joseph, D.:* Role of rate of ingestion of diet on regulation of intermediate metabolism (meal-eating vs. 'nibbling'). Metab. clin. Exp. *9:* 492–500 (1960).

169 *Suttie, J.W.:* Effect of dietary fluoride on the pattern of food intake in the rat and the development of a programmed pellet dispenser. J. Nutr. *96:* 529–535 (1968).

170 *Hollifield, G. and Parson, W.:* Metabolic adaptations to a 'stuff and starve' feeding program. I. Studies of adipose tissue and liver glycogen in rats limited to a short daily feeding period. J. clin. Invest. *41:* 245–249 (1962).

171 *Leveille, G.A.:* Glycogen metabolism in meal-fed rats and chicks and the time sequence of lipogenic and enzymatic adaptive changes. J. Nutr. *90:* 449–460 (1966).

172 *Leveille, G.A.:* Lipogenesis in adipose tissue of meal-fed rats. A possible regulatory role of alpha-glycerophosphate formation. Canad. J. Physiol. *45:* 201–214 (1966).

173 *Leveille, G.A.:* Influence of dietary fat and protein of metabolic and enzymatic activities in adipose tissue of meal-fed rats. J. Nutr. *91:* 25–34 (1967).

174 *O'Hea, E.K. and Leveille, G.A.:* Influence of feeding frequency on lipogenesis and enzymatic activity of adipose tissue and on the performance of pigs. An. Aci. *28:* 336–341 (1969).

175 *Freedland, R.A. and Harper, A.E.:* Initiation of glucose 6-phosphatase adaptation in the rat. J. Nutr. *89:* 429–434 (1966).

176 *Chakrabarty, K. and Leveille, G.A.:* Influence of periodicity of eating on the activity of various enzymes in adipose tissue, liver and muscle of the rat. J. Nutr. *96:* 76–82 (1968).

177 *Szepesi, B. and Freedland, R.A.:* Dietary effects on rat liver enzymes in meal-fed rats. J. Nutr. *96:* 382–390 (1968).

178 *Potter, V.R.; Gebert, R.A.; Pitot, H.C.; Peraino, C.; Lamar, C., Jr.; Lesher, S., and Morris, H.P.:* Systematic oscillations in metabolic activity in rat liver and in hepatomas. I. Morris hepatoma No. 7793, Cancer Res. *26:* 1547–1560 (1966).

179 *Wurtman, R.J. and Axelrod, J.:* Daily rhythmic changes in tyrosine transaminase activity of the rat liver. Proc. nat. Acad. Sci., Wash. *57:* 1594–1598 (1967).

180 *Wurtman, R.J.; Shoemaker, W.J., and Larin, F.:* Mechanism of the daily rhythm in hepatic tyrosine transaminase activity: Role of dietary tryptophan. Proc. nat. Acad. Sci., Wash. *59:* 800–807 (1968).

181 *Watanabe, M.; Potter, V.R., and Pitot, H.C.:* Systematic oscillations in liver of normal and adrenalectomized rats on controlled feeding schedules. J. Nutr. *95:* 207–227 (1968).

182 *Zigmond, M.J.; Shoemaker, W.J.; Larin, F., and Wurtman, R.J.:* Hepatic tyrosine transaminase rhythm: Interaction of environmental lighting, food consumption and dietary protein content. J. Nutr. *98:* 71–75 (1969).

183 *Baril, E.F. and Potter, V.R.:* Systematic oscillations of amino acid transport in liver from rats adapted to controlled feeding schedules. J. Nutr. *95:* 228–237 (1968).

184 *Potter, V.R.; Watanabe, M.; Pitot, H.C., and Morris, H.P.:* Systematic oscillations in metabolic activity in rat liver and hepatomas. Survey of normal diploid and other hepatoma lines. Cancer Res. *29:* 55–59 (1969).

185 *Whittle, E.D. and Potter, V.R.:* Systematic oscillations in the metabolism of orotic acid in the rat adapted to a controlled feeding schedule. J. Nutr. *95:* 238–246 (1968).

186 *Peraino, B.; Lamar, C., Jr., and Pitot, H.C.:* Studies on the induction and repression of enzymes in rat liver. IV. Effect of cortisone and phenobarbital. J. biol. Chem. *241:* 2944–2948 (1966).

187 *Fitch, C.D.; Hsu, C., and Dinning, J.S.:* Some factors affecting kidney transamidinase activity in rats. J. biol. Chem. *235:* 2362–2369 (1960).

188 *Walker, J.B.:* Metabolic control of creatine biosynthesis. II. Restoration of transamidinase activity following creatine repression. J. biol. Chem. *236:* 493–498 (1961).

189 *Revel, M. and Hiatt, H.H.:* The stability of liver messenger RNA. Proc. nat. Acad. Sci., Wash. *51:* 810–818 (1964).

190 *Scarano, E.; Petrocellis, B., and Augusti-Tocco, G.:* Studies on the control of enzyme synthesis during the early embryonic development of the sea urchin. Biochim. biophys. Acta *87:* 174–176 (1964).

191 *Gross, P.R.; Malkin, L.I., and Moyer, W.A.:* Templates for the first proteins of embryonic development. Proc. nat. Acad. Sci., Wash. *51:* 407–414 (1964).

192 *Garren, L.D.; Richardson, A.P., Jr., and Crocco, R.M.:* Studies on the role of ribosomes in the regulation of protein synthesis in hypophysectomized and thyroidectomized rats. J. biol. Chem. *242:* 650–656 (1967).

193 *Wool, I.G. and Moyer, A.N.:* Effect of actinomycin and insulin on the metabolism of isolated rat diaphragm. Biochim. biophys. Acta *91:* 248–256 (1964).

194 *Krause, R.L. and Sokoloff, L.:* Effects of thyroxine on initiation and completion of protein chains of hemoglobin *in vitro.* J. biol. Chem. *242:* 1431–1438 (1967).

195 *Schimke, R.T.:* On the control of protein synthesis in animal tissues; in *Smith* Ontog. Immunity Proc. Develop. Immunol. Workshop, 2nd ed., p. 18 (Univ. of Florida Press, Gainsville 1966).

196 *Munro, H.N.:* Role of amino acid supply in regulating ribosome function. Fed. Proc. *27:* 1231 (1968).

197 *Soeiro, R. and Amos, H.:* mRNA half-life measured by use of actinomycin D in animal cells-a caution. Biochim. biophys. Acta *129:* 406–409 (1966).

198 *Szepesi, B. and Freedland, R.A.:* Dietary regulation of pyruvate kinase synthesis in rat liver. j. Nutr. *95:* 591–602 (1968).

199 *Szepesi, B. and Freedland, R.A.:* Further studies on the dietary control of pyruvate kinase activity. Proc. Soc. Biol., NY *132:* 489–491 (1969).

200 *Tanaka, T.; Harano, Y.; Morimura, H., and Mori, R.:* Evidence for the presence of two types of pyruvate kinase in rat liver. Biochem. Biophys. Res. Com. *21:* 55–60 (1965).

201 *Swick, R.W.; Rexroth, A.K., and Stange, J.L.:* The metabolism of mitochondrial proteins. III. The dynamic state of rat liver mitochondria. J. biol. Chem. *243:* 3581–3587 (1968).

202 *Kim, Y.S.:* Half-life of alanine aminotransferase and of total soluble protein in livers of normal and glucocorticoid-treated rats. Mol. Pharmacol. *5:* 105–108 (1969).

203 *Szepesi, B. and Freedland, R.A.:* Time-course of enzyme adaptation. II. Rate of change in two urea cycle enzymes. Life Sci. *8:* 1067–1072 (1969).

204 *Wunner, W.H.; Bell, J., and Munro, H.N.:* The effect of feeding with a tryptophan-free amino acid mixture on rat-liver polysomes and ribosomal ribonucleic acid. Biochem. J. *101:* 417–428 (1966).

Authors' addresses: Dr. *R.A. Freedland,* Department of Physiological Sciences, School of Veterinary Medicine, University of California, *Davis, CA 95616;* Dr. *B. Szepesi,* US Department of Agriculture, Agricultural Research Service, Human Nutrition Research Division, *Beltsville, MD 20705* (USA)

II. Regulatory Mechanisms of Enzyme Synthesis and Degradation

Enzyme Synthesis and Degradation in Mammalian Systems, pp. 141–164
(Karger, Basel 1971)

Regulation of Protein Activity and Turnover Through Specific Modifications in Structure

K. Lemone Yielding[1]

Laboratory of Molecular Biology, University of Alabama in Birmingham, The Medical Center, Birmingham, Ala.

Contents

I.   Introduction

It is the purpose of this discussion to direct attention to the central role of the control of protein structure in biological regulation. No attempt will be made to provide a thorough literature survey, with the realization that much exciting work in this important and rapidly expanding field cannot be cited. The references were selected to serve as illustrative examples of the concepts discussed rather than to provide complete documentation or to establish priority of discovery. Many scientists have grappled with this problem in the past, and it represents a major focus of contemporary activity. Several extremely interesting enzyme systems have been completely omitted from consideration in this discussion and the reader may consult several excellent reviews for more detailed information (7, 25, 68, 114).

It is generally accepted that proteins, and enzymes in particular, serve as the effector molecules for biological processes, and therefore provide an appropriate

1 Recipient of USPHS career development award GM-22698, and project grant AM-08274.

focus for the study of biological regulation. The potential for specific differences in protein structure is practically unlimited, thus satisfying the need of biological systems for variation in function and control. Thus, proteins serve as catalysts, contractile elements, structural units, carrier systems, regulators of osmotic pressure and specific ion concentrations, and specific components in discriminating biological barriers such as membranes. Recent findings have also proposed a role for proteins in memory storage. Although regulation is usually discussed in relation to enzymes, the principles involved and the present discussion are equally pertinent to the other classes of proteins.

In the general problem of enzyme regulation, 3 types of control have been considered: (1) Modification of the *activity* or efficiency of the enzyme (protein) molecules; (2) Changes in the *concentration* of enzyme (protein) as determined by synthesis and degradation; and (3) *Availability of enzyme substrate and removal of products.* In most discussions, these processes are considered as isolated and independent processes, although they obviously must be coordinated in the intact organism. As will be discussed below, the mechanism of regulation through specific modification in enzyme structure extends in a quite general way to contribute to each of these aspects of control.

In regulation we are concerned with the regulatory *signal*, the *receptor* for this signal, and the *transduction* of this signal into a *biological response*, since any theory of regulation must encompass all these factors. The *signal* for a biological response may be defined in chemical terms as a change in concentration of some existing component or the appearance of something new in the internal or external cellular milieu. It may be a single signal or multiple signals involving one or multiple receptors. The signal may serve as an appropriate 'feed back' or 'feed forward' circuit, or it may be more general, such as the case with hormonal or drug control or in regulation of collateral or seemingly distant processes. It may function to assist biologic function, or it may serve in a deleterious way as with toxins or inhibitory drugs. In each instance, however, the signal is chemical and the receptor must be exceedingly discriminating in order to function appropriately. Protein molecules, by virtue of the precise nature of their 3 dimensional structure, provide ideal *receptors* for control signals. Since, in many instances, proteins are also biological effectors, the *transduction* of the signal into biological effects may occur simply through changing the properties of the receptor protein. This can lead directly to a change in a biological process or it may result in the generation of a 'secondary' signal. The present discussion is centered around the concept that the *transduction* process involves a change in the 3 dimensional structure of the receptor molecule (protein).

Although, as will be shown, neither the concern with enzyme regulation nor, in fact, the general concepts expressed in current discussions are completely new; regulation can be examined critically and productively at the present time because of several important insights and developments. First of all, protein structure can now be discussed clearly. The aspects to be considered are primary structure (amino acid sequence); secondary structure (hydrogen bonding of the peptide bonds along the back bone); and tertiary structure (the complex 3 dimensional 'folding' arising from interactions of the amino acid side chains, also known as 'conformation'). The interchain interactions of protein molecules which leads to multichain complexes is termed quarternary structure. According to our current understanding, the secondary, tertiary, and quarternary structure of a protein result from the specific interactions of the aminoacid side chains. These interactions are unique for the sequential arrangement of amino acids provided by each primary structure, so that each protein serves as its own ordering mechanism and the interactions of each amino acid side chain contribute to the final state (33, 34, 109). Any modification, then, of the availability of the amino acid side chains, either by alteration of the primary structure or through interactions of the side chain with the environment can result in a change in the 3 dimensional structure of a protein. Thus, an enzyme may no longer be considered simply as a small catalytic site carried on a large and functionless protein molecule since the entire molecule serves to establish and maintain the proper 3 D configuration. Secondly, the 'induced fit' hypothesis of *Koshland* (67) established that catalytic sites may be flexible and the catalytic process may depend on induction of an exquisitely precise configuration of the catalytic region. Obviously, any change in enzyme tertiary structure could modify enzyme action based on these concepts. Finally, contemporary understanding of protein and solvent structure coupled with the extensive development of technology now make it possible to study directly the interaction between *signal* and *receptor* protein and the resulting changes in properties, so that the biology of regulation may be correlated with the chemistry.

In this discussion 3 types of modifications in enzyme structure will be presented which may be related to changes in enzyme function: (1) Non-covalent changes; (2) Covalent modification of amino acid side chains; and (3) Modification of the protein backbone through limited hydrolysis. Although the mechanism for enzyme modification differs somewhat in these three cases, each in fact provides a means of altering the three dimensional properties of the protein molecule, with resulting changes in function.

II. *Enzyme Regulation Through Non-covalent Changes in Enzyme Structure*

Within the past 10 years, the general concept that many enzymes may be regulated through specific, reversible, non-covalent interactions with small molecules has gained wide acceptance. The essential features of such regulation are that the binding of a small molecule (signal) to non-substrate sites provides a metastable control system in which enzyme conformation can be altered directly through a simple binding equilibrium. Thus, appropriate non-substrate molecules can exert control over enzyme action simply, directly, and reversibly. The enzyme molecule serves both as the *receptor* and the *transducer,* and signal strength (small molecule concentration) is coupled directly to the magnitude of response. Multiple signals can be integrated on the same enzyme simply by the presence of multiple binding sites, and a variety of control situations can be provided (stimulation, inhibition, changes in 'V_{max}' and/or 'K_m', and change in substrate specificity). This type of mechanism has attracted wide interest under the name of 'allosteric regulation' which was introduced as a general name for binding of small molecules to non-substrate sites with resultant changes in enzyme (protein) conformation and activity (90). More recently, this term has been used in a more restricted sense to describe such systems which adhere to a particular quantitative model (91) which may or may not be completely general (8, 42, 66, 69, 82, 107, 108, 154). This and other models will be discussed below.

History. Although it is generally believed that the idea of non-substrate regulation through changes in enzyme conformation is a recent insight in enzymology, it is apparent that the essential features of this model have been discussed for many years. For example, in 1934 *Anson* and *Mirsky* reported that both trypsin and hemoglobin could exist in an equilibrium between native and 'reversibly denatured' forms, and that this equilibrium could be influenced by the presence of small molecules (4, 5). Furthermore, it was proposed by *Johnson et al.* and *Brown et al.* that such reversible alterations in protein structure represented a significant mechanism for the biologic regulation of certain enzymic processes *in vivo* (17, 59–63). This concept also provided the basis for understanding 'breaks' in the straight line region of an Arrhenius plot of an enzyme catalyzed reaction (59). Thus, the idea that enzymes and proteins in general serve both as receptors and effectors for drug action was deduced from the idea that drugs could attach themselves specifically to enzymes, alter their properties, and thus modify the physiology of the cell as a whole (23, 48). Moreover, *Botts and Morales* (14) presented equations describing the expected kinetic behavior of 'modified' enzymes in 1952.

Attention has also been directed for many years at the problem of enzyme regulation by metabolic intermediates or products. For example, it was recognized in the late 1930's that phosphorylase could be activated by nucleotides which did not serve as substrates for the phosphorylation reaction (27). Inhibition of an enzyme by a non-substrate molecule arising later in the same metabolic pathway was also recognized as a possible control mechanism of importance by *Dische* who, in 1941 reported that phosphoglycerate specifically inhibited the phosphorylation of glucose in homogenates from erythrocytes (32). Unfortunately, his observations were largely unnoticed until quite recently. This phenomenon of biological control by metabolic products began to receive more general attention beginning in the mid 1950's when it was recognized clearly that metabolic products could govern the rates of their own biosynthesis (1, 104, 133, 147). This type of regulation is undoubtedly a general phenomenon both in lower and higher organisms (25, 114). Feedback control in lower organisms has been an especially useful model for studying mechanisms for enzyme regulation because of its clear relation to biological systems. The concept of a separate (non-catalytic) binding site which could act as the receptor for an inhibitor (signal), although proposed or implied in a number of control situations, was clearly established as a distinct, genetically (and therefore structurally) determined feature of feedback control. First, mutants have been obtained in which catalytic and regulatory function of an enzyme are found to vary independently (19, 96). Secondly, *Gerhart and Pardee* clearly demonstrated *in vitro* that aspartate transcarbamylase could be desensitized to the effects of feedback inhibitors without impairing its catalytic activity (43, 44). Subsequent demonstrations that various types of treatment may selectively abolish the effects of activators or inhibitors on a variety of enzymes has strengthened this concept that enzymes may have *distinct regulation sites apart from catalytic loci* (2, 12, 19, 36, 81, 82, 83, 84, 99, 106, 126). In most systems it cannot be concluded whether desensitization results from blocking of a binding site or through preventing a change in enzyme configuration. For aspartate transcarbamylase, however, *Gerhart and Schachman* showed that the catalytic and regulatory sites are on different protein chains (46).

Frieden provided clear evidence for changes induced by small molecules in enzyme tertiary structure accompanying changes in catalytic activity, in his studies with crystalline glutamate dehydrogenase (37, 38, 39) in which he showed that the state of aggregation of the enzyme molecule could be altered by nucleotide inhibitors and stimulators. Although the precise details of his model have been revised somewhat to fit more extensive data accumulated in the ensuing years, his early experiments were the beginning of a modern surge of insight into enzyme regulation and provided clear evidence for regulator binding sites and small molecule evoked changes in enzyme structure. The biological implications of regulatory effects on this enzyme are not clear, but it has served as an important model for conceptual development (11–13, 38–40, 56, 121, 122, 129, 132, 140, 148, 155). Multiple regulatory reagents have been studied with the enzyme including steroid hormones (149, 151), GTP (40), amino acids (150), metals (155), thyroid hormone (140), and a variety of aromatic hydrocarbons (152), all of which promote changes in enzyme conformation, catalytic activity, and relative substrate specificity. It appears, therefore, that this enzyme has several regulator sites. The independence of the catalytic site and the regulatory effects were demonstrated by the fact that treatment with organic mercurials leads to desensitization (12), while conversely, the catalytic activity can be abolished with acetic anhydride with retention of the effects of regulatory reagents on enzyme conformation (148). These multiple observations prompted the consideration and adoption of a general model for enzyme and other protein regulation by hormones and other metabolically significant molecules; i.e. an enzyme may possess, in addition to its catalytic site, multiple stereospecific loci concerned with regulation of its catalytic activity (38, 39, 129). These sites bind specific regulatory reagents, or signals, which exert their effects by reversibly changing enzyme conformation. These multiple sites may act antagonistically or in concert. The subunit structure and association – dissociation behavior have also figured prominently in discussion of this and other examples of enzyme regulation. Attempts have also been made to extend the features of this model to account for some types of control of enzyme assembly and stability, since both the enzyme cofactor and the multiple reagents serve either to increase or decrease enzyme denaturation in solution (13, 31, 40, 51, 55, 153). This enzyme, which is apparently composed of identical subunits (6), is a particularly interesting model for study because of the variety of types of regulatory effects it shows with clear relationship to conformational changes, and because it is conveniently available as the crystalline protein. It also has served to show the multifunctional nature of control, in keeping with the variety of demands for regulation which exist in the intact cell. In particular it offers the opportunity for determining the relationships between various types of binding sites on the same enzyme chain.

The occurrence of binding sites for multiple regulatory ligands (signals) has particular pertinence to the problem of 'cross-linking' or coordinating multiple related metabolic events. Various phenomena have been described in microorganisms, particularly, in which a number of signals may interact to control a single enzymic step. Thus, 'concerted', 'multivalent', 'cooperative', 'cummulative', or 'sequential' feedback imply the combined effects of several signals on a single protein (18, 28, 29, 58, 93, 95, 100, 141).

There are large numbers of enzyme reactions which are regulated by nonsubstrates, and the implication follows that such effects may result from changes in enzyme conformation. The proof of the relationship between changes in activity and enzyme structure must depend on additional experiments in most cases. Although the role of enzyme structure in determining specificity of binding is easily rationalized, its relationship to the catalytic process awaits a more thorough understanding of enzyme structure and mechanism.

Models for reagent control of protein structure and functions. Several useful models (or mechanisms) have been proposed to assist in generalizing our thinking concerning enzyme regulation. Clearly the word 'model' or 'mechanism' has different meanings to different people. It may consist of a mathematical expression *consistent* with observations (but lacking in exclusive relationship to all data) or it may be a general non-mathematical description which does not require the same limiting assumptions required for setting down specific equations or precise descriptions. Of the general models, the earliest was that proposed by *Anson and Mirsky* and studied extensively by *Johnson* and his associates, stating that proteins could reversibly 'denature' in response to environmental conditions. In recent years, the general model of specific reagent binding to non-substrate sites with induced changes in enzyme tertiary structure leading to changes in catalytic activity was applied to glutamate dehydrogenase in mammalian systems (40, 129, 131, 149) and to transcarbamylase (44, 45), and threonine deaminase (21, 22) in bacterial 'feedback' systems. The interplay between O_2 binding sites on hemoglobin also has served to establish a clear dependence of the binding properties of a protein on its structure in solution (143, 145). Hemoglobin, however, does not have separate sites for regulation and function, and thus differs somewhat from our discussion here.

Mathematical models have been provided by several investigators without defining the physical mechanisms involved, and the complexity of such treatments and the multiple assumptions required for their solution illustrate the difficulty in inducing a physical model purely from kinetic data. They are quite useful, however, in providing a framework for considering experimental results. For example, *Botts and Morales* considered modifier binding extensively in 1952. *Frieden,* following his studies on the reagent induced physical and kinetic changes in glutamate dehydrogenase, extended these considerations to show the variety of changes in kinetic parameters which may result from modifier binding (14, 41, 42). Extensive attempts to formulate both a mathematical and a physical model were provided by *Monod et al.* (90, 91) and *Koshland et al.* (66, 69) who proposed a somewhat specific model. Ironically, the 2 models cannot be differentiated unequivocally using conventional enzyme kinetics, and there may be some reality in each of the proposals. *Monod* and his associates introduced the important term 'allosteric' to describe those protein systems which bind regulatory ligands to sites other than those having catalytic or other functional properties, and attributed the different functional properties of the regulated proteins to changes in conformation, as had been proposed by others. They recognized the general importance of the concept, and have succeeded in promoting its wide acceptance. Subsequent to their first review, they also proposed a rather specific quantitative model based on interactions between protein subunits and the preservation of molecular symmetry. The 'allosteric transition' was explained on the basis of an equilibrium between 2 or more forms of an enzyme, and pictured allosteric ligands as interacting preferentially and in a concerted manner with one or the other of the enzyme forms, as shown in figure 1a, for 2 different ligands. (Recently, *Changeux* has described the consequences of non-exclusive binding of ligand [107].) This model might be paraphrased as a 'pull' model since the ligands influence the equilibrium between enzyme states by 'sequestering' the enzyme in ligand bound forms. Obviously, the different conformations may exhibit a variety of differences in functional properties and the effects of L_A and L_B are antagonistic. Strictly speaking, the ligand influences the distribution of species but not the equilibrium since its presence introduces additional species E_{L_A} and E_{L_B} into the equilibrium expression. Based on this model they predicted co-operative binding for both substrates and allosteric reagents based on subunit interactions and preservation of symmetry, and provided a system of definitions for describing regulatory effects in this context. They also predicted that allo-steric enzymes will all be comprised of subunits. Various interpretations of this model have resulted in 2 widespread conclusions which are not always warrant-ed. First, that an 'allosteric' system always shows cooperative kinetics for sub-strate and/or regulatory ligand binding; and secondly, that an enzyme showing cooperative kinetics is 'allosteric' and must consist of subunits. Such cooperative phenomena clearly have important implications for regulation, but it has been shown that regulation by this mechanism of conformation change may not always be reflected in cooperativity of conventional enzyme kinetic curves (8, 154). Furthermore, cooperative kinetics can be generated by other models. For example, based on the induced fit hypothesis of *Koshland*, a simple model can be devised in which enzyme relaxation following dissociation of products from an enzyme can be rate limiting and can result in non-linear kinetics (31, 139). *Sweeny and Fisher* (125) have generalized that sigmoidal kinetics serve only to eliminate ordered sequence models of enzyme action, and show that such kine-tics can be displayed with substrate binding if there is more than one pathway leading to binding of a single site per enzyme. Thus, kinetic studies are not adequate evidence on which to base a mechanism for a particular enzyme.

1a 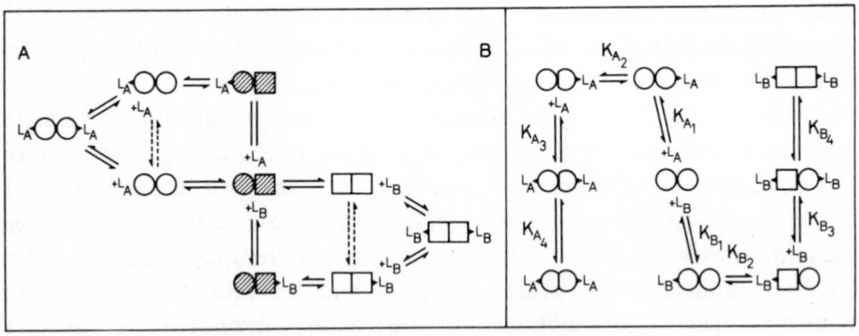   1b

Fig. 1a. 'Pull' model for ligand induced change in protein conformation (adapted from *Monod et al.* [90, 91]). The effects of ligands (L_A and L_B) are superimposed on a pre-existing equilibrium between forms A and B of the enzyme. Subunit interactions and the resulting conformational restraints dictated by molecular symmetry cause the cross hatched species to be unfavored.

Fig. 1b. 'Push' model for ligand induced conformational change (adapted from *Kosh-land et al.* [66, 69]). The extent to which interactions between subunits are altered by ligand will determine the extent to which $K_{A_1} K_{A_2}$ and $K_{B_1} K_{B_2}$ differ from $K_{A_3 A_4}$ and $K_{B_3} K_{B_4}$ and, therefore, the extent of positive (or negative) cooperativity.

Koshland's model for ligand control of protein structure and function explains the effects of reagents on the basis of changes in the equilibrium between different states of the protein molecule; and rationalizes non-linear behavior of ligand binding and substrate binding through sequential binding to protein subunits. This proposal may be paraphrased as a 'push' hypothesis, since it implies that the ligand binding *results* in a conformational change rather than simply selecting a pre-existing state. The essential features of this model are shown in figure 1b (66, 69).

For either type of model, the role of single ligands may be explained rather simply, and although certain predictions differ (8) the implications for regulation would be essentially the same. In each instance, regulation results from changes in enzyme conformation. The effects of multiple ligand binding can be rationalized with either model, but the predictions might differ somewhat. For example, with 'cumulative' inhibition resulting from 2 or more ligands, the *Monod* model would predict a pre-existing equilibrium between the active and inactive enzyme forms with the 'cumulative' inhibitors each acting to stabilize the latter. The *Koshland* model permits a progressive change in conformation with the addition of each additional ligand. With any model the occurrence of multiple regulator effects may make it necessary to evoke *multiple* changes in enzyme conformation. Studies, for example with glutamate dehydrogenase clearly show that a transition between only 2 states will not account for all the changes described with all regulatory reagents (132).

It is highly desirable to keep our models for regulation, and the language used for their description, as general as possible. A number of papers have been published describing *possible* models based on minimal deviations from those under discussion or representing special limiting cases. Some confusion arises from our description of what constitutes a change in structure. In the present discussion I have been most careful to say that *regulators work by evoking changes in structures,* without qualification. Clearly, tertiary structure changes are implied, but discussions of secondary, tertiary, and 'quarternary' structure would be appropriate. It has yet to be determined what is the extent of structure change that accompanies such regulatory transitions. It is difficult, if not impossible, to separate these three aspects of structure, particularly the side chain interactions which determine both tertiary and quarternary structure. Thus, models for 'allosteric' effects based *only* on polymerization of subunits (without 'isomerization' – change in tertiary structure) as proposed by *Nichol,* would seem unlikely (94). On the other hand, polymerization or aggregation may serve as a sensitive index of changes in tertiary structure (11, 40) and may be an important parameter of control (54).

Relatively little attention has been directed toward the physical basis for the transitions in protein structure produced by regulatory ligands. *Monod* refers to 'taut' and 'relaxed' state for proteins in relation to allosteric changes, but the physical basis for this classification is rather arbitrary. In fact, the thermo-dynamics of the allosteric transition have, for the most part, not been investigated and it is likely that a variety of effects will be discovered. It is most interesting in relation to this present volume to consider the effects of regulatory ligands on protein stability. Several models have been studied rather well. Glutamate dehydrogenase has again served a useful role. With this enzyme, reagents may lead either to a considerable increase or decrease in stability to heat denaturation (13, 31, 40, 51, 55, 153). In other instances, allosteric effects have been shown to change the rate at which enzymes are inactivated by proteo-lytic enzymes (116, 146). There is obviously a great need to study the stability of enzymes *in vivo* in relation to the effects of regulatory ligands.

Yet an additional area of inquiry into enzyme regulation by ligand evoked changes in structure is largely unexplored. Much is said about the *concentration* of enzymes in cells, but little about their *availability*. Certain enzymes are present in cells in quite large concentrations and the question of what makes them available to substrates is unsolved. Structure transitions resulting from binding of regulatory ligands may well act in these instances to influence enzyme particle size, solubility, membrane binding (85) and absorption to other macromolecules. Certain 'inhibitor' studies *in vitro* might even serve to facilitate enzyme action *in vivo* through increased availability.

Regulation through structure transitions may be generalized to other protein functions. This mechanism provides the means by which specific small molecules activate or deactivate repressors (see below), modify membrane permeability, or change the function of any biological protein (even, perhaps, 'memory' proteins).

Thus, proteins may be regulated due to the presence of specific binding sites for regulatory ligands of various types. Although this discussion has centered around 'small molecules' as regulatory ligands, macromolecules clearly could interact with proteins to produce the same type of effects. Examples may be cited in which nucleic acids (86, 97) or polysaccharides (88, 135) function as the regulatory signal, or specific regulatory binding proteins may be involved in enzyme regulation (10, 87). The occurrence of subunits is of particular importance to regulation. The subunit structure may assist in the amplification process for ligand interactions, or may provide the specific binding site(s) for the regulators. In proteins comprised of non-identical subunits the subunits themselves might be considered as regulatory ligands. Hemoglobin serves as an example of this type of regulation, the O_2 saturation curve for the β subunits is clearly modified by the presence of the a subunits. Similarly, with aspartate transcarbamylase, the presence of the regulatory subunit changes the activity of the catalytic subunit (46). It also appears that in certain multienzyme complexes, binding of ligand to one of the enzymes can result in regulation of the other enzymes in the complex (9).

III. Enzyme Regulation Through Covalent Changes in Structure

The principles underlying enzyme regulation through changes in enzyme structure are even better established in the case of covalent modification. As a control mechanism, covalent changes differ from the allosteric mechanism in several important features: First, regulation may involve a second control system which receives the signal and responds by modifying the enzyme; secondly, the regulatory change is either irreversible, or, if reversible, requires still another control component which can receive the 'reverse' signal and reverse the protein modification. Thus, covalent regulation is considerably more complicated than 'allosteric' changes, but the principles relating functional properties to changes in tertiary structure are the same. Clearly, covalent modification may also result in a protein form whose sensitivity to non-covalent regulatory ligands is changed, thus also sensitizing it to the direct action of such regulators (16, 79, 80, 111). In general, 2 types of covalent changes must be considered: Modification of the peptide backbone structure; and modification of individual amino acid side chain groups.

Hydrolytic activation of precursor proteins. Synthesis of proteins in the inactive state (zymogens or proenzyme) followed by hydrolytic activation at the time and site of need provides a means of controlling both activity and protein turnover. The proenzyme may be activated either by mixing with an appropriate hydrolytic enzyme, or by the activation of a hydrolytic enzyme by an environmental signal as is the case with blood clotting mechanisms. This general type of enzyme regulation has been reviewed recently by *Ottesen* (98). The most extensive studies on these proenzyme-enzyme transformations have been done with the pancreatic enzymes chymotrypsinogen and trypsinogen, and considerable structural information is available concerning the activation process. With each, a specific region of the protein backbone is excised, leading to a stable activated enzyme. The complete structures of these proteins are known and the changes in properties of the catalytic site which are involved in this activation serve as important models for understanding the structure changes attending other enzyme regulatory processes. From the viewpoint of the present discussion it is pertinent that the changes in conformation (ranging from very subtle to large changes in molecular aggregation) in addition to changing catalytic properties uniformly lead to alterations in susceptibility of the proteins to subsequent denaturation.

The blood clotting mechanism is interesting because of the extensive nature of the proenzyme-enzyme relationships. Some 8 or more inactive precursors circulate, functionless, until the first factor is activated by a change in environment. Subsequent to this first event all are activated sequentially by a 'cascade' of proteolytic transformations, in which each modified protein serves to activate the next component in the series (30). The complement system provides another example of a sequence of biological events resulting from triggering a cascade of proteolytic transformations at the time and site of need. This system was also reviewed recently (92).

Recent studies showing that the a and β chains of insulin are derived by proteolysis from a single larger protein are of particular pertinence to the present discussion even though this protein is not an enzyme (24, 117–120). Whereas the other cases of limited proteolysis involve delivery of inactive precursors to the site of biological function where they are actived, the proinsulin – insulin conversion appears not to be simply part of the secretory process for the hormone, but results in production of circulating protein hormone which is thermodynamically unstable, and can be cleared quickly from the organism. Rapid turnover would be most desirable for such a hormone whose level must be subject to rapid fluctuations for metabolic control. The disulfide bonds which apparently form in proinsulin serve as an 'activation' barrier to denaturation, but when broken in the active insulin result in a molecule which cannot be renatured to a substantial extent. This means that rapid turnover and fluctuations in the level of circulating hormone can be accomplished readily. Thus, certain enzymes (digestive, clotting factors, complement), and insulin are produced in stable forms which are then modified to unstable but active states at the time of need. Following their function, they may be eliminated rapidly or inactivated. In this way rapid control may be exerted over both activity and turnover.

Covalent modification of specific amino acid side chains. The covalent introduction of a ligand onto specific amino acid side chains of a protein can provide a relatively stable modification in protein functional properties. In some instances this process is reversible, but it is noteworthy that this is relatively slow and requires still another enzyme which must itself be regulated. Thus, while 'allosteric' regulation provides for rapid fluctuations in biological activity through changes in side chain environment based on simple equilibria, covalent modification of side chain function permits slower and more sustained responses to regulatory signals. In the latter instance, the reversal of the response requires a secondary signal and a secondary effector, such as another enzyme. For purposes of this discussion, only limited examples of this type of regulation will be considered.

Phosphorylase. It is quite well known that glycogen phosphorylase is interconvertable between 2 different molecular forms (26, 50, 74), by means of specific covalent modification. This modification results in changes both in its catalytic activity and in its response to regulatory nucleotides. Thus, phosphorylase a is a tetramer with a molecular weight of 495,000 while phosphorylase b has a molecular weight of 242,000 (65). Phosphorylase b, which requires AMP for catalytic activity (16, 53), is converted to phosphorylase a by a specific kinase which catalyzes the phosphorylation of 2 serine residues by ATP (72, 73, 75, 76). Although the phosphorylation is accompanied by dimerization to the tetramer, phosphorylase *a* may be maintained in the dimer configuration in solutions of high ionic strength, and it appears that the phosphorylation rather than dimerization is the essential step in the modification of the enzyme's catalytic properties (136, 137, 138). Similarly, the conversion of phosphorylase *a* to *b* involves a specific enzyme (phosphorylase phosphatase) catalyzed modification of enzyme structure with release of the serine bound phosphate (49, 142). These interconversions between enzyme forms are accompanied by at least 2 sets of changes in properties. First, the enzyme is activated with respect to catalytic activity as a consequence of the covalent modification, and secondly, there is an extensive change in the response of the enzyme to noncovalent modification by a variety of nucleotides. The enzyme responsible for covalent modification of phosphorylase (phosphorylase kinase) is also activated by a molecular signal in the form of cyclic 3',5'—AMP, thus providing another key control point in the overall sequence. Much important work has been done with phosphorylase which cannot be reviewed here because of space limitations. Other reviews may be consulted for details (16, 74).

Another example of enzyme regulation resulting from phosphorylation is the inactivation by an ATP specific kinase of the pyruvate dehydrogenase complex from kidney, heart and liver mitochondria (79, 80). In fact, it has been suggested that cyclic nucleotide dependent protein kinases may represent a control point of general importance (77, 89).

Another useful model for the study of both covalent and noncovalent modification of enzyme properties is that of glutamine synthetase in *E. coli.* In this instance, adenylation of the enzyme converts it between 2 forms which differ substantially in their response to feedback control (111, 112). Thus, this covalent modification serves chiefly to limit 'allosteric' control.

These systems, therefore, offer more than one mode of regulation. Regulatory signals can operate 'allosterically' either directly on the enzyme in question or on another specific 'control' enzyme which controls its activity through covalent modification.

IV. Integrated Control of Enzyme Activity

It is particularly noteworthy that all models for regulation must take into account the need for coordinating multiple metabolic events, for example, with interrelated but non-sequential events with converging and branching systems, and with systems having controlled time response of the regulatory effect. In some instances, a single enzyme step must be sensitive to multiple signals which must be coordinated. In other instances, a single signal must be amplified to influence multiple steps. At other times, there needs to be a shift in the sensitivity to control a particular enzyme step. It is also obvious that all these effects may need to be expressed over a short time course or may need to be sustained.

The total integration, therefore, of biological events can be accomplished only through the appropriate and simultaneous application of all regulatory principles. As mentioned earlier, the allosteric mechanism is particularly suitable for coordinating multiple signals. Thus, multiple regulator sites may exist which act in concert either synergistically, additively, or antagonistically to modify enzyme structure and activity. The time course for allosteric regulation, however, is relatively short and could lead to rapid and brief fluctuations in enzyme function. Covalent control, on the other hand, has a more sustained time course but requires that the signal not work directly on the enzyme itself but through another effector system. Thus a signal may act as a simple 'allosteric' effector on a modifying enzyme (for example on a kinase or an adenylating enzyme) which in turn produces a sustained effect on the metabolic step which is to be controlled by promoting covalent changes in the target enzyme. These can alter both catalytic activity and the responsiveness of the enzyme to direct control by other allosteric regulators. The covalent change may be prolonged, or may be under fine control by a second enzyme which reverses the covalent modification. Studies with phosphorylase have provided a most interesting model for hormonal (and drug) regulation which embodies some of these features (105, 124). This model proposes that hormones can work in this enzyme system (and others) by means of stimulated synthesis (mechanism not specified) of cyclic 3',5'—AMP which in turn activates the phosphorylase *b* kinase system resulting in increased levels of 'active' phosphorylase. Thus, the hormone influences the enzyme through the intermediary of another signal 'receiving' and 'transducing' system. The cyclic 3',5'—AMP has been referred to as a 'secondary messenger' in hormone action. It would appear that at least 2 'allosteric' steps and 1 covalent modification are involved in this hormonal effect.

An additional feature of probable importance in providing integrated control of complex pathways is the occurrence of multiple enzymes (isozymes) which can catalyze the same reaction, but differ in regulatory properties. These have been discussed extensively in relation to feedback control in microorganisms (113, 115), and examples of differences have also been cited, in higher organisms (64, 123, 127). Isozymes are discussed elsewhere in this volume.

V.   Role of the Regulation of Protein Structure in Control of Protein Concentrations

Protein levels depend on the combined effects of synthesis and degradation (110). Since the regulation of neither of these events has been thoroughly worked out, it is difficult to predict the influence of any particular perturbation on the final state of a protein system. Nevertheless, changes in tertiary structure evoked by regulatory signals must be given careful scrutiny as a means of control.

Effects of alterations in protein structure on the rate of protein synthesis. Protein synthesis, worked out largely in bacterial systems has been generally accepted as proceeding according to the well known scheme of transcription and translation. A specific messenger RNA is transcribed as dictated by the DNA. A given code word of this messenger is translated by ribosomes which oppose the specific amino acid charged transfer RNA molecules, so that the polypeptide is assembled in the correct sequence. A polypeptide is formed which then assumes the most stable configuration dictated thermodynamically by the resulting intrachain and environmental interactions. This picture is complicated in the case of subunit proteins, particularly when the subunits are not identical. Although the thermodynamic principles are established for assembly of secondary, tertiary and quarternary structure, the kinetics for this process are largely unknown. The elegant studies on the regulation of protein synthesis in bacteria leading to the generally accepted *Jacob-Monod* model (57) of induction and repression at the gene level, may not provide an adequate understanding of higher organisms. From the multiple steps involved in overall protein synthesis, the rate limiting step cannot be determined. It seems very likely that different steps may be limiting for different proteins. Recently, *Tomkins et al.* (128), from studies of tyrosine transaminase in hepatoma cells, have presented a model for regulation in mammalian cells involving repression and derepression at the translational level. In their studies, it also appeared that rate determination was different in different phases of the cell cycle. If assembly of protein subunits or folding of individual subunits are rate limiting steps, it is easy to see how specifically evoked changes in conformation could modify the rate of active protein formation. This could result either through the 'allosteric' type mechanism or by covalent modification of the protein in question. In the case of proteins exported from the cell, structure changes could also influence the rate of secretion. It is thought, for example, that covalent attachment of carbohydrate residues to protein side chains may play a role in secretion from the cell. Experiments on the effects of known allosteric modifiers on the rates of synthesis of specific proteins would be most interesting.

There may also be less direct effects of structure modification on protein synthesis. For example, studies on the repression of the enzymes for histidine biosynthesis in *Salmonella typhimurium* have shown that the state of the feedback sensitive (allosteric) site on the first enzyme in the pathway determines the kinetic pattern of repression since different patterns are obtained in the presence and absence of feedback inhibitor. Furthermore, mutants with altered feedback sensitivity show differences in repression (70, 71).

The additional question is whether a protein may serve to feedback in its own biosynthesis directly, in which case changes in structure could also have important direct effects on biosynthesis by this mechanism.

Apart from the question of the effects of structural changes in a protein on its own biosynthesis, is the question of the mode of action of inducers and co-repressors. The nature of repressor has been studied most extensively for λ phage and for the *lac* operon, and each instance appears to be a multi-subunit protein with multiple binding sites for inducer (78, 103). Inducer is thought to act allosterically to modify the combining properties of repressor. Furthermore, the 'catabolite' repression by glucose that is seen with β-galactosidase in *E. coli* has been correlated with the ability of a glucose metabolite, fructose 1,6-di-phosphate to bind to the enzyme and alter its physical properties (15, 47). A role for inactivation in catabolite repression of yeast malate dehydrogenase has also been suggested (35).

In summary, the role of 'control' evoked changes of enzyme structure in regulating enzyme synthesis directly has not been determined, but remains an exciting possibility. A role for structure changes in determining repressor action on the other hand seems to be highly likely. Thus, both 'allosteric' and covalent changes may be involved in regulation of protein synthesis.

Effects of alterations in protein structure on the rate of protein degradation. Relatively little is known of the mechanism *in vivo* for protein degradation, in spite of rather extensive knowledge of the steps involved in protein synthesis. In an ideal environment, a protein could remain stable almost indefinitely with minimal turnover. Similarly, in an unchanging environment a protein should turn over at a constant rate. Either situation would be inconsistent with the need for the regulation of biological activities required for homeostasis in the living organism. It is well established that biological variation can be accompanied by changes in protein turnover. Since the precise mechanism is not known for protein turnover, the key question of how regulatory signals exert their effects remains unanswered. It seems most likely, however, that regulation of protein structure (and, therefore, stability) will play a central role in this system, particularly since it seems unlikely that each protein will have a specific degradative pathway which could serve to receive and respond to the regulatory signal. It may be that some proteins will have specific degradative mechanisms (101, 102) but this seems unlikely as a general rule.

It is well known that protein structure changes can result in drastic alteration in stability. Many native proteins are relatively resistant to proteolysis, suggesting the requirement for an initial change in conformation resulting in lability of the protein. *Grisolia* (52) reviewed extensively the concept that substrates and cofactors can either stabilize or labilize proteins to denaturation and/or proteolysis and provided a list of more than 30 enzymes in which substantial effects had been documented. In our own and other studies with glutamate dehydrogenase it has become apparent that 'allosteric' reagent can either increase or decrease enzyme stability, and it is now generally accepted that one type of evidence for reagent induced change in tertiary structure is a change in stability. Assuming that the rate of proteolysis is determined by the concentration of susceptible substrate, it seems likely that changes from the native, stable form of a protein by covalent modification or 'allosteric' mechanisms may well represent a rate limiting step in turnover. Thus, the 'turnover' signal could operate directly on the protein in the case of the 'allosteric' mechanism, or could serve directly to stimulate or inhibit enzymes concerned with generation of control ligands or with covalent modification of the protein in question. Studies of turnover in response to known 'allosteric' modifiers will be most interesting.

VI. Summary

The concept of regulation of the biological activity of proteins through specific modifications in their structure is firmly established as a general biological principle, although many details of mechanisms are still to be elucidated. These changes can occur non-covalently by the 'allosteric' mechanism, by covalent modification of protein side chains, or through alteration in primary structure. Each of these events, in addition to regulating the activity of protein molecules, may also serve an important role in regulating protein synthesis and turnover. The final relationships between biological control signals, protein structure, and protein turnover are crucial issues toward which much attention is directed.

VII. References

1   Adelberg, E.A. and Umbarger, H.E.: Isoleucine and valine metabolism in E. coli. V. Alpha-ketoisovaleric acid accumulations. J. biol. Chem. 205: 475–482 (1953).

2   Ahlfors, C.E. and Mansour, T.E.: Studies on heart phosphofructokinase desensitization of the enzyme to adenosine triphosphate inhibition. J. biol. Chem. 244: 1247–1251 (1969).

3   Anderson, S.R. and Weber, G.: Multiplicity of binding by lactate dehydrogenases. Biochemistry 4: 1948 (1965).

4   Anson, M.L. and Mirsky, A.E.: The equilibrium between active native trypsin and inactive denatured trypsin. J. gen. Physiol. 17: 393–396 (1934).

5   Anson, M.L. and Mirsky, A.E.: The equilibria between native and denatured hemoglobin in salicylate solutions and the theoretical consequences of the equilibrium between native and denatured protein. J. gen. Physiol. 17: 399–408 (1933–34).

6   Appella, E. and Tomkins, G.M.: The subunits of bovine liver glutamate dehydrogenase: demonstration of a single peptide chain. J. molec. Biol. 18: 77–89 (1966).

7   Atkinson, D.E.: Regulation of enzyme activity. Ann. Rev. Biochem. 35: 85–124 (1966).

8   Atkinson, D.E.; Hathaway, J.A., and Smith, E.C.: Kinetics of regulatory enzymes. Kinetic order of the yeast diphosphopyridine nucleotide isocitrate dehydrogenase reaction and a model for the reaction. J. biol. Chem. 240: 2682–2690 (1965).

9   Bailin, G. and Lukton, A.: Allosteric properties of a phosphorylase a-glutamic-pyruvic transaminase complex. Biochim. biophys. Acta 128: 317–326 (1966).

10 *Bechet, J. and Wiame, J.M.:* Indication of a specific regulatory binding protein for ornithinetranscarbamylase in saccharomyces cerevisiae. Biochem. biophys. Res. Comm. *21:* 226–234 (1964).

11 *Bitensky, M.W.; Yielding, K.L., and Tomkins, G.M.:* Reciprocal changes in alanine and glutamate dehydrogenase activities after exposure of crystalline bovine L-glutamate dehydrogenase to organic mercury. J. biol. Chem. *240:* 663–667 (1965).

12 *Bitensky, M.W.; Yielding, K.L., and Tomkins, G.M.:* The reversal by organic mercurials of 'allosteric' changes in glutamate dehydrogenase. J. biol. Chem. *240:* 668–673 (1965).

13 *Bitensky, M.W.; Yielding, K.L., and Tomkins, G.M.:* The effect of allosteric modifiers on the rate of denaturation of glutamate dehydrogenase. J. biol. Chem. *240:* 1077–1082 (1965).

14 *Botts, J. and Morales, M.:* Analytical description of the effects of modifiers and of enzyme multivalency upon the steady state catalyzed reaction rate. Trans. Faraday Soc. *49:* 696–707 (1953).

15 *Brewer, M.E. and Moses, V.:* Metabolite-promoted heat lability of beta-galactosidase and its relation to catabolite repression. Nature, Lond. *214:* 272–273 (1967).

16 *Brown, D.H. and Cori, C.F.:* The enzymes, vol. 5, p. 707, *Boyer, Lardy and Myrbach* (eds.) (Academic Press, New York 1961).

17 *Brown, D.E.; Johnson, F.H., and Marshland, D.A.:* The pressure, temperature relations of bacterial luminescence. J. cell. comp. Physiol. *20:* 151–168 (1942).

18 *Caskey, C.T.; Ashton, D.M., and Wyngaarden, J.B.:* The enzymology of feedback inhibition of glutamine phosphoribosylpyrophosphate amidotransferase by purine ribonucleotides. J. biol. Chem. *239:* 2570–2579 (1964).

19 *Changeux, J.P.:* The feedback control mechanism of biosynthetic L-threonine deaminase by L-isoleucine. Cold Spr. Harb. Symp. quant. Biol. *26:* 313–318 (1961).

20 *Changeux, J.P.:* Effet des analogues de la L-thréonine et de la L-isoleucine sur la L-thréonine désaminase. J. molec. Biol. *4:* 220–225 (1962).

21 *Changeux, J.P.:* Allosteric interactions on biosynthetic L-threonine deaminase from E. coli K12. Cold Spr. Harb. Symp. quant. Biol. *28:* 497–504 (1963).

22 *Changeux, J.P.:* Sur les propriétés allostériques de la L-thréonine désaminase de biosynthèse. IV. Le phénomène de désensibilization. Bull. Soc. Chim. biol. *47:* 115 (1965).

23 *Clark, A.J.:* in *Heffters* Handbuch der experimentellen Pharmakologie, Ergänzungswerk, p. 4 (J. Springer, Berlin 1937).

24 *Clark, J.L. and Steiner, D.F.:* Insulin biosynthesis in the rat: Evidence for two proinsulins. Proc. nat. Acad. Sci., Wash. *62:* 278–285 (1969).

25 *Cohen, G.N.:* Regulations of enzyme activity in microorganisms. Ann. Rev. Microbiol. *19:* 106–126 (1965).

26 *Cori, G.T.:* The effect of stimulation and recovery on the phosphorylase *a* content of muscle. J. biol. Chem. *158:* 333–339 (1945).

27 *Cori, G.T.; Colowick, S.P., and Cori, C.F.:* The action of nucleotides in the disruptive phosphorylation of glycogen. J. biol. Chem. *123:* 381–389 (1938).

28 *Datta, P. and Gest, H.:* Control of enzyme activity by concerted feedback inhibition. Proc. nat. Acad. Sci., Wash. *52:* 1004–1009 (1964).

29 *Datta, P. and Gest, H.:* Alternative patterns of end-product control in biosynthesis of amino-acids of the aspartate family. Nature, Lond. *203:* 1259–1261 (1964).

30 *Davie, E.W. and Ratnoff, O.D.:* In *Neurath*'s The proteins, vol. 3, pp. 360–444 (Academic Press, New York 1965).

31 *DiPrisco, G. and Strecker, H.J.:* Studies on the effect of ionic compounds on the stability of GDH. Biochim. biophys. Acta *122:* 413–422 (1966).

32 *Dische, Z.:* Interdependence of various enzymes of the glucolytic system and the automatic regulation of their activity within the cells. (1) Inhibition of the phosphorylation of glucose in red corpuscles by monophosphoglyceric and diphosphoglyceric acids-state of the diphosphoglyceric acid and the phosphorylation of glucose. Trav. Membres. Soc. Chim. Biol. *23:* 1140–1148 (1941).

33 *Epstein, C.J.; Goldberger, R.F.; Young, D.M., and Anfinsen, C.B.:* A study of the factors influencing the rate and extent of enzymic reactivation during reoxidation of reduced ribonuclease. Arch. Biochem. Suppl. 1: 223–231 (1962).

34 *Epstein, C.J.; Goldberger, R.F., and Anfinson, C.B.:* The genetic control of tertiary protein structure: Studies with model systems. Cold Spr. Harb. Symp. quant. Biol. *28:* 439–449 (1963).

35 *Ferguson, J.J., Jr.; Boll, M., and Holzer, H.:* Yeast malate dehydrogenase: Enzyme inactivation in catabolite repression. Europ. J. Biochem. *1:* 21–25 (1967).

36 *Forest, P.B. and Kemp, R.G.:* Alteration of the allosteric properties of phosphofructokinase by modification of a single thiol group. Biochem. biophys. Res. Comm. *33:* 763–768 (1968).

37 *Frieden, C.:* The dissociation of glutamic dehydrogenase by reduced diphosphopyridine nucleotide (DPNH). Biochim. biophys. Acta *27:* 431–432 (1958).

38 *Frieden, C.:* Glutamic dehydrogenase. I. The effect of coenzyme on the sedimentation velocity and kinetic behavior. J. biol. Chem. *234:* 809–814 (1959).

39 *Frieden, C.:* Glutamic dehydrogenase. II. The effect of various nucleotides on the association-dissociation and kinetic properties. J. biol. Chem. *234:* 815–820 (1959).

40 *Frieden, C.:* Glutamic dehydrogenase. V. The relation of enzyme structure to the catalytic function. J. biol. Chem. *238:* 3286–3299 (1963).

41 *Frieden, C.:* Treatment of enzyme kinetic data. I. The effect of modifiers on the kinetic parameters of single substrate enzymes. J. biol. Chem. *239:* 3522–3531 (1964).

42 *Frieden, C.:* Treatment of enzyme kinetic data. II. The multisite case: comparison of allosteric models and a possible new mechanism. J. biol. Chem. *242:* 4045–4052 (1967).

43 *Gerhart, J.C. and Pardee, A.B.:* Separation of feedback inhibition from activity of aspartate transcarbamylase (ATCase). Fed. Proc. *20:* 224 (1961).

44 *Gerhart, J.C. and Pardee, A.B.:* The enzymology of control by feedback inhibition. J. biol. Chem. *237:* 891–896 (1962).

45 *Gerhart, J.C. and Pardee, A.B.:* The effect of the feedback inhibition, CTP, on subunit interactions in aspartate transcarbamylase. Cold Spr. Harb. Symp. quant. Biol. *28:* 491–496 (1963).

46 *Gerhart, J.C. and Schachman, H.K.:* Distinct subunits for the regulation and catalytic activity of aspartate transcarbamylase. Biochemistry *4:* 1054–1062 (1965).

47 *Gest, H. and Mandelstam, J.:* Heat denaturation of B-galactosidase: A possible approach to the problem of catabolite repression and its site of action. Nature, Lond. *211:* 72–73 (1966).

48 *Goldstein, A.:* The mechanism of enzyme-inhibitor-substrate reactions. Illustrated by the cholinesterase-physostigmine-acetylcholine system. J. gen. Physiol. *27:* 529–580 (1944).

49 *Graves, D.J.; Fischer, E.H., and Krebs, E.G.:* Specificity studies on muscle phosphorylase phosphatase. J. biol. Chem. *235:* 805–809 (1960).

50 *Green, A.A. and Cori, G.T.:* Crystalline muscle phosphorylase. I. Preparation, properties, and molecular weight. J. biol. Chem. *151:* 21–29 (1943).

51 *Grisolia, S.; Fernandez, M.; Amelunxen, R., and Quijada, C.L.:* Glutamate dehydrogenase inactivation by reduced nicotinamide-adenine dinucleotide phosphate. Biochem. J. *85:* 568–576 (1962).

52 *Grisolia, S.:* The catalytic environment and its biological implications. Physiol. Rev. *44:* 657–712 (1964).

53 *Helmreich, E. and Cori, C.F.:* The role of adenylic acid in the activation of phosphory-lase. Proc. nat. Acad. Sci., Wash. *51:* 131–138 (1964).

54 *Huang, C.Y. and Frieden, C.:* Rates of GDP induced and GTP-induced depolymeriza-tion and isomerization of the bovine liver glutamate dehydrogenase-coenzyme com-plex: A possible controlling factor in metabolic regulation. Proc. nat. Acad. Sci., Wash. *64:* 338–344 (1969).

55 *Inagaki, M.:* Denaturation and inactivation of enzyme proteins. XII. Thermal inactiva-tion and denaturation of glutamic acid dehydrogenase and the effect of its coenzyme on these processes. J. Biochem. *46:* 1001–1010 (1959).

56 *Iwatsubo, M. and Pantaloni, D.:* Mécanisme d'action de la L-glutamate déshydrogénase et rôle des effecteurs ADP and GTP. Etude de la phase initiale rapide. C.R. Acad. Sci. Se. D. *264:* 1200–1203 (1967).

57 *Jacob, F. and Monod, J.:* Genetic regulatory mechanisms in the synthesis of proteins. J. molec. Biol. *3:* 318–356 (1961).

58 *Jensen, R.A.; Nasser, D.S., and Nester, E.W.:* Comparative control of a branch-point enzyme in microorganisms. J. Bact. *94:* 1582–1593 (1967).

59 *Johnson, F.H.; Brown, D., and Marsland, D.:* A basic mechanism in the biological effects of temperature, pressure and narcotics. Science *95:* 200–203 (1942).

60 *Johnson, F.H.; Brown, D., and Marsland, D.:* Pressure reversal of the action of certain narcotics. J. cell. comp. Physiol. *20:* 269–276 (1942).

61 *Johnson, F.H.; Eyring, H., and Kearns, W.:* A quantitative theory of synergism and antagonism among diverse inhibitors, with special reference to sulfanilamide and ure-thane. Arch. Biochem. *3:* 1–31 (1943–44).

62 *Johnson, F.H.; Eyring, H.; Steblay, R.; Chaplin, H.; Huber, C., and Gherardi, G.:* The nature and control of reactions in bioluminescence with special reference to the mechanism of reversible and irreversible inhibitions by hydrogen and hydroxyl ions, temperature, pressure, alcohol, urethane, and sulfanilamide in bacteria. J. gen. Physiol. *28:* 463–537 (1944–45).

63 *Johnson, F.H. and Schneyer, L.:* The quinine inhibition of bacterial luminescence. Amer. J. trop. Med. *24:* 163–175 (1944).

64 *Katsunuma, T.; Temma, M., and Katunuma, N.:* Allosteric nature of a glutaminase isozyme in rat liver. Biochem. biophys. Res. Comm. *32:* 433–437 (1968).

65 *Keller, P.J. and Cori, G.T.:* Enzymic conversion of phosphorylase *a* to phosphorylase *b.* Biochim. biophys. Acta *12:* 235–238 (1953).

66 *Kirtley, M.E. and Koshland, D.E., Jr.:* Models for cooperative effects in proteins con-taining subunits. Effects of interacting ligands. J. biol. Chem. *242:* 4192–4206 (1967).

67 *Koshland, D.E., Jr.:* Application of a theory of enzyme specificity to protein synthesis. Proc. nat. Acad. Sci., Wash. *44:* 98–104 (1958).

68 *Koshland, D.E., Jr. and Neet, K.E.:* The catalytic and regulatory properties of en-zymes. Ann. Rev. Biochem. *37:* 359–410 (1968).

69 *Koshland, D.E., Jr.; Nemethy, G., and Filmer, D.:* Comparison of experimental binding data and theoretical models in proteins containing subunits. Biochemistry *5:* 365–385 (1966).

70 *Kovach, J.S.; Berberich, M.A.; Venetianer, P., and Goldberger, R.F.:* Repression of the histidine operon: Effect of the first enzyme on the kinetics of repression. J. Bact. *97:* 1283–1290 (1969).

71 *Kovach, J.S.; Phang, J.M.; Ference, M., and Goldberger, R.F.:* Studies on repression of the histidine operon. II. The role of the first enzyme in control of the histidine system. Proc. nat. Acad. Sci., Wash. *63:* 481–488 (1969).

72 *Krebs, E.G. and Fischer, E.H.:* The phosphorylase *b* to *a* converting enzyme of rabbit skeletal muscle. Biochim. biophys. Acta *20:* 150–157 (1956).

73 *Krebs, E.G. and Fischer, E.H.:* The role of metals in the activation muscle phosphory-lase. Ann. NY Acad. Sci. *88:* 378–384 (1960).

74 *Krebs, E.G. and Fischer, E.H.:* Molecular properties and transformations of glycogen phosphorylase in animal tissues. Adv. Enzymol. *24:* 263–358 (1962).

75 *Krebs, E.G.; Graves, D.H., and Fischer, E.H.:* Factors affecting the activity of muscle phosphorylase *b* kinase. J. biol. Chem. *234:* 2867–2873 (1959).

76 *Krebs, E.G.; Kent, A.B., and Fischer, E.H.:* The muscle phosphorylase *b* kinase re-action. J. biol. Chem. *231:* 73–83 (1958).

77 *Kuo, J.F. and Greengard, P.:* An adenosine 3',5'-monophosphate-dependent protein kinase from escherichia coli. J. biol. Chem. *244:* 3417–3419 (1969).

78 *Lieb, M.:* Allosteric properties of the repressor. J. molec. Biol. *39:* 379–382 (1969).

79 *Linn, T.C.; Pettit, F.H.; Hucho, F., and Reed, L.J.:* Alpha keto acid dehydrogenase complexes. XI. Comparative studies of regulatory properties of the pyruvate dehydro-genase complexes from kidney, heart, and liver mitochondria. Proc. nat. Acad. Sci., Wash. *64:* 227–234 (1969).

80 *Linn, T.C.; Pettit, F.H., and Reed, L.J.:* Alpha keto acid dehydrogenase complex from beef kidney mitochondria by phosphorylation and dephosphorylation. Proc. nat. Acad. Sci., Wash. *62:* 234–241 (1969).

81 *Lorenson, M.Y. and Mansour, T.E.:* Studies on heart phosphofructokinase binding properties of native enzyme and of enzyme desensitized to allosteric control. J. biol. Chem. *244:* 6420–6431 (1969).

82 *Maeba, P. and Sanwal, B.D.:* The allosteric threonine deaminase of Salmonella. Kinetic model for the native enzyme. Biochemistry *5:* 525–535 (1966).

83 *Mankovitz, R. and Segal, H.L.:* Dissociation produced loss of regulatory control of homoserine dehydrogenase of rhodospirillum rubrum. II. Some properties of the re-gulatable and nonregulatable forms. Biochemistry *8:* 3765–3767 (1969).

84 *Martin, R.G.:* The first enzyme in histidine biosynthesis: The nature of feedback inhibition by histidine. J. biol. Chem. *238:* 257–268 (1963).

85 *Masters, C.J.; Sheedy, R.J.; Winzor, D.J., and Nichol, L.W.:* Reversible adsorption of enzymes as a possible allosteric control mechanism. Biochem. J. *112:* 806–808 (1969).

86 *Mehler, A.H. and Mitna, S.K.:* The activation of arginyl transfer ribonucleic acid syn-thetase by transfer ribonucleic acid. J. biol. Chem. *242:* 5495–5499 (1967).

87 *Messenguy, F. and Wiame, J.M.:* The control of ornithinetranscarbamylase activity by arginase in saccharomyces cerevisial. FEBS Letters *3:* 47–49 (1969).

88 *Metzger, B.; Helmreich, E., and Glaser, L.:* The mechanism of activation of skeletal muscle phosphorylase A by glycogen. Biochemistry *57:* 994–1001 (1967).

89 *Miyamoto, E.; Kuo, T.F., and Greengard, P.:* Cyclic nucleotide-dependent protein kinases. III. Purification and properties of adenosine 3',5'-monophosphate-dependent protein kinase from bovine brain. J. biol. Chem. *244:* 6395–6402 (1969).

90 *Monod, J.; Changeux, J.P., and Jacob, F.:* Allosteric proteins and cellular control systems. J. molec. Biol. *6:* 306–329 (1963).

91 *Monod, J.; Wyman, J., and Changeux, J.P.:* On the nature of allosteric transitions: a plausible model. J. molec. Biol. *12:* 88–116 (1965).

92 *Muller-Eberhard, J.:* Complement. Ann. Rev. Biochem. *38:* 398–414 (1969).

93 *Nester, E.W. and Jensen, R.A.:* Control of aromatic acid biosynthesis in B. subtilis: sequential feedback inhibition. J. Bact. *91:* 1594–1598 (1966).

94 *Nichol, L.W.; Jackson, W.J., and Winzor, D.J.:* A theoretical study of the binding of small molecules to a polymerizing protein system. A model for allosteric effects. Bio-chemistry *6:* 2449–2456 (1967).

95 *Nierlich, D.P. and Magasanik, B.:* Regulation of purine ribonucleotide synthesis by end product inhibition. J. biol. Chem. *240:* 358–365 (1965).

96 *O'Donovan, G.A. and Ingraham, J.L.:* Cold-sensitive mutants of E. coli resulting from increased feedback inhibition. Proc. nat. Acad. Sci., Wash. *54:* 451–457 (1965).

97 *Ohta, T.; Shimada, I., and Imahoni, K.:* Conformational change of tyrosyl-RNA synthetase induced by its specific transfer RNA. J. molec. Biol. *26:* 519–524 (1967).

98 *Ottesen, M.:* Induction of biological activity by limited proteolysis. Ann. Rev. Biochem. *36:* 55–76 (1967).

99 *Patte, J.C.; Lebras, G.; Loviny, T., and Cohen, G.N.:* Rétro-inhibition et répression de l'homosérine déshydrogenase d'escherichia coli. Biochim. biophys. Acta *67:* 16–30 (1963).

100 *Paulus, H. and Gray, E.:* Multivalent feedback inhibition of aspartokinase in bacillus polymyxa. J. biol. Chem. *239:* 4008–4009 (1964).

101 *Rechcigl, M., Jr.:* In vivo turnover and its role in the metabolic regulation of enzyme levels. Enzymologia *34:* 23–39 (1968).

102 *Rechcigl, M., Jr. and Heston, W.E.:* Genetic regulation of enzyme activity in mammalian system by the alteration of the rates of enzyme degradation. Biochem. biophys. Res. Commun. *27:* 119–124 (1967).

103 *Riggs, A.D. and Bourgeois, S.:* On the assay, isolation and characterization of the *lac* repressor. J. molec. Biol. *34:* 361–364 (1968).

104 *Roberts, R.B.; Abelson, P.H.; Cowie, D.B.; Bolton, E.T., and Britton, R.J.:* Studies on the biosynthesis in E. coli – Carnegie Inst. Wash. D.C. Publ. No. 607 (1955).

105 *Robison, G.A.; Butcher, R.W., and Sutherland, E.W.:* Cyclic AMP. Ann. Rev. Biochem. *37:* 149–174 (1968).

106 *Rosen, S.M. and Rosen, O.M.:* The effect of iodination upon the catalytic and regulatory activities of fructose 1,6-diphosphatase. Biochemistry *6:* 2094–2097 (1967).

107 *Rubin, M.M. and Changeux, J.P.:* On the nature of allosteric transitions: implications of non-exclusive ligand binding. J. molec. Biol. *21:* 265–274 (1966).

108 *Sanwal, B.D.; Stachow, C.S., and Cook, R.A.:* A kinetic model for the mechanism of allosteric activation of nicotinamide-adenine dinucleotide-specific isocitric dehydrogenase. Biochemistry *4:* 410–421 (1965).

109 *Schachman, H.K.:* Considerations of the tertiary structure of proteins. Cold Spr. Harb. Symp. quant. Biol. *28:* 409–430 (1963).

110 *Schimke, R.T.; Sweeney, E.W., and Berlin, C.M.:* The roles of synthesis and degradation in the control of rat liver trytophan pyrrolase. J. biol. Chem. *240:* 322–331 (1965).

111 *Shapiro, B.M.; Kingdon, H.S., and Stadtman, E.R.:* Regulation of glutamine synthetase. VII. Adenyl glutamine synthetase: A new form of the enzyme with altered regulatory and kinetic properties. Proc. nat. Acad. Sci., Wash. *58:* 647–649 (1967).

112 *Shapiro, B.M. and Stadtman, E.R.:* 5'-adenyl-O-tyrosine. The novel phosphodiester residue of adenylylated glutamine synthetase from escherichia coli. J. biol. Chem. *243:* 3769–3771 (1968).

113 *Stadtman, E.R.:* Symposium on multiple forms of enzymes and control mechanisms. II. Enzyme multiplicity and function in the regulation of divergent metabolic pathways. Bact. Rev. *27:* 170–181 (1963).

114 *Stadtman, E.R.:* Allosteric regulation of enzyme activity. Adv. Enzymol. *28:* 41–154 (1966).

115 *Stadtman, E.R.:* The role of multiple enzymes in the regulation of branched metabolic pathways. Ann. NY Acad. Sci. *151:* 516–530 (1968).

116 *Stancel, G.M. and Deal, W.C., Jr.:* Metabolic control and structure of glycolytic enzymes. V. Dissociation of yeast glyceraldehyde-3-PO$_4$-dehydrogenase into subunits by ATP. Biochem. biophys. Res. Comm. *31:* 398–403 (1968).

117 *Steiner, D.F. and Clark, J.L.:* The spontaneous reoxidation of reduced beef and rat proinsulins. Proc. nat. Acad. Sci., Wash. *60:* 622–629 (1968).

118 *Steiner, D.F. and Clark J.L.:* Insulin biosynthesis: Demonstration of 2 proinsulins. Proc. nat. Acad. Sci., Wash. *62:* 278-285 (1969).

119 *Steiner, D.F.; Cunningham, D.; Spigelman, L., and Alen, B.:* Insulin biosynthesis. Evidence for a precursor. Science *157:* 697–700 (1967).

120 *Steiner, D.F. and Oyer, P.E.:* The biosynthesis of insulin and a probable precursor of insulin by a human islet cell adenoma. Proc. nat. Acad. Sci., Wash. *57:* 473–480 (1967).

121 *Sund, H.:* Struktur und Wirkungsweise der Glutaminsäuredehydrogenase. I. Grösse und Gestalt der Glutaminsäuredehydrogenase aus Rinderleber. Acta chem. scand. *17:* 102–106 (1963).

122 *Sund, H.; Pilz, I., and Herbst, M.:* Studies of glutamic dehydrogenase. 5. The X-ray small-angle investigation of beef liver GDH. Europ. J. Biochem. *7:* 517–525 (1969).

123 *Susor, W.A. and Rutter, W.J.:* Some distinctive properties of pyruvate kinase purified from rat liver. Biochem. biophys. Res. Comm. *30:* 14–20 (1968).

124 *Sutherland, E.W. and Rall, T.W.:* The relation of adenosine-3',5'-phosphate and phosphorylase to the actions of catecholamines and other hormones. Pharmacol. Rev. *12:* 265–299 (1960).

125 *Sweeny, J.R. and Fisher, J.R.:* An alternative to allosterism and cooperativity in the interpretation of enzyme kinetic data. Biochemistry *7:* 561–565 (1968).

126 *Taketa, K. and Pogell, B.M.:* Allosteric inhibition of rat liver fructose 1,6-diphosphatase by adenosine 5'-monophosphate. J. biol. Chem. *240:* 651–662 (1965).

127 *Tanaka, T.; Sue, F., and Morumura, H.:* Feed-forward activation and feed-back inhibition of pyruvate kinase type L of rat liver. Biochem. biophys. Res. Comm. *29:* 444–449 (1967).

128 *Tomkins, G.M.; Gelehrter, T.D.; Granner, D.; Martin, D.M., Jr.; Samuels, H.H., and Thompson, E.B.:* Control of specific gene expression in higher organisms. Science *166:* 1474–1480 (1969).

129 *Tomkins, G.M. and Yielding, K.L.:* Regulation of the enzymic activity of glutamic dehydrogenase mediated by changes in its structure. Cold Spr. Harb. Symp. quant. Biol. *26:* 331–341 (1961).

130 *Tomkins, G.M.; Yielding, K.L.; Curran, J.F.; Summers, M.R., and Bitensky, M.W.:* The dependence of the substrate specificity on the conformation of crystalline glutamate dehydrogenase. J. biol. Chem. *240:* 3793–3798 (1965).

131 *Tomkins, G.M.; Yielding, K.L.; Talal, N., and Curran, J.:* Protein structure and biologic regulation. Cold Spr. Harb. Symp. quant. Biol. *28:* 461–471 (1963).

132 *Thompson, W. and Yielding, K.L.:* 8-anilino naphthalene sulfonate binding as a probe for conformational changes induced in GDH by regulatory (allosteric) reagents. Arch. biochem. Biophys. *29:* 399–406 (1968).

133 *Umbarger, H.E.:* Evidence for a negative-feedback mechanism in the biosynthesis of isoleucine. Science *123:* 848 (1956).

134 *Yates, R.A. and Pardee, A.B.:* Pyrimidine biosynthesis in Escherichia coli. J. biol. Chem. *221:* 743–756 (1956).

135 *Wang, J.H.; Shonka, M.L., and Graves, D.J.:* Influence of carbohydrates on phosphorylase structure and activity. I. Activation by preincubation with glycogen. Biochemistry *4:* 2296–2301 (1965).

136 *Wang, J.H. and Graves, D.J.:* Effect of ionic strength on the sedimentation of glycogen phosphorylase a. J. biol. Chem. *238:* 2386–2389 (1963).

137 *Wang, J.H. and Graves, D.J.:* The relationship of the dissociation to the catalytic activity of glycogen phosphorylase a. Biochemistry *3:* 1437–1444 (1964).

138 *Wang, J.H.; Shonka, M.L., and Graves, D.J.:* The effect of glucose on the sedimentation and catalytic activity of glycogen phosphorylase. Biochem. biophys. Res. Comm. *18:* 131–135 (1965).

139 *Weber, G. and Anderson, S.R.:* Multiplicity of binding. Range of validity and practical test of Adair's equation. Biochemistry *4:* 1942–1947 (1965).

140 *Wolff, J.:* The effect of thyroxine on isolated dehydrogenases. II. Sedimentation changes in glutamic dehydrogenase. J. biol. Chem. *237:* 230–235 (1962).

141 *Woolfolk, C.A. and Stadtman, E.R.:* Cumulative feedback inhibition in the multiple end product regulation of glutamine synthetase activity in E. Coli. Biochem. biophys. Res. Comm. *17:* 313–319 (1964).

142 *Wosilait, W.D. and Sutherland, E.W.:* Relationship of epinephrine and glucagon to liver phosphorylase; enzymatic inactivation of liver phosphorylase. J. biol. Chem. *218:* 469–481 (1956).

143 *Wyman, J.:* Heme proteins; *Anson and Edsall* in Adv. Prot. Chem. *4:* 407–531 (1948).

144 *Wyman, J.:* Allosteric effects in hemoglobin. Cold Spr. Harb. Symp. quant. Biol. *28:* 483–489 (1963).

145 *Wyman, J., Jr. and Allen, D.W.:* The problem of the heme interactions in hemoglobin and the basis of the Bohr effect. J. Polymer Sci. *7:* 499–518 (1951).

146 *Yang, S.T. and Deal, W.C., Jr.:* Metabolic control and structure of glycolytic enzyme. VII. Destabilization and inactivation of yeast glyceraldehyde-3-PO_4-dehydrogenase by adenosine phosphates and chymotrypsin. Biochemistry *8:* 2814–2820 (1969).

147 *Yates, R.A. and Pardee, A.B.:* Control of pyrimidine biosynthesis in Escherichia coli by a feedback mechanism. J. biol. Chem. *221:* 757–770 (1956).

148 *Yielding, K.L.:* Persistance of the effects of regulatory reagents on conformation of catalytically inactive glutamate dehydrogenase. Biochem. biophys. Res. Comm. *29:* 424–429 (1967).

149 *Yielding, K.L. and Tomkins, G.M.:* Structural alterations in crystalline glutamatic dehydrogenase induced by steroid hormones. Proc. nat. Acad. Sci., Wash. *46:* 1483–1488 (1960).

150 *Yielding, K.L. and Tomkins, G.M.:* An effect of L-leucine and other essential amino acids on the structure and activity of glutamic dehydrogenase. Proc. nat. Acad. Sci., Wash. *47:* 983–989 (1961).

151 *Yielding, K.L. and Tomkins, G.M.:* Studies on the interaction of steroid hormones with glutamic dehydrogenase. Recent. Progr. Hormone Res. *18:* 467–489 (1962).

152 *Yielding, K.L. and Tomkins, G.M.:* The inhibition of the glutamic dehydrogenase by 1, 10-phenanthroline and its analogs. Biochem. biophys. Acta *62:* 327–331 (1962).

153 *Yielding, K.L. and Tomkins, G.M.:* The regulation of enzyme structure and function by steroid hormones. Amer. J. Med. *33:* 1–3 (1962).

154 *Yielding, K.L.; Tomkins, G.M.; Holt, B.B.; Summers, M.R., and Gaudin, D.:* Alterations in the structure and activity of glutamate dehydrogenase as a model for enzyme regulation. Excerpta medica international congress series No. 132, pp. 503–507 (1966). (Proc. of the 2nd internat. congr. on hormonal steroids.)

155 *Yielding, K.L.; Tomkins, G.M., and Trundle, D.:* Metal ion requirement for pyridine nucleotide induced disaggregation of glutamic dehydrogenase. Biochem. biophys. Acta *77:* 703–705 (1963).

Author's address: Dr. *K.L. Yielding,* Laboratory of Molecular Biology, University of Alabama in Birmingham, The Medical Center, *Birmingham, AL 35233* (USA)

Enzyme Synthesis and Degradation in Mammalian Systems, pp. 165–199
(Karger, Basel 1971)

Regulatory Mechanisms of Enzyme Synthesis: Enzyme Induction[1]

Thomas D. Gelehrter

Division of Medical Genetics, Department of Medicine
University of Washington, Seattle, Wash.

Contents

I. Introduction

Multicellular differentiated organisms are characterized by the presence of diverse cell types, which differ quantitatively and qualitatively in their complement of structural and catalytic proteins. Since all somatic cells in a multicellular organism appear to possess the same set of genetic information (65, 130), these differences in protein complement must reflect the variable expression of a constant genetic complement. Thus each cell must express only a part of its genetic information, and different cell types express different portions of their genome. Similarly, a given cell type may express different components of its genetic information during various stages of its differentiation and development, and during different phases of the cell cycle; and the adult differentiated cell

1 This review was aided by US Public Health Service Special Fellowship 1 FO3
AM43624-01 from the National Institute of Arthritis and Metabolic Diseases.

may modify its pattern of gene expression in response to a variety of environmental stimuli. Thus, an understanding of the regulation of specific gene expression is fundamental to our understanding of development and differentiation, and to the mechanisms by which a cell responds to a variety of physiologic and pharmacologic stimuli.

II. Regulatory Mechanisms of Protein Synthesis in Microorganisms

In recent years a great deal has been learned about the physiologic regulation of gene expression in the relatively simpler prokaryotic organisms, the bacteria and their viruses. Quantitative control of specific enzyme activity operates at two levels: first, by regulation of enzyme activity (discussed in the preceding chapter of this volume); and second, by regulation of enzyme biosynthesis. Genetic analysis of mutations affecting protein synthesis revealed the existence of a 'double genetic determinism' (77) regulating induced enzyme synthesis in bacteria. Mutations affecting structural genes, whose nucleotide sequences encode information for the amino acid sequences of specific polypeptides, result in an alteration in the structure of a protein without affecting its rate of synthesis. Mutations in distinct and separate regulatory genes result in alterations in the rate of synthesis of a protein without affecting its structure. (It should be noted that mutations in a structural gene could theoretically alter the rate of synthesis or degradation of the protein and thereby affect the amount of the protein present in the cell [76]. In fact, *Yoshida* [199, 200] has demonstrated this phenomenon in man.)

The studies of *Jacob and Monod* and their coll. (77) on the induction of the *lac* operon in *E. coli* have led to a model for the regulation of induced enzyme synthesis, which is presented in simplified outline form below, and in figure 1. Details of this and other models are discussed in two excellent, critical reviews (34, 116).

The synthesis of the proteins (E_1, E_2 and E_3) encoded by a given operon (defined as a cluster of contiguous structural genes (S_{G_1}, S_{G_2}, S_{G_3}) showing coordinate expression, and their associated controlling sites (R_G and O) is regulated by the concentration of labile, polycistronic messenger ribonucleic acid (mRNA) coding for these proteins. The concentration of this mRNA is determined by the rate at which the structural genes (S_{G_1}, S_{G_2}, S_{G_3}) are transcribed into RNA. This in turn is controlled by the repressor (R), an allosteric protein which is the product of a separate regulatory gene (R_G). The repressor binds specifically to the operator (O), a region of DNA at one extremity of the operon, so as to prohibit transcription of its associated structural genes (fig. 1a). Inducers are effector molecules which can interact with the repressor in such a way as to alter its configuration and prevent its binding to the operator, thereby allowing transcription of the structural genes to occur (fig. 1b). In a repressible, as opposed to inducible, operon, effector molecules, called co-repressors, activate rather than inhibit the repressor, and therefore turn off mRNA synthesis.

Although this model was based initially on indirect though impressive genetic experiments, it has been confirmed recently by direct biochemical studies (50, 156), in which the repressor has been isolated, characterized, and shown to bind specifically to both the inducers and the operator region of the *lac* operon.

The most important features of the model are that regulation of specific protein synthesis operates at the genetic level, and that it operates by controlling the rate of synthesis of a labile polycistronic mRNA. Thus a gene is active, i.e. being expressed, only when its corresponding mRNA is being synthesized. The basic ideas of the operon model have been found applicable in a variety of inducible systems studied in microorganisms (34, 116), but many of the details of regulation remain to be clarified. Furthermore, the details of regulation of one operon need not apply universally to all other operons. Studies of the biosynthesis of histidine (159) and of other repressible biosynthetic pathways have suggested that the primary site of regulation may be at the level of translation of the mRNA rather than at the level of transcription (116, 177, 185). Furthermore, in the arabinose pathway (174) there is evidence that regulation is positive rather than negative as it is in the lactose system.

Not surprisingly, however, the elegance and clarity of the *Jacob and Monod* model have led to its extensive application to the problems of gene regulation and especially enzyme induction in higher organisms. It is essential to consider whether such an application is justified, or useful. In both eukaryotic cells as well as prokaryotic cells, it is known that DNA is the hereditary material, and that genetic information is expressed via the transcription of an intermediate mRNA which is then translated by ribosomes, transfer RNA, etc. In addition, the genetic code is essentially the same in mammalian and bacterial cells (111). Furthermore, there is evidence in Diptera for direct gene activation (18), and in at least fungi (27, 63) and maize (131) for operon-like units. In fungi, these operons may be under either positive or negative control (63).

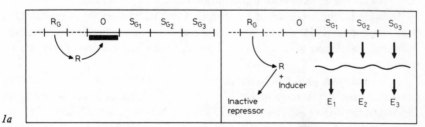

Fig. 1a and 1b. Schematic model for enzyme induction in bacterial cells. Modified from (77). See text for explanation.

III. Differences Between Microorganisms and Mammals
Relevant to Gene Regulation

The problems of adaptation posed to microorganisms and to the cells of higher organisms are quite different (121, 185). The rapidly dividing free-living bacterium must adapt rapidly to wide swings in its environment in order to maintain its rapid growth in changing conditions. Selection is for maximal growth rate. In the differentiated metazoan organism, various physiologic functions are more or less permanently delegated to certain cells. Many of these cell types divide infrequently or not at all. The environment of individual cells within the organism is relatively stable. Thus selection might be expected to favor optimal function at the cellular level rather than growth, and intercellular coordination of physiologic functions at the level of the organism.

Furthermore, there are fundamental differences between microorganisms and higher animals with regard to the organization of the genetic material and to macromolecular synthesis.

1. Unlike the bacterial chromosome, which is a single molecule of naked DNA, the chromosomes of higher organisms are highly complex structures containing RNA and various proteins in addition to DNA (125).

2. Mammalian cells contain at least a thousand times more DNA than bacteria (9), organized in several chromosomes within a distinct cellular structure, the nucleus.

3. The DNA of higher organisms contains large portions of repetitive nucleotide sequences (9).

4. Large portions of the genome are more or less permanently inactivated; and in mammalian somatic cells an entire chromosome (the X-chromosome) is apparently inactivated (110).

5. There is as yet no evidence in mammals for the physical organization of genes for related functions in operons.

6. Consistent with the above, only monocistronic mRNA's have been found in the polyribosomes of animal cells (97), whereas in bacteria, polycistronic mRNA's are the units of both transcription and translation (115).

7. In bacteria, the processes of transcription and translation are physically (12) and functionally (75) related, whereas in higher organisms the sites of transcription and translation are physically separate, in the nucleus and cytoplasm respectively.

8. In animal cells, unlike bacteria, large portions of the genome are transcribed into extremely labile, large molecules of RNA which are degraded without ever leaving the nucleus (74, 164, 165, 173). The function of this RNA is unknown. Some of it may serve as precursor to functional mRNA; and an intranuclear regulatory role has also been suggested recently (10). In any case, these data would suggest that models of regulation at the level of gene transcription alone

would be insufficient to explain the fine control of protein synthesis in animal cells.

9. Finally, in rapidly dividing bacteria, mRNA's are very labile, while ribosomes and proteins essentially do not turn over (72), whereas in animal cells, the mRNA population appears relatively more heterogeneous and more stable (148, 155, 194), while ribosomes and proteins do turn over (7, 109, 151, 168). In fact, control of protein turnover is obviously an important regulatory mechanism in animal cells ([168], see Chapter by *Rechcigl,* pp. 236–310).

Taken together, these considerations suggest that regulation of gene expression in animal cells might be considerably more complex than in bacteria, involve different steps in the processing of genetic information, and might affect any of the following physiologic events: (1) Synthesis of an mRNA intermediate complimentary to the DNA template, i.e. 'transcription', (2) selective stabilization or protection from degradation of a sequence of the RNA within the nucleus, (3) transport of such RNA from the nucleus to the cytoplasm, (4) association of the RNA with ribosomes to make polysomal aggregates, (5) the initiation, elongation, and termination of peptide synthesis, i.e. 'translation', (6) the release and folding of peptide products, (7) association of peptide subunits, and finally, (8) the degradation of the finished protein. The complexity of gene expression in animal cells and the many potential regulatory sites should caution against the uncritical application of models derived from lower organisms.

IV. Enzyme Induction in Higher Organisms

A major experimental approach to the problem of the modulation of gene expression in animal cells, as in microorganisms, has been the study of enzyme induction. We may define enzyme induction as an adaptive increase in the number of molecules of a specific enzyme secondary to either an increase in the rate of synthesis of the enzyme or a decrease in the rate of its degradation. As noted earlier, alteration of the rate of degradation of a protein represents an important control mechanism in animal cells and will be the subject of a subsequent chapter. The significance to enzyme induction, e.g. hepatic tryptophan pyrrolase, of stabilization of enzymes by substrates or cofactors has been well reviewed recently by *Greengard* (58) and by *Schimke* (168, 170). Therefore, for the purposes of this discussion I will consider enzyme induction to mean a selective increase in the rate of synthesis of a specific enzyme, a definition analogous to that accepted in microbial systems (19).

A wide variety of experimental systems has been employed in the study of enzyme induction, including the intact animal (83, 90), the isolated perfused organ (5, 51, 66), organ culture (124, 193), and tissue culture (169, 182). Similarly, a great range of inducing agents has been used including hormones

(99, 137), diet (143), drugs (20, 43), virus infection (128), substrates (102, 167), vitamins (69, 138), gasses (67) and even mechanical obstruction (81).

The 'induction' of a great many enzymes has been studied. In the case of most of these only an increase in enzyme activity has been demonstrated, which must be considered at best only a first approximation to establishing a change in cellular content of the enzyme. Inhibition of enzyme induction by inhibitors of protein synthesis such as puromycin and cycloheximide establishes only that protein synthesis is required for induction to occur. It does not establish enhanced *de novo* synthesis of the enzyme under study, although it is frequently interpreted in this way. This distinction is illustrated by the careful studies of *Griffin and Cox* on the induction by prednisolone of alkaline phosphatase in HeLa cells. When HeLa cells are grown in the presence of the glucocorticoid, prednisolone, alkaline phosphatase activity, after a lag of 15 to 24 h, increases to 5 to 15 times the control level. This increase is inhibited by puromycin, and by Actinomycin D, if the latter is added during the first 8 h of steroid treatment (60). Careful immunochemical and enzymologic studies have established, however, that the concentration of alkaline phosphatase protein is the same in fully induced and basal cells, but that the induced enzyme differs in its enzymatic properties (61). The authors suggest that the hormone mediates a conformational change in the enzyme during its synthesis resulting in increased catalytic activity, without increasing the rate of synthesis of the enzyme.

In relatively few systems has there been convincing evidence provided for an increase in the rate of synthesis of enzyme molecules. Furthermore, few experimental systems have been studied in sufficient depth as yet to enhance our understanding of the molecular mechanisms underlying enzyme induction. Therefore, rather than attempting to review superficially the many reports of putative enzyme induction, I have chosen to discuss in some detail two major, but different, experimental models of enzyme induction, the induction of hepatic tyrosine aminotransferase (TAT) by glucocorticoids, and the induction of hepatic microsomal aryl hydroxylase by polycyclic hydrocarbons, in order (1) to try to draw some inferences regarding the basic mechanism of enzyme induction, and (2) to illustrate some experimental and conceptual approaches to the problem of specific gene regulation.

The reader is referred to other chapters in this volume for a detailed discussion of some other well-studied examples of enzyme induction and of the ontogeny of inducible enzymes. Also, *Knox and Greengard* (91) have extensively reviewed the induction of several enzymes of nitrogen metabolism (including hepatic tryptophan pyrrolase) in a recent symposium. Other major examples of enzyme induction which cannot be discussed here include (1) the induction of hepatic alanine aminotransferase by glucocorticoids (172), (2) hepatic and HeLa cell arginase by arginine (166, 167), (3) HeLa cell alkaline phosphatase by prednisolone (60), (4) embryonic chick retina glutamine synthetase by glucocorticoids (124, 154), and (5) hepatic δ-aminolevulinic acid synthetase by sex steroids and barbiturates (52, 117). In addition, the specific induction of some non-catalytic proteins has also been investigated, including (1) the induction of chick oviduct avidin synthesis by progesterone (139, 140), (2) hepatic and HeLa cell apo-ferritin by iron (14, 30), (3) rat mammary gland casein by insulin, hydrocortisone, and prolactin (191), (4) hemoglobin by erythropoetin (93, 142) and (5) zinc-transport by glucocorticoids (23).

V. Hormonal Induction of Hepatic Tyrosine Aminotransferase

The administration of the glucocorticoid hormone, cortisol, to rats results in an increased activity of several hepatic enzymes, including tryptophan pyrrolase (or tryptophan oxygenase) (89), alanine alpha-ketoglutarate transaminase (172), tyrosine aminotransferase (104), tryptophan a-ketoglutarate transaminase (16), several urea cycle enzymes (166) and serine dehydrase (143). The effects of steroid are selective, however, in that the activity of several other transaminases, including aspartic-a-ketoglutarate transaminase (157) and histidine-pyruvate- and phenylalanine-pyruvate-transaminase (105) are not increased. The qualitative and quantitative enzymatic response to steroid administration is markedly affected by the turnover rates of the enzymes themselves (7), the nature, dose, route of administration and duration of administration of the steroid, and by the age, and hormonal and nutritional state of the recipient animal. Thus it is not surprising that an almost bewildering array of observations has been reported concerning enzyme induction in intact animals (see review by *Nichols and Rosen* (137).

Since the initial report of *Lin and Knox* (104), the hormonal induction of hepatic tyrosine aminotransferase, E.C.2.6.1.5, (TAT) has been extensively studied as a model of enzyme induction in higher organisms. TAT catalyses the first and rate-limiting step in the catabolism of tyrosine to fumarate and aceto-acetate (104). The enzyme is found primarily in liver, and to a lesser extent in kidney (104); 85–90 % of the hepatic activity is localized in the cytosol fraction (83). TAT is also found in the brain, predominantly associated with the mitochondrial fraction; its activity can be altered by agents which affect catecholamine metabolism (49).

The intraperitoneal administration of hydrocortisone (e.g. 2–5 mg/100 g rat) to intact or adrenalectomized rats results, after a lag of 1–2 h, in a rapid, 4- to 5-fold increase in the activity of hepatic TAT which is maximal within about 6 h, and then declines with a half-life of about 2.5 h (87). The increase in TAT activity can be inhibited by inhibitors of protein synthesis (87), and it has been demonstrated immunochemically that the increase can be attributed entirely to a steroid-mediated increase in the rate of TAT synthesis (83, 87) (fig. 2). Approximately 6 h after administration of the incuding steroid, the rate of TAT synthesis falls rapidly to the basal level, presumably due to metabolism and/or excretion of the hormone. Repeated administration of steroid prevents the decline in enzyme activity (64). Total hepatic protein synthesis is also increased by steroids but to a considerably smaller degree (35).

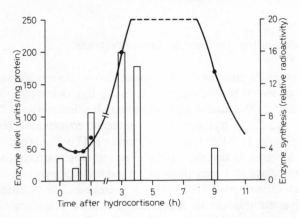

Fig. 2. Immunochemical analysis of the rate of synthesis of hepatic TAT after induction with hydrocortisone. Adrenalectomized rats were given hydrocortisone (2 mg/100 g) at zero time. Three or more animals at each point were given 30 μCi of ^{14}C-leucine 10 min before they were killed and the enzyme was then isolated immunochemically. The points represent enzyme activity plotted against the left ordinate. The bars represent mean values of relative enzyme radioactivity (cpm in isolated enzyme x 10^3/cpm/mg total soluble protein) and are plotted against the right ordinate. Figure 2 in *Kenney et al.* (87).

The steroid-mediated induction of TAT is prevented by the prior or simultaneous administration of actinomycin D (AMD) (39, 56), an inhibitor of DNA-directed RNA synthesis (153). Furthermore, several groups have demonstrated that the cortisol-stimulated induction of TAT is associated with rather striking increases in the activity of RNA polymerase, and in nuclear RNA labeling which precede the increase in TAT activity (5, 6, 39, 59). Total hepatic RNA is also increased in amount (35). These findings have been interpreted to indicate that steroids activate genes directly to stimulate the synthesis of mRNA's for the induced enzymes (5, 25, 82). In other words, the mechanism of enzyme induction in the rat is thought to be analogous to that in *E. coli.*

The available data do not support such a simple interpretation, however. First of all, it is important to emphasize that the inhibition of enzyme induction by inhibitors of RNA synthesis means only that RNA synthesis is required for enzyme induction to occur, and not that enhanced mRNA synthesis is the mechanism by which induction occurs. Accumulation of mRNA could occur without change in its rate of synthesis, simply by protecting mRNA's from degradation (see below). Furthermore, since the great majority of RNA synthesized in the animal cell nucleus is degraded without entering the cytoplasm, it is probably meaningless to draw inferences about specific, functional (i.e. polysome-associated) mRNA's by examining the nuclear products of transcription (165).

Secondly, the magnitude of the increase in RNA synthesis is too large to be interpreted simply as the increased synthesis of mRNA's for the induced enzymes. Indeed, density gradient centrifugation, base composition analysis, and differential phenol-extraction of the RNA demonstrates that the synthesis of all classes of RNA, ribosomal (rRNA) and transfer RNA (tRNA) as well as putative mRNA, appears to be stimulated by steroids (88). Thus one must explain the apparently specific induction of a limited number of enzymes in terms of an apparently non-specific stimulation of the synthesis of all classes of RNA by glucocorticoids. Several laboratories (28, 36, 96) have claimed to demonstrate directly the cortisol-stimulated synthesis of specific TAT-mRNA, but these reports must be viewed with some caution. It is as yet impossible in animal cells to define specific mRNA's by hybridization to the DNA of specific genes (with the exception of integrated viral genes) as has been accomplished in microorganisms (2, 70); and only in the case of hemoglobin (98) has a mammalian mRNA definitely been identified by its template activity *in vitro*.

Thirdly, it is possible that the marked increase in RNA synthesis is a separate effect of the steroid treatment, unrelated to the enzyme induction. In fact, the 2 effects of steroids can be dissociated in the newborn rat. Prior to the 13th day of postnatal life, glucocorticoids stimulate TAT synthesis as in the adult, but have no effect on RNA polymerase activity (6). Furthermore, as will be discussed below, steroid-mediated induction of TAT in tissue culture (45) and organ culture (193) is not associated with any increase in overall RNA synthesis.

Two lines of evidence point to regulation of TAT synthesis at some steps in protein synthesis beyond gene transcription. The first is the 'paradoxical' superinduction of TAT by actinomycin D, reported by *Garren* and co-workers (40). As described above, about 6 h after cortisol administration, at which time the rate of general protein synthesis is increasing, the rate of TAT synthesis (and tryptophan pyrrolase as well) falls rapidly to the basal rate. The administration of AMD (or 5-fluorouracil) at or after this time paradoxically 'reactivates' the synthesis of TAT and tryptophan pyrrolase, and increases the level of these enzymes. Puromycin, an inhibitor of protein synthesis, prevents the superinduction caused by AMD. These data argue that the decline in TAT activity observed 6 h or more after steroid administration is not due to the disappearance of a labile TAT mRNA, but rather to the specific repression of the translation of a stable TAT template. These authors suggested that TAT synthesis might be regulated at the level of translation by a cytoplasmic repressor. As will be discussed later, such 'paradoxical' effects of AMD may be quite common in animal cells. The interpretation of *Garren*'s data has been made somewhat difficult, however, by the difficulty in reproducing the effect in intact animals (120), and by the fact that inhibitors of protein and RNA synthesis may interfere with the degradation of hepatic TAT *in vivo* (64, 85, 101, 168), although they do not interfere with the degradation of tryptophan pyrrolase.

A second line of evidence suggesting control at a translational (or strictly speaking, post-transcriptional) level comes from observations of the repression of TAT synthesis by growth hormone (84). When adrenalectomized rats are subjected to severe stress, TAT levels fall with a $t_{1/2}$ of 2.5 h, and immunochemical studies have confirmed that TAT synthesis has virtually stopped under these conditions. The repression is specific since general protein synthesis is unaffected. AMD does not affect the basal level of TAT (confirming *Garren*'s finding that TAT mRNA must be relatively stable), but does inhibit the stress-induced repression. Therefore, the repression must act by inhibiting translation of the stable TAT template, though initiation of the repression requires RNA synthesis. Stress does not cause repression of TAT in the hypophysectomized rat and small doses of growth hormone (somatotropin) produce the effect, although the response is very erratic and may be indirect (84).

A variety of other factors are also known to affect TAT activity level in rat liver. Casein hydrolysate causes a several-fold increase in TAT level (but not tryptophan pyrrolase) which is inhibited by AMD and puromycin, and is additive to the effects of cortisol (158).

The pancreatic hormones, insulin and glucagon, both cause a rapid, though short-lived (maximal effect at 2 to 3 h after administration) induction of TAT activity (73). The rate of synthesis of TAT, assayed immunochemically, increases within 30 min of hormone treatment and entirely accounts for the increase in TAT activity. Although glucagon and insulin have antagonistic biologic effects, both hormones induce TAT with the same kinetics, and are both inhibited by AMD. The effects of insulin and glucagon are not additive with each other, but each is additive with hydrocortisone.

The $N^6, O^{2'}$-dibutyryl analogue of cyclic adenosine $3',5'$-monophosphate (cAMP) also increases the rate of synthesis of TAT with the same kinetics as glucagon and insulin. The effect of cAMP on TAT synthesis is additive to that of hydrocortisone, but not to insulin or glucagon. Cyclic AMP also increases the activity of hepatic phosphoenolpyruvate carboxykinase, which is unaffected by hydrocortisone, but not that of tryptophan pyrrolase, which is induced by hydrocortisone (192).

Finally, TAT activity shows marked circadian rhythmicity, even in the adrenalectomized or hypophysectomized rat (17, 197).

VI. Hepatoma Cells in Tissue Culture

Although a great deal has been learned from studies such as those described above, the difficulties of experimentation in the 'enormously complex environment of the living animal' (86) are readily apparent. The variables of circadian rhythm, nutritional status, the problems of circulation and distribution of inducers and substrates, of uncontrollable changes in extracellular environment, and the influences of extrahepatic tissues all make experimentation in the whole animal a complicated and difficult endeavor. The use of the isolated perfused liver system (66) represents an important step toward a more workable experimental system, as does the short-term organ culture of fetal liver (193). A major step in the development of a flexible, simpler experimental system, however, was the observation that glucocorticoids induced an increase in the activity of TAT in a rat hepatoma in tissue culture (147).

It is obviously of considerable advantage to be able to study enzyme induction in a well-defined, experimentally-controllable and manipulable system. For this reason *Thompson* and his coll. (182) established in continuous tissue culture a line of rat hepatoma cells derived from the Morris hepatoma 7288C (182), which they designated HTC (hepatoma tissue culture). These cells resemble liver cells in morphology, and exhibit several characteristic hepatic constituents such as TAT, alanine aminotransferase and alcohol dehydrogenase, though they lack other characteristic liver proteins such as tryptophan pyrrolase and albumin. They are heteroploid in chromosome constitution with a modal chromosome number of about 65. HTC cells grow in either monolayer or suspension culture with a population-doubling time of about 24 h, and can be cloned with high efficiency (182).

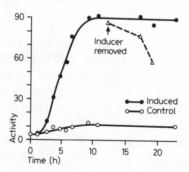

Fig. 3. Induction of TAT in HTC cells by dexamethasone phosphate. A culture of HTC cells, grown to a density of about 8×10^5 cells/ml, was resuspended in fresh medium and divided into two portions. To the first portion, dexamethasone phosphate was added to a final concentration of $5 \times 10^{-5} M$. The other portion was used as a control. Enzyme activity is expressed as milliunits enzyme/mg cell protein. Figure 2 in *Tomkins et al.* (188).

The addition of 10^{-5} *M* dexamethasone phosphate, a synthetic glucocorticoid, to either growing or stationary cultures of HTC cells results in a marked increase in the activity of TAT (fig. 3). After a lag of 1 to 2 h, enzyme activity increases rapidly, reaching a new steady state after 8 to 10 h, 5 to 15 times higher than the basal activity. The induced level of TAT is maintained as long as the inducer is present, but falls rapidly if inducer is removed and the medium replaced with steroid-free medium (182, 188) (fig. 3). The striking similarity of the induction phenomenon *in vivo* and *in vitro* suggests that TAT induction in HTC cells is a valid model for the normal biological control mechanism even though HTC cells are malignant, and clearly different from normal hepatocytes in a number of other respects.

Induction of TAT can be prevented by inhibiting protein synthesis with puromycin or cycloheximide (182), suggesting that enzyme induction involves new enzyme synthesis. To confirm this, TAT was purified from rat liver (71) and antibody prepared to it (53). It could be shown that the steroid-mediated increase in TAT activity is accompanied by an increase in the number of enzyme molecules, and further that the increase in enzyme activity is entirely accounted for by an increase in the rate of synthesis of TAT, as determined by immunochemical techniques (53). Consistent with these observations, it was shown that the turnover of TAT is not affected by the inducing steroid (53, 145); nor, under the conditions of the experiments, does the degradation of TAT require RNA or protein synthesis (3). Thus, TAT degradation is not altered by AMD or cyclohexmide in HTC cells (3) in contrast to the situation reported *in vivo* (85).

The steroidal induction of TAT in HTC cells is quite specific. General protein synthesis is not affected (186), and thus far only one other cellular property, an alteration in cell surface resulting in increased 'stickiness' of the cells, is known to be induced (4). Neither growth nor DNA synthesis (145) are required for induction to occur, and steroids do not affect the growth of TAT cells in the concentrations used (182, 186).

In striking contrast to the findings in rat liver, glucocorticoids do not affect overall RNA synthesis in HTC cells, under conditions in which TAT is induced 5- to 15-fold (45); nor is there any change in the turnover of rapidly-labeled RNA (45), or any increase in RNA polymerase activity (48). These data suggest (as noted above) that the gross changes in RNA synthesis evoked by steroid *in vivo* are not an essential part of the induction of TAT.

The induction of TAT in HTC cells is inhibited by AMD in concentrations (0.1–0.2 μg/ml) which inhibit RNA synthesis by 85 to 95 %, suggesting that RNA synthesis is required for induction to occur (145, 182). Furthermore, it has been shown by inhibitor studies that the amount of enzyme induction obtained is proportional to the amount of presumptive mRNA which accumulates (145). Consistent with these indirect studies, direct examination of HTC cell RNA by double-labeling techniques has shown the accumulation of a small amount of rapidly-labeled, non-ribosomal, polysome-associated RNA in steroid-treated cells. This RNA has some of the characteristics ascribed to mRNA, but for the reasons discussed earlier it is not possible to equate it with specific mRNA (45). Both indirect (145, 187) and direct (45) approaches suggest further that presumptive TAT mRNA can accumulate in the absence of protein synthesis.

Removal of the inducing steroid from fully-induced HTC cells, or antagonism of the inducing steroid with non-inducing steroids (162) results in a rather prompt decline in the rate of TAT synthesis (162, 188). This phenomenon is similar to that observed in bacteria, in which inducer must be present continually to maintain the induced rate of specific mRNA synthesis (77). In bacteria, therefore, inhibition of RNA synthesis produces the same result as removal of the inducer (77, 100). If the enzyme induction in HTC cells is explicable in terms of the *Jacob-Monod* model, i.e. inducer acts by stimulating mRNA synthesis, then inhibition of TAT mRNA synthesis by AMD should mimic the results of removing inducer. In fact, however, this is not observed. If transcription of the TAT gene is inhibited by 0.1 to 0.2 µg/ml AMD in either basal (145) or induced (47, 188) HTC cells, TAT activity is maintained at its previous level for several hours and then falls only very gradually. Since AMD does not stabilize TAT under these experimental conditions (3), these findings argue that (1) the TAT message is relatively stable, at least the presence of inducer, and (2) that the presence of inducer is required for 'translation' of the TAT message rather than for its synthesis.

Another observation, incompatible with the *Jacob-Monod* model, is the finding that the addition to basal (145) or induced HTC cells of higher concentrations of AMD (1—5 µg/ml), which inhibit RNA synthesis by more than 95 %, 'paradoxically' increases TAT activity (182, 183, 188), as had been observed earlier in the rat (40). This increase is blocked by inhibition of protein synthesis (183, 188) and can be shown by immunochemical techniques to reflect increased synthesis of TAT (183, 188). General protein synthesis is not affected. *Reel and Kenney* (152) have confirmed the 'superinduction' of TAT by AMD in both HTC cells and in the H-35-Reuber rat hepatoma cell line, but have argued that AMD blocks degradation of TAT in their experiments. Under nutritional 'step-down' conditions, which these authors may have used, TAT turnover is markedly enhanced, and this enhancement is blocked by AMD (3); but under our experimental conditions TAT degradation is clearly not affected by AMD (3, 188), and superinduction must reflect an increased rate of TAT synthesis.

Although interpretations based on inhibitor data may always be somewhat suspect, and though AMD may have various effects on animal cells (175), it is felt that superinduction is the result of virtually complete inhibition of RNA synthesis. Structural derivatives of actinomycin which do not inhibit RNA synthesis do not superinduce; inhibitors of DNA synthesis such as hydroxyurea similarly do not superinduce TAT nor interfere with its superinduction by AMD

(183). Other inhibitors of RNA synthesis such as mitomycin C and 5-flurouracil have also been found to cause superinduction, although results with the latter agent have been inconsistent thus far (183). The recent finding of an inhibitor of RNA synthesis, a-amanitin, which specifically interferes with mammalian RNA polymerase, rather than binding to the DNA template (78), should allow a further test of this interpretation.

When RNA synthesis is virtually completely blocked in fully-induced HTC cells, TAT synthesis is not only increased ('superinduction'), but it becomes constitutive, i.e. the steroid inducer may be removed, but enzyme synthesis continues at the elevated rate (183, 188).

These results are clearly not compatible with a model of enzyme induction in which the inducer acts only at the gene level to stimulate mRNA synthesis. In fact there is no compelling evidence in HTC cells necessitating control at a transcriptional level, but the observations noted above do require some kind of regulation at a post-transcriptional level. Therefore, *Tomkins* and his coll. have presented a model in which inducing steroids have a single action; to antagonize a post-transcriptional repressor which both inhibits messenger translation and promotes messenger degradation (188) (fig. 4). In this model, the synthesis of TAT is regulated by at least two genes: the structural gene for TAT (S_G) and a separate regulatory gene (R_G). During the inducible phases of the cell cycle (see below), we propose that the structural gene is transcribed continuously into its mRNA, which is then translated to synthesize TAT. The regulatory gene also produces its product, the repressor, at a constant rate. We assume that the repressor is a protein, though it could also be an RNA. We propose further that the repressor is quite labile; and that its synthesis is relatively less sensitive to inhibition by AMD than is the TAT mRNA; i.e. requires higher concentrations of AMD to inhibit its synthesis. (There is evidence in both bacterial [92, 149] and mammalian cells [144] that the transcription of different genes exhibits differential sensitivity to inhibition by AMD.)

The repressor is assumed to interact reversibly with the TAT message such that translation of the message (now in the 'MR' configuration) is inhibited.

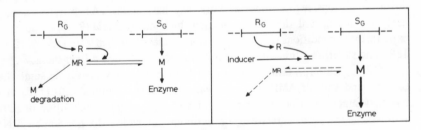

Fig. 4. Schematic model for TAT induction in HTC cells. Modified from *Tomkins et al.* (188). See text for explanation.

Since we postulate the TAT message can only be degraded at a significant rate in the MR form, the repressor both inhibits messenger translation and enhances its degradation (fig. 4a). Finally, we propose that inducing steroids directly or indirectly antagonize the repressor such that TAT message (now in the M form) is translated and its degradation prevented. Since TAT message synthesis is continuous, the concentration of the mRNA increases under these conditions (fig. 4b).

Thus, the model satisfactorily explains the following observations described in more detail above: (1) Glucocorticoid hormones stimulate the rate of enzyme synthesis; (2) Enzyme-specific mRNA accumulates in the presence of inducer (even though its rate of synthesis is unchanged), and it can accumulate even when protein synthesis is inhibited; (3) RNA synthesis is necessary for induction to occur, and the amount of induction obtained is proportional to the amount of mRNA accumulated; but continued RNA synthesis is not required for the maintenance of the induced rate of synthesis since, in the presence of the inducer, the accumulated mRNA is kept in the translatable and stable M configuration; (4) When RNA synthesis is virtually completely inhibited, enzyme synthesis becomes constitutive since the labile repressor rapidly disappears; (5) In the presence of RNA synthesis the constant presence of the inducing steroid is required to maintain the induced rate of enzyme synthesis. Removal of the inducer allows the free repressor to inhibit mRNA translation rapidly, followed by the degradation of the message.

The model suggests, therefore, that the rapid decrease in enzyme synthesis following removal of inducer is due to repression of translation of the mRNA rather than disappearance of a very labile mRNA, and predicts that after removal of inducer the message, and hence enzyme synthesizing capacity, could be 'rescued' by antagonizing or inhibiting the repressor. When RNA synthesis is inhibited by AMD at various intervals after removing inducer from previously induced cells, just such a 'rescue' or restoration of TAT synthesis is observed (188). Thus a significant pool of inactive message exists in the cells after removal of inducer, which can be activated by antagonizing the repressor. Since repressed messenger seems to disappear more rapidly than messenger in the M from, it appears that repressor somehow makes the message more labile.

Another major line of support for the model comes from the study of TAT induction in synchronized cells. TAT can be synthesized during all phases of the cell cycle, but can be induced by steroids only during S (period of DNA synthesis) and latter 2/3 of Gl (interval between mitosis and onset of DNA synthesis) (112). Since steroids are believed to induce TAT by antagonizing repressor, it has been suggested that during the non-inducible phases of the cycle (mitosis, first 1/3 of G1, and G2 [interval between end of DNA synthesis and mitosis]) repressor is not made, as the transcription of both the TAT and repressor genes is repressed by a steroid-insensitive mechanism (113). (This mechanism may not be entirely specific, since during mitosis at least, essentially all gene transcription is turned off.) Pre-existing TAT message continues to be translated, and is stable during this period since repressor is not present (113).

The model would then predict that TAT synthesis should be constitutive during G2, mitosis, and the beginning of G1, but that it should decline rapidly during G1 when repressor begins to be made, unless repressor synthesis is blocked with AMD, or its action antagonized by inducing steroid. As predicted, previously-induced HTC cells, collected in mitosis, continue to synthesize TAT at the fully induced rate during mitosis and early in G1 even in the absence of the steroid inducer (113). At about the third hour of G1 (at the same time in the cell cycle when TAT becomes inducible in uninduced cultures) TAT synthesis falls rapidly unless steroid is added to the cells. This fall can also be prevented by the addition of 1–5 μg/ml AMD (113). As in random cells, AMD does not alter the turnover of TAT under these conditions, and it has been demonstrated immunochemically that the high level of TAT activity reflects a high rate of TAT synthesis (113). Similar experiments have shown that when the inducing steroid is removed during the G2 phase of the cell cycle from previously-induced cells, TAT synthesis becomes constitutive, consistent with the prediction that repressor is not made during the G2 period (114).

Recently we have reported further evidence for post-transcriptional regulation of TAT in HTC cells. Although TAT can be induced (about 5-fold) in HTC cells incubated with glucocorticoids in a chemically-defined serum-free medium, the presence of 5 % bovine serum markedly enhances the extent of enzyme induction (46). The addition of dialyzed bovine serum to cells previously induced in the absence of serum causes a 2- to 3-fold further increase in the rate of synthesis of TAT, as measured by radio-immunoprecipitation techniques, within 2 h. AMD in concentrations sufficient to block completely hormonal induction of TAT did not significantly inhibit this effect of serum, suggesting that serum also acts at a step in protein synthesis beyond gene transcription. Serum was also found to stimulate overall protein synthesis by about 40 %, and to cause a shift in ribosome distribution toward polysomal aggregates (46).

Since certain serum requirements in cultured animal cells can be satisfied by insulin (103, 171, 181), and since insulin and glucagon have been reported to induce TAT in rat liver (73), the effect of insulin on TAT synthesis in HTC cells was also examined. The addition of 0.1 U/ml insulin to cells previously induced with steroids in serum-free medium causes a rapid 2- to 3-fold further increase in the rate of TAT synthesis, which is almost completely insensitive to AMD (47). The stimulation of TAT synthesis by insulin is probably not secondary to effects on glucose or amino acid transport, since the magnitude of the effect is independent of the concentration of either glucose or amino acids in the medium, and occurs in the absence of each component. Insulin, like serum, also stimulates a modest increase in overall protein synthesis and shift in ribosome distribution.

However, when maximally effective concentrations of both serum and insulin are added simultaneously to previously induced HTC cells, there is an additive stimulation of TAT activity, suggesting that these 2 agents affect different aspects of TAT synthesis (47).

The presence of the inducing steroid is required to permit the maximum effect of both serum and insulin on TAT synthesis, but both enhance TAT synthesis in the absence of the steroid, suggesting that they can circumvent the action of the repressor in some way (47). Unlike the glucocorticoids, insulin does not cause the accumulation of TAT message, nor the sustained induction of TAT (47). Thus it appears that three humoral agents, glucocorticoids, insulin, and a macromolecular factor in serum, all regulate the synthesis of one enzyme, presumably by affecting different post-transcriptional aspects of enzyme synthesis.

In preliminary experiments we have observed that purified glucagon can also stimulate TAT synthesis (48). This is of particular interest since HTC cells lack measurable levels of adenyl cyclase activity, and cyclic adenosine 3',5'-monophosphate (cAMP) (54). Consistent with the possibility that glucagon affects TAT synthesis by a mechanism not involving the cyclic AMP-adenyl cyclase system is our observation that $N^6,O^{2'}$-dibutyryl cyclic AMP has no effect on TAT synthesis (48).

Evidence compatible with the existence of a cytoplasmic repressor has also been obtained using quite a different approach, namely that of viral-mediated cell fusion (68). Binucleate heterokaryons, containing one nucleus from an HTC cell, and one nucleus from cells of a diploid rat liver line (22) which lacks inducible TAT, have in every case lacked inducible TAT as assayed histochemically (184). This result is at least consistent with the presence in the non-inducible cell line of a steroid-insensitive repressor capable of repressing TAT synthesis directed by the HTC cell nucleus. Using comparable techniques, Davidson has reached similar conclusions with respect to the regulation of melanin synthesis in rodent cells (26), but Littlefield, examining the regulation of folate reductase in hamster cells, failed to find evidence for such repression (108).

The exact mechanism of post-transcriptional regulation is, of course, unknown. From the considerations discussed earlier, there are many potential regulatory sites between the transcription of the gene and completion of the translation process. Some of the most reasonable possibilities are (1) selection and protection from degradation of nascent mRNA's in the nucleus, (2) transport of mRNA from nucleus to cytoplasm, (3) association of mRNA with ribosomal units to form polyribosomes, (4) activation or inactivation of ribosomes by association of regulatory molecules with ribosomes or bound message, or (5) alteration of initiation or termination of protein synthesis. Since binding of mRNA's to ribosomes appears to protect them from degradation by nucleases (13), and is also required for messenger translation, we have suggested as one possible mechanism that the repressor might attach to specific sequences at the 5' end of the messenger, preventing polyribosome formation (188). The observation that steroids increase the number of polyribosomes in HTC cells, even in the presence of AMD, is consistent with such a hypothesis (48). Obviously, the other alternatives listed are also possible, and there is evidence that estrogens, for example, may selectively regulate the transport of potential messenger RNA from the nucleus to the cytoplasm of rabbit uterine cells (15).

How universally our proposed model can be applied to enzyme induction in animal cells is unknown. Post-transcriptional control of protein synthesis of a general or non-specific (i.e. affecting overall protein synthesis) kind appears to operate in unfertilized eggs (123) and in mitotic cells (161), in the mechanism of insulin action on protein synthesis in muscle (196), and the serum stimulation of protein synthesis in primary chick embryo cells (176).

Evidence suggesting more specific post-transcriptional control has been obtained in a number of experimental systems. The synthesis of hemoglobin in reticulocytes, for example, occurs in the absence of RNA synthesis. The synthesis of early proteins in sea urchin development is resistent to AMD (62), as is the increase in rat liver gulonolactone hydroxylase activity induced by somatotropin (178). The synthesis of a number of other proteins including estrogen-induced hepatic phosphoprotein in male chickens (57), cortisol-induced glutamine synthetase in chick embryo retina (124, 154), and UDP-galactose polysaccharide transferase during slime mold development (179), all become AMD-resistant after a period of initial sensitivity to the antibiotic, suggesting that these proteins are synthesized from stable templates.

In addition, there are now a number of reports in which AMD appears to enhance the synthesis of specific proteins. The synthesis of chick oviduct avidin, assayed immunochemically, is enhanced by AMD (139), as is the incorporation of ^{14}C-acetate into rat liver cholesterol (118). Both effects are inhibited by puromycin. AMD also increases incorporation of amino acids into lens crystallins in highly differentiated lens fiber cells (141). The activities of arginase in non-growing Chang liver cells (31), of deoxycytidylate aminohydrolase in early sea urchin embryos (163), and of leucocyte alkaline phosphatase in rabbit white cells (132) are all increased by inhibition of RNA synthesis with actinomycin. Rat adipose tissue clearing factor lipase activity is enhanced 10-fold by AMD, and the enhancement is prevented by puromycin (195). *Nebert and Gelboin* (see below) have demonstrated a puromycin-sensitive AMD superinduction of aryl hydroxylase activity in hamster fetus cells in culture (133). AMD has been found to enhance interferon synthesis stimulated by double-stranded RNA in rabbit kidney cells (190), and to restore protein synthesis and rhythmic contractions in cultured rat heart cells which have ceased to beat (129). Interestingly, cortisol has a similar effect on the cultured rat cells. Recently, *Ambrose* has reported that low concentrations of AMD paradoxically enhance the secondary antibody response by rabbit lymph node fragments *in vitro* (1). Prompted by *Garren's* report suggesting a putative translational repressor (40), *Ambrose* has sought for and described a non-dialyzable 'antibody-inhibitory material' in his cultures which apparently blocks antibody synthesis. Finally, when AMD in relatively high doses is added to organ cultures of chick embryo retinas 4 h after the addition of cortisol, glutamine synthetase synthesis becomes constitutive (124).

Thus there is considerable evidence from very diverse experimental systems that AMD may paradoxically enhance the activity or synthesis of a number of specific proteins, at least consistent with the model presented for post-transcriptional control. Unfortunately, however, too few inducible enzyme systems have been investigated in sufficient detail to provide convincing evidence for or against the role of specific post-transcriptional repressors. In several of the experimental systems noted above (57, 124, 133, 154, 179), as in HTC cells, the induced synthesis of a specific protein is initially AMD-sensitive, but subsequently becomes AMD-insensitive. Inducer is required during both phases, however, unless RNA synthesis has been completely inhibited by high concentrations of AMD. These findings could be explained by a model in which the inducer acts both at the level of transcription and translation. The model proposed for the regulation of TAT synthesis in HTC cells (fig. 4) thus represents a minimal model in which all the observed phenomena can be explained by a single site of inducer action.

Regulation at the level of gene transcription must also be important in metazoan organisms. Evidence for transcriptional control of specific gene expression has been obtained cytologically in Diptera (18) and in the case of ribosomal RNA in Amphibia (11). However, from what is known of the nature of the RNA synthesized in the nuclei of higher animals (165), it is difficult to envision how the fine adjustments required in enzyme induction can be made at the level of transcription. It seems much more likely that transcriptional controls may be most important in the sequential activation and repression of large numbers of genes which occurs during development and differentiation. These changes, also often hormonally effected (180), are generally stable, if not irreversible, and of large magnitude (55). Various simple (121) and more complex (10) variations of the basic Jacob and Monod model have been proposed to explain such regulation. Post-transcriptional regulation might be expected in the finer modulation of the differentiated state, as in specific enzyme induction.

It would be unlikely theoretically and unjustified experimentally to state that the model for steroidal induction of TAT is sufficient to explain all examples of enzyme induction in mammals. On the other hand it seems equally clear that the most profitable approach to an understanding of the basic mechanisms of enzyme induction and gene regulation in animal cells will come from a detailed study of a limited number of favorable model systems. The study of enzyme induction in cultured cells such as HTC cells offers such a unique opportunity. The experimenter can work with a pure, cloned population of functioning cells in an environment which can be rigidly controlled and easily manipulated. Somatic cell genetic techniques hopefully may allow one to begin genetic analyses (see below) as has been so successful in microorganisms. Although such *in vitro* systems are artificial in the sense of their isolation from other tissue influences, this is more than compensated for by the increased versatility of the isolated cell system. The analysis of TAT induction in HTC cells is a representative example of the advantages of such an approach. The advantages and disadvantages of cultured cells for the study of enzyme regulatory mechanisms in animal cells have been extensively discussed in a recent review (169).

VII. Induction of Hepatic Drug-metabolizing Enzymes

More than 200 drugs, insecticides, carcinogens and other chemicals are known to stimulate the activity of a variety of hepatic microsomal drug-metabolizing enzymes (see reviews by *Conney* [20] and *Gelboin* [43]). The enzymes affected are mostly of the mixed oxygenase type, including those involved in C- and N-hydroxylation, N-, S-, and O-demethylation, and glucuronide formation (43). The pharmacologic actions of the effector or inducing compounds are diverse. There is no apparent relation between their structure or activity and their ability to induce; nor, for instance, between their carcinogenicity and inducing capacity (20). Just as in the case of hormonal induction of soluble hepatic enzymes, there is enormous variation in drug-mediated microsomal enzyme induction with respect to the experimental animals' sex, age, nutritional and hormonal state, and species and strain of origin (20, 43). The latter finding obviously stresses the significance of genetic factors in microsomal enzyme induction. These considerations have been pursued actively in the case of pharmacogenetic studies in man (95, 126, 189), but have received relatively less attention in investigations in experimental animals (24, 119, 126).

Induction of drug-metabolizing microsomal enzymes results in accelerated metabolism of their substrates, either drugs or endogenous body constituents such as hormones, and a consequent alteration in the intensity and duration of action of these substrates. The physiologic consequences of the enzyme induction, therefore, depend on the relative activities of the substrate compound and its metabolites (20).

The implications of microsomal enzyme induction in human medicine are very great, particularly in such areas as drug toxicity and tolerance, carcinogenesis, and in the treatment of diseases resulting from certain enzyme deficiencies (21, 198).

Conney has suggested that microsomal enzyme inducers fall into two major classes, characterized by phenobarbital, and by the polycyclic hydrocarbon, 3-methylcholanthrene. The two types of inducers differ in their course and intensity of induction; their specificity or lack of it, and in the metabolic pathways they affect (20). The administration of polycyclic hydrocarbons to the rat causes an increase in hepatic protein content, proliferation of the agranular endoplasmic reticulum, and the induction of a variety of microsomal enzymes both in liver and in other tissues including kidney, intestine and lung (44, 134). The enzymes affected are different from those induced by glococorticoids, and are unaffected by glucocorticoid administration (44).

The induction of these enzymes is prevented by puromycin treatment, suggesting that the increase in activity reflects *de novo* synthesis of the enzymes (42). Since these microsomal enzymes have not been solubilized and purified, however, no immunochemical studies have been possible to confirm this interpretation. Inhibition of RNA synthesis with AMD also blocks the enzyme induction suggesting that RNA synthesis is also required (42).

In an effort to circumvent the complexities of studying this phenomenon in intact animals, *Gelboin* and his coll. have begun to investigate the induction of aryl hydrocarbon hydroxylase (benzo[a]pyrene hydroxylase) by its substates, the polycyclic hydrocarbons, in short-term cultures of fetal hamster cells (133). Aryl hydroxylase is a multicomponent enzyme system of the microsomal fraction of the cell, which converts a variety of polycyclic hydrocarbons to a variety of phenolic derivaties. The enzyme requires NADPH and molecular oxygen, and involves the microsomal electron transport chain of which the heme protein, P-450, is a part. The enzyme is inducible in secondary (i.e. first subculture) cell cultures derived from the whole hamster fetus, as well as in cells from fetal lung, gut, liver, or limb (133).

Following the addition of 13 μM 1,2-benz(a)anthracene (BA), a relatively weak carcinogenic hydrocarbon, to growing monolayer cultures of whole fetal hamster cells, aryl hydroxylase activity, after a lag of about 30 min, increases linearly for approximately 24 h to a maximum level, 15 to 30 times higher than the basal level of activity (fig. 5a). Various other polycyclic hydrocarbons also evoke this response, though less dramatically. Glucocorticoids, however, are without effect on this enzyme. BA, at the concentrations used in these experiments, does not affect growth or morphology of the hamster cells (133).

As in the intact animal, the induction of aryl hydroxylase *in vitro*, is also inhibited by cycloheximide and puromycin, inhibitors of protein synthesis (133). In the absence of immunochemical studies, however, this result cannot be equated with a demonstration of *de novo* synthesis of the enzyme upon induction. Aryl hydroxylase induction is also blocked by AMD (fig. 5a), confirming that RNA synthesis is also required (133). Furthermore, in experiments comparable to those performed in HTC cells (146), these authors have shown

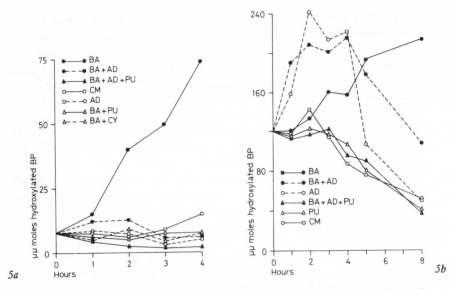

Fig. 5a. The effects of 0.4 μ*M* actinomycin D (AD), 60 μ*M* puromycin (PU), and 3.5 μ*M* cycloheximide (CY) on aryl hydroxylase induction in fetal hamster cells in cuture by 13 μ*M* BA. The responses to inducer alone (BA) and control medium alone (CM) are also illustrated. The antibiotics and the inducer were added simultaneously. At these concentrations of antibiotics, actinomycin D prevented more than 90 % of RNA synthesis, and puromycin and cycloheximide prevented more than 95 % of protein synthesis. Figure 9 in *Nebert and Gelboin* (133).

Fig. 5b. The effects of 0.4 μ*M* actinomycin D (AD), 60 μ*M* puromycin (PU) and control medium (CM) on hydroxylase activity of previously induced fetal hamster cells. The cultures had been treated with 13 μ*M* BA for 25 h, at which time the medium of all dishes was replaced with fresh medium containing the indicated additions. Under these conditions, actinomycin D inhibited RNA synthesis by more than 90 % after 30 min incubation, and by 95 to 98 % after 90 min incubation. Figure 10 from *Nebert and Gelboin* (133).

that the initial phase of the microsomal oxygenase induction appears to involve the accumulation of 'induction-specific-RNA', and that this RNA can accumulate in the absence of protein synthesis (135).

Removal of the inducer from previously induced hamster cells results in a decline in aryl hydroxylase activity with a half-life of 3 to 4 h (fig. 5b). Interruption of protein synthesis with puromycin (in the presence or absence of the inducer) causes an essentially identical decline in enzyme activity (133). Inhibition of RNA synthesis with 0.4 μ*M* AMD (10 μg/ml), however, causes a prompt and striking 'paradoxical' increase in arylhydroxylase activity, whether or not inducer is present. This increase in enzyme activity requires concomitant protein synthesis, and is prevented by puromycin (133, 136) (fig. 5b). These observations are strikingly similar to those described earlier in HTC cells (182, 183, 188). Finally, the addition of AMD to induced hamster cells at various intervals from zero to as long as 6 h after removing the inducer stimulated a significant increase in hydroxylase activity (136), a finding again exactly comparable to that reported in HTC cells (188).

Nebert and Gelboin have interpreted these observations to indicate that the inducing hydrocarbons act at the level of gene transcription to 'deplete' a putative transcriptional repressor, resulting in the synthesis and accumulation of induction-specific mRNA (44, 135). Once this accumulation has occurred (and it can occur in the absence of protein synthesis, so it is 'translation-independent' [135]), enzyme induction then becomes AMD-insensitive and hence 'transcription-independent' (135). Finally, the authors suggest that the newly synthesized mRNA accumulates at a nuclear site bound to DNA, from which it can be displaced by AMD, accounting for the 'paradoxical' stimulation of enzyme activity by the inhibitor (44, 135).

The above model, however, does not account for the requirement for the continued presence of the inducer in order to maintain the induced level of hydroxylase activity. Since the induced level of aryl hydroxylase does not depend on continued RNA synthesis (it is 'transcription-independent'), the requirement for inducer suggests that it must act as some post-transcriptional step in enzyme synthesis or processing. Secondly, there is no evidence in animal cells that newly synthesized mRNA accumulates in the nucleus associated with DNA, although this presumably is possible. Finally, the experimental observations reported in these studies are strikingly similar to those described in quite a different experimental system, namely the glucocorticoid induction of TAT in HTC cells (188), and are entirely consistent with the model for enzyme induction (fig. 4) presented earlier.

VIII. Genetic Approaches to Enzyme Induction

The analysis of enzyme induction in *E. coli* was based largely on genetic studies (77), and the Jacob-Monod model derived from investigations of several regulatory mutations affecting enzyme synthesis (77). In contrast to the success of genetic analyses in bacteria, one is struck by the relative paucity of genetic approaches to enzyme induction in animal cells. Although genetic studies are clearly more difficult in higher organisms that in prokaryotes, genetic approaches to the regulation of protein synthesis in mammals have been undertaken, though rarely in the investigation of enzyme induction.

Using inbred strains of mice, *Ganschow and Paigen* have demonstrated that the structure and intracellular location of hepatic β-glucuronidase are under the control of separate genes (37). *Rechcigl and Heston* (150) and *Ganschow and*

Schimke (38) have reported that separate genes control the catalytic activity and rate of degradation (and hence cellular concentration) of hepatic catalase in mice. Recently, *Blake* (8) has described inter-strain differences in the qualitative and quantitative aspects of TAT induction in inbred mice, suggesting that an important beginning has been made in the genetic exploration of enzyme induction.

Ideally, one would like to have available regulatory mutants in cell culture. Unfortunately, no regulatory mutants for TAT have been discovered thus far among HTC cells; but it is worth noting that in HTC cells, regulatory mutations have been mimicked by inhibitors (AMD) and certain phases of the cell cycle; e.g. enzyme synthesis is constitutive in mitotic cells, or in the absence of RNA synthesis. Quantitative variations in certain enzymes, however, have been described in established cell lines (169), and techniques for the induction and selection of nutritional mutants have been devised (80), but thus far no definitely proven regulatory mutants are available. Fortunately, mouse tissues explanted to tissue culture frequently retain many of their enzymatic markers (160), so that it may be possible to study *in vitro* the genetic differences in enzyme complement (79) and in enzyme induction demonstrated *in vivo* (8, 122).

Another important source of possible enzyme regulatory mutants may be the large number of inherited enzyme alterations in man (107). At present, few if any of these 'inborn errors' have been convincingly established to result from regulatory mutations (29, 33), although this remains an attractive possibility.

Providing that suitable regulatory mutants can be obtained, genetic studies can be performed with somatic cells, utilizing the techniques of cell fusion (68) and hybridization (32, 106). Some examples of this approach to the problem of quantitative regulation of enzyme synthesis have already been noted (26, 108, 184). Some of the problems and possibilities of somatic cell genetic approaches have been discussed in recent reviews (41, 94).

IX. Summary

The problems of biochemical regulation posed to the highly-differentiated metazoan organism are clearly different from those faced by the free-living, rapidly-dividing prokaryotic cell. Furthermore the organization of the genetic material in higher organisms, and the physiology of macromolecular synthesis also differ in significant ways. From these considerations, it should be apparent that the control of gene expression in mammalian cells, as exemplified by enzyme induction, might also be expected to differ from the regulatory mechanisms demonstrated so elegantly in microorganisms.

The greater complexity of higher animals also means that the problems of investigating regulatory mechanisms are correspondingly greater. The very considerable difficulties in trying to get at the mechanism of enzyme induction in intact animals are a good illustration of this problem. It seems, therefore, that progress in understanding the mechanisms of enzyme induction will most probably come from the study, in depth, of a relatively few, particularly favorable experimental systems. The information gained, in a relatively short time, from studies of TAT induction in HTC cells, and of aryl-hydroxylase in fetal hamster cells illustrate the value of such studies.

From the investigation of such experimental systems, a model for enzyme induction in animal cells has been presented. How universally this model may be applied to enzyme induction in higher organisms is unknown at present. It seems unlikely that this, or any single model, could account successfully for all examples of a biologic phenomenon which we must expect to be very heterogeneous. From the considerations presented earlier, one would expect regulatory controls to operate at a number of different steps in the flow of genetic information. On the other hand, there is evidence that the model proposed above is applicable to a number of quite diverse enzyme systems. Perhaps its most important feature is its emphasis on post-transcriptional control, and its demonstration that regulation of gene expression need *not* operate at the genetic or transcriptional level. There is good evidence that post-transcriptional controls must be quite universal among animal cells. With currently available methodology, unfortunately, it is almost impossible to demonstrate, directly, transcriptional control of specific enzyme synthesis in animal cells. The nuclear products of transcription cannot as yet be equated with specific functional (cytoplasmic) messenger RNA's, nor can they be hybridized selectively with the isolated DNA coding for specific proteins. Transcriptional controls undoubtedly exist in animal cells, but their role in enzyme induction remains to be established firmly. Inhibition of enzyme induction by inhibitors of RNA synthesis is not sufficient evidence from which to draw such conclusions.

The paucity of genetic approaches to the problems of enzyme induction in higher organisms stands in striking contrast to the success of such approaches in microorganisms. Hopefully, the selection of regulatory mutants or variants *in vivo* and in cultured cells will open the way to a better understanding of enzyme induction.

The advantages of relatively simpler experimental model systems in which to explore the molecular basis for enzyme induction have been stressed. Once progress has been made by such an approach, however, investigators must face the far more difficult problem of trying to fit their observations into the integrated physiology of the whole organism.

Acknowledgment

I wish to thank Dr. *E. Brad Thompson* for critically reviewing the manuscript, and especially to thank Dr. *Gordon M. Tomkins* for introducing me to this exciting field and for encouragement and stimulation.

X. References

1 *Ambrose, C.T.:* Regulation of the secondary antibody response *in vitro*. Enhancement by actinomycin D and inhibition by a macromolecular product of stimulated lymph node cultures. J. exp. Med. *130:* 1003–1010 (1969).

2 *Attardi, G.; Naono, S.; Rouviere, J.; Jacob, F., and Gros, F.:* Production of messenger RNA and regulation of protein synthesis. Cold Spr. Harb. Symp. quant. Biol. *28:* 363–372 (1963).

3 *Auricchio, F.; Martin, D., and Tomkins, G.:* Control of degradation and synthesis of induced tyrosine aminotransferase studied in hepatoma cells in culture. Nature, Lond. *224:* 806–808 (1969).

4 *Ballard, P. and Tomkins, G.M.:* Hormone induced modification of the cell surface. Nature, Lond. *224:* 344–345 (1969).

5 *Barnabei, O. and Sereni, F.:* Cortisol-induced increase of tyrosine-α-ketoglutarate transaminase in the isolated perfused rat liver and its relation to ribonucleic acid synthesis. Biochim. biophys. Acta *91:* 239–247 (1964).

6 *Barnabei, O.; Romano, B.; diBitonto, G.; Tomasi, V., and Sereni, F.:* Factors influencing the glucocorticoid-induced increase in RNA polymerase activity of rat liver nuclei. Arch. Biochem. *113:* 478–486 (1966).

7 *Berlin, C.M. and Schimke, R.T.:* Influence of turnover rates on the responses of enzymes to cortisone. Molec. Pharmacol. *1:* 149–156 (1965).

8 *Blake, R.:* Personal communication.

9 *Britten, R.J. and Kohne, D.E.:* Repeated sequences in DNA. Science *161:* 529–540 (1968).

10 *Britten, R.J. and Davidson, E.H.:* Gene regulation for higher cells: A theory. Science *165:* 349–357 (1969).

11 *Brown, D.D. and Littna, E.:* RNA synthesis during the development of Xenopus laevis, the South African clawed toad. J. molec. Biol. *8:* 669–687 (1964).

12 *Byrne, R.; Levin, J.G.; Bladen, H.A., and Nirenberg, M.W.:* The *in vitro* formation of a DNA ribosome complex. Proc. nat. Acad. Sci., Wash. *52:* 140–148 (1964).

13 *Castles, J.J. and Singer, M.F.:* Degradation of polyuridylic acid by ribonuclease II: Protection by ribosomes. J. molec. Biol. *40:* 1–17 (1969).

14 *Chu, L.L.H. and Fineberg, R.A.:* On the mechanism of iron-induced synthesis of apoferritin in HeLa cells. J. biol. Chem. *244:* 3847–3854 (1969).

15 *Church, R.B. and McCarthy, B.J.:* Unstable nuclear RNA synthesis following estrogen stimulation. Biochim. biophys. Acta *199:* 103–114 (1970).

16 *Civen, M. and Knox, W.E.:* Pattern of adaptive control of levels of rat liver tryptophan transaminase. Science *129:* 1672–1673 (1959).

17 *Civen, M.; Ulrich, R.; Trimmer, B.M., and Brown, C.B.:* Circadian rhythms of liver enzymes and their relationship to enzyme induction. Science *157:* 1563–1564 (1967).

18 *Clever, U.:* Regulation of chromosomal function. Ann. Rev. Genet., vol. 2, pp. 11–30 (Annual Reviews, Palo Alto 1968).

19 *Cohn, M.; Monod, J.; Pollock, M.R.; Spiegelman, S., and Stanier, R.Y.:* Terminology of enzyme formation. Nature, Lond. *172:* 1096 (1953).

20 *Conney, A.H.:* Pharmacological implications of microsomal enzyme induction. Pharmacological Reviews, vol. 19, pp. 317–366 (Williams and Wilkins, Baltimore 1967).

21 *Conney, A.H.:* Drug metabolism and therapeutics. New Engl. J. Med. *280:* 653–660 (1969).

22 *Coon, H.:* Clonal culture of differentiated rat liver cells. J. Cell Biol. *39:* 29a (1968).

23 *Cox, R.P.:* Hormonal stimulation of zinc uptake in mammalian cell cultures. Molec. Pharmacol. *4:* 510–521 (1968).

24 *Cram, R.L.; Juchau, M.R., and Fouts, J.R.:* Differences in hepatic drug metabolism in various rabbit strains before and after pretreatment with phenobarbital. Proc. Soc. exp. Biol., N.Y. *108:* 872–875 (1965).

25 *Davidson, E.H.:* Hormones and genes. Sci. Amer. *212:* 36–45 (1965).

26 *Davidson, R.; Ephrussi, B., and Yamamoto, K.:* Regulation of melanin synthesis in mammalian cells, as studied by somatic hybridization. I. Evidence for negative control. J. cell. Physiol. *72:* 115–128 (1968).

27 *Douglas, H.C. and Hawthorne, D.C.:* Regulation of genes controlling synthesis of the galactose pathway enzymes in yeast. Genetics *54:* 911–916 (1966).

28 *Drews, J. and Brawerman, G.:* Alterations in the nature of ribonucleic acid synthesized in rat liver during regeneration and after cortisol administration. J. biol. Chem. *242:* 801–808 (1967).

29 *Dreyfus, J.C.:* The application of bacterial genetics to the study of human genetic abnormalities. Progress in Medical Genetics, vol. 6, pp. 169–200 (Grune & Stratton, New York 1969).

30 *Drysdale, J.W.:* Regulation of ferritin synthesis; in *San Pietro, Lamborg and Kenney* Regulatory mechanisms for protein synthesis in mammalian cells, pp. 431–466 (Academic Press, New York/London 1968).

31 *Eliasson, E.E.:* Regulation of arginase activity in Chang liver cells in the absence of net protein synthesis. Biochem. biophys. res. Comm. *27:* 661–667 (1967).

32 *Ephrussi, B.:* Hybridization of somatic cells and phenotypic expression; in Developmental and metabolic control mechanisms and neoplasia, pp. 486–503 (Williams & Wilkins, Baltimore 1965).

33 *Epstein, C.:* Structural and control gene defects in hereditary diseases in man. Lancet *ii:* 1068 (1964).

34 *Epstein, W. and Beckwith, J.R.:* Regulation of gene expression. Ann. Rev. Biochem., vol. 37, pp. 412–436 (Annual Reviews, Palo Alto 1968).

35 *Feigelson, P. and Feigelson, M.:* Studies on the mechanism of cortisone action; in *Litwack and Kritchevsky* Actions of hormones on molecular processes, pp. 218–233 (J. Wiley & Sons, New York 1964).

36 *Finkel, R.; Henshaw, E., and Hiatt, H.:* Early changes in liver cytoplasmic RNA of hydrocortisone-treated rats. Molec. Pharmacol. *2:* 221–226 (1966).

37 *Ganschow, R. and Paigen, K.:* Separate genes determining the structure and intracellular location of hepatic glucuronidase. Proc. nat. Acad. Sci., Wash. *58:* 938–945 (1967).

38 *Ganschow, R.E. and Schimke, R.T.:* Independent genetic control of the catalytic activity and the rate of degradation of catalase in mice. J. biol. Chem. *22:* 4649–4658 (1969).

39 *Garren, L.D.; Howell, R.R., and Tomkins, G.M.:* Mammalian enzyme induction by hydrocortisone: The possible role of RNA. J. molec. Biol. *9:* 100–108 (1964).

40 *Garren, L.D.; Howell, R.R.; Tomkins, G.M., and Crocco, R.M.:* A paradoxical effect of actinomycin D: The mechanism of regulation of enzyme synthesis by hydrocortisone. Proc. nat. Acad. Sci., Wash. *52:* 1121–1129 (1964).

41 *Gartler, S.M. and Pious, D.A.:* Genetics of mammalian cell cultures. Humangenetik *2:* 83–114 (1966).

42 *Gelboin, H.V. and Blackburn, N.R.:* The stimulatory effect of 3-methylcholanthrene on benzpyrene hydroxylase activity in several rat tissues: Inhibition by actinomycin D and puromycin. Cancer Res. *24:* 356–360 (1964).

43 *Gelboin, H.V.:* Carcinogens, enzyme induction, and gene action; in Advances in cancer research, vol. 10, pp. 1–81 (Academic Press, New York 1967).

44 *Gelboin, H.V.:* Effect of carinogens on gene action; in Exploitable molecular mechanisms and neoplasia, pp. 285–311 (Williams and Wilkins, Baltimore 1968).

45 *Gelehrter, T.D. and Tomkins, G.M.:* The role of RNA in the hormonal induction of tyrosine aminotransferase in mammalian cells in tissue culture. J. molec. Biol. *29:* 59–76 (1967).

46 *Gelehrter, T.D. and Tomkins, G.M.:* Control of tyrosine aminotransferase synthesis in tissue culture by a factor in serum. Proc. nat. Acad. Sci., Wash. *64:* 723–730 (1969).

47 *Gelehrter, T.D. and Tomkins, G.M.:* Post-transcriptional control of tyrosine aminotransferase synthesis by insulin. Proc. nat. Acad. Sci., Wash. *66* (June, 1970, in press).

48 *Gelehrter, T.D.* (unpublished data).

49 *Gibb, J.W. and Webb, J.G.:* Effects of reserpine, α-methyltyrosine, and L-3,4,-dihydroxy-phenylalanine on brain tyrosine transaminase. Proc. nat. Acad. Sci., Wash. *63:* 364–369 (1969).

50 *Gilbert, W. and Müller-Hill, B.:* Isolation of the lac repressor. Proc. nat. Acad. Sci., Wash. *56:* 1891–1898 (1966). – The lac operator is DNA. Proc. nat. Acad. Sci., Wash. *58:* 2415–2421 (1967).

51 *Goldstein, L.; Stella, E.J., and Knox, W.E.:* The effect of hydrocortisone on tyrosine-α-ketoglutarate transaminase and tryptophan pyrrolase activities in the isolated, perfused rat liver. J. biol. Chem. *237:* 1723–1726 (1962).

52 *Granick, S. and Kappas, A.:* Steroid control of porphyrin and heme biosynthesis: A new biological function of steroid hormone metabolites. Proc. nat. Acad. Sci., Wash. *57:* 1463–1467 (1967).

53 *Granner, D.K.; Hayashi, S.; Thompson, E.B., and Tomkins, G.M.:* Stimulation of tyrosine aminotransferase synthesis by dexamethasone phosphate in cell culture. J. molec. Biol. *35:* 291–301 (1968).

54 *Granner, D.; Chase, L.R.; Aurbach, G.D., and Tomkins, G.M.:* Tyrosine aminotransferase: Enzyme induction independent of adenosine 3',5'-monophosphate. Science *162:* 1018–1020 (1968).

55 *Green, H. and Todaro, G.J.:* The mammalian cell as differentiated organism; in Ann. Rev. Microbiology, vol. 21, pp. 573–600 (Annual Reviews, Palo Alto 1967).

56 *Greengard, O.; Smith, M.A., and Acs, G.:* Relation of cortisone and synthesis of ribonucleic acid to induced and developmental enzyme formation. J. biol. Chem. *238:* 1548–1551 (1963).

57 *Greengard, O.; Gordon, M.; Smith, M.A., and Acs, G.:* Studies on the mechanism of diethylstilbesterol-induced formation of phosphoprotein in male chickens. J. biol. Chem. *239:* 2079–2082 (1964).

58 *Greengard, O.:* The quantitative regulation of specific proteins in animal tissues; words and facts. Enzym. biol. clin. *8:* 81–96 (1967).

59 *Greenman, D.L.; Wicks, W.D., and Kenney, F.T.:* Stimulation of ribonucleic acid synthesis by steroid hormones. J. biol. Chem. *240:* 4420–4426 (1965).

60 *Griffin, M.J. and Cox, R.:* Studies on the mechanism of hormonal induction of alkaline phosphatase in human cell cultures. I. Effects of puromycin and actinomycin D. J. Cell Biol. *29:* 1–9 (1966).

61 *Griffin, M.J. and Cox, R.P.:* Studies on the mechanism of hormone induction of alkaline phosphatase in human cell cultures. II. Rate of enzyme synthesis and properties of base level and induced enzymes. Proc. nat. Acad. Sci., Wash. *56:* 946–953 (1966).

62 *Gross, P.R.; Malkin, L.I., and Moyer, W.A.:* Templates for the first proteins of embryonic development. Proc. nat. Acad. Sci., Wash. *51:* 407–414 (1964).

63 *Gross, S.R.:* Genetic regulatory mechanisms in the fungi. Ann. Rev. Gen., vol. 3, pp. 395–424 (Annual Reviews, Palo Alto 1969).

64 *Grossman, A. and Mavrides, C.:* Studies on the regulation of tyrosine aminotransferase in rats. J. biol. Chem. *242:* 1938–1405 (1967).

65 *Gurdon, J.B.:* Transplanted nuclei and cell differentiation. Sci. Amer. *219:* 24–35 (1968).

66 *Hager, C.B. and Kenney, F.T.:* Regulation of tyrosine-α-ketoglutarate transaminase in rat liver. J. biol. Chem. *243:* 3296–3300 (1968).

67 *Hakami, N. and Pious, D.A.:* Regulation of cytochrome oxidase in human cells in culture. Nature, Lond. *216:* 1087–1090 (1967).

68 *Harris, H.; Sidebottom, E.; Grace, D.M., and Bramwell, M.E.:* The expression of genetic information: A study with hybrid animal cells. J. Cell Sci. *4:* 499–525 (1969).

69 *Haussler, M.R. and Nagode, L.A.:* Induction of specific intestinal brush border enzymes by vitamin D. J. Cell Biol. *43:* 51a (1969).

70 *Hayashi, M.; Spiegelman, S.; Franklin, N., and Luria, S.E.:* Separation of the RNA message transcribed in response to a specific inducer. Proc. nat. Acad. Sci., Wash. *49:* 729–736 (1963).

71 *Hayashi, S.; Granner, D.K., and Tomkins, G.M.:* Tyrosine aminotransferase, purification and characterization. J. biol. Chem. *242:* 3998–4006 (1967).

72 *Hogness, D.S.; Cohen, M., and Monod, J.:* Studies on the induced synthesis of β-galactosidase in *Escherichia coli.* Biochim. biophys. Acta *16:* 99–116 (1955).

73 *Holten, D. and Kenney, F.T.:* Regulation of tyrosine α-ketoglutarate transaminase in rat liver. VI. Induction by pancreatic hormones. J. biol. Chem. *242:* 4372–4377 (1967).

74 *Houssais, J.F. and Attardi, G.:* High molecular weight non-ribosomal-type nuclear RNA and cytoplasmic messenger RNA in HeLa cells. Proc. nat. Acad. Sci., Wash. *56:* 616–623 (1966).

75 *Imamoto, F.; Ito, J., and Yanofsky, C.:* Polarity studies with the tryptophan operator. Cold Spr. Harb. Symp. quant. Biol. *31:* 235–249 (1966).

76 *Itano, H.:* Genetic regulation of peptide synthesis in hemoglobins. J. Cell Physiol. *67:* 65–76 (1966).

77 *Jacob, F. and Monod, J.:* Genetic regulatory mechanisms in the synthesis of proteins. J. molec. Biol. *3:* 318–356 (1961).

78 *Jacob, S.T.; Sajdel, E.M., and Munro, H.N.:* Specific action of α-amanitin on mammalian RNA polymerase protein. Nature, Lond. *225:* 60–62 (1970).

79 *Kandutsch, A.A. and Coleman, D.L.:* Inherited metabolic variations; in *Green* Biology of the laboratory mouse, 2nd ed., pp. 377–386 (McGraw-Hill, New York 1966).

80 *Kao, F. and Puck, T.T.:* Genetics of somatic mammalian cells. VII. Induction and isolation of nutritional mutants in Chinese hamster cells. Proc. nat. Acad. Sci., Wash. *60:* 1275–1281 (1968).

81 *Kaplan, M.M. and Righetti, A.:* Induction of liver alkaline phosphatase by bile duct ligation. Biochim. biophys. Acta *184:* 667–669 (1969).

82 *Karlson, P.:* Regulation of gene activity by hormones. Humangenetik *6:* 99–109 (1968). – New concepts on the mode of action of hormones. Perspect. Biol. Med. *6:* 203–213 (1963).

83 *Kenney, F.T.:* Induction of tyrosine-α-ketoglutarate transaminase in rat liver. III. Immunochemical analysis. J. biol. Chem. *237:* 1610–1614 (1962).

84 *Kenney, F.T.:* Regulation of tyrosine-α-ketoglutarate transaminase in rat liver. V. Repression in growth hormone-treated rats. J. biol. Chem. *242:* 4367–4371 (1967).

85 *Kenney, F.T.:* Turnover of rat liver transaminase: Stabilization after inhibition of protein synthesis. Science *156:* 525–528 (1967).

86 *Kenney, F.T.; Holten, D., and Albritton, W.L.:* Hormonal regulation of liver enzyme synthesis. Nat. Cancer Inst. Monogr. No. 27, pp. 315–323 (1966).

87 *Kenney, F.T.; Reel, J.R.; Hager, C.B., and Wittliff, J.L.:* Hormonal induction and repression; in *San Pietro, Lamborg and Kenney* Regulatory mechanisms for protein synthesis in mammalian cells, pp. 119–142 (Academic Press, New York/London 1968).

88 *Kenney, F.T.; Wicks, W.D., and Greenman, D.L.:* Hydrocortisone stimulation of RNA synthesis in induction of hepatic enzymes. J. cell. comp. Physiol. *66:* 125–136 (1965).

89 *Knox, W.E. and Auerbach, V.H.:* The hormonal control of tryptophan peroxidase in the rat. J. biol. Chem. *214:* 307–313 (1955).

90 *Knox, W.E.; Auerbach, V.H., and Lin, E.C.C.:* Enzymatic and metabolic adaptations in animals. Physiol. Rev. *36:* 164–252 (1956).

91 *Knox, W.E. and Greengard, O.:* The regulation of some enzymes of nitrogen metabolism – An introduction to enzyme physiology. Adv. Enzymol. *3:* 247–313 (1965).

92 *Kodawaki, K.; Hosada, J., and Maruo, B.:* Effects of actinomycin D and 5-fluorouracil on the formation of enzymes in Bacillus subtilis. Biochim. biophys. Acta *103:* 311–318 (1965).

93 *Krantz, S.B. and Goldwasser, E.:* On the mechanism of erythropoietin-induced differentiation. II. Effect on RNA synthesis. Biochim. biophys. Acta *103:* 325–332 (1965).

94 *Krooth, R.S.; Darlington, G.A., and Velaquez, A.A.:* The genetics of cultured mammalian cells. Ann. Rev. Genet., vol. 2, pp. 141–164 (Annual Reviews, Palo Alto 1968).

95 *LaDu, B.N.:* Pharmacogenetics. Toxicol. app. Pharmacol. *7:* 27–38 (1965).

96 *Lang, N.; Herrlich, P., and Sekeris, C.E.:* On the mechanism of hormone action. VIII. Induction by cortisol of a messenger RNA coding for tyrosine-a-ketoglutarate transaminase in an *in vitro* system. Acta Endocrin. *57:* 33–44 (1968).

97 *Latham, H. and Darnell, J.E.:* Distribution of mRNA in the cytoplasmic polyribosomes of the HeLa cell. J. molec. Biol. *14:* 1–12 (1965).

98 *Laycock, D.G. and Hunt, J.A.:* Synthesis of rabbit globin by a bacterial cell free system. Nature, Lond. *221:* 1118–1122 (1969).

99 *Lee, K.L. and Miller, D.N.:* Induction of mitochondrial a-glycerophosphate dehydrogenase by thyroid hormone. Arch. Biochem. *120:* 638–645 (1967).

100 *Leive, L.:* Some effects of inducer on synthesis and utilization of B-galactosidase messenger RNA in actinomycin sensitive *Escherichia coli.* Biochem. biophys. res. Comm. *20:* 321–327 (1965).

101 *Levitan, I.B. and Webb, T.E.:* Modification by 8-azaguanine of the effects of hydrocortisone on the induction and inactivation of tyrosine transaminase of rat liver. J. biol. Chem. *244:* 341–347 (1969).

102 *Lieber, C.S. and DeCarli, L.M.:* Ethanol oxidation by hepatic microsomes: Adaptive increases after ethanol feeding. Science *162:* 117–118 (1968).

103 *Lieberman, I. and Ove, P.:* Growth factors for mammalian cells in culture. J. biol. Chem. *234:* 2754–2758 (1959).

104 *Lin, E.C.C. and Knox, W.E.:* Adaptation of the rat liver tyrosine-a-ketoglutarate transaminase. Biochim. biophys. Acta *26:* 85–88 (1957).

105 *Lin, E.C.C. and Knox, W.E.:* Specificity of adaptive response of tyrosine-a-ketoglutarate transaminase in the rat. J. biol. Chem. *233:* 1186–1189 (1958).

106 *Littlefield, J.W.:* The use of drug resistant markers to study the hybridization of mouse fibroblasts. Exp. Cell Res. *41:* 190–196 (1966).

107 *Littlefield, J.W.:* Control mechanisms in animal cell cultures. Arch. Biochem. *125:* 410–415 (1968).

108 *Littlefield, J.W.:* Hybridization of hamster cells with high and low folate reductase activity. Proc. nat. Acad. Sci., Wash. *62:* 88–95 (1969).

109 *Loeb, J.N.; Howell, R.R., and Tomkins, G.M.:* Turnover of ribosomal RNA in liver. Science *149:* 1093–1095 (1965).

110 *Lyon, M.F.:* Chromosomal and subchromosomal inactivation. Ann. Rev. Genet., vol. 2, pp. 31–52 (Annual Reviews, Palo Alto 1968).

111 *Marshall, R.E.; Caskey, C.T., and Nirenberg, M.:* Fine structure of RNA codewords recognized by bacterial, amphibian, and mammalian transfer RNA. Science *155:* 820–826 (1967).

112 *Martin, D.; Tomkins, G.M., and Granner, D.:* Synthesis and induction of tyrosine aminotransferase in synchronized hepatoma cells in culture. Proc. nat. Acad. Sci., Wash. *62:* 248–255 (1969).

113 *Martin, D.W.; Tomkins, G.M., and Bressler, M.A.:* Control of specific gene expression examined in synchronized mammalian cells. Proc. nat. Acad. Sci., Wash. *63:* 842–849 (1969).

114 *Martin, D.W. and Tomkins, G.M.:* The appearance and disappearance of the post-transcriptional repressor of tyrosine aminotransferase synthesis during the HTC cell cycle. Proc. nat. Acad. Sci., Wash. *65:* 1064–1068 (1970).

115 *Martin, R.G.:* The one operon-one messenger theory of transcription. Cold Spr. Harb. Symp. quant. Biol. *28:* 357–361 (1963).

116 *Martin, R.G.:* Control of gene expression. Ann. Rev. Genet., vol. 3, pp. 181–216 (Annual Reviews, Palo Alto 1969).

117 *Marver, H.S.; Collins, A.; Tschudy, D.P., and Rechcigl, M., Jr.:* δ-Aminolevulinic acid synthetase. II. Induction in rat liver. J. biol. Chem. *241:* 4323–4329 (1966).

118 *deMatteis, F.:* Stimulation of liver cholesterol synthesis by actinomycin D. Biochem. J. *109:* 775–785 (1968).

119 *Meier, H. and Fuller, J.L.:* Responses to drugs; in *Green* Biology of the laboratory mouse, 2nd ed., pp. 447–455 (McGraw-Hill, New York 1966).

120 *Mishkin, E.P. and Shore, M.L.:* Inhibition by actinomycin D of the induction of tryptophan pyrrolase by hydrocortisone. Biochim. biophys. Acta *138:* 169–174 (1967).

121 *Monod, J. and Jacob, F.:* General conclusions: Teleonomic mechanisms in cellular metabolism, growth and differentiation. Cold Spr. Harb. Symp. quant. Biol. *26:* 389–401 (1961).

122 *Monroe, C.B.:* Induction of tryptophan oxygenase and tyrosine aminotransferase in mice. Amer. J. Physiol. *214:* 1410–1414 (1968).

123 *Monroy, P.; Maggio, R., and Rinaldi, A.M.:* Experimentally induced activation of the ribosomes of the unfertilized sea urchin egg. Proc. nat. Acad. Sci., Wash. *54:* 107–111 (1965).

124 *Moscona, A.A.; Moscona, M.H., and Saenz, N.:* Enzyme induction in embryonic retina: The role of transcription and translation. Proc. nat. Acad. Sci., Wash. *61:* 160–167 (1968).

125 *Moses, M. and Coleman, J.R.:* Structural patterns and the functional organization of chromosomes; in *Locke* The role of chromosomes in development, pp. 11–50 (Academic Press, New York/London 1964).

126 *Motulsky, A.G.:* Pharmacogenetics; in Progress in medical genetics, vol. 3, pp. 49–74 (Grune & Stratton, New York/London 1964).

127 *McAuslan, B.R.:* The induction and repression of thymidine kinase in the poxvirus-infected HeLa cell. Virology *21:* 383–389 (1963).

128 *McAuslan, B.R.:* Regulation of enzymes induced by animal viruses. Nat. Cancer Inst. Monogr. No. 27, pp. 211–219 (1966).

129 *McCarl, R.L. and Shaler, R.C.:* The effects of actinomycin D on protein synthesis and beating in cultured rat heart cells. J. Cell Biol. *40:* 850–854 (1968).

130 *McCarthy, B.J. and Hoyer, B.H.:* Identity of DNA and diversity of messenger RNA molecules in normal mouse tissues. Proc. nat. Acad. Sci., Wash. *52:* 915–922 (1964).

131 *McClintock, B.:* Genetic systems regulating gene expression during development; in *Locke* Control mechanisms in developmental processes, pp. 84–112 (Academic Press, New York/London 1967).

132 *McCoy, E.E. and Ebadi, M.:* The paradoxical effect of hydrocortisone and actinomycin on the activity of rabbit leucocyte alkaline phosphatase. Biochem. biophys. res. Comm. *26:* 265–271 (1967).

133 *Nebert, D.W. and Gelboin, H.V.:* Substrate-inducible microsomal aryl hydroxylase in mammalian cell culture. I. Assay and properties of the induced enzyme. J. biol. Chem. *243:* 6242–6249 (1968). – II. Cellular responses during enzyme induction. J. biol. Chem. *243:* 6250–6261 (1968).

134 *Nebert, D.W. and Gelboin, H.V.:* The *in vivo* and *in vitro* induction of aryl hydrocarbon hydroxylase in mammalian cells of different species, tissues, strains, and developmental and hormonal states. Archiv. Biochem. *134:* 76–89 (1969).

135 *Nebert, D.W. and Gelboin, H.V.:* The role of RNA and protein synthesis in microsomal aryl hydrocarbon hydroxylase induction in cell cultures: The independence of transcription and translation. J. biol. Chem. *245:* 160–168 (1970).

136 *Nebert, D.W. and Bausserman, L.L.:* Fate of inducer during induction of aryl hydrocarbon hydroxylase activity in mammalian cell culture. II. Intracellular polycyclic hydrocarbon content during enzyme induction and decay. Molec. Pharmacol. (submitted for publication).

137 *Nichol, C.A. and Rosen, F.:* Adaptive changes in enzymatic activity induced by glucocorticoids; in *Litwack and Kritchevsky* Actions of hormones on molecular processes, pp. 234–256 (J. Wiley & Sons, New York 1964).

138 *Olson, R.E.; Kipfer, R.H., and Li, L.F.:* Adv. Enz. Reg. *7:* 83 (1969).

139 *O'Malley, B.W.: In vitro* hormonal induction of a specific protein (avidin) in chick oviduct. Biochemistry *6:* 2546–2551 (1967).

140 *O'Malley, B.W. and Kohler, P.O.:* Studies on steroid regulation of synthesis of a specific oviduct protein in a new monolayer culture system. Proc. nat. Acad. Sci., Wash. *58:* 2359–2366 (1967).

141 *Papaconstantinou, J.:* Molecular aspects of lens cell differentiation. Science *156:* 338–346 (1967).

142 *Paul, J. and Hunter, J.A.:* Synthesis of macromolecules during induction of haemoglobin synthesis by erythropoietin. J. molec. Biol. *42:* 31–41 (1969).

143 *Peraino, C. and Pitot, H.C.:* Studies on the induction and repression of enzymes in rat liver. II. Carbohydrate repression of dietary and hormonal induction of threonine dehydrase and ornithine δ-transaminase. J. biol. Chem. *239:* 4308–4313 (1964).

144 *Perry, R.P.:* The nucleus and the synthesis of ribosomes. Nat. Canc. Inst. Monogr. No. 18, pp. 325–340 (1964).

145 *Peterkofsky, B. and Tomkins, G.M.:* Effect of inhibitors of nucleic acid synthesis on steroid-mediated induction of tyrosine aminotransferase in hepatoma cell cultures. J. molec. Biol. *30:* 49–61 (1967).

146 *Peterkofsky, B. and Tomkins, G.M.:* Evidence for the steroid-induced accumulation of tyrosine aminotransferase messenger RNA in the absence of protein synthesis. Proc. nat. Acad. Sci., Wash. *60:* 222–228 (1968).

147 *Pitot, H.C.; Peraino, C.; Morse, P.A., and Potter, V.R.:* Hepatomas in tissue culture compared with adapting liver *in vivo.* Nat. Canc. Inst. Monogr. No. 13, pp. 229–245 (1964).

148 *Pitot, H.C.; Peraino, C.; Lamar, C., and Kennan, A.L.:* Template stability of some enzymes in rat liver and hepatoma. Proc. nat. Acad. Sci., Wash. *54:* 845–851 (1965).

149 *Pollock, M.R.:* The differential effect of actinomycin D on the biosynthesis of enzymes in Bacillus subtilis and Bacillus cereus. Biochim. biophys. Acta *76:* 80–93 (1963).

150 *Rechcigl, M., Jr. and Heston, W.E.:* Genetic regulation of enzyme activity in mammalian systems by the alteration of the rates of enzyme degradation. Biochem. biophys. res. Comm. *27:* 119–124 (1967).

151 *Rechcigl, M., Jr.:* Relative role of synthesis and degradation in the regulation of catalase activity; in *San Pietro, Lamborg and Kenney* Regulatory mechanisms for protein synthesis in mammalian cells, pp. 399–415 (Academic Press, New York/London 1968).

152 *Reel, J.R. and Kenney, F.T.:* 'Superinduction' of tyrosine transaminase in hepatoma cell cultures: Differential inhibition of synthesis and turnover by actinomycin D. Proc. nat. Acad. Sci., Wash. *61:* 200–206 (1968).

153 *Reich, E.; Franklin, R.M.; Shatkin, A.J., and Tatum, E.L.:* Action of actinomycin D on animal cells and viruses. Proc. nat. Acad. Sci., Wash. *48:* 1238–1245 (1962).

154 *Reif-Lehrer, L. and Amos, H.:* Hydrocortisone requirement for the induction of glutamine synthetase in chick-embryo retinas. Biochem. J. *106:* 425–430 (1968).

155 *Revel, M. and Hiatt, H.H.:* The stability of liver messenger RNA. Proc. nat. Acad. Sci., Wash. *51:* 810–818 (1964).

156 *Riggs, A.D. and Bourgeois, S.:* On the assay, isolation and characterization of the lac repressor. J. molec. Biol. *34:* 361–364 (1968).

157 *Rosen, F.; Roberts, N.R.; Budnick, L.E., and Nichol, C.A.:* An enzymatic basis for the gluconeogenic action of hydrocortisone. Science *127:* 287–288 (1958).

158 *Rosen, F. and Milholland, R.J.:* Effects of casein hydrolysate on the induction and regulation of tyrosine-α-ketoglutarate transaminase in rat liver. J. biol. Chem. *243:* 1900–1907 (1968).

159 *Roth, J.R.; Silbert, D.F.; Fink, G.R.; Voll, M.J.; Anton, D.; Hartman, P.E., and Ames, B.N.:* Transfer-RNA and the control of the histidine operon. Cold Spr. Harb. Symp. quant. Biol. *31:* 383–392 (1966).

160 *Ruddle, F.H.:* Isozymic variants as genetic markers in somatic cell populations *in vitro.* Nat. Cancer Inst. Monogr. No. 29, pp. 9–13 (1968).

161 *Salb, J.M. and Marcus, P.I.:* Translational inhibition in mitotic HeLa cells. Proc. nat. Acad. Sci., Wash. *54:* 1353-1358 (1965).

162 *Samuels, H.H. and Tomkins, G.M.:* J. molec. Biol. (in press).

163 *Scarano, E.; de Petrocellis, B., and Augusti-Tocco, G.:* Studies on the control of enzyme synthesis during early embryonic development of the sea urchins. Biochim. biophys. Acta *87:* 174–176 (1964).

164 *Scherrer, K.; Marcaud, L.; Zajdele, F.; London, I.M., and Gros, F.:* Patterns of RNA metabolism in a differentiated cell: A rapidly labeled, unstable 60S RNA with messenger properties in duck erythroblasts. Proc. nat. Acad. Sci., Wash. *56:* 1571–1578 (1966).

165 *Scherrer, K. and Marcaud, L.:* Messenger RNA in Avian erythroblasts at the transcriptional and translational levels and the problem of regulation in animal cells. J. Cell Physiol. *72:* 181–212 (1968).

166 *Schimke, R.T.:* Studies on factors affecting the levels of urea cycle enzymes in rat liver. J. biol. Chem. *238:* 1012–1018 (1963).

167 *Schimke, R.T.:* Enzymes of argine metabolism in cell culture: Studies on enzyme induction and repression. Nat. Cancer Inst. Monogr. No. 13, pp. 197–216 (1963).

168 *Schimke, R.T.:* Protein turnover and the regulation of enzyme levels in rat liver. Nat. Cancer Inst. Monogr. No. 27, pp. 301–314 (1966).

169 *Schimke, R.T.:* The study of enzyme regulatory mechanisms in cultured cells; in *Tritsch* Axenic mammalian cell reactions, pp. 181–217 (Marcel Dekker, New York/London 1969).

170 *Schimke, R.T.; Sweeney, E.W., and Berlin, C.M.:* The roles of synthesis and degradation in the control of rat liver tryptophan pyrrolase. J. biol. Chem. *240:* 322–331 (1965).

171 *Schwartz, A.G. and Amos, H.:* Insulin dependence of cells in primary culture: Influence on ribosome integrity. Nature, Lond. *219:* 1366–1367 (1968).

172 *Segal, H. and Kim, Y.S.:* Environmental control of enzyme synthesis and degradation. J. cell. comp. Physiol. *66:* 11–22 (1965).

173 *Shearer, R.W. and McCarthy, B.J.:* Evidence for ribonucleic acid molecules restricted to the cell nucleus. Biochemistry *6:* 283–289 (1967).

174 *Sheppard, D. and Englesberg, E.:* Further evidence for positive control of the *L*-arabinose system in gene araC. J. molec. Biol. *25:* 443–454 (1967).

175 *Singer, M.F. and Leder, P.:* Messenger RNA: An evaluation. Ann. Rev. Biochem., vol. 35, pp. 195–230 (Annual Reviews, Palo Alto 1966).

176 *Soeiro, R. and Amos, H.:* Arrested protein synthesis in polysomes of cultured chick embryo cells. Science *154:* 662–665 (1966).

177 *Stent, G.S.:* The operon: On its third anniversary. Science *144:* 816–820 (1964).

178 *Stubbs, D.W. and Haufrect, D.B.:* Effects of actinomycin D and puromycin on induction of gulonolactone hydrolase by somatotrophic hormone. Arch. Biochem. *124:* 365–371 (1968).

179 *Sussman, M.:* Protein synthesis and the temporal control of genetic transcription during slime mold development. Proc. nat. Acad. Sci., Wash. *55:* 813–818 (1966).

180 *Tata, J.R.:* Regulation of protein synthesis by growth and developmental hormones; in *San Pietro, Lamborg and Kenney* Regulatory mechanisms for protein synthesis in mammalian cells, pp. 299–321 (Academic Press, New York/London 1968). – Hormonal regulation of growth and protein synthesis. Nature, Lond. *219:* 331–337 (1968).

181 *Temin, H.M.:* Studies on carcinogenesis by Avian sarcoma viruses. VI. Differential multiplication of uninfected and of converted cells in response to insulin. J. Cell Physiol. *69:* 377–384 (1967).

182 *Thompson, E.B.; Tomkins, G.M., and Curran, J.F.:* Induction of tyrosine-α-ketoglutarate transaminase by steroid hormones in a newly established tissue culture line. Proc. nat. Acad. Sci., Wash. *56:* 296–303 (1966).

183 *Thompson, E.B.; Granner, D.K., and Tomkins, G.M.:* Superinduction of tyrosine aminotransferase by actinomycin D in HTC cells. J. molec. Biol. (in press).

184 *Thompson, E.B. and Gelehrter, T.D.:* (unpublished data).

185 *Tomkins, G.M. and Ames, B.N.:* The operon concept in bacteria and higher organisms. Nat. Cancer Inst. Monogr. No. 27, pp. 221–234 (1966).

186 *Tomkins, G.M.; Thompson, E.B.; Hayashi, S.; Gelehrter, T.; Granner, D., and Peterkofsky, B.:* Tyrosine transaminase induction in mammalian cells in tissue culture. Cold Spr. Harb. Symp. quant. Biol. *31:* 349–360 (1966).

187 *Tomkins, G.M.; Gelehrter, T.D.; Granner, D.K.; Peterkofsky, B., and Thompson, E.B.:* Regulation of gene expression in mammalian cells; in Exploitable molecular mechanisms and neoplasia, pp. 229–250 (Williams & Wilkins, Baltimore 1968).

188 *Tomkins, G.M.; Gelehrter, T.D.; Granner, D.; Martin, D.; Samuels, H.H., and Thompson, E.B.:* Control of specific gene expression in higher organisms. Science *166:* 1474–1480 (1969).

189 *Vesell, E. and Page, J.G.:* Genetic control of phenobarbital-induced shortening of plasma antipyrene half-lives in man. J. clin. Invest. *48:* 2202–2208 (1969).

190 *Vilček, J.; Rossman, T.G., and Varacalli, F.:* Differential effects of actinomycin D and puromycin on the release of interferon induced by double-stranded RNA. Nature, Lond. *222:* 682–683 (1969).

191 *Voytovich, A.E. and Topper, Y.J.:* Hormone-dependent differentiation of immature mouse mammary gland *in vitro.* Science *158:* 1326–1327 (1967).

192 *Wicks, W.D.; Kenney, F.T., and Lee, K-L:* Induction of hepatic enzyme synthesis *in vivo* by adenosine $3',5'$-monophosphate. J. biol. Chem. *244:* 6008–6013 (1969).

193 *Wicks, W.D.:* Induction of tyrosine-a-ketoglutarate transaminase in fetal rat liver. J. biol. Chem. *243:* 900–906 (1968).

194 *Wilson, S.H. and Hoagland, M.B.:* Physiology of rat liver polysomes. Biochem. J. *103:* 556–566 (1967).

195 *Wing, D.R. and Robinson, D.S.:* Clearing-factor lipase in adipose tissue. Biochem. J. *106:* 667–676 (1968).

196 *Wool, I.G.; Stirewalt, W.S.; Kurihara, K.; Low, R.B.; Bailey, P., and Oyer, D.:* Mode of action of insulin in the regulation of protein biosynthesis in muscle; in Recent progress in hormone research, vol. 24, pp. 139–208 (Academic Press, New York 1968).

197 *Wurtman, R.J. and Axelrod, J.:* Daily rhythmic changes in tyrosine transaminase activity of the rat liver. Proc. nat. Acad. Sci., Wash. *57:* 1594–1598 (1967).

198 *Yaffee, S.J.; Levy, G.; Matsuzawa, T., and Balich, T.:* Enhancement of glucuronide-conjugating capacity in a hyperbilirubinemic infant due to apparent enzyme induction by phenobarbital. New Engl. J. Med. *275:* 1461–1466 (1966).

199 *Yoshida, A.; Stamatoyannopoulos, G., and Motulsky, A.G.:* Negro variant of glucose-6-phosphate dehydrogenase deficiency (A⁻) in man. Science *155:* 97–99 (1967).

200 *Yoshida, A. and Motulsky, A.G.:* A pseudocholinesterase variant (E cynthiana) associated with elevated plasma enzyme activity. Amer. J. hum. Genet. *21:* 486–498 (1969).

Author's address: *Thomas D. Gelehrter,* M.D., Division of Medical Genetics, Department of Medicine, University of Washington, *Seattle, WA 98105* (USA). Present address: Division of Medical Genetics, Departments of Medicine and Pediatrics, Yale University School of Medicine, *New Haven, CT 06510* (USA)

Enzyme Synthesis and Degradation in Mammalian Systems, pp. 200–215
(Karger, Basel 1971)

Enzyme Repression[1]

O. Hänninen

Department of Physiology and Department of Biochemistry, University of Turku, Turku

Contents

I.   Introduction

The specific switching off and on of enzyme biosynthesis by a few repressor molecules at the DNA-level is a very economical way of regulating the cell metabolism. The blocking of enzyme synthesis at the translational level costs a large amount of high energy equivalents spent in transcription of DNA and in the transport of messenger RNA to the cytoplasm. The lowering of unnecessarily high enzyme levels by active degradation means an additional loss of energy already used for amino acid activation and for the translation of messenger

1 This review was aided by a grant from the National Research Council for Natural Sciences, Finland and US Public Health Service (AM-06018-07).

RNAs to numerous protein copies. Thus these other mechanisms used in mammalian cells to lower enzyme levels are very wasteful.

Enzyme repression will be defined in this context as a decrease or complete abolition of the transcription of a specific gene or functional genetic unit. This is reflected as a decrease of the corresponding protein or proteins in the cell. The definition used by microbiologists (14) according to which 'repression refers to inhibition of the synthesis of specific enzymes in response to the addition of a compound presumed to increase the effector concentration in the cell', is not applicable to the mammalian cells due to the active control of the translation and messenger transfer. *Weber* and collaborators (86) have suggested the use of the term 'suppression' to describe repression in mammalian cells in contrast to that in microorganisms. They want in this way to place the endogenous and extracellular (hormonal) repressing effectors into different categories. Unfortunately, suppression already has a specific meaning in molecular biology and should not be used in this second way.

The studies on the enzyme regulation in microorganisms have shown that the expression of genes (or gene groups) is controlled by repressing effectors, corepressors, which bind to specific regulator molecules, aporepressors. Aporepressors are products of specific regulator genes. The complex (or the aporepressor itself in inducible systems) has a high affinity towards the operator region of the operon in question. Only two aporepressor species have been isolated and characterized, the *lac*-operon aporepressor of *Escherichia coli*, a tetrametric protein (molecular weight 150,000) binding with double stranded *lac*-DNA and *lac*-effectors, which, when present, abolish the affinity of the repressor towards the *lac*-DNA (20), and λ-phage repressor, a protein with molecular weight 30,000 (60). Their properties fit nicely to the *Jacob-Monod* model (33) of the regulation of enzyme synthesis described above. Whether similar regulator genes, aporepressors and operons as in microorganisms exist in mammalian cells cannot be decided at the present moment due to the scanty experimental data.

According to *Britten*'s and *Davidson*'s theory (7) of gene regulation the genome in higher organisms contains gene batteries, sets of protein coding producer (structural) genes and receptor (operator) genes. One set (operon) may contain one or several receptor genes. The function of gene batteries is controlled by specific activator RNAs produced by integrator genes when activated by sensor genes. Several sensor genes control the same gene battery. There may be one or many integrator genes activated by a single sensor gene and sending activator RNA(s) to control each a set of gene batteries. This theory resembles the regulon concept in bacteria (44).

Mammalian cells like the cells of other eukaryotes contain much more DNA than the cells of prokaryotes. Apparently not all DNA is active in the current regulation of cell metabolism. The expression of certain genes is long-term repressed and inactive due to gene packaging. Other genes are in an active state and at least partly expressed due to a labile repression. In prokaryotes all genes are unpacked and either fully expressed or rapidly derepressable (inducible).

The studies of repression in mammalian cells are hampered much more than the studies of derepression by the diversity and many levels of control in protein metabolism. The existence of significant translational control (see *Pitot* in this volume) and the regulation by protein degradation (*Rechcigl* in this volume) indicate that the simple recording of decreasing enzyme levels does not necessarily indicate a decreased transcription of the corresponding DNA sequence due to a specific effector. The only fool proof evidence for a repression would be the demonstration of a decrease or abolition of the synthesis of a specific messenger RNA, which is impossible at the present moment. Compounds specifically blocking the template activity of DNA (39), inhibiting RNA polymerase (43) or ribosomal function (82) can be used in the studies of derepression to differentiate the effects of effectors at the transcriptional and translational levels, but they are less useful in the case of repression studies.

II. Repression Caused by Metabolites

In microorganisms repression is usually found in anabolic pathways like those for the biosynthesis of amino acids, vitamins, purines and pyrimidines. In some cases the substrates cause a derepression and the pathway products, a repression of the enzyme biosynthetic pathway (e.g. mandelate catabolism in *Pseudomonas*, which is controlled by multivalent and sequential derepression and repression) (75). In mammalian cells similar examples are known, which can be explained by means of specific repression, although the data published are not usually conclusive.

A. Amino Acid Metabolism

The first indication of repression in mammalian cells was the demonstration of decreased glutamyl transferase levels by glutamine in HeLa cell cultures (47).

The first case of enzyme repression described in living mammals was the decrease of arginine-glycine transamidinase levels in rat kidneys when feeding the animals with high creatine diet (16, 78a). Since chicks are more sensitive to effects of creatine they have been used to clarify the mechanism in detail (78b). The level of arginine-glycine transamidinase can be lowered very effectively by the addition of either creatine or precursor guanidinoacetic acid to the diet. The restoration of enzyme levels after derepression can be blocked by feeding ethionine, the effect of which can be overcome by methionine. This suggests that enzyme synthesis is needed to increase the enzyme activity during derepression (78c). Enzyme repression can be achieved even in chick embryos by injecting creatine into eggs. Also arginine has some repressing activity (79). Guanidinoacetic acid methyltransferase is not sensitive to creatine repression (78b).

It appears that arginine biosynthesis in mammalian cells is controlled by enzyme repression as in microorganisms (22). Both arginosuccinate synthetase and argininosuccinase are repressed in L-cell strain of mouse fibroblasts as in HeLa- and KB-cells, when they are grown in the presence of excess arginine. The enzyme levels increase 3- to 15-fold after cultivating them in the presence of growth limiting concentrations of citrulline or arginine (66). This suggests that the genes of these enzymes might be located in the same functional genetic unit. The arginase activity behaves inversely, which might indicate that arginine is capable of causing a derepression of arginase (66).

The biosynthesis of proline from glutamic acid is repressible in mouse fibro-blast L-cells. Exogenous proline added to culture medium represses the for-mation of Δ'-pyrroline-5-carboxylic acid reductase (62).

Serine biosynthesis is also repressable in mammalian cells. Serine in culti-vation medium lowers the levels of 3-phosphoglycerate dehydrogenase and phosphoserine phosphatase in KB-cells (58). Cysteine administration can prevent the increase of these two enzymes in rat liver, when the animals are kept on low protein diet to increase the enzyme levels (15).

L-Cysteine, *L*-cystine and mercaptopropionic acid cause a repression of alka-line phosphatase by some unknown mechanism in human skin fibroblasts *in vitro* (11). Inorganic phosphate does not cause a repression in cultivated mam-malian cells although it is a corepressor in microorganisms (52).

B.   Biosynthesis of Heme and Cytochromes

The initial and rate limiting enzyme in the porphyrin biosynthesis is δ-aminolevulinic acid synthetase located in mitochondria. Its biosynthesis can be induced by foreign compounds like allylisopropylacetamide and hormones like pregnandiol. This increased synthesis can be blocked in chick liver cultures by several metalloporphyrins, including heme. It is possible that heme and steroids are the natural effectors which control the enzyme synthesis under physiological conditions by binding to a corepressor of δ-aminolevulinic acid synthetase gene (23, 36). The glucuronic acid conjugates of the active steroids and these steroids in the presence of exogenous UDPglucuronic acid are unable to cause an induc-tion (36). The administration of hemin to rats causes irregular oscillations in δ-aminolevulinic acid synthetase levels in the liver (83).

Steroids have also been proposed to act as the natural inducers of the cytochrome P-450 cored drug metabolizing enzymes (10). Their absence causes an extensive repression of this enzyme machinery, e.g. in adrenalectomized rats (61). The administration of foreign compounds such as phenobarbitone causes high increase of these enzymes in rat liver microsomes. The administration of heme prevents the phenobarbitone mediated induction of cytochrome P-450 and cytochrome b_5 and that of δ-aminolevulinic acid synthetase in rat liver. It also prevents the phenobarbitone stimulated increases of microsomal protein, phospholipid and NADPH-cytochrome c reductase. Hematin and methalbumin are also potent effectors like heme itself, but bilirubin, peroxidized heme and inorganic iron are all ineffective in comparable concentrations (49).

The data above suggest that heme (or heme containing compounds) might act as corepressors in the biosynthesis of many microsomal and mitochondrial proteins.

After becoming barbitone dependent rats show drug hypersensivity due to a low level of various drug metabolizing enzymes and a low response to phenobarbitone induction (76). Furthermore, the inductive response of rats to barbitone decreases with increasing age (40). These findings suggest that earlier derepressions of the drug metabolizing enzymes might result in some kind of permanent memory in the liver. The successive derepression — repression cycles of cytochrome P-450 and glucuronolactone dehydrogenase of rat liver induced by phenobarbitone with intervals of one month repeated up to three times elicit, however, a similar response every time. If the cycles are repeated after an interval of only four days, a significantly lower response is obtained during the third than during the first cycle. These results indicate that previous cycles do not cause any memory effects in cytochrome P-450 or glucuronolactone dehydrogenase synthesis, if the interval is long enough to permit a reconstitution after a derepression phase (32).

C. Nucleotide Metabolism

Cultured mammalian cells are able to synthesize purines and pyrimidines from precursors. Cultivation of human carcinoma HeLa cells in the presence of adenine depresses the incorporation of glycine into adenine and guanine (65). Purine biosynthesis is also lowered in L-strain cells of mouse fibroblasts in the presence of adenine or guanosine. This suggests an end product repression, which is not, however, due to a repression of inosinicase, IMP-dehydrogenase or adenylosuccinase but of some so far unidentified enzyme (45). Aspartate transcarbamylase, an enzyme of pyrimidine biosynthesis, can be repressed by uridine in Sarcoma-180 cells, when grown in the presence of 6-azauridine to cause a pyrimidine deficiency and aspartate transcarbamylase derepression (13).

It is also possible that the degradative enzymes are repressible. The cultivation of mouse fibroblasts in the presence of nicotinamide adenine dinucleotide lowers very rapidly the NADase activity in cells. This probably is not due entirely to the suggested and confirmed inhibition of the enzyme, but also to repressed enzyme synthesis, since low enzyme levels persist for several cell generations even in the absence of NAD in the culture medium (42).

D. Lipid Biosynthesis

The biosynthesis of cholesterol from acetate in liver is very sensitive to dietary cholesterol content. Biosynthesis is nearly completely prevented in rats and also in man, when the cholesterol content of diet increases over 0.5 %. The blockage is mediated by a lipoproteincholesterol complex, which acts on β-hydroxy-β-methylglutaryl-CoA conversion to mevalonate. The inhibition is rapidly effective (2.5 h) *in vivo*, and it has been explained to be due to the specific feedback inhibition of the respective reductase (69), although an inhibition of enzyme synthesis has not been experimentally excluded. During cholesterol feeding the conversion of mevalonate to cholesterol is also markedly depressed, which has been explained by assuming a repression of corresponding enzyme machinery (70).

E. Carbohydrate Metabolism

In the carbohydrate pathways repressions caused by pathway metabolites are not common. For example the biosynthesis of glucose by gluconeogenic enzymes is not directly controlled by the end product. Glucose causes a repression or derepression of these enzymes only indirectly by controlling the release of hormones (86).

In the glucuronic acid pathway UDP glucuronyltransferase of embryonic liver is repressed by some unknown compound(s) excreted from chorioallantoic membrane (71). Of the various pathway enzymes only hepatic microsomal uronolactonase appears to be decreased (repressed) after the administration of D-glucaro-1, 4-lactone to rats, and none after the administration of D-glucuronolactone (fig. 1) (30b). Both of these metabolites are powerful feedback inhibitors of some glucuronic acid pathway enzymes *in vitro*: D-glucaro-1, 4-lactone of β-glucuronidase and D-glucuronolactone of glucuronate reductase (30a).

The administration of many chemical compounds foreign to the body causes a derepression of some enzymes of the glucuronic acid pathway, e.g. UDP glucuronyltransferase. One model compound, cinchophen, when administered to rats, causes also an extensive depression of hepatic L-gulonolactone oxidase, renal *myo*-inositol oxygenase, and a less marked depression of hepatic and renal UDPglucuronyl pyrophosphatase, renal glucuronic acid 1-phosphatase, renal glucuronate reductase and hepatic, renal and mucosal (small intestine) 3-hydro-xyacid dehydrogenase. These depressions together with the induction and in-creased levels of UDPglucose dehydrogenase and UDP glucuronyltransferase in the liver, kidney and small intestinal mucosa and glucuronolactone dehydrogenase in the liver and small intestinal mucosa all direct the flux in the glucuronic acid pathway to support and enhance the biosynthesis of glucuronides. The structural genes of the pathway enzymes are located in different genome sections (30a).

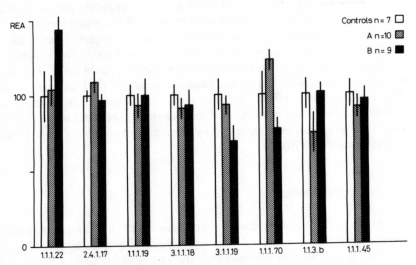

Fig. 1. The effect of *D*-glucuronolactone (**A**) and *D*-glucaro-1,4-lactone (**B**) adminis-trations on the activity of various enzymes of the glucuronic acid pathway in the rat liver. Metabolites were given intraperitoneally 5 days at a dosage level of 7 mmoles/kg of body weight. 1.1.1.22 UDPglucose dehydrogenase, 2.4.1.17 UDP glucuronyltransferase, 1.1.1.19 glucuronate reductase, 3.1.1.18 aldonolactonase, 3.1.1.19 uronolactonase, 1.1.1.70 glucu-ronolactone dehydrogenase, 1.1.3.b gulonolactone oxidase and 1.1.1.45 3-hydroxyacid dehydrogenase. REA is the relative enzyme activity, when the mean activity of the controls has been taken as 100. The number of rats and the standard errors of the means are given.

F. Catabolite Repression

Carbohydrates, especially glucose or even more generally any compound which can serve efficiently as a source of intermediary metabolites and of energy, have a repressive effect on the biosynthesis of many enzymes in different pathways in microorganisms. This effect has been called the *glucose effect or more generally catabolite repression* (46). Similar effects have been observed in mammalian tissues.

Sucrose administration inhibits effectively the allylisopropylacetamide induced increase of hepatic δ-aminolevulinic acid synthetase in rats (77). Glucose administration inhibits effectively the casein induced synthesis of threonine dehydratase and ornithine ketoacid aminotransferase (57), but this has been shown in subsequent studies to be due to the inhibition of translation and not due to that of transcription (54). Glucose administration causes a cessation of serine dehydratase synthesis, but the amount of enzyme protein bound to ribosomes is much increased which might suggest an inhibition of enzyme release (56). A similar mechanism may be behind other glucose effects of mammalian cells, and it has also been described in microorganisms during glucose repression (25). The glucose and fructose blockage of the fasting-mediated increase of hepatic dimethylaminoazobezene reductase probably takes place at the translational level (34). The increase of hepatic phosphoenolpyruvate carboxykinase during fasting returns to normal levels after refeeding with glucose or glycerol containing diets. Insulin may be involved in this effect (87). Glucose administration counteracts the hormone or casein hydrolysate induced increases of several aminotransferases and some other amino acid catabolizing enzymes in rat liver, but the mechanism has not been clarified (55, 64).

III. Hormonal Repression

In multicellular organisms like mammals the hormones have an important role in the regulation of metabolism. The release of hormones is regulated by small molecular weight metabolites or by other hormones. It is very probable that in many cases these effectors affect the hormone biosynthesis via repression and/or derepression, but this has not been verified. In the peripheral target tissues hormones are mostly metabolic activators either at the level of enzyme catalysis (e.g. cyclic AMP mediated effects) or at the level of enzyme biosynthesis.

Glucocorticoids induce the synthesis of hepatic gluconeogenic enzymes, puryvate carboxylase, phosphoenolpyruvate kinase, fructose diphosphatase and glucose-6-phosphatase (86). The key enzymes of glycolysis, glucokinase, phosphofructokinase and pyruvate kinase in the rat liver are induced by insulin as is glycogen synthetase (85). On the other hand, insulin causes a repression of gluconeogenic enzymes, and it even prevents the effect of glucocorticoids in the biosynthesis of gluconeogenic enzymes shown by the decrease of glucocorticoid induced RNA biosynthesis and enzyme levels after insulin administration (84).

A selective repression of hepatic tyrosine aminotransferase, based on the inhibition of RNA synthesis, has been demonstrated in rats. The synthesis is completely blocked by moderate doses of growth hormone (37). The growth hormone effect probably explains the repression of tyrosine aminotransferase synthesis during stress, since this stress repression is seen only in the intact but not in hypophysectomized rats (38).

IV. Long-term Repression

Mammalian cells contain about three orders of magnitude more DNA than prokaryote cells. Only a small fraction of native mammalian nuclear DNA is, however, able to act as a template in RNA synthesis (3). In adult mammalian hepatocyte less than 5 % of the genome is expressed (6).

A. Redundancy

The high amount of DNA in mammalian cells is not only due to the addition of new genes not found in prokaryotes, but also to high redundancy. In fact more than one third of the DNA of higher organisms is made up of repeated sequences, which recur anywhere from a thousand to a million times per cell. Only about 60 % of calf DNA does not exhibit repeated sequences (8). The redundant sequences may be either tandem or probably even distributed in different genome segments. This redundancy is best known in the case of ribosomal DNA (63). Certain DNA sequences may also be amplified and serve as templates of 'slaves', which may remain either in the neighbourhood of their 'masters' (9), or they may be transferred outside of the chromosomes as ribosomal genes in amphibian oocytes (18, 51). It is not known whether a similar amplification occurs in mammalian tissues.

The multiple copies of certain genes may provide more templates of important RNAs and proteins, which are of importance during development, e.g. by providing more ribosomes for protein synthesis (53) and for the specific functions carried out by the highly specialized cells. It is possible that the remarkably stable graded specialization of the mucosal cells in the small intestine is due to graded masking of redundant genes (1, 31).

During development and cellular differentiation there is an orderly progression of gene activation and masking. This leads to a situation in which fully differentiated cells, despite their common genotype, express different genes as shown by RNA hybridization experiments (21a) and by the distribution of various enzymes in different tissues (12). Even functionally related enzymes may be absent in some tissues and present in others (30a). The inactivation of genes is not necessarily, however, irreversible, and it can be reversed; e.g. by hormones (24). This process needs complicated regulatory mechanisms.

B. Chromatin Proteins

The long-term switching off of unused genome regions in the differentiated mammalian tissues apparently takes place by stable masking of the genes and supercoiling of DNA with the aid of histones and nonhistone chromatin proteins.

Quantitatively the different histone fractions are the most important chromatin proteins, since the total amount of histones is approximately equal to the amount of DNA (for recent reviews 19, 35). *Stedman and Stedman* (74) already have suggested that histones might play a role in the control of gene activity. Native calf thymus nucleohistone is a poor primer for RNA synthesis *in vitro* and probably also *in vivo*. The removal of histones increases 300–400 % the RNA synthesis, while the restoration of histones to the nuclei results in an immediate inhibition of nucleotide incorporation (2, 3). Histones inhibit the transcription by forming complexes with DNA (lysine-rich histones) and by acting also directly on RNA polymerase (arginine-rich histones) (28, 72, 73). Avian erythrocyte nuclei are completely inactive, and they contain a specific histone, not found in any other bird tissue, which replaces the lysine-rich histones, when the nuclei become inactive (35).

The primary structure of calf thymus histone fraction IV has been described (41). Its N-terminal part contains many positive charges, and it is probably involved with the binding to DNA. This portion contains a unique lysyl-residue, which can be acetylated. The enzymatic acetylation of histone has been suggested to be necessary for the reopening of the masked genome sequences. The acetylation is much increased, when RNA synthesis increases, e.g. after partial hepatectomy in the liver (2, 59).

The probable function of histones is to act as unspecific repressors by taking part in the organization of chromosomes and supercoiling the appropriate DNA sequences which leads to gene packaging (19, 35). The repressed chromatin occurs as condensed masses of heterochromatin while active chromatin occurs as extended euchromatin (17).

In addition to histones the chromatin contains many types of nonhistone proteins. More nonhistone proteins are found in the template active than repressed chromatin of calf thymus cells (17). The nonhistone proteins may be responsible for the organ-specific template activity of chromatin (21b).

Most of the nonhistone proteins of chromatin are acidic. These acidic proteins are very heterogeneous – and include DNA polymerase. These proteins can form insoluble complexes with the histone subfractions (80, 81), and they combine also with DNA, but these complexes have the same template activity as pure DNA (48). The turnover of the acidic proteins is rapid (26) and some of them rapidly incorporate phosphate (5).

Chromatin also contains significant amounts of ribonucleoproteins. They are necessary for the reconstitution of functional chromatin *in vitro* (4, 29). The ribonucleic acid chain of calf thymus chromatin ribonucleoprotein consists of 40 nucleotides which include 3–4 dihydrouridylic acid residues per chain (28, 68). The protein part of the molecule is acidic at least in chick embryo chromatin (29). The turnover of this ribonucleoprotein is probably very high, since nuclear RNA has in general a very short half-life (27, 67).

The properties of the acidic nonhistone chromatin proteins suggest that they might contain the specific aporepressors in mammalian cells. In one case a specific uptake of a corepressor by these proteins has been demonstrated. Estradiol-17β is stereospecifically bound with high affinity (equilibrium constant 2×10^{-8} molar) by a nonhistone protein of calf endometrial chromatin (50).

V. Summary

Mammalian cells contain about three orders of magnitude more DNA than the prokaryote cells. Mammalian DNA exhibits a high redundancy. A major part of the DNA in the mammalian cells is template inactive due to a long-term repression caused by gene masking and DNA supercoiling, in which the chromatin proteins, especially histones, are important. This type of repression is very stable under various physiological conditions in fully differentiated cells. A minor part of the DNA is template active. Genes of these areas are expressed either fully or they are derepressable.

The biosynthesis of many enzymes in mammalian cells is repressable by simple metabolites like pathway endproducts in much the same way as in microorganisms. Examples are given from the metabolism of amino acids, heme and cytochromes, nucleotides, lipids and carbohydrates. Catabolite repression, which is very important in the control of the biosynthesis of many enzymes in microorganisms, can also be demonstrated in mammalian cells. Some hormones, which mostly act as stimulators of metabolism, work also by enzyme repression.

In some cases the repression can be explained on the basis of the microbial repression model. Development and differentiation which lead to the expression of different portions of the genome in different cell types of the same organism having the same genotype by specific long-term repression need, however, additional regulatory mechanisms to control this complicated progress.

VI. References

1   *Aitio, A., Hänninen, O., and Hietanen, E.:* The stability of the metabolic gradient in the rat small intestine. Abstracts, Meeting of the Scandinavian Physiological Society, Helsinki, pp. 6–7 (1970).

2   *Allfrey, V.G.:* Control mechanisms in ribonucleic acid synthesis. Cancer Res. *26:* 2026–2040 (1966). – Some observations on histone acetylation and its temporal

relationship to gene activation; in *San Pietro, Lamborg and Kenney* Regulatory mechanisms for protein synthesis in mammalian cells, pp. 65–100 (Academic Press, New York 1968).

3   *Allfrey, V.G. and Mirsky, A.E.:* Evidence for the complete DNA-dependence of RNA synthesis in isolated thymus nuclei. Proc. nat. Acad. Sci., Wash. *48:* 1590–1596 (1962).

4   *Bekhor, I.; Kung, G.M., and Bonner, J.:* Sequence-specific interaction of DNA and chromosal protein. J. molec. Biol. *39:* 351–364 (1969).

5   *Benjamin, W. and Gellhorn, A.:* Acidic proteins of mammalian nuclei: isolation and characterization. Proc. nat. Acad. Sci., Wash. *59:* 262–268 (1968).

6   *Bresnick, E. and Madix, J.C.:* Activation of chromatin by 3-methylcholanthrene; in *Gillette, Conney, Cosmides, Estabrook, Fouts and Mannering* Microsomes and drug oxidations, pp. 431–449 (Academic Press, New York 1969).

7   *Britten, R.J. and Davidson, E.H.:* Gene regulation for higher cells: A theory. Science *165:* 349–357 (1969).

8   *Britten, R.J. and Kohne, D.E.:* Repeated sequences in DNA. Science *161:* 529–540 (1968).

9   *Callan, H.G.:* The organization of genetic units in chromosomes. J. Cell Sci. *2:* 1–7 (1967).

10 *Conney, A.H.:* Pharmacological implications of microsomal enzyme induction. Pharmacol. Rev. *19:* 317–366 (1967).

11 *Cox, R.P. and MacLeod, C.M.:* Repression of alkaline phosphatase in human cell cultures by cysteine and cystine. Proc. nat. Acad. Sci., Wash. *49:* 504–510 (1963). – Regulation of alkaline phosphatase in human cell cultures. Cold Spr. Harb. Symp. quant. Biol. *29:* 233–251 (1964).

12 *Dixon, M. and Webb, E.C.:* Comparative enzyme biochemistry; Enzymes, 2nd ed., pp. 636–659 (Longmans, Green and Co, London 1964).

13 *Ennis, H.L. and Lubin, M.:* Capacity for synthesis of a pyrimidine biosynthetic enzyme in mammalian cells. Biochim. biophys. Acta *68:* 78–83 (1963).

14 *Epstein, W. and Beckwith, J.R.:* Regulation of gene expression. Annu. Rev. Biochem. *37:* 411–436 (Annual Reviews, Palo Alto 1968).

15 *Fallon, H.J.; Hackney, E.J., and Byrne, W.L.:* Serine biosynthesis in rat liver. Regulation of enzyme concentrations by dietary factors. J. biol. Chem. *241:* 4157–4167 (1966).

16 *Fitch, C.D.; Hsu, C., and Dinning, J.S.:* Some factors affecting kidney transamidinase activity in rats. J. biol. Chem. *235:* 2362–2364 (1960).

17 *Frenster, J.H.:* Nuclear polyanions as derepressors of synthesis of ribonucleic acid. Nature *206:* 680–683 (1965).

18 *Gall, J.G.:* Differential synthesis of the genes for ribosomal RNA during amphibian oogenesis. Proc. nat. Acad. Sci., Wash. *60:* 553–560 (1968). – The genes for ribosomal RNA during oogenesis. Genetics *61:* suppl. 1, 121–132 (1969).

19 *Georgiev, G.P.:* Histones and the control of gene action. Annu. Rev. Genet. *3:* 155–180 (Annual Reviews, Palo Alto 1969).

20 *Gilbert, W. and Müller-Hill, B.:* The *lac* operator is DNA. Proc. nat. Acad. Sci., Wash. *58:* 2415–2421 (1967).

21 *Gilmour, R.S. and Paul, J.:* (a) Restriction of deoxyribonucleic acid template activity in chromatin is organ specific. Nature *210:* 992–993 (1966). – (b) The nature of the specific restriction of template activity in the chromatin of animal cells. Biochem. J. *104:* 27P–28P (1967).

22 *Gorini, L.; Gundersen, W., and Burger, M.:* Genetics of regulation of enzyme synthesis in the arginine biosynthesis pathway of *Escherichia coli.* Cold Spr. Harb. Symp. quant. Biol. *26:* 173–182 (1961).

23 *Granick, S.:* The induction *in vitro* of the synthesis of δ-aminolevulinic acid synthetase in chemical porphyria: A response to certain drugs, sex hormones, and foreign compounds. J. biol. Chem. *241:* 1359–1375 (1966).

24 *Hamilton, T.H. and Teng, C.-S.:* Regulation by estrogen of synthesis of chromatin-directed RNA and of non-histone chromatin proteins. Genetics *61:* suppl. 1, 381–390 (1969).

25 *Hauge, J.G.; MacQuillan, A.M.; Cline, A.L., and Halvorson, H.O.:* The effect of glucose repression on the level of ribosomal-bound beta-glucosidase. Biochem. biophys. Res. Commun. *5:* 267–269 (1961).

26 *Holoubek, V. and Crocker, T.T.:* DNA-associated acid proteins. Biochim. biophys. Acta *157:* 352–361 (1968).

27 *Houssais, J.-F. and Attardi, G.:* High molecular weight nonribosomal-type nuclear RNA and cytoplasmic messenger RNA in HeLa cells. Proc. nat. Acad. Sci., Wash. *56:* 616–623 (1966).

28 *Huang, R.C. and Bonner, J.:* Histone, a suppressor of chromosomal RNA synthesis. Proc. nat. Acad. Sci., Wash. *48:* 1216–1222 (1962). – Histone-bound RNA, a component of native nucleohistone. Proc. nat. Acad. Sci., Wash. *54:* 960–967 (1965).

29 *Huang, R.C. and Huang, P.C.:* Effect of protein-bound RNA associated with chick embryo chromatin on template specificity of the chromatin chick embryo. J. molec. Biol. *39:* 365–378 (1969).

30 *Hänninen, O.:* (a) On the metabolic regulation in the glucuronic acid pathway in the rat tissues. Ann. Acad. Sci. fenn. A 11, 142 (1968). – (b) Enzyme repression and derepression by pathway metabolites in the glucuronic acid cycles (in press).

31 *Hänninen, O. and Hartiala, K.:* Studies on inorganic phosphate, inorganic pyrophosphatase and alkaline nonspecific phosphomonoesterase levels in the gastrointestinal tract of the rat. Acta chem. scand. *19:* 817–822 (1965).

32 *Hänninen, O.; Kivisaari, E., and Antila, K.:* Repeated derepression cycles of the glucuronolactone dehydrogenase and cytochrome P-450 in the rat liver induced by phenobarbitone administration. Biochem. Pharmacol. *18:* 2203–2210 (1969).

33 *Jacob, F. and Monod, J.:* Genetic regulatory mechanisms in the synthesis of proteins. J. molec. Biol. *3:* 318–356 (1961).

34 *Jervell, K.F.; Christoffersen, T., and Mörland, J.:* Studies on the 3-methylcholanthrene induction and carbohydrate repression of rat liver dimethylaminoazobenzene reductase. Arch. Biochem. *111:* 15–22 (1965).

35 *Johns, E.W.:* The histones, their interactions with DNA, and some aspects of gene control; in *Wolstenholme and Knight* Homeostatic regulators, pp. 128–143 (Churchill, London 1969).

36 *Kappas, A. and Granick, S.:* Steroid induction of porphyrin synthesis in liver cell culture. II. The effects of heme, uridine diphosphate glucuronic acid, and inhibitors of nucleic acid and protein synthesis on the induction process. J. biol. Chem. *243:* 346–351 (1968).

37 *Kenney, F.T.:* Regulation of tyrosine alfa-ketoglutarate transaminase in rat liver. V. Repression in growth hormone-treated rats. J. biol. Chem. *242:* 4367–4371 (1967).

38 *Kenney, F.T. and Albritton, W.L.:* Repression of enzyme synthesis at the translational level and its hormonal control. Proc. nat. Acad. Sci., Wash. *54:* 1693–1698 (1965).

39 *Kersten, H. and Kersten, W.:* Inhibitors acting on DNA and their use to study DNA replication and repair; in *Bücher and Sies* Inhibitors tools in cell research; pp. 11–31 (Springer Verlag, Berlin 1969).

40 *Klinger, W. and Kramer, B.:* Untersuchungen zum Mechanismus der Enzyminduktion bei Ratten und Mäusen. Acta biol. med. germ. *15:* 707–711 (1965).

41 *Lange, R.J. de; Fambrough, D.M.; Smith, E.L., and Bonner, J.:* Calf and pea histone IV. II. The complete amino acid sequence of calf thymus histone IV; presence of e-N-acetyllysine. J. biol. Chem. *244:* 319–334 (1969).

42 *Lieberman, I.:* The mechanism of the specific depression of an enzyme activity in cells in tissue culture. J. biol. Chem. *225:* 883–898 (1957).

43 *Lill, U.; Santo, R.; Sippel, A., and Hartman, G.:* Inhibitors of the RNA polymerase reaction; in *Bücher and Sies* Inhibitors tools in cell research; pp. 48–59 (Springer Verlag, Berlin 1969).

44 *Mass, W.K. and Clark, A.J.:* Studies on the mechanism of repression or arginine biosynthesis in *Escherichia coli.* II. Dominance or repressibility in diploids. J. molec. Biol. *8:* 365–370 (1964).

45 *McFall, E. and Magasanik, B.:* The control of purine biosynthesis in cultured mammalian cells. J. biol. Chem. *235:* 2103–2108 (1960).

46 *Magasanik, B.:* Catabolite repression. Cold Spr. Harb. Symp. quant. Biol. *26:* 249–254 (1961).

47 *Mars, R., de:* The inhibition by glutamine of glutamyl transferase formation in human cells in culture. Biochim. biophys. Acta *27:* 435–436 (1958).

48 *Marushige, K.; Brutlog, D., and Bonner, J.:* Properties of chromosomal nonhistone protein of rat liver. Biochemistry *7:* 3149–3155 (1968).

49 *Marver, H.S.:* The role of heme in the synthesis and repression of microsomal protein; in *Gillette, Conney, Cosmides, Estabrook, Fouts and Mannering* Microsomes and drug oxidations; pp. 495–511 (Academic Press, New York 1969).

50 *Maurer, H.R. and Chalkley, G.R.:* Some properties of a nuclear binding site of estradiol. J. molec. Biol. *27:* 431–441 (1967).

51 *Miller, O.L., Jr. and Beatty, B.R.:* Extrachromosomal nucleolar genes in amphibian oocytes. Genetics *61:* suppl. 1, 133–143 (1969).

52 *Nitowsky, H.M. and Herz, F.:* Alkaline phosphatase in cell cultures of human origin. Nature *189:* 756–757 (1961).

53 *Pavan, C. and Cunha, A.B. da:* Gene amplification in ontogeny and phylogeny of animals. Genetics *61:* suppl. 1, 290–304 (1969).

54 *Peraino, C.; Lamar, C., Jr., and Pitot, H.C.:* Studies on the mechanism of carbohydrate repression in rat liver. Adv. Enzyme Regulation, vol. 4, pp. 199–217 (Pergamon Press, 1966).

55 *Pestana, A.:* Dietary and hormonal control of enzymes of amino acid catabolism in liver. Europ. J. Biochem. *11:* 400–404 (1969).

56 *Pitot, H.C. and Jost, J.-P.:* Studies on the regulation of the rate of synthesis and degradation of serine dehydratase by amino acids and glucose *in vivo;* in *San Pietro, Lamborg and Kenney* Regulatory mechanisms for protein synthesis in mammalian cells; pp. 283–298 (Academic Press, New York 1968).

57 *Pitot, H.C. and Peraino, C.:* Carbohydrate repression of enzyme induction in rat liver. J. biol. Chem. *238:* PC 1910–PC 1912 (1963).

58 *Pizer, L.I.:* Enzymology and regulation of serine biosynthesis in cultured human cells. J. biol. Chem. *239:* 4219–4226 (1964).

59 *Pogo, B.G.T.; Pogo, A.O., and Allfrey, V.G.:* Histone acetylation and RNA synthesis in rat liver regeneration. Genetics *61:* suppl. 1, 373–379 (1969).

60 *Ptashne, M.:* Specific binding or the λ-phage repressor to λ-DNA. Nature *214:* 232–234 (1967).

61 *Remmer, H.:* Die Wirkung der Nebennierenrinde auf den Abbau von Pharmaka in den Lebermikrosomen. Naturwissenschaften *45:* 522–523 (1958).

62 *Rickenberg, H.V.:* In discussion to the paper of *Schimke.* Studies on adaptation of urea cycle enzymes in the rat. Cold Spr. Harb. Symp. quant. Biol. *26:* 366 (1961).

63 *Ritossa, F.M. and Scala, G.:* Equilibrium variations in the redundancy of rDNA in *Drosophila melanogaster.* Genetics *61:* suppl. 1, 305–317 (1969).

64 *Sahib, M.K. and Murti, C.R.K.:* Induction of histidine-degrading enzymes in protein-starved rats and regulation of histidine metabolism. J. biol. Chem. *244:* 4730–4734 (1969).

65 *Salzman, N.P. and Sebring, E.D.:* Utilization of precursors for nucleic acid synthesis by human cell cultures. Arch. Biochem. *84:* 143–150 (1959).

66 *Schimke, R.T.:* Enzymes of arginine metabolism in mammalian cell culture. I. Repression of arginine-succinate synthetase and argininosuccinase. J. biol. Chem. *239:* 136–145 (1964).

67 *Shearer, R.W. and McCarthy, B.J.:* Evidence for ribonucleic acid molecules restricted to the cell nucleus. Biochemistry *6:* 283–289 (1967).

68 *Shih, T.Y. and Bonner, J.:* Chromosomal RNA of calf thymus chromatin. Biochim. biophys. Acta *182:* 30–35 (1969).

69 *Siperstein, M.D. and Fagan, V.M.:* Studies on the feedback regulation of cholesterol synthesis. Adv. Enzyme Regulation, vol. 2, pp. 249–264 (Pergamon Press, Oxford 1964).

70 *Siperstein, M.D. and Guest, M.:* Studies on the site of the feedback control of cholesterol synthesis. J. clin. Invest. *39:* 642–652 (1960).

71 *Skea, B.R. and Nemeth, A.M.:* Factors influencing premature induction of UDP-glucuronyltransferase activity in cultured chick embryo liver cells. Proc. nat. Acad. Sci., Wash. *64:* 795–798 (1969).

72 *Sonnenberg, B.P. and Zubay, G.:* Nucleohistone as a primer for RNA synthesis. Proc. nat. Acad. Sci., Wash. *54:* 415–420 (1965).

73 *Spelsberg, T.C.; Tankersley, S., and Hnilica, L.S.:* The interaction of RNA polymerase with histones. Proc. nat. Acad. Sci., Wash. *62:* 1218–1225 (1969).

74 *Stedman, E. and Stedman, E.:* Cell specificity of histones. Nature *166:* 780–781 (1950).

75 *Stevenson, I.L. and Mandelstam, J.:* Induction and multi-sensitive end-product repression in two converging pathways degrading aromatic substances in *Pseudomonas fluorescens.* Biochem. J. *96:* 354–362 (1965).

76 *Stevenson, I.H. and Turnbull, M.J.:* Hepatic drug-metabolizing enzyme activity and duration of hexobarbitone anaesthesia in barbitone-dependent and withdrawn rats. Biochem. Pharmacol. *17:* 2297–2305 (1968).

77 *Tschudy, D.P.; Welland, F.H.; Collins, A., and Hunter, G.:* The effect of carbohydrate feeding on the induction of δ-aminolevulinic acid synthetase. Metabolism *13:* 396–406 (1964).

78 *Walker, J.B.:* (a) Repression of arginine-glycine transamidinase activity by dietary creatine. Biochem. biophys. Acta *36:* 574–575 (1959). – (b) Metabolic control of creatine biosynthesis. I. Effect of dietary creatine. J. biol. Chem. *235:* 2357–2361 (1960). – (c) Metabolic control of creatine biosynthesis. II. Restoration of transamidinase activity following creatine repression. J. biol. Chem. *236:* 493–498 (1961).

79 *Walker, M.S. and Walker, J.B.:* Repression of transamidinase activity during embryonic development. J. biol. Chem. *237:* 473–476 (1962).

80 *Wang, T.Y.:* Nonhistone chromatin proteins from calf thymus and their role in DNA biosynthesis. Arch. Biochem. *122:* 629–634 (1967).

81 *Wang, T.Y. and Johns, E.W.:* Study of the chromatin acidic proteins of rat liver: heterogeneity and complex formation with histones. Arch. Biochem. *124:* 176–183 (1968).

82 Vazquez, D.; Staehelin, T.; Celma, M.L.; Battaner, E.; Fernandez-Munoz, R., and Monro, R.E.: Inhibitors as tools elucidating ribosomal function; in Bücher and Sies Inhibitors tools in cell research, pp. 100–123 (Springer Verlag, Berlin 1969).

83 Waxman, A.D.; Collins, A., and Tschudy, D.P.: Oscillations of hepatic δ-aminolevulinic acid synthetase produced in vivo by heme. Biochem. biophys. Res. Commun. 24: 675–683 (1966).

84 Weber, G.; Singhal, R.L., and Srivastava, S.K.: Insulin: Suppressor of biosynthesis of hepatic gluconeogenic enzymes. Proc. nat. Acad. Sci., Wash. 53: 96–104 (1965).

85 Weber, G.; Singhal, R.L.; Stamm, N.B.; Lea, M.A., and Fisher, E.A.: Synchronous behavior pattern of key glycolytic enzymes, glucokinase, phosphofruktokinase and pyruvate kinase. Adv. Enzyme Regulation, vol. 4, pp. 59–81 (Pergamon Press, Oxford 1966).

86 Weber, G.; Singhal, R.L.; Stamm, N.B., and Srivastava, S.K.: Hormonal induction and suppression of liver enzyme biosynthesis. Fed. Proc. 24: 745–754 (1965).

87 Yang, J.W.; Shrago, E., and Lardy, H.A.: Metabolic control of enzymes involved in lipogenesis and gluconeogenesis. Biochemistry 3: 1687–1692 (1964).

Author's address: O. Hänninen, Ph.D., M.D., Department of Physiology, Kiinamyllynk 10, Turku 3 (Finland)

Enzyme Synthesis and Degradation in Mammalian Systems, pp. 216–235
(Karger, Basel 1971)

Translational Regulation of Enzyme Levels in Liver [1]

H.C. Pitot, J. Kaplan and A. Čihák [2]

McArdle Laboratory, Departments of Oncology and Pathology,
University of Wisconsin, Medical School, Madison, Wis.

Contents

Although classically the presentation of the Operon concept by *Jacob and Monod* (21) argued that all regulation of genetic expression should be at the transcriptional level, sufficient experimental evidence has now accumulated, both from microbial systems and in mammals, to demonstrate rather conclusively that the amount of an enzyme present in a cell or tissue may be the direct result of effects at the translational level of genetic expression. In this short review we will present evidence from data compiled in our own and other laboratories for the existence of the translational regulation of the rate of enzyme synthesis in mammalian liver. In addition, some possible interrelationships between the translational regulation of enzyme synthesis and enzyme degradation will be presented.

1 Part of the work originating from this laboratory was supported by grants from the National Cancer Institute (CA-7175) and the American Cancer Society (P-314).
2 Eleanor Roosevelt International Fellow in Cancer Research (1968–69).

I. Translational Regulation – Definition and Evidence for its Existence

In viewing the dogma of molecular biology as it is presently held (figure 1), the primary potential sites for the regulation of genetic expression are at the level of mRNA synthesis -1- (transcription) and of protein or enzyme synthesis -5- (translation). In addition, there are obviously a number of other possible sites of regulation including transportation, i.e. the transport of mRNA from the nucleus to the cytoplasm, -2- the formation of actively translating polysomes -3- and post translational events, -6-7-8- i.e. the conversion of an inactive to an active form (trypsinogen to trypsin), the allosteric activation or inhibition of enzyme activity, or the *in vivo* degradation of the synthesized enzyme.

In the past decade there have been several examples of long-lived mRNA existing in various bacterial cells. In 1965 *Coleman and Elliott* (8) described the stimulation of extracellular ribonuclease formation in *B. subtilis* by the addition of Actinomycin D to the culture. Stimulation of enzyme synthesis by the antibiotic was dependent on the concentration of the inhibitor used; however, levels of Actinomycin which inhibited RNA synthesis more than 60 % had virtually no effect on ribonuclease synthesis. As we shall see later, the phenomenon of the Actinomycin D stimulation of enzyme levels has been described in mammalian tissues in several instances although its exact mechanism is still subject to question. Later studies by *McClatchy and Rickenberg* (32) demonstrated that in *Salmonella typhimurium* the synthesis of flagellin occurred in the apparent ab-

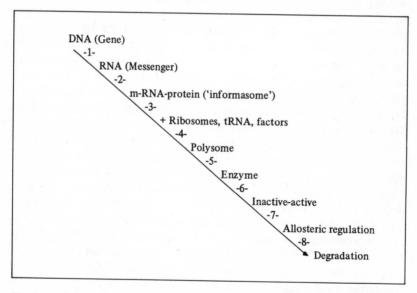

Fig. 1. Biochemical pathways of genetic expression from DNA to protein degradation.

sence of mRNA synthesis which was inhibited either by Actinomycin D or tryptophan starvation. In 1966 *Yudkin* (71) demonstrated that the synthesis of penicillinase in constitutive strains of *B. licheniformis* was much less sensitive to Actinomycin than the synthesis of the enzyme in inducible strains. However, this difference in sensitivity to the antibiotic was not seen in another strain, *B. cereus* (72).

The presence of stable messenger RNA's during oogenesis and early embryogenesis is now well known (2). Most of this work has been carried out in amphibians, invertebrate animals and plants. In addition, as indicated in *Berman*'s review (3), the development of mRNA template stability in differentiating systems has been described in a number of laboratories. In mammalian systems one of the most widely studied translational systems is that of hemoglobin synthesis in the reticulocyte (1). Numerous studies on its regulation have been carried out and very recently *Lockard and Lingrel* (30) have described the synthesis of hemoglobin *in vitro* utilizing an isolated mRNA fraction, ribosomes and the appropriate factors. Translational variation in hemoglobin structure had earlier been reported by *von Ehrenstein* (64).

Other than the hemoglobin studies, the predominant evidence for the existence of stable templates in mammals is the lack of inhibition of translation or protein synthesis by the administration of Actinomycin D and related inhibitors of mRNA synthesis. Reviews of examples of this phenomenon have been published (47, 52). On the basis of much of the data cited in these earlier reviews the majority of enzyme syntheses in mammalian systems appeared to be regulated at the transcriptional level. However, instances of translational regulation then became apparent with the results and interpretation largely based on the insensitivity of enzyme synthesis to Actinomycin D administration. Early studies by *Revel and Hiatt* (48) demonstrated that Actinomycin D administration *in vivo* to rats, while inhibiting RNA synthesis, had little or no effect on cytoplasmic amino acid incorporation *in vivo* or *in vitro*. On the basis of these studies, they suggested that the bulk of the cytoplasmic messenger fraction in liver is stable for 40 h or more. Later studies by *Wilson and Hoagland* (67) showed that, in liver, polysome decay is probably a reasonably accurate reflection of mRNA stability. These authors concluded that in rat liver cytoplasm, mRNA stability varies quite widely with a relatively stable class of RNA messages having a half-life of at least 80 h whereas another group, the majority, decayed with an apparent half-life of 3 to 3 1/2 h. By use of electrophoretic separations, these authors (68) were able to demonstrate that the proportion of albumin synthesized in livers of animals treated with Actinomycin D increased remarkably compared to untreated animals, suggesting that albumin synthesis comprised the primary protein synthesized on stable mRNA. *Bresnick et al.* (5) had earlier demonstrated that the half-life of survival of the template RNA for deoxythymidine kinase in 24-hour regenerating and control adult liver was 7.5

and 3.0 h respectively. The enzyme itself turned over with a half-life of 3.7 and 2.6 h in regenerating and control livers respectively. These authors suggested that the regulation of the lifetime of the deoxythymidine kinase mRNA may be a major controlling factor in the regulation of liver regeneration.

Since the levels of Actinomycin that were utilized in the experiments cited above were lethal to the animals, the work of *Prosky et al.* (45) was of interest in that it demonstrated that considerably lower levels of Actinomycin D which blocked RNA synthesis by only 50 % were accompanied by less than a 5 % inhibition of cytoplasmic protein synthesis in liver. These data tended to support the earlier work suggesting that many templates in liver cytoplasm were in fact stable. However, a complication of these studies may be cited in the work of *Stewart and Farber* (59) who demonstrated that Actinomycin D administration *in vivo,* in contrast to ethionine administration, led to a marked breakdown of hepatic nuclear ribonucleic acid. Thus, one might suggest that at the high doses of Actinomycin one is dealing with several other problems, including an accelerated breakdown of RNA in addition to an inhibition of the synthesis of RNA. These studies, however, do not preclude or invalidate the fact that protein synthesis still occurs in rat liver in the face of almost complete inhibition of RNA synthesis. The work of *Blobel and Potter* also supported this concept (4) in that they demonstrated that administration of Actinomycin D, while causing a marked breakdown in hepatic polysomes, did not result in a complete destruction of the entire polysome profile. Recent studies by *Sarma et al.* (53) have confirmed this work in normal liver and extended it by demonstrating that Actinomycin D administration causes a complete breakdown of free polysomes, but has relatively little effect on those bound to membranes of the endoplasmic reticulum.

II. The 'Paradoxical' Effect of Actinomycin D
   as an Expression of a Translational Phenomenon

As indicated earlier, the studies of *Coleman and Elliott* (8) showed that Actinomycin D actually stimulated the synthesis of ribonuclease by certain bacteria. Although this effect was inexplicable at the time, a related effect was known in that Actinomycin D added to mammalian cells *in vitro* infected with an RNA virus, for example polio virus, actually stimulated the synthesis of the virus (47). This was explained on the basis that Actinomycin D inhibited DNA-directed RNA synthesis in the mammalian cell allowing all of the associated metabolites utilized in this process to be geared to RNA-directed RNA synthesis of the viral RNA. The stimulation of enzyme activity by Actinomycin D was noted by *Peraino and Pitot* (38) in the case of the dietary induction of threonine dehydrase. *Garren et al.* (13) demonstrated that the administration of Actinomycin D *in vivo* several hours after the initiation of tryptophan pyrrolase induction by hydrocortisone in adrenalectomized rats actually caused a further increase in the level of the enzyme. These data have been contested by some authors (35). Later data by *Tomkins'* group (40, 61) indicated that addition of Actinomycin D inhibited the induction of tyrosine aminotransferase, if administered simultaneously with a steroid to cultured hepatoma cells, but stimulated an increased level of the enzyme when added several hours after the initiation of induction. Recent studies by *Reel and Kenney* (46) have substantiated these data *in vitro* and by means of specific immunochemical analyses have suggested that this increase or 'paradoxical' effect of Actinomycin D is the result of a decrease in the rate of degradation of the enzyme combined with a steadily decreasing rate of synthesis. Similar paradoxical effects of Actinomycin have been noted in cultured cells (16, 36) as well as *in vivo* (19, 33, 50).

Since Actinomycin D inhibits DNA-directed RNA synthesis, it is reasonable to assume that the 'paradoxical' effects seen with enzyme synthesis in the presence of high levels of the antibiotic occur independent of nuclear RNA synthesis. Recently *Tomkins et al.* (62) have postulated the existence of a repressor which acts during certain phases of the cell cycle to stimulate mRNA degradation, but only when the translation of the mRNA is inhibited by the repressor. The control of tyrosine aminotransferase synthesis in cultured cells may be explained by this model; however, as in bacterial systems, it is quite improbable that all regulatory systems in mammalian cells can be explained in this way.

III. Translational Regulation of Enzyme Synthetic Rate

As indicated above, most of the experiments that have been carried out in mammalian systems suggesting the existence of translational regulation have based their evidence on the use of inhibitors of RNA synthesis, particularly Actinomycin D, as well as inhibitors of protein synthesis such as puromycin and cycloheximide. In bacterial systems the control of the rate of protein synthesis at the polysomal level has been described for several systems (2). In addition, the role of membranes in controlling protein synthesis, the regulation of ribosome attachment to mRNA, and the hormonal effects on ribosomal activity have been described (2). Experiments designed to investigate specifically the rate of synthesis of an enzyme under environmental conditions have been less frequent. In 1967 *von der Decken* (65) demonstrated that when rats were maintained on a protein-free diet for several days and then allowed to eat a high protein diet for 14 h, the sedimentation patterns of polysomes were unchanged as a result of dietary alteration whereas their ability to incorporate amino acids was markedly enhanced. No significant differences in RNA/protein of the microsomes or amounts of polysomes were observed. These results were interpreted as indi-

cating a translational regulation of gene expression in response to dietary alterations. That polysomes can reform during periods of Actinomycin D blockage of
RNA synthesis has recently been shown by *Stewart and Farber* (58). By administration of the methionine analogue, ethionine, polysome breakdown *in vivo*
occurred. Later administration of adenine and methionine resulted in a re-forma-
tion of polysome patterns *in vivo*. When extremely high doses of Actinomycin D
were administered together with adenine and methionine, polysome re-forma-
tion still occurred essentially as in those animals not treated with the antibiotic.
These data would serve to further indicate a mechanism for translational regulation at the level of formation of polysomes. In studies by *Wilson et al.* (68)
quoted earlier, the relative rate of albumin synthesis in the livers of animals
treated with Actinomycin D more than doubled in comparison with untreated
animals. The data of *Reel and Kenney* (46) on tyrosine transaminase synthesis *in
vitro* demonstrated that enzyme synthesis as judged by immunochemical techniques continued after the addition of Actinomycin D to the medium of hepatoma cells cultured *in vitro*. However, the rate of synthesis of the enzyme
decreased exponentially with a half-life of approximately 3 h. This was true
despite the increase in the total number of units of tyrosine transaminase in the
culture.

The translational regulation of ferritin synthesis in rat liver and HeLa cells
has been described. *Drysdale et al.* (10) demonstrated that iron administration to
protein-deprived rats resulted in increased synthesis of rat liver apoferritin.
Earlier studies by these authors had shown that iron stimulation of ferritin
synthesis was also resistant to Actinomycin and an increased synthesis of ferritin
could be demonstrated *in vitro*. More recent studies by *Chu and Fineberg* (7)
have demonstrated that iron added to HeLa cell cultures resulted in a synthesis
of ferritin within 10 min, this synthesis being unaffected by Actinomycin and in
some instances actually enhanced by the addition of the antibiotic. On the other
hand, the stimulation of leucine incorporation into ferritin in rat liver slices is
inhibited by addition of Actinomycin D to the slice incubation media *in vitro*
(70).

Earlier studies in this laboratory (42) demonstrated that increases in the
activities of serine dehydratase and ornithine transaminase became resistant to
the effects of Actinomycin D several hours after the initiation of induction.
Similarly, the induction of the enzyme also became resistant to the administration of 5-fluoroorotic acid and to high levels of gamma radiation (44). Later
studies from this laboratory (43) demonstrated that the induction of serine
dehydratase became resistant to the administration of Actinomycin D shortly
after the initiation of induction by administration of an amino acid mixture.
This resistance continued from approximately 1 1/2 h after the initiation of
induction by a single dose of casein hydrolysate or amino acids for another 6 to
7 h at which time any further increase in the enzyme was inhibited by admin-

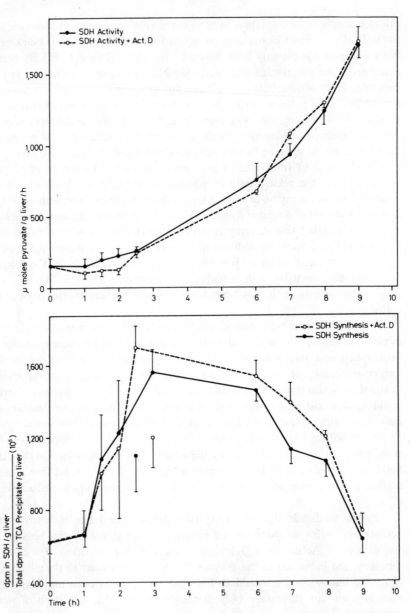

Fig. 2. Serine dehydratase levels and rates of synthesis after administration of a single dose of casein hydrolysate at zero time followed by actinomycin D (1 mg/kg) at each of the time points indicated. The upper portion of the figure denotes enzyme activity and the lower portion the pulse-labeled rate of synthesis carried out in a manner similar to that seen in table I (43).

istration of the antibiotic. Recent studies in this laboratory (figure 2) utilizing pulse labeling with a radioactive amino acid and subsequent quantitative immunochemical precipitation of serine dehydratase from liver extracts of induced animals confirmed the suggestion that the synthesis of serine dehydratase is initially sensitive to the antibiotic, becoming resistant from 1 1/2 to 2 h after the initiation of induction and remaining so for another 6 to 7 h when it again becomes sensitive to the effects of the antibiotic. These data have been interpreted as indicating that the mRNA template for serine dehydratase is synthesized upon induction of the enzyme by a mixture of amino acids and then stabilized within 2 h after the initiation of induction. The stable mRNA template for the enzyme then remains in a functioning translating capacity independent of further RNA synthesis for approximately 6 more hours before new RNA synthesis is required for further translational capacities. On the basis of this, one may further speculate that during the period of template stability, regulation of the rate of synthesis of the enzyme occurs at the translational level or some level beyond the synthesis and transport of mRNA from nucleus to cytoplasm.

Data to confirm this speculation were obtained by *Jost et al.* (24). Utilizing immunochemical techniques and pulse labeling *in vivo*, these authors were able to demonstrate both an increase and decrease in the rate of synthesis of the enzyme serine dehydratase during periods of template stability. The data obtained are seen in part in table I. From these data it can be noted that there is an increase in the rate of synthesis of serine dehydratase from the 6-hour point at which time the system is entirely resistant to the effects of Actinomycin and thus presumably operating only at the translational level to the 10-hour point. The increase is more than a doubling in the rate of synthesis during this time period. In addition, a more dramatic effect is noted when glucose is administered at the 6-hour point. There is a marked inhibition of the rate of synthesis of the enzyme while the activity of the enzyme as measured at the 10-hour point still is relatively high, approaching that seen at the 6-hour point. When the correction for the incorporation of radioactivity into total soluble protein is made, it may be seen that the rate of synthesis of serine dehydratase is essentially the same as that of the control animals on 0 % protein when the experiment was initiated. These data argue rather strongly for the fact of translational regulation of the rate of synthesis of serine dehydratase in rat liver.

Another earlier study from this laboratory which was interpreted as indicating translational control was carried out by *Cho-Chung and Pitot* (6). The data from this experiment are seen in table II. Tryptophan pyrrolase was induced by the administration of tryptophan to adrenalectomized rats. As can be seen from the table and in confirmation of the earlier work of *Greengard et al.* (15), administration of Actinomycin D did not inhibit the tryptophan-induced increase in tryptophan pyrrolase activity in adrenalectomized animals. However, it should be noted that there is almost a doubling of the rate of synthesis of tryptophan pyrrolase when tryptophan is administered to the adrenalectomized animals which were maintained on a low protein diet prior to the beginning of the experiment. This increase in the rate of synthesis occurs even in the presence of Actinomycin D. Since there were no significant differences in the total counts incorporated into supernatant protein, this conclusion seems to be reasonable and again indicates a regulation of the rate of synthesis of an enzyme completely independent of mRNA synthesis.

From these studies it is thus apparent that the rate of enzyme synthesis may be regulated by environmental means during periods wherein enzyme synthesis occurs independent of RNA synthesis. The mechanism of this regulation is as yet obscure. However, studies from several laboratories (22, 66) have now indicated that cyclic nucleotides may be involved in the regulation of synthetic rates of enzymes. In particular, the recent studies of *Jost et al.* (23) demonstrated that the administration of cyclic AMP, or its dibutyryl derivative, induces the synthesis of serine dehydratase. This synthesis rapidly becomes resistant to the effects of

Table I. Regulation of serine dehydratase (SDH) synthesis in rat liver

	SDH Activity[1] (units)	SDH Synthesis (dpm)	Total soluble protein synthesis (dpm)
Control – 0 Time	4	66	3.2×10^5
+ Amino acids – 6 h	60	836	–
+ Amino acids – 10 h	126	1,816	3.2×10^5
+ Amino acids + glucose – 10 h	6	41	3.2×10^5
+ Amino acids (+ glucose at 6 h) n 10 h	44	150	4.9×10^5

1 See *Jost et al.* (24) for details. All values are expressed per gram liver. The times noted are the times of sacrifice. In the last group, amino acids were given at 0 time and amino acids + glucose at 6 h with their sacrifice at 10 h.

Table II. Translational regulation of tryptophan pyrrolase (TP) synthesis

	TP Activity[1] (units)	TP Synthesis (dpm)	Total soluble protein synthesis (dpm)
Control + NaCl	2.0	620	2,400
+ Tryptophan	6.0	1,140	2,300
+ Tryptophan + Actinomycin D	7.0	1,190	2,300

1 See *Cho-Chung and Pitot* (6) for details of the experiment.

Actinomycin D although it is apparent that the initiation of induction still remains sensitive to this antibiotic. Whether or not cyclic AMP is capable of inducing changes in the rate of enzyme synthesis during periods of Actinomycin D resistance has yet to be determined.

IV. Translational Regulation of Enzyme Degradative Rate-Enzyme Turnover

It has been almost 30 years since *Schoenheimer* and his ass. (56) demonstrated the existence of the dynamic turnover of body proteins in the multicellular organism by means of isotopes. Later studies by *Miller* and his ass. (34) demonstrated that fasting of animals resulted in a decrease of a number of hepatic enzymes. This was interpreted as demonstrating the functional significance of the dynamic turnover that had been described by *Schoenheimer*. On the other hand, later studies by *Pitot* (41) clearly demonstrated that prolonged fasting of rats resulted in a dramatic increase in the total and specific activity of the enzyme, threonine dehydrase. These studies demonstrated that not all enzymes in liver decreased in amount upon fasting, but rather that there was a selective degradation of some enzymes while synthesis of others occurred. These studies were most significant in indicating, therefore, that the dynamic turnover of body proteins was not a generalized phenomenon, but could be controlled, presumably by environmental mechanisms.

That the degradation of an enzyme may be controlled by environmental means was first shown conclusively by *Schimke* and his ass. (54). In these early studies utilizing a non-reutilizable labeled amino acid, C^{14}-guanidino-L-arginine, *Schimke* demonstrated that whereas fasting increased the actual arginase content of the liver, the radioactivity of arginase itself was unchanged. In contrast, in changing animals from a high to a low protein diet, the arginase activity decreased as did the radioactivity in the isolated arginase. Thus, by changing the dietary conditions, one could alter the rates of degradation of this enzyme. In later studies *Schimke et al.* (55) demonstrated by immunochemical means that tryptophan administration to adrenalectomized animals maintained on a laboratory chow diet resulted in a stabilization of the enzyme, tryptophan pyrrolase. Although the rate of synthesis of the enzyme was not changed by tryptophan administration under these conditions, the degradative rate of enzyme was markedly diminished. As indicated earlier in studies by *Hoagland* and his ass. (68), the synthesis of serum albumin appears to occur on a stable mRNA template. *Kirsch et al.* (28) demonstrated that when animals were depleted of dietary protein, the catabolic rate of albumin in the liver did not change for 6 days although its synthetic rate had already decreased significantly during this period. After that time the catabolic rate did decrease. Similarly, on protein repletion the catabolic rate gradually increased although the synthetic rate increased exponentially for 2 days, then leveled off. From these data it would appear that the rate of synthesis and degradation of serum albumin may well be controlled by entirely different mechanisms. As pointed out by *Tracht et al.* (63), one of the regulatory mechanisms for albumin synthesis may be the rate of delivery of the protein into the circulation from the liver cell. However, in the experiments of *Kirsch et al.* (28), changes in plasma and albumin concentration were not nearly as abrupt as those seen in the degradative or synthetic rates. Studies by *Drysdale and Munro* (9) demonstrated that administration of iron to animals after leucine-C^{14} administration resulted in an inhibition of the loss of radioactivity from liver ferritin. These authors suggest that iron-rich ferritin molecules are less susceptible to intrinsic degradation than iron-poor molecules.

The significance of the studies by *Schimke* on the tryptophan stabilization of tryptophan pyrrolase (55) can be further viewed in the light of the earlier studies of *Greengard et al.* (15) who showed that this stabilization was not affected by Actinomycin D although being inhibited by inhibitors of protein synthesis. Thus, the stabilization phenomenon still represented enzyme synthesis occurring on a stable template. Later studies by *Greengard,* extending her earlier

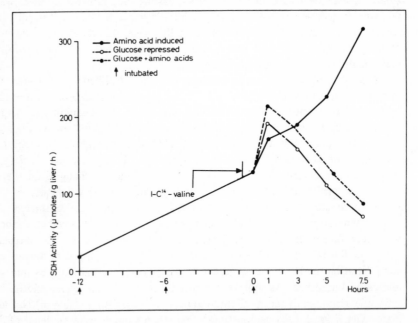

Fig. 3. The amino acid induction and glucose repression of serine dehydratase (SDH). Animals were pre-induced with amino acids up to the zero time point and 1-^{14}C-valine administered 45 min prior to this point. At zero time animals were divided into 3 groups, one of which received amino acids, another glucose, and the third glucose plus amino acids in appropriate doses (24).

studies, suggested that co-factors, in particular pyridoxine, when administered to animals could result in a stabilization and increased levels of tyrosine transaminase if the vitamin was given to deficient animals (14). However, the studies of *Holten et al.* (18) showed that pyridoxine administration to adrenalectomized rats resulted in a significant increase in the rate of synthesis of tyrosine transaminase, this increase being sensitive to the effects of Actinomycin D. *Khairallah and Pitot* (27) showed that pyridoxine administration to adrenalectomized animals caused a three-fold increase in the activity of serine dehydratase, but essentially no change in the rate of synthesis of the enzyme. By prelabeling the enzyme *in vivo* with C^{14}-valine, these authors also demonstrated that when pyridoxine was administered to the animal, the half-life of decay of serine dehydratase radioactivity *in vivo* was doubled as compared to the control group. These authors concluded that in fact pyridoxine administration was resulting in enzyme stabilization much like that described by *Schimke* for tryptophan on tryptophan pyrrolase. Thus, it appears that both co-factors and substrates are capable of stabilizing an enzyme *in vivo* and preventing the degradation which normally occurs as a mechanism for controlling the level of the enzyme.

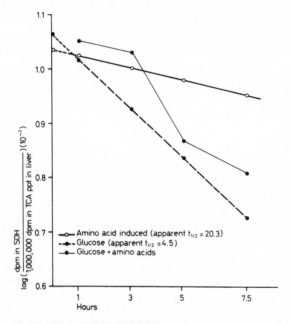

Fig. 4. The turnover of serine dehydratase relative to the total trichloroacetic acid-precipitable proteins in liver. The experiment is essentially that seen in figure 3 except that the points represent the radioactivity of the serine dehydratase quantitatively precipitated by its specific antibody. The counts in the enzyme are divided by the counts in the total soluble protein to correct for pool size (24).

Studies by *Jost et al.* (24) also directed themselves to this point, but in a somewhat different fashion. In these studies (figure 3) animals were prelabeled with C^{14}-valine after serine dehydratase had been previously induced to a constant rate of synthesis by administration of a mixture of amino acids. Shortly thereafter groups of animals were given a further dose of amino acids, or glucose, or both. As can be seen in the figure, those animals to which amino acids were administered continued to synthesize the enzyme, whereas glucose administration, even in the presence of the amino acids, resulted in a decay of enzyme activity. Figure 4 shows that the decay in the radioactivity of serine dehydratase in animals given glucose with or without amino acids was much greater than that in animals given amino acids alone. These studies indicate that the administration of glucose during period of translational regulation (figure 2) of the synthesis of serine dehydratase results in not only a complete inhibition of the synthesis of the enzyme, but also its decay even in the face of the obviously stabilizing influence of the inducing agent, an amino acid mixture. Thus, it would appear that glucose inhibition of enzyme synthesis also acts to affect enzyme degradation under these conditions in preventing the stabilization of the 'substrate'. As might be expected from combining the data of figures 3 and 4, the specific radioactivity of the enzyme decreased in the amino acid induced animals, but remained constant in those given glucose alone.

Table III. Stabilization of tyrosine-α-ketoglutarate transaminase in rat liver after L-tryptophan administration

Application	Tyrosine transaminase (μmoles/g liver per hour)	%	Tyrosine transaminase synthesis (dpm/g liver)	Total protein synthesis (dpm/g liver)
Intact				
Control	69.2 ± 7.0 (21)	100	4,360 ± 560 (5)	228,800 ± 16,700
Cortisone	591.2 ± 23.3 (20)	854.3	13,040 ± 860 (5)	293,800 ± 6,900
L-Tryptophan	350.4 ± 26.6 (16)	506.3	4,760 ± 960 (8)	220,600 ± 21,300
Adrenalectomized				
Control	64.7 ± 7.4 (18)	100	4,560 ± 600 (5)	192,200 ± 4,300
Cortisone	590.6 ± 41.2 (16)	912.8	12,720 ± 840 (5)	270,300 ± 20,700
L-Tryptophan	285.6 ± 22.8 (13)	441.4	3,920 ± 300 (7)	177,200 ± 8,400

Holtzman male intact and adrenalectomized rats (200 g) were kept 3 days on 12.5 % protein diet. Intubation of L-tryptophan (20 mg/100 g; 5 ml of 0.8 % saline solution) and i.p. injection of cortisone (5 mg per animal) were performed after 12 h of starvation. 45 min before killing L-leucine-4,5-H^3 (50 μC/15 μmoles per animal) was injected. The animals were killed 5 h after administration of cortisone or tryptophan.

In recent studies in this laboratory, it has become evident that tryptophan plays a wider role in the stabilization of enzyme synthesized *in vivo* than was previously thought. As can be seen from the data of table III the administration of tryptophan *in vivo* results in an increase in the total enzyme present in the liver with no change in its rate of synthesis both in intact and adrenalectomized animals. In contrast, the administration of cortisone dramatically increases the rate of synthesis of the enzyme both in intact and adrenalectomized animals. The tryptophan effect on tyrosine aminotransferase activity has previously been shown by *Rosen and Nichol* (49).

V. Possible Mechanisms of Translational Regulation of Enzyme Synthesis and Degradation

Obviously, the key question to be answered is the mechanism of translational regulation, which includes, some investigators believe, the mechanism of enzyme turnover in resting cells. This question perhaps may be broken down into two parts, first an understanding of the mechanism of the change in the rates of synthesis and degradation, and second the mechanism of enzyme degradation itself.

Studies from this laboratory (26) have argued that cyclic nucleotides play a major role in translational regulation in the case of glucose repression and probably of amino acid induction of serine dehydratase. This was initially thought to involve an effect of the cyclic nucleotide on the release of the finished polypeptide chain from the polysomal unit. Later studies from this laboratory (*Soling and Kaplan,* unpublished) have not supported this initial contention. In bacterial systems the role of cyclic nucleotides in the regulation of genetic expression suggests an action both at the transcriptional and the translational level of genetic expression (37, 39). Recent studies by *Jost* (23) demonstrated the alteration of cyclic nucleotide levels in liver after the administration of amino acids or glucagon and show that the induction of serine dehydratase indicates a role for cyclic AMP in the regulation of the synthesis of this enzyme. Furthermore, the administration of cyclic AMP can itself induce the synthesis of several enzymes in rat liver *in vivo* and *in vitro* (22, 66). Thus, there is ample evidence of the role of cyclic AMP in the regulation of enzyme synthesis, but exactly what that role is still remains obscure. Recent studies in this laboratory by *Inoue* demonstrated that the enzyme serine dehydratase occurs in two isozymic forms (20). These forms may be separated by electrophoresis. The more electro-negative form is regulated in its synthesis by glucagon, whereas the more electro-positive form by steroids. One may thus postulate that the glucagon, and therefore presumably cyclic AMP, regulated form is phosphorylated and that this phosphorylation is under the direct or indirect control of cyclic AMP. If the phosphorylation occurs during protein synthesis, the earlier studies on the release of the enzyme from polysomes *in vitro* as stimulated by addition of cyclic AMP may be in this fashion. Furthermore, the addition of a phosphate group to the enzyme may alter its rate of degradation.

As demonstrated in this review, the amino acid tryptophan may act to stabilize enzymes and inhibit the rate of degradation whereas glucose administration appears to stimulate the rate of enzyme degradation. The effect of glucose is to over-ride the stabilizing effect of amino acids on the enzyme serine dehydratase. Preliminary investigations in this laboratory have indicated that the administration of cycloheximide during glucose repression does not serve to stabilize the enzyme and prevent its further degradation. This result is unlike that described by *Kenney* (25) on the *in vivo* stabilization of tyrosine transaminase after the administration of cycloheximide. The effects of tryptophan on the stabilization of enzymes other than tryptophan pyrrolase and, as described here, tyrosine transaminase, have not been shown unequivocally. On the other hand, several enzymes are known to be 'induced' by tryptophan and these examples may in fact represent an inhibition of degradation rather than an increased rate of synthesis (12, 49, 51). That tryptophan has a peculiar effect in protein synthesis in the liver has been shown by several authors. *Munro* and his ass. (11, 69) have demonstrated that the administration of tryptophan-deficient amino acid mixtures causes a breakdown of polysomes in liver. Administration of tryptophan alone or in mixture with amino acids results in a normal polysome pattern with heavy aggregates and few monosomes. Furthermore, as shown by *Sidransky et al.* (57), tryptophan omission from an inducing amino acid mixture results in a low level of incorporation of radioactive amino acids into protein *in vitro*, whereas inclusion of tryptophan with the mixtures results in a much greater incorporation of label *in vitro*. These studies indicate a peculiar role for the amino acid tryptophan in protein synthesis in the liver, possibly at the level of translation.

Recent studies by *Levitan and Webb* (29) have demonstrated that the administration of purine and pyrimidine analogues results in an increase in the activity of tyrosine transaminase after cortisone administration. These authors suggest that this effect is due to an inhibition of the degradation of the enzyme and have reported preliminary evidence to support this contention. In this laboratory similar results have been carried out as seen from the data of table IV on 2 pyrimidine analogues, 5-fluoroorotate and 5-azacytidine. It can be seen that in the case of tyrosine transaminase there is a dramatic increase in the level of the enzyme, but essentially no change in its rate of synthesis when the pyrimidine analogues are administered. On the other hand, cortisone administration causes a marked increase in the level and rate of synthesis of the enzyme.

Thus, it is apparent that analogues of pyrimidines and purines which may be incorporated into RNA (17, 31) are capable of inhibiting enzyme degradation.

Table IV. Effect of cortisone, 5-fluoroorotate (5-FOA) and 5-azacytidine (5-AC) on tyrosine transaminase activity and synthesis *in vivo*[1]

Treatment (4 h)	Tyrosine transaminase activity (μmoles/g per h)	Tyrosine transaminase synthesis (dpm/g)
Control	84 ± 15.2	2,380 ± 240
+ Cortisone	474 ± 18.9	8,805 ± 470
+ 5-FOA	208 ± 28.3	2,780 ± 260
+ 5-AC	158 ± 31	2,320 ± 280

1 Holtzman male adrenalectomized rats (200 g) maintained for 3 days on a 12.5 % protein diet were starved for 14 h. Cortisone (5 mg/rat), 5-FOA (50 mg/kg), and 5-AC (15 mg/kg) were injected intraperitoneally 4 h prior to sacrifice. 45 min prior to sacrifice *L*-leucine-4,5-H^3 (50 μC/15 μmoles/rat) was injected intraperitoneally. Enzyme activity was determined by the method of *Pitot et al.* (44) and tyrosine transaminase radioactivity by the immunochemical technique of *Jost et al.* (24). Values are ± standard error of the mean.

Furthermore, tryptophan administration also results in a decrease or inhibition of the rate of degradation of several enzymes. Although it is difficult to draw any generalized conclusions, one may suggest that RNA turnover is involved in enzyme degradation as well as enzyme synthesis. Several other authors (25) have suggested that enzyme degradation is a process akin to enzyme synthesis. These studies would support this contention and even suggest that some RNA which turns over in the cytoplasm is involved in enzyme degradation. At the present time the best known example of an RNA which is rapidly broken down and resynthesized in the cytoplasm is the terminal nucleotide, C-C-A, of soluble RNA. Since this terminal trinucleotide involves both purines and pyrimidines, it is possible that analogues may inhibit this turnover and also inhibit enzyme degradation. This suggests a major role for transfer RNA in the regulation of enzyme degradation. Furthermore, the fact that tryptophan itself is capable of enzyme stabilization suggest that tryptophanyl-tRNA may be one of the major factors in regulating the rate of enzyme degradation.

VI. Summary and Conclusions

This short review has not attempted to comprehensively cover the subject of translational regulation in mammalian systems, but rather has presented several possible conclusions from some of the data which have accumulated to date. The possibility that tryptophanyl-tRNA is a major factor in the regulation of enzyme degradation may be easily tested in a number of *in vivo* and *in vitro* systems. The mechanism of the rate of synthesis of enzymes at the translational level, although still unknown, may in several instances involve cyclic nucleotides, particularly cyclic AMP, as major mediators in this process. The possible role of 'repressors' in translational regulation has been postulated. Only future experimentation can answer these questions and determine the validity of the models proposed.

VII. References

1 *Baglioni, C. and Colombo, B.:* Control of hemoglobin synthesis. Cold Spr. Harb. Symp. quant. Biol. *29:* 347–356 (1964).

2 *Berman, M.:* A survey of translational control in bacteria and vertebrate cells. Bull. Inst. Cell. Biol., Univ. Conn. *86:* 1–13 (1967).

3 *Berman, M.:* Translational control in higher organisms. Bull. Inst. Cell. Biol., Univ. Conn. *87:* 1–22 (1967).

4 *Blobel, G. and Potter, V.R.:* Studies on free and membrane-bound ribosomes in rat liver. II. Interaction of ribosomes and membranes. J. molec. Biol. *26:* 293–301 (1967).

5 *Bresnick, E.; Williams, S.S., and Mosse, H.:* Rates of turnover of deoxythymidine kinase and of its template RNA in regenerating and control liver. Cancer Res. *27:* 469–475 (1967).

6 *Cho-Chung, Y.S. and Pitot, H.C.:* Regulatory effects of nicotinamide on tryptophan pyrrolase synthesis in rat liver *in vivo.* Europ. J. Biochem. *3:* 401–406 (1968).

7 *Chu, L.L.H. and Fineberg, R.A.:* On the mechanism of iron-induced synthesis of apo-ferritin in HeLa cells. J. biol. Chem. *244:* 3847–3854 (1969).

8 *Coleman, G. and Elliott, W.H.:* Extracellular ribonuclease formation in *Bacillus subtilis* and its stimulation by actinomycin D. Biochem. J. *95:* 699–706 (1965).

9 *Drysdale, J.W. and Munro, H.N.:* Regulation of synthesis and turnover of ferritin in rat liver. J. biol. Chem. *241:* 3630–3637 (1966).

10 *Drysdale, J.W.; Olafsdottir, E., and Munro, H.N.:* Effect of ribonucleic acid depletion on ferritin induction in rat liver. J. biol. Chem. *243:* 552–555 (1968).

11 *Fleck, A.; Shepherd, J., and Munro, H.N.:* Protein synthesis in rat liver: Influence of amino acids in diet on microsomes and polysomes. Science *150:* 628–629 (1965).

12 *Foster, D.O.; Ray, P.D., and Lardy, H.A.:* A paradoxical *in vivo* effect of L-tryptophan on the phosphoenolpyruvate carboxykinase of rat liver. Biochemistry *5:* 563–569 (1966).

13 *Garren, L.D.; Howell, R.R.; Tomkins, G.M., and Crocco, R.M.:* A paradoxical effect of actinomycin D: The mechanism of regulation of enzyme synthesis by hydrocortisone. Proc. nat. Acad. Sci., Wash. *52:* 1121–1129 (1964).

14 *Greengard, O.:* The regulation of apoenzyme levels by coenzymes and hormones. Adv. Enz. Reg. *2:* 277–288 (1964).

15 *Greengard, O.; Smith, M.A., and Acs, G.:* Relation of cortisone and synthesis of ribonucleic acid to induced and developmental enzyme formation. J. biol. Chem. *238:* 1548–1551 (1963).

16 *Griffin, M.J. and Cox, R.P.:* Studies on the mechanism of hormonal induction of alkaline phosphatase in human cell cultures. J. Cell Biol. *29:* 1–9 (1966).

17 *Heidelberger, C.:* Fluorinated pyrimidines. Progr. nucl. Acid Res. molec. Biol. *4:* 1–50 (1965).

18 *Holten, D.; Wicks, W.D., and Kenney, F.T.:* Studies on the role of vitamin B_6 derivatives in regulating tyrosine-a-ketoglutarate transaminase activity *in vitro* and *in vivo*. J. biol. Chem. *242:* 1053–1059 (1967).

19 *Homoki, J.; Beato, M., and Sekeris, C.E.:* 'Paradox' effect of cortisol and actinomycin D on RNA polymerase activity of rat liver nuclei. FEBS Letters *1:* 275–278 (1968).

20 *Inoue, H. and Pitot, H.C.:* Regulation of the synthesis of serine dehydratase isozymes. Adv. Enz. Reg. (in press).

21 *Jacob, F. and Monod, J.:* Genetic-regulatory mechanisms in the synthesis of proteins. J. molec. Biol. *3:* 318–356 (1961).

22 *Jost, J.-P.; Hsie, A.W., and Rickenberg, H.V.:* Regulation of the synthesis of rat liver serine dehydratase by adenosine 3′,5′-cyclic monophosphate. Biochem. biophys. Res. Comm. *34:* 748–754 (1969).

23 *Jost, J.-P.; Hsie, A.; Hughes, S.D., and Ryan, L.:* Role of adenosine 3′,5′-cyclic monophosphate in the induction of hepatic enzymes. I. Kinetics of the induction of rat liver serine dehydratase by adenosine 3′,5′-cyclic monophosphate. J. biol. Chem. (in press).

24 *Jost, J.-P.; Khairallah, E.A., and Pitot, H.C.:* Studies on the induction and repression of enzymes in rat liver. V. Regulation of the rate of synthesis and degradation of serine dehydratase by dietary amino acids and glucose. J. biol. Chem. *243:* 3057–3066 (1968).

25 *Kenney, F.T.:* Turnover of rat liver tyrosine transaminase: Stabilization after inhibition of protein synthesis. Science *156:* 525-528 (1967).

26 *Khairallah, E.A. and Pitot, H.C.:* 3′,5′-cyclic AMP and the release of polysome-bound proteins *in vitro*. Biochem. biophys. Res. Comm. *29:* 268–274 (1967).

27 *Khairallah, E.A. and Pitot, H.C.:* Studies on the turnover of serine dehydrase: Amino acid induction, glucose repression and pyridoxine stabilization; in *Yamada, Katsunuma and Wada* Symposium on pyridoxal enzymes, pp. 159–164 (Maruzen, Tokyo 1968).

28 *Kirsch, R.; Frith, L.; Black, E., and Hoffenberg, R.:* Regulation of albumin synthesis and catabolism by alteration of dietary protein. Nature, Lond. *217:* 578–579 (1968).

29 *Levitan, I.B. and Webb, T.E.:* Modification by 8-azaguanine of the effects of hydrocortisone on the induction and inactivation of tyrosine transaminase of rat liver. J. biol. Chem. *244:* 341–347 (1969).

30 *Lockard, R.E. and Lingrel, J.B.:* The synthesis of mouse hemoglobin β-chains in a rabbit reticulocyte cell-free system programmed with mouse reticulocyte 9S RNA. Biochem. biophys. Res. Comm. *37:* 204–212 (1969).

31 *Matthews, R.S.F.:* Biosynthetic incorporation of metabolite analogues. Pharmacol. Rev. *10:* 359–406 (1958).

32 *McClatchy, J.K. and Rickenberg, H.V.:* Heterogeneity of the stability of messenger ribonucleic acid in *Salmonella typhimurium*. J. Bact. *93:* 115–121 (1967).

33 *McCoy, E.E. and Ebadi, M.:* The paradoxical effect of hydrocortisone and actinomycin on the activity of rabbit leucocyte alkaline phosphatase. Biochem. biophys. Res. Comm. *26:* 265–271 (1967).

34 *Miller, L.L.:* Changes in rat liver enzyme activity with acute inanition. Relation of loss of enzyme activity in liver protein loss. J. biol. Chem. *172:* 113–121 (1948).

35 *Mishkin, E. Patricia and Moris, L. Shore:* Inhibition by actinomycin D of the induction of tryptophan pyrrolase by hydrocortisone. Biochim. biophys. Acta *138:* 169–174 (1967).

36 *Nebert, D.W. and Gelboin, H.V.:* Substrate-inducible microsomal aryl hydroxylase in mammalian cell culture. J. biol. Chem. *243:* 6250–6261 (1968).

37 *Pastan, Ira and Perlman, R.L.:* Stimulation of tryptophanase synthesis in *Escherichia coli* by cyclic 3′,5′-adenosine monophosphate. J. biol. Chem. *244:* 2226–2232 (1969).

38 *Peraino, C. and Pitot, H.C.:* Studies on the induction and repression of enzymes in rat liver. II. Carbohydrate repression of dietary and hormonal induction of threonine dehydrase and ornithine-δ-transaminase. J. biol. Chem. *239:* 4308–4313 (1964).

39 *Perlman, R.L. and Pastan, I.:* Regulation of β-galactosidase synthesis in *Escherichia coli* by cyclic adenosine 3′,5′-monophosphate. J. biol. Chem. *243:* 5420–5427 (1968).

40 *Peterkofsky, Beverly and Tomkins, G.M.:* Effect of inhibitors of nucleic acid synthesis on steroid-mediated induction of tyrosine aminotransferase in hepatoma cell cultures. J. molec. Biol. *30:* 49–61 (1967).

41 *Pitot, H.C.:* Studies on the control of protein synthesis in normal and neoplastic rat liver. Ph.D. Dissertation, Tulane University, 1959.

42 *Pitot, H.C. and Peraino, C.:* Studies on the induction and repression of enzymes in rat liver. I. Induction of threonine dehydrase and ornithine-δ-transaminase by oral intubation of casein hydrolysate. J. biol. Chem. *239:* 1783–1788 (1964).

43 *Pitot, H.C.; Peraino, C.; Lamar, C., Jr., and Kennan, A.L.:* Template stability of some enzymes in rat liver and hepatoma. Proc. nat. Acad. Sci., Wash. *54:* 845–851 (1965).

44 *Pitot, H.C.; Peraino, C.; Lamar, C., Jr., and Lesher, S.:* Effect of gamma radiation on dietary and hormonal induction of enzymes in rat liver. Science *150:* 901–903 (1965).

45 *Prosky, L.; Roberts, B., Jr.; O'Dell, R.G., and Imblum, R.L.:* Differential effects of actinomycin D on nucleic acid and protein synthesis in rat liver. Arch. Biochem. Biophys. *126:* 393–398 (1968).

46 *Reel, J.R. and Kenney, F.T.:* 'Superinduction' of tyrosine transaminase in hepatoma cell cultures: Differential inhibition of synthesis and turnover by actinomycin D. Biochemistry *61:* 200–205 (1968).

47 *Reich, E. and Goldberg, I.H.:* Actinomycin and nucleic acid function. Prog. nucl. Acid Res. molec. Biol. *3:* 183–234 (1964).

48 *Revel, M. and Hiatt, H.H.:* The stability of liver messenger RNA. Proc. nat. Acad. Sci., Wash. *51:* 810–818 (1964).

49 *Rosen, F. and Nichol, C.A.:* Studies on the nature and specificity of the induction of several adaptive enzymes responsive to cortisol. Adv. Enz. Reg. *2:* 115–135 (1964).

50 *Rosen, F.; Raina, Prem Nath; Milholland, R.J., and Nichol, C.A.:* Induction of several adaptive enzymes by actinomycin D. Science *146:* 661–663 (1964).

51 *Rotschild, M.A.; Oratz, M.; Mongalli, J.; Fishman, L., and Schreiber, S.S.:* Amino acid regulation of albumin synthesis. J. Nutr. *98:* 395–403 (1969).

52 *Samuels, L.D.:* Actinomycin and its effects. New Engl. J. Med. *271:* 1252–1258, 1301–1308 (1964).

53 *Sarma, D.S.R.; Reid, J.M., and Sidransky, H.:* The selective effect of actinomycin D on free polyribosomes of mouse liver. Biochem. biophys. Res. Comm. *36:* 582–588 (1969).

54 *Schimke, R.T.:* The importance of both synthesis and degradation in the control of arginase levels in rat liver. J. biol. Chem. *239:* 3808–3817 (1964).

55 *Schimke, R.T.; Sweeney, E.W., and Berlin, C.M.:* The roles of synthesis and degradation in the control of rat liver tryptophan pyrrolase. J. biol. Chem. *240:* 322–331 (1965).

56 *Schoenheimer, R.:* The dynamic state of body constituents (Harvard Univ. Press, Cambridge, Mass. 1964).

57 *Sidransky, H.; Sarma, D.S.R.; Bongiorno, M., and Verney, E.:* Effect of dietary tryptophan on hepatic polyribosomes and protein synthesis in fasted mice. J. biol. Chem. *243:* 1123–1132 (1968).

58 *Stewart, E.A. and Farber, E.:* Reformation of functional liver polyribosomes from ribosome monomers in the absence of RNA synthesis. Science *157:* 67–69 (1967).

59   *Stewart, Gloria A. and Farber, E.:* The rapid acceleration of hepatic nuclear ribonucleic acid breakdown by actinomycin but not by ethionine. J. biol. Chem. *243:* 4479–4485 (1968).

60   *Szepesi, B. and Freedland, R.A.:* Control of tyrosine-α-ketoglutarate transaminase synthesis in rat liver: Studies on superinduction in force-fed rats. J. Nutr. *97:* 255–259 (1969).

61   *Thompson, E. Brad; Tomkins, G.M., and Curran, J.F.:* Induction of tyrosine-α-ketoglutarate transaminase by steroid hormones in a newly established tissue culture line. Proc. nat. Acad. Sci., Wash. *56:* 296–303 (1966).

62   *Tomkins, G.M.; Gelehrter, T.D.; Granner, D.; Martin, D., Jr.; Samuels, H.H., and Thompson, E.B.:* Control of specific gene expression in higher organisms. Science *166:* 1474–1480 (1969).

63   *Tracht, M.E.; Tallal, L., and Tracht, D.E.:* Intrinsic hepatic control of plasma albumin concentration. Life Sci. *6:* 2621–2628 (1967).

64   *Ehrenstein, E. von:* Translational variations in the amino acid sequences of the α-chain of rabbit hemoglobin. Cold Spr. Harb. Symp. quant. Biol. *31:* 705–714 (1966).

65   *Decken, A. von der:* Evidence for regulation of protein synthesis at the translational level in responses to dietary alterations. J. Cell Biol. *33:* 657–663 (1967).

66   *Wicks, W.D.:* Induction of hepatic enzymes by adenosine 3′,5′-monophosphate in organ culture. J. biol. Chem. *244:* 3941–3950 (1969).

67   *Wilson, S.H. and Hoagland, M.B.:* Physiology of rat liver polysomes – The stability of messenger ribonucleic acid and ribosomes. Biochem. J. *103:* 556–566 (1967).

68   *Wilson, S.H.; Hill, H.Z., and Hoagland, M.B.:* Physiology of rat liver polysomes – Protein synthesis by stable polysomes. Biochem. J. *103:* 567–572 (1967).

69   *Wunner, W.H.; Bell, J., and Munro, H.N.:* The effect of feeding with a tryptophan-free amino acid mixture on rat-liver polysomes and ribosomal ribonucleic acid. Biochem. J. *101:* 417–427 (1966).

70   *Yoshino, Y.; Manis, J., and Schachter, D.:* Regulation of ferritin synthesis in rat liver. J. biol. Chem. *243:* 2911–2917 (1968).

71   *Yudkin, M.D.:* Protein synthesis by long-lived messenger ribonucleic acid in bacteria. Biochem. J. *100:* 501–506 (1966).

72   *Yudkin, M.D.:* Lifetime of messenger ribonucleic acid for penicillinase synthesis in several strains of bacilli. Biochem. J. *108:* 675–677 (1968).

Authors' addresses: *Henry C. Pitot*, M.D., Ph.D., McArdle Laboratory, University of Wisconsin, Medical School, *Madison, WI 53706; Joel Kaplan*, Ph.D., General Electric Research and Development Center, *Schenectady, N.Y.* (USA); *Alois Čihák*, RNDr., Ph.D., Institute for Organic Chemistry and Biochemistry, Czechoslovak Academy of Sciences, *Prague* (Czechoslovakia)

Enzyme Synthesis and Degradation in Mammalian Systems, pp. 236–310
(Karger, Basel 1971)

Intracellular Protein Turnover and the Roles of Synthesis and Degradation in Regulation of Enzyme Levels

Miloslav Rechcigl, Jr.

Research and Institutional Grants Division, Agency for International Development,
U.S. Department of State, Washington, D.C.

Contents

I. Introduction

The idea of intracellular protein being in a state of dynamic equilibrium in which the proteins are continually being broken down and replenished by re-synthesis is not new, nor has it been universally accepted without a challenge. Although generally recognized today, its validity was still seriously doubted only a few years ago (9, 78, 126).

Numerous studies carried out in the last two decades have conclusively demonstrated a great heterogeneity of turnover rates between proteins of different animal tissues (133, 211). One might argue that most of such studies have involved measurement of the turnover of the total protein of a given organ which is the mean value for turnover of the multitude of different proteins present within the tissues of which the organ is composed. The foregoing conclusion is, however, fully supported by a number of careful studies on the turnover rates of individual proteins, notably, those of the muscle and the liver. More recently, unequivocal evidence for the existence of turnover has been provided by the kinetic studies of enzyme synthesis and degradation.

The purpose of this chapter is to briefly review the development of the concept of protein turnover and to examine the recent literature with particular emphasis on the turnover of enzymes in the mammalian tissues, and the relative role of enzyme synthesis and degradation in the maintenance and regulation of enzyme levels.

II. The Development of the Concept of Protein Turnover

A. Early Theories of Protein Metabolism

Relative constancy which the animal maintains with respect to its nitrogen content, under diverse physiological conditions, has led to numerous hypotheses and vigorous polemics among the early investigators.

One of the earliest views, forwarded in 1867 by *Carl von Voit* (224), maintained that the ingested protein after becoming *circulating protein* was catabolized by the living tissues without actually becoming an integral part of them.

Voit's views were severely criticized by *Edward F.W. Pflüger* (146) who firmly believed that food protein must first become an integral part of the living protoplasm before it could be utilized. While *Voit* assumed that tissue proteins are relatively inert, *Pflüger* was of the opinion that only proteins which are present in cells and which are not in solution are subject to catabolic changes.

Another theory was advanced by *Max Rubner* (176) who maintained that the study of metabolism must be considered in association with the energy exchanges. While other investigators were content with two classes, *Rubner* postulated the existence of four types of proteins, according to the degree of stability of their metabolism, i.e. *Organeiweiss, Meliorationseiweiss, Übergangs-eiweiss,* and *Vorratseiweiss.*

B. *Folin*'s Theory of Endogenous and Exogenous Protein Metabolism

As a result of the chemical analyses of urine, *Otto Folin* (43) advanced a new theory which differentiated between two forms[1] of catabolism which were essentially independent and quite different from each other. One type, charac-terized by the urinary excretion of urea and inorganic sulfate, was thought to be extremely variable and to depend on the composition of the diet. This was *exogenous* or *intermediate metabolism.* The second type was thought to be relatively constant and to lead to the urinary excretion of creatinine and neutral sulfur, and to a lesser extent of uric acid and ethereal sulfates. This second type of metabolism which was supposed to reflect *wear and tear* of the tissues, *Folin* termed *tissue* or *endogenous metabolism.*

The doctrine of *Folin* regarding the two forms of metabolism had been generally accepted until about 1940 when new advances had been made using the radioactive isotopes. Up to this time, however, there was a consensus among the scientists that the major portion of the ingested amino acids, in an animal maintained in nitrogen equilibrium, was not used for protein synthesis, but was catabolized and the nitrogen excreted in the form of urinary urea.

C. The Dynamic State of Body Constituents

Although *Schoenheimer* and his collaborators are recognized as initiating the idea of *dynamic state of body constituents,* it was actually *Borsook and Keighley* (13) who first seriously challenged *Folin*'s concept of the static endo-genous metabolism. On the basis of their experiments these authors put forward the view that the rate of breakdown and concomitant resynthesis of protein in animal organism in nitrogen balance is far greater than was envisaged in *Folin*'s theory of endogenous and exogenous protein metabolism. It is questionable, however, whether their concept of *continuous nitrogen metabolism,* which was based primarily on indirect evidence, would have been accepted had it not been for the dramatic findings with the application of the isotope techniques to the study of protein metabolism.

---

1 In reality these correspond quite closely to the *wear and tear* and to the *dynamic quotas of protein metabolism* proposed by *Rubner* (176).

The pioneering work of *Schoenheimer*'s group (189) with ^{15}N-labeled amino acids cut away all the support of *Folin*'s theory. Their experiments definitely disproved the view that, when the nitrogen intake of an animal equals its nitrogen output, the nitrogen that is excreted stems mainly from the dietary protein. Since there could not have been a net gain in the body protein under these conditions, an amount of tissue protein equivalent to that newly formed must have been degraded. In view of this work, tissue proteins could not be regarded as inert structural substances but as labile compounds, which are continuously being catabolized and resynthesized, regardless of whether the food is available to the organism or not.

D.   Turnover in Growing Bacteria

The concept of the general dynamic state of tissue proteins has been universally recognized until several studies on the induction of β-galactosidase in *Escherichia coli* indicated that the proteins of this organism might be stable components of the cell (78, 108, 174). In one of these studies, *Hogness et al.* (78) demonstrated that bacteria containing ^{35}S-labeled protein contributed no radioactivity to β-galactosidase which was subsequently induced and isolated. Conversely, labeled β-galactosidase lost no measurable radioactivity in growing cultures, either in the presence or absence of the inducer. Generalizing from such findings, these investigators questioned not only the universality of the dynamic state but also the interpretation of *Schoenheimer*'s results with whole animals, suggesting that the dynamic state might not be due to intracellular turnover, but to *replacement of older by younger cells in the tissues, and by the replacement of protein lost by secretion.*

Be that as it may, the conclusions of *Hogness et al.* (78) have forced a reappraisal of the concept of the dynamic state, which gave a stimulus for a number of new exciting studies both in bacterial and mammalian systems.

E.   Turnover in Nongrowing Bacteria

It soon became apparent that the reported 'stability' of proteins in microorganisms is confined only to some bacteria, and then only when they were in a state of rapid growth.

Mandelstam (121) made a direct comparison of protein stability in growing and nongrowing cells, using a strain of *Escherichia coli* requiring leucine and threonine. The total loss of radioactivity from the growing cells prelabeled with ^{14}C-leucine was less than 4 %, over half of which took place in the first 30 min while the cells were still in lag phase; after this first hour no detectable turnover of protein was seen. By contrast protein degradation in the nongrowing cells continued linearly at the rate of 4—5 %/h for the duration of the experiment.

Rates of similar magnitude were reported by *Borek et al.* (12). In *Bacillus cereus, Urbá* (221) found a rate of degradation of 7 %/h in nongrowing cells and 1.4 %/h in growing cells. In yeast, the rate of protein degradation in growing cells was only 4 % of that in resting cells (71, 72). Superficially, it would seem that the dissimilarity in stability of protein in growing and nongrowing cells is a general phenomenon among the microorganisms.

It is of interest to note, however, that *Fox and Brown* (45) not too long ago reported, by direct measurement, a protein-degradation rate of 2.7 % per generation in logarithmically growing *Escherichia coli*. This value is significantly greater than that estimated by other investigators who used indirect methods (e.g. from the appearance of various compounds which might be derived from protein degradation), and is not very different from the rate estimated by *Mandelstam* (121) and *Borek et al.* (12) for nongrowing *Escherichia coli*.

According to the recent studies of *Pine* (147), the intracellular turnover of protein in non-proliferating *Escherichia coli* is kinetically a heterogeneous process comprising the breakdown and resynthesis of broad spectra of protein populations. In extension of these studies, *Pine* (148) also examined protein breakdown in this organism under varying conditions of growth and nutritional deprevidation. Optimal breakdown rates, estimated over a short time periods of protein synthesis and decay were found invariant and unrepressible by any condition of growth or starvation. Surprising, although not unexpected, were *Pine*'s findings (148) regarding the selection imposed by the physiological state of the cell for breakdown of limited populations of proteins. During the process of growth, 1/2—3/4 of the protein susceptible in the resting state was progressively spared, and breakdown was continued in populations that were more selected and more frequently regenerated. According to these studies as well as subsequent work of *Willetts* (236, 237) in growing *Escherichia coli* there is approximately 1 % of total protein being degraded at a fixed rate of 20 %/h. This rate is maintained even during nongrowing stages although the number of proteins subject to degradation is increased.

In view of these experiments it is clear that the earlier view concerning the lack of turnover in growing bacteria is no longer tenable. The protein turnover evidently plays a significant role in all states of microbial growth.

F. Turnover in Mammalian Cells

Hogness et al. (78) pointed out earlier that the interpretation of the incorporation data in the case of the mammalian systems involves some inherent ambiguities which are evident when one considers the degree of homogeneity of the cellular population. Thus, mammalian systems consist of heterogeneous population of cells in which some cells grow and multiply, some die and lyse, others secrete large amounts of proteins, while others appear to remain very stable. In contrast, the bacterial system consists of a 'homogeneous' population of cells in which there is no observable cell lysis or secretion of proteins from the cells and where all the cells are placed in an identical environment. To avoid these complications, a number of investigators examined the turnover rates of mammalian cells in tissue culture.

A definite evidence for protein turnover was derived from studies with peritoneal exudates. *Moldave* (127) observed that prelabeled Ehrlich ascites cells, implanted intraperitoneally in a dialysis membrane in mice, lose radioactivity to the medium under the conditions which allowed cell division but no net growth, even though their total protein content remained constant. In the comparable experiments of *Greenlees and LePage* (66), radioactivity was continuously released from the ^{14}C-glycine labeled cells of the TA3 ascites tumor during the growth in the peritoneal cavity of the mouse. Similarly, *Forssberg and Revesz* (44) showed that during growth of two Ehrlich ascites tumors the release of previously incorporated radioactivity from total proteins is different when two different amino acids are used as labels. The loss of radioactivity from multiplying ascites cells was interpreted as evidence for an intracellular degradation of the protein molecule.

Occurrence of turnover was also demonstrated by *Harris and Watts* (73) in their studies on the *in vitro* incorporation of ^{14}C-valine into the protein of resting rabbit macrophages recovered from a peritoneal exudate.

Furthermore, *Eagle et al.* (33) have shown that serially cultured human cells incorporated ^{14}C-phenylalanine into protein in the absence of net synthesis, at a rate of 1 %/h. In contrast to the initial reports on bacteria and yeast, the turnover also occurred, in growing cultures, in which it was evidenced by incorporation in excess of that represented by the net increase in protein. In subsequent studies, it was shown that the incorporation was largely an intracellular process, and only in small part due to cell death or protein secretion, followed by degradation and resynthesis.

Similar turnover was reported in Walker carcinoma 256 and human skin epithelium in tissue culture by *Jordan et al.* (89). The constancy of catabolic rate, reported by these investigators, and reversely, the afore-mentioned finding of *Eagle et al.* (33) that amino acid incorporation continues at a constant rate in the absence of net synthesis implies that most of the cell proteins are involved, and was not due to the rapid turnover of a relatively small portion. There is, as yet, no information as to whether this average rate masks important differences in the turnover of individual proteins.

The contradictory conclusions of *King et al.* (102) concerning the lack of turnover of cell protein in L cells maintained in logarithmic growth were subsequently questioned and presumably disproved (90).

G. Turnover in the Living Animal

The literature on the turnover of proteins in animal systems is so volumi-
nous that it is virtually impossible to even attempt to summarize it in so little
available space. The animal systems which involve heterogenous populations of
cells present, of course, additional problems, concerning the interpretation of
the data. We shall, therefore, limit ourselves to a few selected examples in which
the turnover rate was demonstrated beyond any doubt.

Swick (206), for example, estimated the rate of renewal of liver proteins
from the measurement of the incorporation of $^{14}CO_2$ into the guanidine group
of arginine in adult rats continuously exposed to isotope for periods of 1 to
8 days. The calculated half-life of liver proteins ranged from 2 to 4 days, depending upon the duration of the exposure. Clearly, this renewal could hardly be
accounted for by the replacement of cells, since the life span of liver cells has
been shown to be at least 4 to 5 months (19, 119a, 206).

Another evidence for the occurrence of the intracellular turnover was fur-
nished by *Buchanan* (19) who measured the total carbon turnover in rats and
mice by feeding the animals a uniformly labeled diet. In some organs there was
nearly complete replacement by dietary carbon during the 50-day experimental
period. It was generally believed that adult brain tissue does not undergo cellular
replacement, yet, in these experiments, 70 % of the total carbon present was
replaced by dietary carbon within 7 weeks.

These as well as other studies (133, 211) have conclusively demonstrated a
great heterogeneity of turnover rates between proteins of different tissues. One
might argue, of course, that most of these studies have involved measurement of
the turnover of the total protein of a given organ, which is the mean value for
turnover of the multitude of different proteins present within the tissues of
which the organ is composed. The above conclusion is, however, fully supported
by a number of careful studies on the turnover rates of individual proteins,
notably those of the muscle, collagen and of the plasma. More recently, unequivocal evidence for the existence of turnover was provided by the kinetic studies
of enzyme synthesis and degradation. We shall examine these studies in more
detail in the next section.

III. Turnover of Enzymes in the Muscle Tissue

The first direct evidence for the turnover of enzymes in mammalian systems
was provided by the striking experiments of *Velick* and his collaborators (75,
200, 222). ^{14}C-labeled amino acids were injected into adult rabbits, and several
enzyme proteins were isolated from the muscle, with subsequent estimation of

the specific activities of the relevant amino acid residues obtained from the recrystallized enzymes.

In one of these studies (200) it was observed that the specific activity of each amino acid in aldolase was 1.8 times the specific activity of the corresponding amino acid in the glyceraldehyde-3-phosphate dehydrogenase. The constant ratio of the specific activities of different amino acids residues in the two proteins was interpreted as meaning that the two enzymes are synthesized from the same pool of amino acid precursors and that they turn over metabolically *at different rates.*

In another study on muscle aldolase and phosphorylase, *Heimberg and Velick* (75) injected different amino acids serially at intervals into a single animal, before the isolation of the enzymes, in order to obtain some information concerning time relationships in protein synthesis. The specific activities of the amino acids in aldolase at different intervals after injection were comparable with the results of earlier experiments (200) in which the amino acids had been administered in a single injection. Maintenance of the specific activity levels of the amino acid residues of aldolase over an extended period while turnover was still occurring has been explained as a consequence of a relatively slow protein turnover rate. It was further observed that, although the individual specific activities varied widely from one amino acid to the other, the ratio of the specific activity of each amino acid in aldolase to that of the corresponding amino acid in phosphorylase was essentially constant and equal to unity. It was concluded, therefore, that both aldolase and phosphorylase are in steady state equilibrium with the same pools of free amino acids and that the two proteins turn over metabolically *at the same rate.*

In the last paper of the series (222) aldolase, glyceraldehyde-3-phosphate dehydrogenase, myosin, L-meromyosin, H-meromyosin, actin, tropomyosin, and serum albumin were isolated from the muscle of rabbit after a single injection of phenylalanine-3-^{14}C. Pure phenylalanine and tyrosine, labeled metabolically, were isolated from hydrolysates of these proteins and the specific radioactivities of the amino acids were determined. From the specific activity ratios, phenylalanine to tyrosine, it appeared that H-meromyosin, actin and glycolytic enzymes were synthesized from the same amino acid precursor pools. The specific activities of amino acids in L- and H-meromyosins were significantly different, suggesting that myosin subunits are synthesized independently and turn over at different rates.

Making certain assumptions, i.e. that the specific activities of the free amino acids in the liver and muscle are identical and that half-life of rabbit serum albumin is four days, *Velick* (222) calculated the apparent half-lives of phenylalanine in the individual muscle proteins. His estimates showed that the half-life of glyceraldehyde-3-phosphate dehydrogenase was approximately 100 days, H-meromyosin, 80 days, actin, 67 days, aldolase, 50 days, phosphorylase,

50 days, tropomysin, 27 days, and L-meromyosin, 20 days. In these studies, the examined proteins are presumably found within the same cells, which are in a steady state; therefore, lysis of old cells and regeneration cannot account for the observed differences in the turnover rates of individual muscle proteins.

More recently, *Schapira et al.* (179) questioned the validity of *Velick*'s assumption that the free amino acid pool is identical in the liver and the muscle, and suggested that a computation of the lifetime of a muscle protein based on the identity of the free pools is likely to be too long. Using a direct method of estimating protein turnover by following the exponential decrease of ^{14}C-glycine prelabeled rat muscle aldolase, *Shapira et al.* (179) calculated that the half-life of this protein is 20 days.

IV. Synthesis and Degradation of Specific Enzymes in the Liver

A. Tryptophan Oxygenase

Abundant literature has accumulated since the discovery of the *adaptation phenomenon* (105) in the liver tryptophan oxygenase (pyrrolase), the enzyme catalyzing the oxidation of L-tryptophan to L-formylkynurenine. The activity of the enzyme can be greatly increased by injections of cortisone, as well as by substrate administration; this has been shown to result from an increased amont of newly synthesized enzyme protein, rather than activation of an inactive precursor (42).

Tryptophan oxygenase (TPO) has another distinction as well. In addition to being the first demonstrated *inducible* enzyme in the mammalian system, it is also the first liver enzyme whose turnover *in vivo* has been thoroughly investigated by the use of inhibitors. *Feigelson, Dashman and Margolis* (40) studied the kinetics of the decreased TPO activity subsequent to both substrate and cortisone induction. Their results have shown that the decay in enzyme concentration follows the *kinetics of a first-order reaction* with a calculated $t_{1/2}$ of 2.4 and 2.2 h following tryptophan and cortisone injection, respectively. The kinetics of enzyme synthesis and degradation did not seem to be influenced by a large extra-hepatic tumor (Walker 256 carcinoma) nor by purine analogs 8-azaguanine and 6-mercaptopurine.

In conformity with these experiments, *Nemeth* (132) reported that in the presence of puromycin, which was used to stop protein synthesis, the endogenous level of TPO activity fell to half-value in approximately 2 to 4 h. The elevated level of TPO after injection of the substrate also fell to half-value in about the same time, suggesting that the turnover rates of the enzyme before and after induction are approximately the same. Later, *Schimke, Sweeney and Berlin* (185) reported a value of 2.5 h, determined by following the rate of loss of radioactivity from the prelabeled enzyme. The same figure was obtained by *Garren et al.* (49) following a decay curve from high levels induced with hydrocortisone. Since, in the latter experiments, the administration of puromycin during the hydrocortisone-induction had no effect on the rate of enzyme decline, it was concluded that the changes in the enzyme levels were due entirely to *changes in the rate of enzyme synthesis.*

Several studies using inhibitors of protein and RNA synthesis, as well as immunochemical evidence, indicated that there is a difference in the mode of action of hormonal and substrate inducers of TPO (27, 65). This conclusion is also supported by the experiments of *Schimke, Sweeney and Berlin* (185) on the roles of synthesis and degradation in the control of TPO. An analysis of the time course of changing enzyme levels indicated that *hydrocortisone administration increases the rate of enzyme synthesis,* whereas *tryptophan administration decreases the rate of degradation of the enzyme.* This was based on the finding that tryptophan alone caused a *linear increase* in levels of TPO to six times those of controls, while hydrocortisone alone resulted in an *exponential increase,* reaching a maximum 7 to 8 times the control levels. Furthermore, the simultaneous administration of hydrocortisone and tryptophan produced a linear increase in TPO at a rate 7 times that produced by tryptophan alone, resulting in a 40- to 50-fold increase. In the latter case, the enormous increase in enzyme was presumably a result of both an *increased rate of synthesis* and *cessation of enzyme degradation.* The above hypothesis that whereas hydrocortisone increases the rate of the enzyme synthesis, tryptophan may be acting by preventing the naturally occurring degradation of the enzyme, has been supported by the studies on the *in vivo* incorporation of isotopic amino acids into, and loss from, TPO isolated by precipitation with specific antiserum (185).

Seglen and Jervell (196) recently developed a simple perfusion system which allows efficient induction of TPO by glucocorticoids. This has permitted the authors to determine the respective roles of enzyme synthesis and degradation in the induction process as compared to the basal state. The amount of enzyme increased 3- to 4-fold during 4-hour perfusion following the addition of glucocorticoids. In the presence of cycloheximide the synthesis of TPO was blocked, and the enzyme was degraded with a half-life of approximately 2 h. This value is similar to previously reported half-life figures for this enzyme. The half-life measured at various times during glucocorticoid induction was identical with that under the basal, non-induced conditions. It may thus be concluded that *glucocorticoids act only to increase the rate of synthesis of this enzyme without affecting its rate of degradation.*

Additional evidence for stabilization of TPO by its substrate has been provided by observations (49, 185) that tryptophan administration can prevent the characteristic rapid fall in enzyme activity which occurs several hours after the removal of the inducer. The concomitant administration of general inhibitors of protein synthesis, such as puromycin or cycloheximide, did not alter the tryptophan effect under these conditions.

Of the various tryptophan analogues tested, only a-methyltryptophan seems to impart stability similar to L-tryptophan (186). A good correlation exists, however, between the abilities of tryptophan analogues to stabilize TPO *in vivo* and to induce the enzyme in adrenalectomized rats.

An alternative explanation that both the tryptophan and cortisone induction of TPO may be associated with an increased rate of synthesis was voiced by *Greengard* (61) based on indirect evidence obtained by the use of metabolic inhibitors.

The stabilization of TPO by its substrate may possibly have to be reanalyzed in view of the recently developed assay for the enzyme (107) measuring the active, reduced holoenzyme, as well as the inactive apoenzyme and oxidized holoenzyme, which might have been underestimated in earlier studies. It has been suggested that tryptophan may bring about *in vivo* stabilization of TPO by activating the native apoenzyme to reduced holoenzyme (64, 106).

In this connection, there is special interest in the studies of *Marver et al.* (125) on the tryptophan-mediated induction of ALA-synthetase, the rate limiting enzyme in heme biosynthesis. This effect is thought to result from increased binding of 'free' or dissociable heme to the enzyme TPO making less heme available for apparent repression of ALA-synthetase. The authors believe that reciprocal control mechanisms exist between hemoprotein and ALA-synthetase participating in *coordinate synthesis* of the heme and apoenzyme moieties of TPO.

The existence of activation phenomenon with respect to TPO was originally brought to light independently by *Greengard and Feigelson* (60) and by *Chytil* (21). It is not yet certain whether the stimulatory effect of the microsomes of adult-rat liver on TPO, as described by the former investigators, is due, or in any way related, to the same activator found by *Chytil* in fetal liver supernatant. The afore-mentioned studies, as well as subsequent work by *Greengard and Feigelson,* have been amply reviewed (41, 64). We shall, therefore, concentrate on the findings of *Chytil* and his collaborators, which are of special interest, since they offer, in addition, a new insight into the question of enzyme regulation in development.

During a study on the development of TPO in the rat, it was found that fetal liver, though devoid of any detectable activity of this enzyme, contains a heat-stable activator of the enzyme (21). Subsequently, it was demonstrated that TPO activity in the soluble protein fraction (105,000 x g liver supernatant) can be increased by the addition of boiled extracts from other organs (brain, lung, etc.) of adult rats (22). TPO from liver thus appears to exist in an inactive form which can be readily converted into the active one. The relative amount of inactive enzyme was higher in the cortisone-induced animals than in the tryptophan-treated ones, suggesting the possibility that first formed after induction with cortisone is an inactive protein which is readily converted to active enzyme. It was further shown that the relative amount of the inactive enzyme increases with preincubation of the enzymic preparation at 37 °C (24, 25). The enzymatic activity which is lost during the preincubation period can be completely restored by the addition of cyclic adenosine 3',5'-monophosphate. These results indicate that the conversion of the inactive form into the active form is fully reversible and takes place in the absence of external hematin which was believed to be the unique activator of this enzyme (41). The validity of this conclusion was recently confirmed by *Gray* (59) and *Knox, Piras and Tokuyama* (107). During a study of the degradation of the cyclic 3',5'-adenosine monophosphate by the enzyme preparation, it was found that not only this compound, but also purine ribotides, ribosides, and even purines themselves can reactivate the preincubated TPO preparation (26). Whereas all pyrimidines tested were completely inactive, hypoxanthine and xanthine appeared to be the most effective compounds for restoring the activity of TPO, suggesting a possible role of xanthine oxidase in the activation process. These findings point out again the necessity of re-evaluating the results obtained by measuring the activity of tryptophan pyrrolase *in the high speed supernatants.*

Recently *Chytil* (23) has brought evidence that the conversion of the inactive TPO to the active enzyme in high speed liver supernatants by cyclic adenosine 3',5'-monophosphate is due to the formation of hypoxanthine from the cyclic nucleotide. He was able to identify liver xanthine oxidase to be responsible for the activation of TPO and localize the mode of action of xanthine oxidase on the conversion of the inactive holoenzyme to the active holoenzyme (93). Indeed administration of an inhibitor of xanthine oxidase together with the inducer (corticoid) promotes accumulation of the inactive holotryptophan pyrrolase as was evidenced by immunochemical titration (92, 93).

The idea that activation of apoenzyme to reduced holoenzyme brings about *in vivo* stabilization of TPO by tryptophan does not necessarily conflict with the earlier hypothesis *(vide supra),* since the increase in holo- at the expense of apoenzyme may create enzyme molecules which are resistant to destruction, or it may permit faster synthesis of enzyme protein. Rather, the improved technique for the enzyme determination further underlines the apparent difference existing between hormonal and substate induction. It appears that hydrocortisone causes the accumulation of much TPO activity revealed only by the heat-labile particulate activator (64), whereas, after treatment with tryptophan, there is very little of such latent enzyme or of the inactive apoenzyme (64). Tryptophan, if administered to hydrocortisone-treated rats, converts most of the inactive enzyme (apo- and latent forms) to the active form (holoenzyme).

The process of induction of TPO by corticoids can be resolved in two steps, first of which represents the synthesis *de novo* of inactive holotryptophan pyrrolase and the second involves the activation of these molecules by xanthine oxidase (92).

In this connection, it should be noted that actinomycin D, a powerful inhibitor of RNA synthesis, prevents the cortisone-mediated induction of TPO, but does not interfere with the enzyme induction by its substrate (65). This would suggest that substrate induction does not depend on RNA synthesis, whereas hormonal induction requires new DNA-mediated synthesis of RNA.

Somewhat paradoxical results with actinomycin were reported by *Garren et al.* (49) and *Tomkins et al.* (214). In their studies, actinomycin D blocked the induction of TPO when given simultaneously with hydrocortisone. It also inhibited enzyme synthesis when given 1–2 h after the hormone, but when given 4 or more hours after hydrocortisone, it stimulated enzyme activity instead. 5-Fluorouracil also stimulated enzyme synthesis when given 4 or more hours after hydrocortisone. In order to account for the late response to inhibitors of RNA synthesis, the authors postulated the existence of a *cytoplasmic repressor* which acts by inhibiting *translation* of the messenger RNA (corresponding to the enzyme) rather than interfering with the rate of DNA *transcription* (i.e. messenger RNA synthesis). According to *Tompkins et al.* (214) the following conclusions can be made concerning this mechanism: (a) repression depends on continual synthesis of protein as well as RNA, (b) the 'repressor' has a rapid rate of degradation, (c) repression depends either on a protein or the product of the action of a protein, and (d) it acts to block the translation of a specific mRNA. For further details, the chapter of *Gelehrter* (pp. 165–199) should be consulted.

B.   Tyrosine Aminotransferase

This soluble enzyme (TAT) which catalyzes the reaction between tyrosine and a-ketoglutarate, yielding p-hydroxyphenylpyruvate and glutamate has a number of characteristics comparable to those of tryptophan oxygenase, despite certain differences. Like the latter, TAT can be induced in intact rats by its substrate or by hydrocortisone (116, 117). Unlike tryptophan oxygenase, however, TAT induction is not affected by substrate administration in the absence of functional adrenals; the administration of the substrate together with the hormone, however, elevates the TAT level above that produced by hydrocortisone alone. This was interpreted as a *permissive action of hydrocortisone,* that acts supposedly by allowing an additional substrate induction by the tyrosine.

Kenney and Flora (98) subsequently demonstrated that tryptophan and suspensions of inorganic materials (e.g. Celite, bentonite) were as effective as tyrosine in inducing TAT in intact rats, although ineffective after adrenalectomy. In view of this evidence, the authors concluded that the effect of tyrosine and other agents which cause severe stress is non-specific, and they attributed the TAT induction under such conditions to stimulation of adrenocorticoid secretion. In a subsequent series of papers, *Kenney* showed that the induction by adrenocorticoids of TAT of rat liver involves enzyme synthesis *de novo* and established that the *rate of enzyme synthesis is increased* by induction (94). The possibility that the newly formed enzyme also reflects some enzyme accumulation resulting from the inhibition of enzyme degradation was excluded by comparing the rate of the enzyme decay under induced and under steady-state conditions.

The rate of enzyme degradation in the steady-state, as measured by the loss of the specific radioactivity of the ^{14}C-prelabeled TAT is such that removal of 50 % of the enzyme requires approximately 3 to 4 h (94). These results are in conformity with the figure of 3 h, reported by *Lin and Knox* (117), determined by the rate of the loss of TAT activity after the induced increase has ceased. *Goldstein, Stella and Knox* (53), utilizing a perfusion technique in which enzyme synthesis was blocked by puromycin, found that TAT was degraded 50 % in approximately 2 h, the rate being unaffected by the inducing dose of hydrocortisone.

Simultaneous administration of actinomycin D, an inhibitor of RNA synthesis, prevents the cortisone-induced elevation of TAT just as was reported for tryptophan oxygenase (49, 65). The antibiotic did not cause a fall in the basal level of the enzyme, which has been interpreted to indicate that the basal messenger for TAT is relatively stable. As with tryptophan oxygenase, actinomycin stimulated TAT activity when given more than 4 h after hormone treatment (49, 214); this was the basis for postulating the existence of a *repressor* capable of selectively inhibiting enzyme synthesis at the translational level *(vide supra)*. A similar mechanism appears to be also operating in tissue culture (213). The *superinduction* by actinomycin in culture cells was recently reinvestigated by *Reel and Kenney* (170), using brief *pulse* labeling times to measure synthesis and *chase* to measure turnover. Their results indicate that enzyme synthesis is not stimulated after actinomycin treatment but instead enzyme turnover is effectively blocked. They have therefore concluded that the *superinduction* of tyrosine aminotransferase reflects an actinomycin-mediated blockade of enzyme turnover and not the action of a cytoplasmic repressor. For further discussion and interpretation, the reader should refer to *Gelehrter's* chapter on enzyme induction (pp. 165–199).

Evidence for the existence of *selective repression of enzyme synthesis* has been provided by the studies of *Kenney and Albritton* (97). These authors found that tyrosine and Celite, which are effective inducers of TAT in intact rats, actually *depressed* the level of this enzyme in adrenalectomized animals; this resembles the results reported by *Rosen and Milholland* (172). The data of *Kenney and Albritton* (97) show that in the presence of tyrosine, the decline in enzyme activity follows *first-order kinetics,* the half-life being 2 1/2 h, the normal rate of degradation in the intact animal by radioactive techniques. Loss of enzyme activity at a rate equal to the known rate of enzyme degradation indicates that the rate of enzyme synthesis has been reduced virtually to zero, under these conditions. In pursuing their studies further, the authors found that actinomycin has no significant effect on the base level of TAT at doses sufficient to prevent induction of the enzyme by hydrocortisone, just as was reported by *Garren et al.* (49). Since new templates were not required to maintain the continued synthesis of the enzyme, it was assumed that repression under the stress of inducing agents must have occurred *at the translational* rather than the genetic, or transcriptional, level. Although tyrosine inhibits enzyme synthesis in adrenalectomized animals, this inhibition can readily be removed by the addition of actinomycin. This suggests that although the inducer acts by repression of enzyme synthesis at the translational level, *synthesis of the repressor is dependent upon transcriptional* events requiring the continued synthesis of RNA. The stressing agents are ineffective in hypophysectomized rats suggesting a hormonal factor of pituitary origin in the initiation of repression.

Further studies of *Kenney* (95) revealed that the hypophyseal growth hormone (somatotrophin, STH) is the pituitary compound responsible for eliciting repression of TAT synthesis in the liver. The initial response of the enzyme to the administration of growth hormone is, however, extremely variable. In some cases, the TAT level was already lowered after 1 h, and continued to fall in first-order fashion for several hours thereafter. In other experiments, the enzyme level rose during the first 1 or 2 h after hormone treatment, before it started to fall. This variable response, which occurred in adrenalectomized rats subjected to stress, as well as in growth hormone-treated animals that were either adrenalectomized or hypophysectomized or both, indicates that some capacity for TAT induction probably remains in rats deprived of glucocorticoids. The results could thus be explained if growth hormone treatment resulted in stimulation of this inducing capacity which, being variable in response, might be due to the action of other hormones released in animals treated with growth hormone.

Subsequent studies (79) further showed the existence of two extra-adrenal hormonal factors capable of inducing the enzyme, namely the polypeptide hormones insulin and glucagon. Elevation of the TAT level by either of those hormones is *due entirely to a rapid acceleration of its synthesis* and is clearly distinct from the induction by hydrocortisone. Effects of insulin and glucagon on the rate of transaminase synthesis are not additive, suggesting different mechanisms involved in their induction of the enzyme.

Inhibitors of protein synthesis, like those of RNA synthesis, do not appreciably influence the basal enzyme level, although these drugs will block hormonal induction of the enzyme (65). The apparent discrepancy in the failure of inhibitors of protein synthesis to lower the level of the enzyme which normally must be maintained by continual enzyme synthesis has been reinvestigated by *Kenney*

(96). The rate of TAT turnover in the steady-state or basal condition was determined with increased precision by *label and chase* experiments, using improved immunological techniques to isolate the ^{14}C-labeled enzyme. During an interval in which no turnover of the bulk of the liver soluble proteins could be detected, radioactivity was lost from TAT at a rate indicating a half-life of 1.5 ± 0.3 h. Inhibitors of protein synthesis (cycloheximide or puromycin) did not significantly alter the TAT level over a period of 4 to 6 h. In view of the demonstrated rapid turnover, the failure of cycloheximide to depress the enzyme level must mean either that it does not, in fact, block TAT synthesis, or that *both synthesis and degradation are inhibited to an equivalent extent*. The animals were, therefore, given ^{14}C-amino acids in a brief *pulse* exposure, the labeling time being short relative to the half-life of the enzyme; under these conditions, contribution of turnover is negligible, and the extent of isotope incorporation into the enzyme is a measure of its rate of synthesis. The experiments demonstrated that TAT synthesis was effectively blocked over virtually the entire period of both cycloheximide and puromycin treatment. With enzyme synthesis stopped, the enzyme level would be expected to drop to 10—15 % of the control level, if the normal degradative rate was maintained. Since the enzyme level was unchanged, *Kenney* (96) concluded that degradation, as well as synthesis, must be blocked in the treated animals. This conclusion was confirmed by measurements of TAT turnover in animals treated with cycloheximide after labeling of the enzyme had been completed. More recently, *Reel and Kenney* (170) have also reported that actinomycin D inhibits the turnover of induced TAT in hepatoma cells in culture.

Since the activity of the enzyme is known to be affected by a large number of hormonal agents *in vivo*, *Jervell and Seglen* (87) found it desirable to reexamine the effect of inhibitors of protein synthesis on TAT in the isolated, perfused rat liver. Addition of the synthetic glucocorticoid dexamethasone to the perfusate produced a marked increase in the enzyme. In the presence of cycloheximide at a dose level which inhibits protein synthesis the induction of TAT was blocked and a rapid fall in the enzyme activity resulted. The basal non-induced enzyme level was also lowered by the cycloheximide treatment.

In both cases the half-life of the enzyme was approximately 2—3 h. This is in agreement with previous measurements *in vivo* (94, 117) and indicates that degradation of TAT is unaffected by cycloheximide in the perfused liver. Since an alteration of enzyme synthesis is the only observed effect of cycloheximide on TAT in the perfused liver, indirect mechanisms may thus be responsible for the diverse effects seen *in vivo*. It was proposed (195) that insulin may be the agent responsible for the inhibition of TAT degradation *in vivo*. *Auricchio et al.* (5), on the other hand, suggested that fasting of animals may account for the prevention of enzyme degradation *in vivo* while a nutritional step-down condition may be responsible for a comparable phenomenon seen in cell culture.

It should be noted, however, that *Levitan and Webb* (113) have recently reported that both 8-azaguanine and cycloheximide prevent normal fall in TAT activity from the steroid-induced to the basal level in their perfused liver system. The apparent discrepancy between these and the studies of *Jervell and Seglen* (87) must thus await further study.

Regulatory mechanisms of TAT induction and subsequent inactivation in the presence of inhibitors of protein synthesis was also studied by *Grossman and Mavrides* (69), using a liver biopsy technique. Puromycin, while inhibiting enzyme synthesis when given during the initial phase of induction, caused an unexpected reappearance of enzyme activity, following its administration during the inactivation phase. To interpret this potentiated response, the authors postulated a *repressor* which is presumably formed about 4 h after hormone administration; inhibition of synthesis of this hypothetical repressor supposedly allows the temporary synthesis of the enzyme. Administration of actinomycin D during the early stages of induction curtailed further enzyme synthesis and prevented inactivation of the enzyme already induced. However, when given during the inactivation phase, it had little effect and it did not prevent the disappearance of the induced enzyme activity. An inactivating system is formed, presumably during the induction cycle, which is responsible for the removal of the induced enzyme. The activity of the inactivator appears to be dependent on an extrahepatic factor, since removal of the pituitary gland inhibits the inactivation phase of the induction cycle.

An additional mechanism, termed *cofactor induction,* for control of the TAT level was proposed by *Greengard and Gordon* (63), on the basis of their finding that administration of large doses of pyridoxine causes elevation of this enzyme in the livers of adrenalectomized rats. This assumed *de novo* synthesis of enzyme, as indicated by the inhibitory effect of puromycin, was, according to *Greengard and Gordon* (63), insensitive to actinomycin. Since the unstable TAT apoenzyme under *in vitro* conditions can be stabilized by pyridoxal phosphate, *Holten, Wicks and Kenney* (80) investigated the possibility that the increased enzyme level in adrenalectomized rats following pyridoxine administration is also due to enzyme stabilization *in vivo*. *A priori,* this system seems to be analogous to the substrate-induction of tryptophan pyrrolase by tryptophan *(vide supra),* which lends further credence to the latter hypothesis. However, according to the detailed immunological studies of *Holten, Wicks and Kenney* (80), pyridoxine was ineffective in preventing TAT degradation *in vivo,* but *it increased the rate of enzyme synthesis.* Furthermore, contrary to the earlier results of *Greengard and Gordon* (63), the induction of TAT by pyridoxine was as sensitive to actinomycin as the hormonal induction. This raises the question of whether TAT coenzyme interactions actually play a significant role in this process. There is a strong possibility that pyridoxine acts indirectly by altering the hormonal balance.

C. Alanine Aminotransferase

Also termed as glutamic-alanine (GAT) or glutamic-pyruvic transaminase, this enzyme catalyzes the reaction between *L*-glutamic acid and pyruvic acid to give *a*-ketoglutaric acid and *L*-alanine. Together with glutamic-oxaloacetic (or glutamic-aspartic) transaminase, they play an important role in the transamination reactions leading to the elimination of nitrogen from amino acids, and are an important source of keto acids for Krebs cycle and for gluconeogenesis. Whereas the level of the glutamic-oxalacetic transaminase appears to be relatively stable, this is not true of glutamic-pyruvic transaminase.

Corticoid treatment, as well as various conditions known to increase the rate of gluconeogenesis, such as high protein intake, diabetes, and fasting, produce a marked increase in liver GAT (173). The properties of the enzyme, purified 200- to 300-fold from the livers of normal and corticoid-treated animals, are indistinguishable during the several steps of the purification procedure and in the kinetic parameters studied (190). The corticoid-induced activity is also significantly blocked by the administration of ethionine. These observations, together with the finding that purified enzyme preparations from control and corticoid-treated rats appeared in the same peak, when subjected to chromatography (138) with virtually the same specific activity (190) are consistent with an increased tissue level of the enzyme. *Segal* (190) has subsequently presented direct immunological evidence for changes in the amount of liver GAT as the underlying basis for corticoid-induced alterations in the enzyme's activity.

The mechanism of the corticoid induction has been studied in some detail by *Segal and Kim* (191). It has been shown that administration of prednisolone raises the level of GAT approximately 5-fold over a course of 5 days, at which time the rising curve gradually slopes off to a plateau. The rate of degradation necessary to bring about such a plateau would result in a half-life of 1.2 days. Upon withdrawal of the prednisolone, there is a lag of several days, after which the level of GAT begins to fall, with a rate of decline that approximates a half-life of 3 1/2 days. It would appear from these data that *in the presence of the hormone, the rate of destruction of the enzyme is about 3-fold greater than in its absence.* Since the level of GAT rises 5-fold under the influence of prednisolone at a time when the rate of enzyme destruction is increased approximately 3-fold, the increased GAT activity would thus have to result from a rather enormous increase in the rate of enzyme synthesis.

In view of these discrepant results, it was therefore considered important to derive comparative values by the independent method of disappearance of radioactivity from the labeled proteins (101). Experiments with both L-arginine-(guanido-^{14}C) and uniformly labeled L-leucine-^{14}C have shown no significant difference in the half-life of GAT in normal and glucocorticoid-treated animals. Calculations gave half-lives for the normal animals of 4.3 days with guanidolabeled arginine and 4.2 days with uniformly labeled leucine, respectively, and 4.3 days for the glucocorticoid-treated rats. The values are closely comparable to those determined earlier (3.5 days) from the decline of enzyme activity from the induced to the normal level (191).

That the turnover rate of the induced enzyme does not specifically differ from the basal enzyme is supported by the studies of *Swick, Rexroth and Stange* (208) who have reported similar values in the control animals and in the treated animals on a high protein diet or after administration of corticosteroids. It should be noted, however, that in these studies the half-life of GAT was calculated to be 18 h which is considerably lower than the value obtained by *Kim* (101). Be. that as it may, in comparison to a short-lived rat liver GAT, the half-life of the enzyme in rat muscle is about 20 days (101).

The relatively slow rise of GAT to a new steady-state level, following its induction by a hormone, when compared with tryptophan pyrrolase or tyrosine transaminase, is of some interest and might possibly be the result of a delayed or lower response on the part of the enzyme-forming system. Initial studies (192) did not favor this view since the rate of precursor incorporation into the enzyme was already at a maximum 12 h after initiation of hormone treatment, although the maximum activity level is not reached for 5 days.

To throw some light on this question, *Kim* (100) measured the relative rate of synthesis of GAT at various times after the administration of massive doses of prednisolone or hydrocortisone using the L-leucine-^{14}C pulse-labeling method. The rate of synthesis was unchanged 3.5 h after the administration of prednisolone and 6.5 h after the administration of hydrocortisone. Moreover, the rate was, increased approximately 2-fold after 6.5 h of prednisolone or 5 days of hydrocortisone. Tyrosine aminotransferase and tryptophan oxygenase activities, in comparison, reach a maximum within a few hours after a glucocorticoid administration. This observation indicates that the stimulatory effects of glucocorticoid hormone on the synthesis of liver proteins are *sequential rather than simultaneous.*

D. Catalase

Together with the closely related peroxidases, this heme-containing enzyme has been classed with the hydroperoxidases, which catalyze reactions between hydroxyperoxides and various oxidizable compounds. Despite intensive research, resulting partly from the possible implication of catalase in neoplasia (67, 82), as postulated by the early investigators, we still do not know with certainty whether the physiological function of catalase is a protective catalitic one, or a peroxidatic one. *De Duve and Baudhuin* (28a) recently suggested, on the basis of the high content of this enzyme in the liver microbodies (peroxisomes) which also contain at least three hydrogen peroxide-forming oxidases (82), that catalase acts mainly in its peroxidatic capacity, with its catalatic activity serving as a safety device to stem the overflow of hydrogen peroxide, should the supply of hydrogen donors for the peroxidatic reactions fail to keep up with the production of hydrogen peroxide (82).

Liver catalase can be altered under a variety of experimental conditions, including diet, hormones, drugs, etc. While numerous factors can bring about a severe depression of the enzymatic activity, instances in which the enzyme activity could be elevated above the basal steady-state level are rare in the mammalian systems (82).

The pronounced lowering of catalase activity observed in the liver of cachectic tumor-bearing animals (160, 167) and the rapidity with which its normal level was restored following the surgical removal of the tumor (67), suggest that the enzyme must have a high rate of turnover. The rapidity with which radioactive iron was incorporated into the enzyme molecule (20, 119, 212) accorded with this view.

The early data suggesting that a highly active form of catalase was present in the liver of normal rats but absent in rats bearing the Novikoff hepatoma were later shown to be an artifact of the purification procedure (60). When catalase of livers from tumor-bearing animals was isolated directly from the particulate fractions in which it is located, a highly active enzyme was obtained for which the specific activity was nearly the same as that obtained from normal animals (154). This meant that the lowering of liver catalase in the tumor-bearing host must result from a change in the rate of catalase synthesis or destruction of the enzyme, rather than from an alteration in structure resulting in a lowered specific activity of the enzyme.

New insight into the question of the turnover of liver catalase has been obtained recently in the author's laboratory. Using 3-amino-1,2,4-triazole (AT) and allylisopropylacetamide (AIA), either singly or in combination, *Price, Rechcigl and Hartley* (153), *Price et al.* (154) and *Rechcigl and Price* (165, 166) developed simple but sensitive techniques for measuring the kinetics of catalase synthesis and destruction *in vivo*. AT acts by irreversibly *inhibiting catalase without interfering with its resynthesis,* while AIA acts by *blocking the synthesis of the new enzyme without interfering with the activity* of previously formed catalase.

After the administration of AT, the return of catalase activity is paralleled by a corresponding uptake of ^{59}Fe into the catalase (fig. 1), indicating that the return of catalase results from the formation of a new enzyme rather than a reversal of the inhibitory process which led to the initial fall in the enzyme activity of the new enzyme. These data are in agreement with the findings of *Margoliash, Novogrodsky, and Schejter* (122) who have demonstrated that AT binds to the protein moiety of catalase to form an irreversible complex.

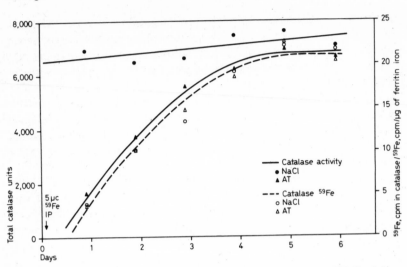

Fig. 1. Liver catalase activity and radioactivity in rats after injection of NaCl solution and AT, followed 2 h later by 5 μc of ^{59}Fe (in the form of ferric ammonium citrate). From *Price et al.* (154).

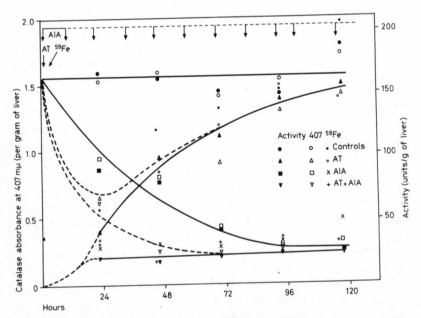

Fig. 2. Relationship of catalase activity to spectra after treatment with 3-amino-1,2,4-triazole (AT) or/and allylisopropylacetamide (AIA), followed later by ^{59}Fe (in the form of ferric ammonium citrate). From *Rechcigl and Price* (166).

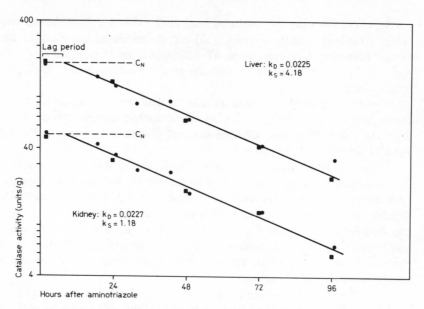

Fig. 3. Kinetics of catalase degradation in liver and kidney, using 3-amino-1,2,4-triazole (AT). Semilogarithmic plot of C_N-C_{AT} *versus* time, where C_N is the normal level of catalase, C_{AT} is the catalase level at various times after the injection of AT, 1 g/kg of body weight, intraperitoneally. Each point represents the average of 5 animals. The k_D is the first order rate constant for catalase decay, as determined by the slope of the solid line drawn to fit the experimental points. The k_S, the rate of catalase synthesis, equals $k_D C_N$. From *Rechcigl* (158).

From kinetic studies on the rate of resumption of catalase activity after the administration of AT, it was calculated that the observed data could be accounted for by the synthesis of 4.8 units (i.e. 28 μg) of catalase per hour per gram of rat liver, if 2.25 % of the catalase molecules present were being destroyed each hour.

By use of AIA to block the formation of new catalase, it was shown that the rate of catalase disappearance was nearly the same as that calculated from the AT data. By isolating catalase from the liver under these conditions, it was shown (fig. 2) that AIA almost completely blocks the incorporation of ^{59}Fe into liver catalase (166). *Schmid, Figen, and Schwartz* (188) reached the same conclusion using ^{14}C-labeled glycine.

There remains a small fraction of catalase, about 8 % of the total, whose synthesis is not blocked by AIA, although it is effectively inactivated by AT (166). This suggests the existence of a second catalase; data indicate that this small pool is turning over very rapidly and that there may be as many catalase molecules synthesized and destroyed per hour in this small, highly active pool as in the main pool of liver catalase.

In agreement with the results of *Price et al.* (154) and *Rechcigl and Price* (166), it has been recently reported (152) that the calculated half-life of catalase derived from rate of recovery from AT inhibition is about 1 1/2 days, as is the apparent half-life of the heme prosthetic groups measured with ^{14}C-δ-amino-levulinic acid.

Comparative studies on the kinetics of catalase synthesis and destruction in the rat liver and the kidney (154, 158) demonstrated identical rates of catalase destruction in both organs, although the rate of catalase synthesis in the liver was four times that of the kidney (fig. 3).

During the first 5 days of starvation (fig. 4), the total liver catalase activity fell progressively as the total weights of the organs, but the rate of catalase synthesis per gram of tissue remained constant (158). Since, during starvation, there is a marked loss of the liver RNA which is essential for protein synthesis, it may be inferred from these data that in the early stages of starvation the animal may control the rate of catalase synthesis by progressively decreasing the number of synthesizing units, but that those synthesizing units which remain may be operating at full activity.

After 5 days, the starving animal enters a second phase in which the rate of catalase synthesis per gram of liver starts to fall, and this continues until the animal's death. As can be seen in figure 5, this decrease in the enzyme activity is

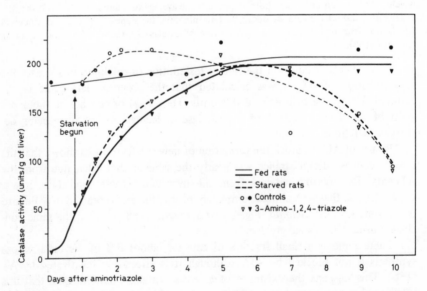

Fig. 4. Time course of changes in the concentration of the hepatic catalase activity in normal and starved rats after injection of 3-amino-1,2,4-triazole. ----= starved rats; ——= fed animals. From *Rechcigl* (158).

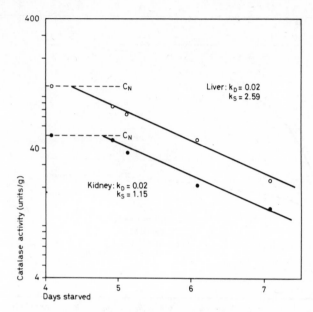

Fig. 5. Kinetics of hepatic and renal catalase degradation during severe starvation, using 3-amino-1,2,4-triazole (AT) and allylisopropylacetamide (AIA). The renal enzyme values are presented in form of a semilogarithmic plot of C_N-C_{AT} *versus* time, and those of the hepatic catalase are plotted as C_{AIA}-C_{AT+AIA}, where C_N is the normal level of tissue catalase at various times after the administration of AT (1 g/kg body weight), C_{AIA} is the catalase level during administration of AIA (200 mg/kg body weight twice daily); and C_{AT+AIA} is the catalase activity of animals given both drugs. From *Rechcigl* (158).

entirely accounted for by the reduction in the rate constant of enzyme synthesis, with no apparent effect on the rate constant for enzyme degradation.

In contrast to those undergoing starvation, animals on a protein-free diet (fig. 6) had a decreased rate of catalase synthesis per gram of tissue (159). Although there was less catalase being synthesized, the *fraction* of enzyme molecules being destroyed per unit of time was within normal limits (fig. 7), thus resulting in a lower concentration of catalase within the liver. However, on the protein-free diet, there was a relatively small decrease in the liver size, in contrast to starvation conditions where a marked decrease occurred.

Similarly, the depression of catalase activity following the feeding of cycasin, a toxic component of cycad nut, was explained (fig. 8) primarily by lower rate of catalase synthesis (164). Possibly of greater biological significance, however, is the observation that cycasin, besides affecting the catalase synthesis, significantly *lowered the rate of catalase destruction* — a rate that appears to be almost 'constant' in liver and kidney of normal rats (155).

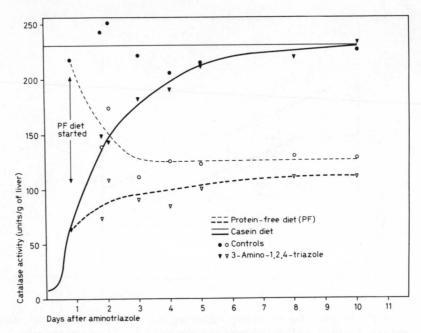

Fig. 6. Time course of changes in the concentration of hepatic catalase activity of protein deficient and protein-fed rats after injection of 3-amino-1,2,4-triazole. ---- = protein-free diet; —— = casein diet. From *Rechcigl* (159).

Since significant levels of catalase were found in several hepatomas (83, 167, 168), it was of interest to study the rates of synthesis and destruction in these tumors. In view of the question as to whether proteins are being 'turned over' during the logarithmic growth of the cells, it is of paramount importance to determine whether the rate constant for catalase destruction of neoplastic tissues is similar to that of the liver and kidney.

Initial studies on tumors were hampered by the finding that many of the examined high-catalase-hepatomas, particularly the HC (high catalase) – hepatoma, did not respond to the inhibitors. Nevertheless, several hepatomas, including 5123 and 7316A, were found in which virtually all catalase was destroyed following the administration of AT, thus making it possible to carry out the kinetic studies on neoplastic as well as normal tissues within the same animal (163).

Using the AT technique, it was found (fig. 9) in the 5123 hepatoma-bearing rats that during each hour, 3.88 units of catalase were being synthesized per gram of the host liver, as compared to 0.92 units in the tumor. The first-order rate constant for catalase destruction, however, was practically the same in both types of tissues. It is of particular interest that the rate constant for catalase

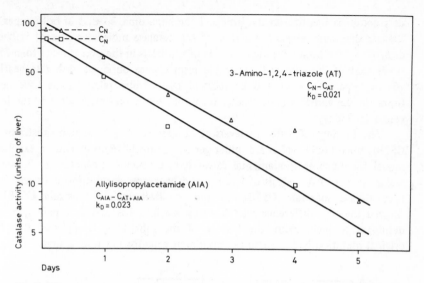

Fig. 7. Kinetics of hepatic catalase degradation during protein deficiency using 3-amino-1,2,4-triazole (AT) and allylisopropylacetamide (AIA). Semilogarithmic plot of C_N-C_{AT} and of C_{AIA}-C_{AT+AIA} *versus* time. From *Rechcigl* (159).

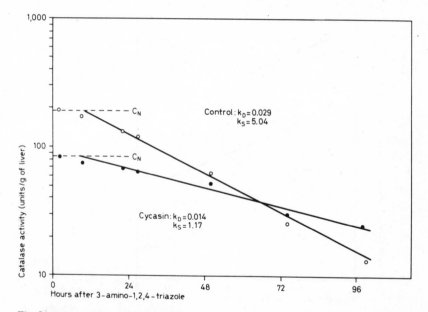

Fig. 8. Kinetics of hepatic catalase synthesis and degradation in rats fed diet with and without 200 mg % cycasin using 3-amino-1,2,4-triazole (AT). Semilogarithmic plot of C_N-C_{AT} *versus* time. From *Rechcigl and Laqueur* (164).

destruction is identical in the liver and the hepatoma, so that in both types of tissues, the same *constant fraction* of the catalase molecules present is being destroyed each hour, although the rate of synthesis in the liver is more than four times that of the tumor. Comparable results were obtained with rats bearing 7316A hepatoma. It should be recalled that similar relationships have been found in the earlier kinetic studies on the turnover rate of catalase in the liver versus the kidney.

The findings of different levels of catalase activity in certain substrains of C57BL mice (161) which are under genetic control (76) provided an excellent model for pursuing fundamental research in biochemical genetics in the mammalian system. An analysis of F_1, F_2, and backcross generations between high-liver-catalase substrain C57BL/He and low-liver-catalase substrain C57BL/6 showed that the difference was due to a single autosomal gene pair with low dominant to high. From the history of the substrains, it appeared that the original mutation had been to the dominant gene for low level activity. There is

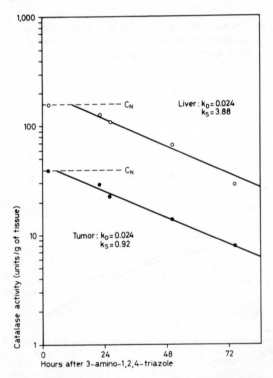

Fig. 9. Kinetics of catalase synthesis and degradation in tissues bearing Hepatoma 5123, using 3-amino-1,2,4-triazole (AT). Semilogarithmic plot of C_N-C_{AT} *versus* time. From *Rechcigl, Hruban and Morris* (163).

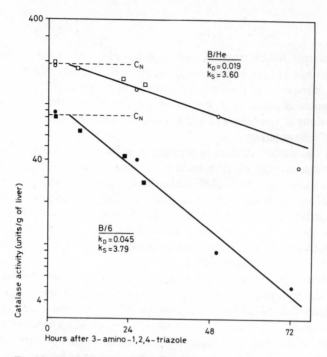

Fig. 10. Kinetics of liver catalase destruction in 2 substrains of C57BL mice. Semiloga-rithmic plot of C_N-C_{AT} *versus* time. Circles and squares denote data from two separate experiments. From *Rechcigl and Heston* (162).

a question concerning the mechanism by which the mutant gene brings about the observed reduction in enzyme activity. Experiments were therefore carried out to investigate the kinetics of catalase synthesis and destruction in the high and low catalase substrains. The data of *Rechcigl and Heston* (162) show that in C57BL/He substrain, there were 3.60 units liver catalase synthesized per hour, as compared with 3.79 units in C57BL/6 substrain (fig. 10). The calculated rate constant for catalase destruction in the former group was 1.9 %/h, while in the latter group it was 4.5 %. It is evident from these studies that an *alteration in the rate of degradation,* as well as in the rate of enzyme synthesis, may play a significant role in controlling the level of catalase. The alteration in the rate of catalase decay in the above studies is a unique finding, since in all rats and mice examined so far in our laboratory under a variety of experimental conditions, the rates of catalase destruction seemed to be almost uniform (155). The finding of *Rechcigl and Heston* (162) that the difference in activity between C57BL/He and C57BL/6 is due to a 2-fold difference in the rate of liver catalase degra-dation, has been recently confirmed by *Ganschow and Schimke* (48).

E. Arginase

The level of this soluble enzyme, hydrolyzing arginine to ornithine plus urea, is known to rise and fall in response to factors that alter the rate of urea excretion, as a consequence of their effect on protein breakdown (181). For this reason, liver arginase is particularly well suited for the comparative studies *in vivo* of contributions of synthesis and degradation of an enzyme with reference to changing metabolic requirements.

Purification and characterization of arginase from livers with widely differing specific activities revealed no differences in the kinetic and physicochemical properties of the enzyme molecule (180). It has been, therefore, concluded that the observed differences in the enzyme activities are due to differences in the amount of a specific enzyme protein rather than to the effect of activators or inhibitors. That the protein synthesis is involved in the increased levels of enzyme is also indicated by the finding that the amino acid analogue, ethionine, prevents increases in arginase activity (181).

In extensive studies on the control of arginase levels in rat liver, *Schimke* (182) followed the incorporation of ^{14}C-guanidino-arginine into arginine. The enzyme was extracted from the liver by a relatively simple procedure that gave approximately 1400-fold purification in five steps, giving a homogeneous preparation which has been shown by immunological analysis to have a single antigenic protein. The use of the ^{14}C-guanidino-*L*-arginine appears to be particularly fortuitous, since 98 % of the ammonia of the guanidine group has been shown to be excreted as urea, and only 2 % is reincorporated into protein upon degradation of the enzyme *in vivo*.

Using these techniques, *Schimke* (182) found that the rate of degradation of arginase in rats on diets containing 8, 30, and 70 % casein seems to be essentially the same under all three dietary conditions, its half-life being 4 to 5 days. Perhaps of greater interest are the *changes in turnover of arginase under changing dietary conditions.* Upon the initiation of fasting, the total content of arginase rises progressively, despite a marked reduction in liver weight. It appears that no radioactivity is lost from arginase during the fasting period, indicating that *prelabeled arginase is not degraded.* The increased content of arginase in the liver results from the continual synthesis of arginase at a relatively constant rate. On change from a high (70 %) to a low (8 %) protein diet, the rate of enzyme degradation is increased, and the rate of synthesis is diminished, thereby decreasing the content of enzyme.

Although arginine fails to augment significantly the levels of arginase in the living animal, there are several reports in the literature which have demonstrated *substrate induction* in cultured mammalian cells (104, 183). In addition, a profound increase in the arginase level under *in vitro* conditions can be brought about by the addition of manganese to the medium. Using an antibody that specifically precipitates arginase, it has been established that these effects are the result of *de novo* synthesis. An analysis of the time course of arginine- and manganese-mediated increases of arginase in $HeLaS_3$ cells (183) indicated that two different mechanisms of induction are involved. Whereas *arginine apparently acts by increasing the rate of enzyme synthesis, the addition of manganese to the medium brings about stabilization of the arginase molecule,* thus reducing its degradation; this leads inevitably to the accumulation of newly synthesized enzyme. The stabilizing effect of manganese was also observed by *Eliasson and Strecker* (35) in Chang's liver cells.

Whereas arginase in cell culture (HeLa, Kb, or L-929 cells were used) is raised by its substrate arginine, the levels of argininosuccinate synthetase and argininosuccinase, the two enzymes concerned with the conversion of citrulline to arginine, are 2- to 15-fold greater when the concentration of arginine in the medium limits growth than when cells are grown in excess arginine (183). This phenomenon is similar to the *arginine repression* that occurs in microorganisms.

The lowering of arginase activity in Chang's liver cells by glucose, as reported by *Eliasson* (34) is thought to be an expression of *catabolite repression* in mammalian cells. Further support for the existence of repression phenomenon in cell cultures was provided by studies in which the response of arginase was investigated with reference to variations in intracellular concentration of the products of the metabolic sequence from arginine to proline (35). Whereas an increase of arginase activity has been produced by increasing the concentration of lysine, leucine, valine, or ornithine, a decrease in the enzyme activity followed the addition of proline, a product of the same reaction sequence. These results suggest that the arginine level is regulated by a *negative feedback control,* in which a product of the reaction chain, initiated by the arginase-catalyzed reaction, depresses the synthesis of the enzyme. Thus, the observed increase in the enzyme activity, following the administration of certain amino acids, and which was shown to result from *de novo* synthesis, could be brought about by lowering the level of metabolic repressor. Conversely, the inhibitory effect of proline might be due either directly or indirectly to the accumulation of a repressor of enzyme synthesis.

F. Glucokinase

This recently-discovered phosphotransferase (ATP:D hexose phosphotransferase), (31, 223, 225), possibly the first and limiting step of glycogen synthesis from glucose, catalyzes the formation of glucose-6-phosphate from glucose and ATP. In contrast to relatively non-specific hexokinase, which also occurs in the liver, glucokinase is specific to glucose and is apparently confined to hepatic parenchymal cells (202).

Whereas hexokinase is not appreciably affected by dietary or hormonal alterations, glucokinase activity falls to very low levels during starvation or in alloxan diabetes (178, 197, 228). The reappearance of glucokinase, on refeeding or after administration of insulin, respectively, can be blocked by the inhibitors of protein synthesis, indicating that the new activity represents *de novo* synthesis of enzyme. A similar inhibitory effect by actinomycin supports this conclusion and further suggests that formation of messenger RNA is also involved.

If insulin is administered to alloxan diabetic rats, the level of glucokinase rises from barely detectable levels to normal in the course of 1 day (202, 203). Withdrawal of the insulin is followed by a disappearance of the glucokinase over a 4-day period. Halving of glucokinase by fasting normal rats takes approximately 1 day as compared to 2 days observed after stopping insulin administration to the alloxanized animals. These figures are in agreement with a half-life of 33 h reported by *Niemeyer* (134) from glucokinase decay after transferring animals from a normal diet to a high-fat carbohydrate-free diet.

Glucose alone is also effective in restoring glucokinase levels after depletion by starvation, while a carbohydrate-free diet leads to glucokinase levels nearly as low as complete starvation. Since in both alloxan diabetes and starvation there is a marked decrease in insulin secretion, whereas these two conditions have opposite effect on blood glucose, *Sols et al.* (203) postulated that insulin rather than glucose is the enzyme inducer. An alternative hypothesis of glucose induction conditioned by a *permissive* role of insulin was advanced by *Niemeyer et al.* (135).

Glucokinase can be elevated in fasted rats by the administration of insulin alone, although the levels are considerably smaller than those obtained by glucose (135, 203). *Sols, Sillero and Salas* (203) are of the opinion, however, that there may be a good induction by insulin even under these conditions, but that the synthetized enzyme is rapidly degraded because of the low glucose concentration, based on the finding that glucose, but not insulin, when added to incubated liver slices, protects glucokinase *in situ*.

This interpretation has been challenged by *Niemeyer et al.* (136) on the grounds that 2-deoxyglucose stabilizes the enzyme *in vitro* but is ineffective *in vivo*. After showing that as much as 6 units of insulin do not modify the low levels of liver enzyme obtained in rats fed a carbohydrate-free diet, and that insulin did not accelerate the enzyme induction by glucose, they proposed instead that the induction of enzyme is being elicited by glucose, insulin being required in a 'permissive' fashion, presumably acting at the inner membranes of liver cells to allow glucose entry into the nucleus. This view has been reinforced by the finding that glucagon prevents the enzyme induction, and that this effect is not counteracted by an excess of insulin. Glucagon may thus act as a physiological repressor while glucose may function both as an inducer in the liver cell and as an agent blocking the release of glucagon from the a-cells of the pancreas (220).

There is an apparent lag of about 2 h from administration of external inducer to detectable increase of glucokinase in liver (203). The results of delayed administration of actinomycin and puromycin indicate that formation of messenger RNA starts within 1 h after the inducer administration, and that completion of active enzyme rapidly follows polypeptide synthesis. Messenger RNA for glucokinase seems fairly stable, its half-life being greater than 8 h.

Inhibition by glucagon as noted by *Niemeyer et al.* (136) may be identical to that of actinomycin, insofar as these inhibitors are effective only when given very early in the response, both presumably interfering with the formation of messenger RNA for the enzyme induction.

A paradoxical effect of actinomycin, reminiscent of the findings of *Garren et al.* (49) with tryptophan pyrrolase *(vide supra),* has been noted on the disappearance of glucokinase by fasting. When actinomycin and fluorophenylalanine were given 8 h after removal of food, 18 h later there was no decrease of glucokinase in the treated animals, while 20 h later there was an actual increase in the enzyme level. In the livers of the control animals, on the other hand, there was the usual moderate decrease in the glucokinase activity.

Interesting studies on the development of glucokinase in the neonatal rat were presented by *Walker and Holland* (227). The enzyme, which is absent from fetal liver, develops gradually, starting about 16 days post-parturition and reaching adult levels 10–12 days later. The inhibitory effects of ethionine or *p*-fluorophenylalanine on this process, as well as those of puromycin and actinomycin, indicate that the appearance of the enzyme represents *de novo* synthesis. Both glucose and insulin are apparently necessary for the normal development of glucokinase, for this is retarded in starved and alloxan-diabetic neonatal animal. The presence of insulin in the pancreas immediately after birth and availability of carbohydrate to the neonatal animal would seem to suggest that there must be some other reason than a lack of substrate and/or insulin to account for the late appearance of the enzyme after birth. Furthermore, initial attempts to hasten glucokinase appearance by the infusion of glucose or insulin were unsuccessful.

However, subsequent studies (226) showed that the ability to synthesize glucokinase develops very rapidly and that the nature of the diet determines the rate of appearance of activity. Thus, feeding a high-glucose diet to weanling rats showed that high hepatic glucokinase could be induced at 18 days of age, i.e. 2 days after development of the enzyme begins. The normal development of glucokinase activity, in contrast, can be retarded in weanling rats by feeding carbohydrate-free, high-fat and high-protein diets.

Many of these physiological properties of glucokinase in the adult animal are also characteristic of phosphofructokinase and pyruvate kinase, the other two key glycolytic enzymes involved in channeling glucose to lactate (232, 234), suggesting that these three enzymes are produced on the same functional genome. Whereas insulin may act as an inducer in case of the glycolytic enzymes, it also acts as a suppressor of the four key gluconeogenic enzymes, i.e. glucose-6-phosphatase, fructose-6-diphosphatase, phosphoenolpyruvate carboxykinase, and pyruvate carboxylase, which are thought to be produced on another functional genome unit (233).

In this connection, mention should be made of the studies of *Freedland, Cunliffe and Zinkl* (46) who observed that while insulin increases glucose-phosphatase and fructose-1,6-diphosphatase activities in response to cortisone treatment, it does not prevent increases in response to diets high in protein or fructose. Thus, the general suppressor theory of insulin postulated for gluconeogenic enzymes (233) awaits further clarification.

G. Serine Dehydratase and Ornithine Aminotransferase

Because of certain similar adaptive properties, and because they have usually been studied concurrently, these two amino acid-catabolizing enzymes will be discussed together. As used here, the term serine dehydratase (SDH) denotes the rat liver enzyme — cystathionine synthetase-*L*-serine and *L*-threonine dehydratase — which catalyzes the deamination of *L*-threonine, as well as of *L*-serine, in addition to the synthesis of cystathionine from *L*-serine and *L*-homocysteine (129). Ornithine δ-aminotransferase (OAT) is a mitochondrial enzyme, catalyzing the conversion of *L*-ornithine to glutamic γ-semialdehyde (143).

Both enzymes can be induced to very high levels in the liver by feeding high protein diet or by oral intubation of casein hydrolysate to protein-depleted rats (149). A similar effect can be achieved by the administration of free amino acid mixtures (141). When administered singly, only tryptophan produces a significant induction. A lack of response following the feeding of the substrate *L*-threonine has already been noted by *Goldstein, Knox and Behrman* (52).

The casein hydrolysate-mediated increase in the enzymes' level can be prevented by puromycin, suggesting that net protein synthesis is involved in their induction. SDH induction by a single dose of casein hydrolysate exhibits a rapid turnover; preliminary studies (149) indicate that the half-life of this enzyme is about 3 h. These authors' studies do not allow estimation of the half-life of OAT although it is apparent from the data on hand that the half-life is quite long. *Swick, Rexroth and Stange* (208) have subsequently estimated the half-life of the latter enzyme to be 19 h.

If actinomycin D, a potent inhibitor of template (messenger) RNA synthesis in mammalian cells, is administered during the initiation of induction of a given enzyme, complete inhibition of enzyme synthesis results. But if the administration of actinomycin is delayed until 12 h after induction is initiated, there is no significant effect on enzyme synthesis. These data indicate that the template

RNA for SDH and OAT must be in a stable form that can continue to be utilized as a template for protein synthesis during the period of actinomycin insensitivity.

With the afore-mentioned assumptions in mind, *Pitot et al.* (150) have estimated the approximate *lifetime of the RNA templates for enzymes*, this being defined as the length of time during which induced synthesis of a given enzyme is insensitive to actinomycin D. Their data show that the actinomycin-resistant period of SDH synthesis is about 6 to 8 h in length, for OAT, 18 to 24 h, while for tyrosine aminotransferase, less than 3 h, and in the case of tryptophan oxygenase, several weeks (151). Surprisingly, in hepatomas, the template lifetimes of most enzymes investigated thus far are significantly different from liver, implicating altered template stability in the genesis of neoplastic disease.

Almost complete suppression of the SDH and OAT induction also occurs if glucose is given along with the casein hydrolysate (144). This *repression*, resembling *glucose effect* in microorganisms, is apparently not the result of decreased absorption of amino acids, nor does it arise from a general depression of protein synthesis in the liver, or an ATP depletion.

The dietary induction of both enzymes, as well as that of tyrosine aminotransferase, does occur, however, even in the presence of glucose, if the animals are pretreated with cortisone (142). The possibility that cortisone in these studies acts as the primary inducer can be excluded on the grounds that cortisone alone is without any effect, and furthermore, earlier studies (149) clearly demonstrated that the dietary induction could proceed in the absence of glucocorticoids. Their experiments with orotic acid labeling of nuclear RNA suggest that glucose administration inhibits the stimulation of nuclear RNA synthesis caused by the administration of casein hydrolysate, a situation comparable to that seen in microorganisms. It is interesting to note, however, that if the administration of glucose is delayed until after stabilization of the template for SDH and OAT has occurred, glucose is still effective in curtailing further enzyme induction. Thus, in contrast to bacteria, where messenger RNA synthesis is prevented during repression, glucose repression in mammalian liver may be occurring at the translational stage in protein synthesis, involving the formation of peptides.

In the cited studies by *Peraino, Lamar and Pitot* (142), insulin neither enhanced nor diminished the repression produced by glucose, which makes it unlikely as the primary agent in the repression phenomenon. The possible involvement of insulin in the regulation of serine dehydratase was suggested by *Ishikawa, Ninagawa, and Suda* (85), who reported a 5- to 6-fold elevation in the enzyme in alloxan diabetic rats, and prevention of this elevation by insulin administration. It was further shown that SDH is readily inducible by glucocorticoids in the absence of insulin, while in its presence hydrocortisone was relatively ineffective. In other studies (46), the reported increase in SDH following cortisol treatment was significantly reduced by insulin.

More recent studies of *Jost, Khairallah and Pitot* (91), utilizing specific antibodies and pulse labeling with ^{14}C-valine, seem to support the thesis that glucose or metabolite thereof regulates the synthesis of the enzyme serine dehydratase by causing a rapid and virtually complete inhibition of the synthesis of this enzyme. The hormone, glucagon, which has been shown to affect the enzyme induction, stimulates a rapid increase in the rate of synthesis in this enzyme. Similar effects have been noted after administration of L-tryptophan. Amino acid induction and glucose repression as well as hormonal induction by both glucagon and hydrocortisone occur in adrenalectomized animals. It is thus apparent that the synthesis of serine dehydratase may be regulated by amino acids and glucose as well as by the hormones, hydrocortisone and glucagon, the latter being the most efficient inducer of the enzyme of those compounds tested.

Inoue and Pitot (84) have recently reported that crystalline serine dehydratase prepared from the livers of rats fed on a high protein diet can be separated into two distinct forms by gel electrophoresis or DEAE cellulose chromatography. The two forms are apparently identical with respect to immunochemical titration and gel diffusion with specific antibody, heat stability, and the Km for serine and threonine. Glucagon administration induces the synthesis of the more electronegative form while cortisone seems to favor the synthesis of the more electropositive form of SDH. Pulse-labeling *in vivo* with ^{3}H-leucine after cortisone administration results in a higher specific radioactivity of the more positive than the more negative form indicating that the latter is not a precursor of the former. From these data it would appear that each enzyme is synthesized independently of the other, one form being induced by corticosteroids and the other by glucagon.

H. δ-Aminolevulinate Synthetase

This mitochondrial hepatic enzyme, which is thought to be rate-controlling in heme biosynthesis (215), catalyzes the condensation of glycine and succinyl coenzyme A to form δ-aminolevulinic acid (ALA). Its usual very low activity can be elevated to high levels by the administration of allylisopropylacetamide (AIA) (217) or by 3,5-dicarbethoxy-1,4-dihydrocollidine (DDC) compounds (56) known to produce experimental porphyria. Excessive excretion of porphyrin precursors which accompanies acute intermittent porphyria (216) is, in fact, the result of marked increase in the level of ALA-synthetase (145, 215).

The AIA-mediated induction of ALA-synthetase in rats is markedly inhibited by the administration of carbohydrate (217), a phenomenon also seen with certain other inducible enzymes (e.g. serine dehydratase) *(vide supra)*. This *glucose effect* is thought to explain the observation that the excretion of porphyrin precursors in experimental and acute intermittent porphyria is reciprocally related to the intake of dietary carbohydrate.

An increase in the activity of ALA-synthetase can be also induced in liver parenchyma cells *in vitro* by a number of chemicals that induce an acute porphyria in animals (55). It has been demonstrated with fluorescent microscopy that this *in vitro* induction is a result of increased synthesis of the enzyme.

The mechanism of induction of the ALA-synthetase has been intensively investigated, using a number of compounds which are known to interfere with different steps of protein and nucleic acid synthesis, both under *in vivo* (74, 124, 130) and *in vitro* (55) conditions.

In their kinetic studies *in vivo*, *Marver et al.* (124) have shown that ALA-synthetase increases markedly after injection of AIA, the total increase being about 8-fold, and the maximum being achieved in about 12 h. They have shown further that the use of puromycin inhibits formation of the enzyme, which would point to a dependence of induction upon protein synthesis. The half-life of ALA-synthetase, measured by the rate of degradation of the induced enzyme following administration of this antibiotic, was estimated to be about 67 to 72 min. Similar value, i.e. 75 min, has been reported by *Stein et al.* (204) with the use of cycloheximide.

The injection of actinomycin D or 5-fluorouracil results in a rapid decline in the level of ALA-synthetase in the liver, and glucose administration is as effective as actinomycin D in blocking the induction of the enzyme. Apparently, the absence of glucose causes a marked stimulation of the formation of ALA-synthetase, while the presence of glucose markedly inhibits its synthesis. This effect might explain the clinical observation that the excretion of porphyrin precursors in acute intermittent porphyria, which is known to be a reflection of the levels of ALA-synthetase, is reciprocally related to the intake of carbohydrate in the diet.

It is of particular interest to note that actinomycin D effectively blocks the induction of ALA-synthetase by AIA. When actinomycin D is administered 14 h after AIA, the ALA-synthetase falls 47 % in 4 h. It is suggested, on the basis of these data, that the messenger RNA for ALA-synthetase is quite unstable and may have a half-life approximating that of the enzyme itself, i.e. 65—75 min.

Although earlier studies in chicken erythrocytes (18) indicated that iron is necessary for the activity of this enzyme, there is no evidence at present that hepatic ALA-synthetase requires iron for its activity. According to *Stein et al.* (204), however, large doses of orally administered chelated iron produce a marked synergistic effect on the induction of hepatic enzyme. The effect appears to be specific for iron citrate, as opposed to other metals in complex with citrate, and specific also in affecting the δ-ALA-synthetase as compared with another induced enzyme, tyrosine aminotransferase. Inorganic iron salts, likewise, did not produce an augmentation of the induction of hepatic enzyme. By a method with the use of cycloheximide, the turnover rate of hepatic ALA-synthetase was measured 17 h and 5 h after induction by the combined administration of ferric citrate and AIA. In comparison, it was found that the half-life (57 and 62 min, respectively) was not longer than that seen during induction by AIA alone (75 min). The data suggest, therefore, that *iron augments the rate of synthesis of the enzyme rather than activating or stabilizing the enzyme.*

Recent studies of *Hayashi, Yoda and Kikuchi* (74) indicate that ALA-synthetase accumulates not only in the mitochondrial fraction but also in the microsomal and soluble fractions when rats are treated with AIA. The enzyme synthesis in both mitochondrial and extra-mitochondrial fractions was inhibited by cycloheximide but was not appreciably affected by chloramphenicol. The induction of ALA-synthetase in either fraction was also suppressed by the administration of mitomycin C, actinomycin D, hemin, or bilirubin. The apparent half-life of the enzyme in the soluble fraction appeared to be much shorter (20 min) than that of the enzyme in the mitochondrial fraction (68 min), suggesting that the ALA-synthetase in the mitochondria is more stable than that in the soluble fraction. Furthermore, a definite difference between these two enzymes was observed on fractionation by ammonium sulfate. On the basis of these data it would appear that ALA-synthetase is first synthesized in the microsomal system and subsequently transferred into mitochondria where it settles, the enzyme being probably modified to some extent before or after entering the mitochondria.

Extensive studies *in vitro* (55) indicate that the level of ALA-synthetase is controlled by *feedback repression* in which heme may be the corepressor. *Granick* (55) has postulated that in the diseased state an ALA-synthetase operator gene is faulty and that porphyrinogenic drugs compete with heme for the aporepressor site on a repressor protein and thereby prevent repression of ALA-synthetase production. The *in vivo* induction of ALA-synthetase by tryptophan provides further evidence for the possible role of heme in the repression phenomenon (125). This is thought to result from the increased binding of 'free' or 'dissociable' heme to apoenzyme of tryptophan pyrrolase, thus making less heme available for repression of ALA-synthetase. Direct evidence for the participation of heme in the *in vivo* repression of hepatic ALA-synthetase has been obtained in subsequent studies by intravenous injection of heme (231).

V. Kinetics of Enzyme Synthesis and Degradation

On the basis of this author's work and that of his associates (154, 165, 166) on the kinetics of catalase synthesis and degradation *in vivo,* a conclusion has been reached that the enzyme is being synthesized at a *constant rate* and that a

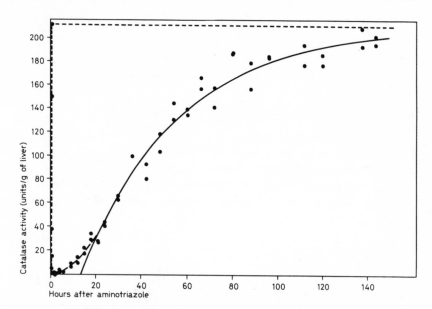

Fig. 11. Disappearance and return of liver catalase activity after injection of 3-amino-1,2,4 triazole (AT). • = experimental data; —— = theoretical curve to fit Equation 2 in text on the assumption of a rate of synthesis of 4.8 units of catalase per hour and a rate of degradation of 2.25 % of the catalase molecules present per hour; ------ = period of inactivation and lag period before experimental data approach theoretical curve; ——— = plateau value of animals not given injections of AT. From *Price et al.* (154).

constant fraction of active molecules present in the tissues is being destroyed per unit time. It now turns out that the same type of relationship, i.e. that *the rate of enzyme synthesis conforms to zero order kinetics whereas that of enzyme degradation to first-order* apparently also holds true for other mammalian enzyme systems, at least the ones that had been examined to date in this regard (7, 182, 185, 192).

In the simplest terms the above relationships can be expressed as follows:

$$\frac{dC}{dt} = k_S - k_D C \tag{1}$$

where C is the enzyme activity at any time t, k_S is the rate constant for enzyme synthesis (i.e. the *amount* of enzyme activity being synthesized per unit time), and k_D, the first-order rate constant for enzyme degradation (i.e. the *fraction* of enzyme molecules being destroyed per unit time).

At the steady state level, when the amount of enzyme synthesized per unit time equals the amount being destroyed, i.e.:

$$\frac{dC}{dt} = 0$$

then:

$$C_t = \frac{k_S}{k_D} (1 - \exp^{-k_D t}). \tag{2}$$

Hence, if k_D is known, k_S can readily be calculated from the equation.

$$k_S = k_D C_N, \tag{3}$$

where C_N is the enzyme activity at the basal plateau.

The experimental evidence for the above model is shown in figure 11 relating to the kinetic studies on the rate of catalase return following the enzyme

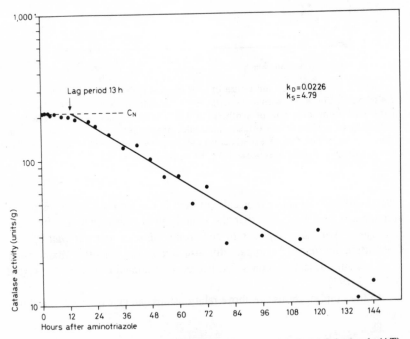

Fig. 12. Kinetics of liver catalase degradation *in vivo* using 3-amino-1,2,4-triazole (AT). Semilogarithmic plot of data in figure 11 presented as C_N-C_{AT} *versus* time, in which C_N is the normal level of liver catalase and C_{AT} is the catalase level at various times after administration of AT. Each point represents the average of 2 animals. The k_D is the first-order constant for catalase destruction, as determined by the slope of the solid line (——) drawn to fit the experimental points. The k_S, the rate constant for catalase synthesis, equals $k_D C_N$. The solid line (——) extrapolates back to the dashed line (– – –), which represents C_N, at 13 h. This is equivalent to the lag period or time required for the excretion of AT, catalase formed during this period being inactived. From *Price et al.* (154).

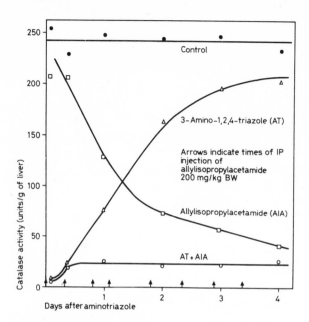

Fig. 13. Liver catalase activity in rats after injection of 3-amino-1,2,4-triazole (AT) and allylisopropylacetamide. From *Price, Rechcigl and Hartley* (153).

irreversible destruction in the presence of 3-amino-1,2,4-triazole (AT). In this figure are depicted the catalase activities of 60 individual rats at various times after the injection of AT. The solid line represents the theoretical curve that was obtained from the above assumptions if 4.8 units or 28 μg of catalase per hour were being synthesized and 2.25 % of the catalase molecules present were being destroyed, with an initial lag period of 13 h for excretion of AT, since catalase formed during this initial period will be inactivated by the drug.

If the reappearance of catalase activity is plotted semilogarithmically, as the difference between C_N, the normal catalase level, and C_{AT}, the catalase level in the aminotriazole treated animals, as shown in figure 12, a reasonably straight line is obtained with a slope equal to k_D, the rate of catalase degradation.

Although the data obtained fit the equations based on the assumptions presented, it is possible that another group of assumptions and equations could be developed that would also fit the data. A separate line of evidence was therefore sought to check the validity of the above assumptions. The critical point was to seek further evidence for the first-order curve for catalase degradation *in vivo.* Such a first-order kinetics would imply that catalase molecules are being destroyed in a random fashion, without regard to their age, so that in a given period of time newly formed enzyme molecules had the same risk of destruction as older ones.

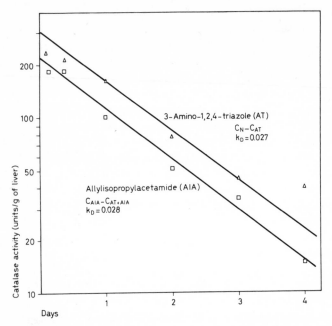

Fig. 14. Kinetics of liver catalase destruction *in vivo* by 3-amino-1,2,4-triazole (AT) and allylisopropylacetamide (AIA). Semilogarithmic plot of C_N-C_{AT} and of C_{AT+AIA} *versus* time, where C_N is the normal level of liver catalase, C_{AT} is the catalase level at various times after the administration of AT (1 g/kg body weight), C_{AIA} is the catalase level during administration of AIA (200 mg/kg body weight twice daily), and C_{AT+AIA} is the liver catalase level of animals given both drugs. From *Price, Rechcigl and Hartley* (153).

It seemed that the best way to test the point was to stop catalase synthesis and then to follow the disappearance of catalase from the liver over the next few days. To do this we used allylisopropylacetamide (AIA), a derivative of Sedormid, which has been shown by *Schmid, Figen and Schwarz* (188) to cause an acute porphyrinuria, a marked elevation in liver porphyrins, and a fall in liver catalase.

The experiment was carried out in a manner which would simultaneously determine the rate constant for catalase disappearance obtained with AIA with the rate constant obtained with AT. For the experiment 1 group of rats was given a single injection of AT, a 2nd group was given AIA, a 3rd group was given both drugs, and the 4th was used as controls. It can be seen from the data plotted in figure 13 that while the catalase activity of the AT group was rising, that in the AIA group was falling. The group of animals given both drugs was of particular interest since it showed whether or not catalase synthesis had been completely blocked by AIA. Instead of staying at the minimal value expected from red cell catalase (2 %) there was an early increase in catalase activity to a low plateau of about 10 % of the normal value. This represents catalase which can be destroyed by AT but whose synthesis is not blocked by AIA.

If the rate at which the AT curve approaches the control values is compared with the rate at which the AIA curve approaches the low plateau of the group given both drugs, in a semilogarithmic plot of the data, as shown in figure 14, it will be seen that the two lines are obtained whose slopes are nearly equal. The slope of these lines is equal to k_D, the rate constant for degradation. The rate constant for catalase destruction determined from the AT data is almost identical with that obtained when catalase synthesis is blocked with AIA. This gives reasonably secure evidence that the initial assumptions are correct and that the newly formed catalase molecule does not have a finite life span, but instead has the same risk of destruction in each unit of time as an older molecule.

VI. Heterogeneity of Enzyme Turnover Rates

The degradation rates of specific enzymes in the rat liver have been tabulated and are shown together with methods employed in their determination in table I. They are expressed in the conventional way as a half-life $(t_{1/2})$ which can be estimated from the enzyme decay curve or derived from the first-order rate decay constant, using the following relationship:

$$t_{1/2} = \frac{0.693}{k_D}$$

where 0.693 is the natural logarithm of 2 (ln 2).

A number of different techniques have been employed in arriving at these rates, some simple, some elaborate, each with certain assumptions and limitations. It is beyond the scope of this chapter to discuss the merits of a particular measurement. No single method is applicable to all enzymes and different techniques may not necessarily bring the same answer. In fact, some methods may give erroneous results, either grossly over-estimating or under-estimating the true half-life of a given enzyme. A case in point are the isotope decay studies of catalase which showed a 2-fold variation in the half-life of the enzyme depending on the type of isotope used (152).

Some methods carry the assumption that the inhibition of protein synthesis in the presence of antibiotics such as puromycin or cycloheximide does not alter the rate of enzyme degradation. Although it has been demonstrated experimentally that such inhibitors do not indeed affect the rate of degradation of a number of enzymes, recent studies of *Kenney* (96) have provided evidence for the alteration of degradation of tyrosine aminotransferase after puromycin or cycloheximide administration.

In order to be sure of a given value, one should therefore, whenever possible, use several different approaches for determining enzyme degradation rates.

Table I lists the degradation rates of various enzymes in the liver of rats. Relatively few measurements have been made on the turnover rates of hepatic enzymes using other species of animals. As to the latter category, this author is aware only of the studies on creatine transamidinase (229) and xanthine dehydrogenase (128) in chicks and on catalase (48, 162), NADPH-cytochrome c reductase (88), and δ-aminolevulinate dehydratase (32) in mice.

It is apparent from table I as well as from the earlier discussion of synthesis and degradation of specific enzymes that a remarkable heterogeneity exists in the turnover rates of individual enzymes. The half-lives of hepatic enzymes vary from as few as 11 min for ornithine decarboxylase to as many as 20 days for NAD glycohydrolase.

Only sporadic data are available on the turnover rates of enzymes of other tissues. Among these are histidine decarboxylase of rat stomach with a half-life of 1.8 to 2.1 h (201), RNA polymerase of rat uterus with a half-life of 1 h (54), and creatine transamidinase of rat kidney with a half-life of 38 h (229). Comparative studies in the author's laboratory (154, 166) on the liver *versus* the kidney catalase in the same animal revealed no significant differences in the rate of enzyme degradation, notwithstanding the large variation existing in the rate of its synthesis.

On the other hand, recent communication from our laboratory (162) on the kinetics of hepatic catalase synthesis and destruction in two closely related substrains of mice set forth evidence for the gene regulation of enzyme activity by the alteration of the rate constant of enzyme degradation. This finding was confirmed by *Ganschow and Schimke* (48).

In view of the question as to whether enzymes (proteins) are being 'turned over' during the logarithmic growth of the cells, we extended our kinetic studies to several rat hepatomas (163). We were particularly interested to ascertain whether the rate constant for catalase destruction of neoplastic tissues differs significantly from that of the host tissues. The rate constants of destruction in both types of tissues, however, were found to be practically identical (table II). In agreement with our conclusions, *Bresnick and Burleson* (14) did not detect a significant difference in the half-life of deoxythymidine kinase of Novikoff ascites hepatoma and that of neonatal liver, these being 3.5 and 3.8 h, respectively. Furthermore, the turnover rate of tyrosine aminotransferase of two distinct hepatoma cell lines (HTC and H35) in culture (5, 170, 171) were found to be analogous to the degradation rate of the enzyme in the liver.

The latter studies, in addition, would seem to suggest that the degradation rate of enzyme is not altered by maintaining cells under tissue culture conditions.

Only a limited number of enzymes have been studied in cell cultures to make any conclusion on this point. The half-life of δ-aminolevulinate synthetase of chick embryo liver cells is about 4–6 h (55) when compared to the half-life of 1 h observed in the rat liver of the intact animal (124). It is not known whether the reported difference in the half-lives of this enzyme is due to actual differences between *in vivo* and *in vitro* conditions or to the species differences.

In addition to the mentioned studies, *Nebert and Gelboin* (131) determined a half-life of 3–4 h for aryl hydrocarbon hydroxylase in hamster embryo cells and *Turner et al.* (219) showed a half-life of 2 h for ribonucleotide reductase in mouse L cells. No comparative values exist, however, for the respective enzymes in tissues under *in vivo* conditions.

In contrast to the above studies which have conclusively demonstrated the presence of enzyme turnover in cultured cells, *Yagil and Feldman* (238), using rat fibroblasts and KB cells, could not detect any decay in three different enzymes (glucose-6-phosphate dehydrogenase, 6-phosphogluconate dehydrogenase and malate dehydrogenase) when protein synthesis as well as RNA synthesis were blocked. The authors therefore concluded that in cell cultures the enzymes are neither continuously degraded nor continuously produced. Their observations might also be explained, however, on the grounds that the normal turnover of the enzymes was altered in the presence of inhibitors of protein synthesis, as shown in the earlier studies of *Kenney* (96).

A few kinetic measurements of enzyme decay have also been done in perfused liver, namely those on tryptophan oxygenase (53, 196) and tyrosine aminotransferase (196). The half-lives of these enzymes have been found to be comparable to those seen in the intact animal.

Some authors measured enzyme decay rates in regenerated tissues. *Bresnick et al.* (16) reported a half-life of 3.7 h for deoxythymidine kinase in regenerated liver as compared to a value of 2.6 h in normal liver or 3.8 h in neonatal liver (14). Aspartate transcarbamylase of both normal and regenerated liver turned over with a half-life of more than 24 h (16).

In contrast to these studies, *Tsukuda and Lieberman* (218) found more than 4-fold difference between the degradation rate of RNA polymerase in the regenerated liver of rats and in the liver of shamoperated controls, the half-life of the former being 2.5–3 h and that of the latter 12 h. The reason for this difference is not known.

As discussed earlier, muscle enzymes have, as a rule, long half-lives (179, 222). This is particularly striking in the comparison of alanine aminotransferase of the muscle with that of the liver, their half-lives being 20 days and 3 days, respectively (193). In the studies of *Fritz et al.* (47), muscle lactate dehydrogenase possessed a half-life of 31 days as compared to 16 days for the enzyme in the liver and 1.6 days in the heart.

It is apparent from the above that different isozymes of a given enzyme may degrade at vastly different rates. To make matters even more complex, it appears than an enzyme turnover may vary depending on its location in the cell. This

Table I. *In vivo* turnover rates of enzymes in the rat liver

Enzyme	Turnover rate		Method of measurement	References
	$t_{1/2}$	k_D		
Ornithine decarboxylase	11 min	3.78	Decay of basal or induced enzyme activity following inhibition of protein synthesis	Russell and Snyder (177)
δ-Aminolevulinate synthetase	67–72 min	0.597	Decay of induced enzyme activity following inhibition of protein synthesis	Marver et al. (124)
δ-Aminolevulinate synthetase	75 min	0.554	Decay of induced enzyme activity following inhibition of protein synthesis	Stein et al. (204)
δ-Aminolevulinate synthetase (mitochondrion)	68 min	0.612	Decay of basal or induced enzyme activity following inhibition of protein synthesis	Hayashi et al. (74)
δ-Aminolevulinate synthetase (soluble)	20 min	2.08	Decay of basal or induced enzyme activity following inhibition of protein synthesis	Hayashi et al. (74)
Tyrosine aminotransferase	3 h	0.231	Decay of induced enzyme activity	Lin and Knox (117)
Tyrosine aminotransferase	3–4 h	0.173– 0.231	Isotope (^{14}C-leucine) decay under basal conditions	Kenney (94)
Tyrosine aminotransferase	2 h	0.347	Decay of induced enzyme activity	Berlin and Schimke (7)
Tyrosine aminotransferase	2.5 h	0.277	Decay of enzyme activity after stress	Kenney and Albritton (97)
Tyrosine aminotransferase	1.7 h	0.408	Decay following inhibition of protein synthesis	Kenney (96)
Tyrosine aminotransferase	1.5 h	0.462	Isotope (^{14}C-leucine) decay	Kenney (96)
Tyrosine aminotransferase	1.7 h	0.408	Decay of induced enzyme activity	Grossman and Mavrides (69)
Tyrosine aminotransferase	>5 h	<0.139	Decay of basal enzyme activity following inhibition of protein synthesis	Grossman and Mavrides (69)
Tryptophan oxygenase	2.3 h	0.301	Decay of induced enzyme activity	Feigelson et al. (40)
Tryptophan oxygenase	2.5 h	0.277	Decay following inhibition of protein synthesis	Goldstein et al. (53)
Tryptophan oxygenase	2–4 h	0.231	Decay of basal or induced enzyme activity following inhibition of protein synthesis	Nemeth (132)

Table I (continued)

Enzyme	Turnover rate		Method of measurement	References
	$t_{1/2}$	k_D		
Tryptophan oxygenase	2.5 h	0.277	Decay of induced enzyme activity following inhibition of protein synthesis	Garren et al. (49)
Tryptophan oxygenase	2.5 h	0.277	Isotope (^{14}C-leucine) decay in steady state	Schimke et al. (185)
Deoxythymidine kinase	2.6 h	0.267	Decay following inhibition of protein synthesis	Bresnick et al. (16)
Serine dehydratase	3 h	0.231	Decay of induced enzyme activity	Pitot and Peraino (149)
Serine dehydratase	4.4 h	0.157	Decay of prelabeled enzyme	Jost et al. (91)
Phosphoenolpyruvate carboxykinase	5.5 h	0.126	Fall in enzyme activity following inhibition of protein synthesis	Shrago et al. (198)
Aniline hydroxylase	5–7 h	0.099–0.139	Rise of induced enzyme to steady state level	Holtzman and Gillette (81)
RNA polymerase	12 h	0.058	Decay of enzyme activity following inhibition of protein synthesis	Tsukuda and Lieberman (218)
Glucokinase	11 h	0.063	Decay of enzyme activity	Sharma et al. (197)
Glucokinase	12–18 h	0.039–0.058	Decay following inhibition of protein synthesis	Salas et al. (178)
Glucokinase	30 h	0.023	Decay of induced enzyme activity	Niemeyer (134)
Dihydroorotase	0.5 days	0.058	Decay of induced enzyme activity	Bresnick et al. (15)
Ornithine aminotransferase	19 h	0.037	Decay of induced enzyme activity	Swick et al. (208)
Ornithine aminotransferase	15 h	0.045	Rate of isotope incorporation	Swick and Peraino (207)
Thymidylate kinase	18 h	0.039	Decay of induced enzyme activity	Hiatt and Bojarski (77)
Phosphoglycerate dehydrogenase	14.6 h	0.047	Decay of basal or induced enzyme activity following inhibition of protein synthesis	Fallon et al. (39)
Phosphoglycerate dehydrogenase	2.3 days	0.013	Rise of induced enzyme activity	Fallon (37)
Phosphoglycerate dehydrogenase	1.9–2.3 days	0.013–0.015	Decay of induced enzyme activity	Fallon (38)

Table I (continued)

Enzyme	Turnover rate		Method of measurement	References
	$t_{1/2}$	k_D		
Catalase	26–32 h	0.022–	Return of enzyme activity after irreversible inhibition	Price et al. (154)
		0.027		Rechcigl and Price (166)
Catalase	26–32 h	0.022–	Isotope (^{59}Fe) incorporation after irreversible inhibition of the enzyme	Price et al. (154)
		0.027		Rechcigl and Price (166)
Catalase	26–32 h	0.022–	Decay following inhibition of protein synthesis	Price et al. (154)
		0.027		Rechcigl and Price (166)
Catalase	1.5 days	0.019	Return of enzyme activity after irreversible inhibition	Poole et al. (152)
Catalase	1.8 days	0.016	Isotope (ALA,^{14}C) decay	Poole et al. (152)
Catalase	2.5 days	0.012	Isotope (guanidino-^{14}C-arginine) decay	Poole et al. (152)
Catalase	3.7 days	0.008	Isotope (^{14}C-leucine) decay	Poole et al. (152)
Aminopyrine demethylase	2 days	0.014	Decay of induced enzyme activity	Argyris and Magnus (2)
Barbiturate side-chain oxidation enzyme	2.6 days	0.011	Decay of induced enzyme activity	Arias and DeLeon (3)
Aspartate transcarbamylase	>24 h	<0.029	Decay of prelabeled enzyme following inhibition of protein synthesis	Bresnick et al. (16)
Aspartate transcarbamylase	2.5 days	0.011	Decay of induced enzyme activity	Bresnick et al. (15)
Amino-azo dye N-demethylase	>24 h	<0.029	Decay of basal or induced enzyme activity following inhibition of protein synthesis	Conney and Gilman (28)
Acetyl coenzyme A carboxylase	50 h	0.014	Isotope (^{3}H-leucine) decay	Majerus and Kilburn (120)
Alanine aminotransferase	1.2 days	0.024	Rise of induced enzyme activity	Segal and Kim (191)
Alanine aminotransferase	3–3.5 days	0.008–	Decay of induced enzyme activity	Segal and Kim (191)
		0.010		Kim (101)
Alanine aminotransferase	18 h	0.039	Decay of induced enzyme activity	Swick et al. (208)
NADPH-cytochrome c reductase	3.0–3.5 days	0.009	Isotope (^{14}C-leucine) decay under basal conditions	Omura et al. 137)

Table I (continued)

Enzyme	Turnover rate		Method of measurement	References
	$t_{1/2}$	k_D		
NADPH-cytochrome c reductase	2.5 days	0.011	Isotope (guanidino-^{14}C-arginine) under basal conditions	*Kuriyama et al.* (109)
NADPH-cytochrome c reductase	48–56 h	0.012–0.014		*Ernster and Orrenius* (36)
NADPH-cytochrome c reductase	2.5–3.0 days	0.010–0.011	Double isotope (^{14}C-leucine and ^{3}H-leucine) decay	*Arias et al.* (4)
α-Glycerophosphate dehydrogenase	4 days	0.008	Decay of induced enzyme activity following inhibition of protein synthesis	*Tarentino et al.* (210)
Malate dehydrogenase	4 days	0.008	Decay of induced enzyme activity following inhibition of protein synthesis	*Tarentino et al.* (210)
Arginase	4–5 days	0.006–0.008	Continuous isotope incorporation	*Schimke* (182)
Arginase	4–5 days	0.006–0.008	Isotope (guanidino ^{14}C-arginine) decay	*Schimke* (182)
Xanthine oxidase	4 days	0.008	Decay of elevated enzyme activity	*Rowe and Wyngaarden* (175)
Lactate dehydrogenase	16 days	0.002	Continuous isotope incorporation	*Fritz et al.* (47)
NAD Glycohydrolase	20 days	0.0015	Double isotope (^{4}C-leucine and ^{3}H-leucine) decay	*Bock et al.* (10)

situation exists at least in the case of δ-aminolevulinate synthetase whose apparent half-life in the soluble fraction was reported to be 20 min, while that of the mitochondrial enzyme was about 68 min (74).

Table II. The rate constants of catalase synthesis and destruction in neoplastic tissues of the rat[1] *(Rechcigl, 157)*

Tissue	k_D	k_S
Hepatoma 5123	0.02	0.92
Host liver	0.02	3.88
Host kidneys	0.02	1.10

1 Inbred rats of the Buffalo/N strain were used.

VII. Alteration of the Rates of Enzyme Synthesis and Degradation

A continuous synthesis and degradation is a *conditio sine qua non* for the occurrence of enzyme turnover. The alteration in the turnover of a given enzyme can thus be brought forth by the alteration in the rate of enzyme synthesis, or degradation, or both. Our understanding of the relative contribution of synthesis and degradation in the control of enzyme levels under various physiological and pathological conditions is still in rudimentary stages.

In contrasting the mammalian with the bacterial systems, it is often pointed out that the ratio of animal cell life span to enzyme turnover is far greater than in bacteria. As a consequence, investigators, rightly or wrongly, have had the tendency of assigning in animal tissues a greater role to the alteration of enzyme degradation than to that of enzyme synthesis.

This view has become especially prevalent following the dramatic demonstration of the tryptophan oxygenase induction through the mechanism of *enzyme stabilization* (decreased breakdown) by its substrate in the presence of a continued enzyme synthesis (186). However, relative scarcity of other cases of this type of mechanism which have been reported so far has been rather disappointing. The situation is further complicated by the demonstration of the role of tryptophan in the activation of the tryptophan oxygenase which may necessitate a re-interpretation of the stabilization phenomenon of this enzyme by its substrate (106).

Be that as it may, there is no justification at the present, *a priori* to hold the view that a differential enzyme synthesis, which is the major control mechanism in bacteria, plays only a minor role in the mammalian systems. In fact, our own studies (155, 156, 158) on the kinetics of catalase synthesis and breakdown *in*

Table III. The effect of the nutritional state on the rate constants of hepatic catalase synthesis and destruction[1] (*Rechcigl,* 156)

Treatment	k_D	k_S
Chow diet, *ad libitum*	0.02	4.3
Moderate starvation	0.02	4.3
Severe starvation	0.02	2.6
Casein diet	0.02	4.3
Protein-free diet	0.02	1.9

1 Rats of the Sprague-Dawley strain were used.

Table IV. The effect of cycasin on the rate constants of hepatic and renal catalase synthesis and degradation in the rat[1]

	Liver		Kidneys	
	k_D	k_S	k_D	k_S
Cycasin-treated rats	0.014	1.2	0.018	1.0
Pair-fed controls	0.029	5.0	0.033	1.6

1 Rats of the Osborne-Mendel strain were used.

Table V. The effect of sex and age on the rate constants of hepatic and renal catalase synthesis and degradation in the rat[1] (*Rechcigl,* 157)

Sex	Age	Body weight g	Liver		Kidneys	
			k_D	k_S	k_D	k_S
Males	young	210	0.023	4.25	0.023	1.14
Females	young	165	0.022	3.00	0.023	0.83
Males	adult	443	0.030	4.43	0.016	0.58
Females	adult	300	0.030	3.33	0.015	0.37

1 The rats of the Sprague-Dawley strain were used.

vivo would seem to indicate that *the rate of synthesis rather than the rate of degradation may be preferential, although by no means the exclusive way of the mammalian organism to control its levels of enzyme activity,* that is to say, at least as far as the catalase is concerned. As can be observed from table III, a severe starvation and a lack of dietary protein, both of which evoke a marked decrease in the enzyme levels, while greatly reducing the rate constant for enzyme synthesis, have no apparent effect on the enzyme degradation rate constant (158, 159).

Moreover, the depression of catalase activity following the feeding of cycasin (table IV) can be accounted for by a lower rate of enzyme synthesis (164). It should be noted that in addition to affecting the rate of synthesis of the enzyme, cycasin also brought about an alteration in the rate constant of enzyme destruction. This alteration occurred, however, in the opposite direction than one would have *a priori* anticipated, i.e. instead of causing an increased enzyme breakdown, cycasin actually caused a substantial lowering in the rate constant of enzyme degradation. The decrease in catalase activity in the cycasin-treated rats would have been even greater had it not been for the concomitant lowering in the rate of enzyme destruction in these animals. We shall see later on that a concurrent decrease in enzyme synthesis and breakdown can also be elicited by the inhibitors of protein synthesis. The observed effects of cycasin in our laboratory may very well be related to this phenomenon.

Similarly, in neoplastic tissues (table II), the modification in the rate of enzyme synthesis rather than that of its destruction accounted for the observed differences between the normal and tumor tissues (163).

Elsewhere, reference has already been made to the finding of almost identical rate constants for catalase degradation in the liver and in the kidney, although the rate of synthesis in the liver is nearly four times that of the kidney (fig. 3).

In a comparable study on the effect of sex, the calculated rate constant for catalase decay in the liver of young male rats did not significantly differ from that of young females, as shown in table V. Similarly, the higher levels of the enzymatic activity in the kidney of male rats when compared to females were found to be due to different rates of enzymes synthesis rather than those of its breakdown. Comparable results were obtained with adult rats. Interestingly enough, the hepatic catalase of the adult animals underwent a greater decay when compared to the younger rats, while the reverse was true in the case of the renal enzyme (table V).

Additional cogent evidence for the regulation of enzyme levels in the mammalian systems by altering the rate constant for enzyme synthesis has been provided from numerous studies by other investigators. Some of these data have been summarized in table VI. Judging from this table, it would seem that the enzymes whose activity is induced or altered in some way in response to hormonal stimulation, or dietary manipulation, increase or decrease their levels by the alteration of their synthetic rate.

It is also manifest, however, from the earlier discussion that, in addition to, or in place of, the profound and ever-changing rates of enzyme synthesis, a modulation of enzyme degradation may take place under certain conditions.

It is a long established fact that the stabilization of enzymes *in vitro* can be brought about by the addition of substrates or other components of enzymatic reactions. The existence of similar phenomena *in vivo* would not therefore be unexpected (68).

A mention has already been made of the *stabilization of tryptophan oxygenase by its substrate* (186). Of the various analogues tested, only a-methyl-tryptophan, besides L-tryptophan, prevents the rapid decay of enzyme activity in animals in which tryptophan pyrrolase had been previously raised to high levels by repeated administration of both hydrocortisone and L-tryptophan. Both L-tryptophan and its methyl derivative are also effective in preventing *in vitro* degradation of the enzyme caused by heat, urea, ethanol, and proteolysis by trypsin, chemotrypsin, and bacterial proteinase. Whereas the degradation of the enzyme can also be prevented by the incubation of liver slices in the presence of L-tryptophan, no such effect is exerted on the total protein degradation. It is apparent, therefore, that L-tryptophan does not affect the enzyme degradation by preventing proteolysis in general but rather that the effect is specific for tryptophan pyrrolase.

It is not known what physical change in the enzyme molecule is imparted during its stabilization by L-tryptophan. Sedimentation studies indicate no apparent alteration in the aggregational state of enzyme. It is to be expected, however, that some *alteration in the conformational state of the enzyme* must have taken place in order to block the access of proteolytic enzymes to critical peptide bonds. Based on kinetic studies with a-methyltryptophan, *Schimke, Sweeney and Berlin* (186) proposed the existence of two substrate-binding sites, one the catalytic site, and a second site which presumably imparts stability to the enzyme.

That the *conformational changes* in protein structure affect their susceptibility to proteolysis has been demonstrated on a number of occasions. These include the protective effect of glucose on trypsin inactivation of hexokinase (6), the resistance of hemoglobin to carboxypeptidase imparted by oxygenation (239), the decrease of serum albumin digestibility by the binding of tryptophan to protein (123), the increase in the trypsin digestibility of glyceraldehyde-3-phosphate dehydrogenase and aldolase upon coupling with p-mercuribenzoate (209) and others.

It had been suggested that the stable form of tryptophan pyrrolase is that in which hematin, the iron-porphyrin cofactor, is bound to apoenzyme, a binding that is facilitated by the presence of tryptophan (42). According to the recent

Table VI. Response and regulation of some enzymes by the alteration of the rate of enzyme synthesis

Enzyme	Source	Modus operandi	Stimulus	Enzyme response	Enzyme synthesis	References
3-P-glycerate dehydrogenase	Rat liver	In vivo	Protein depletion	Increase	Increase	Fallon et al. (39)
Alanine aminotransferase	Rat liver	In vivo	Prednisolone	Increase	Increase	Segal and Kim (191)
Alanine aminotransferase	Rat liver	In vivo	Prednisolone	Increase	Increase	Kim (101)
Alanine aminotransferase	Rat liver	In vivo	Prednisolone	Increase	Increase	Kim (100)
Alanine aminotransferase	Rat liver	In vivo	Hydrocortisone	Increase	Increase	Kim (100)
Alanine aminotransferase	Rat liver	In vivo	Cortisone	Increase	Increase	Berlin and Schimke (7)
Tyrosine aminotransferase	Rat liver	In vivo	Cortisone	Increase	Increase	Berlin and Schimke (7)
Tyrosine aminotransferase	Rat liver	In vivo	Hydrocortisone	Increase	Increase	Kenney (94), Levitan and Webb (114)
Tyrosine aminotransferase	Rat liver	In vivo	Pyridoxine	Increase	Increase	Holten et al. (80)
Tyrosine aminotransferase	Rat liver	In vivo	Insulin	Increase	Increase	Holten and Kenney (79)
Tyrosine aminotransferase	Rat liver	In vivo	Glucogon	Increase	Increase	Holten and Kenney (79)
Tyrosine aminotransferase	Rat liver	Perfusion	Dexamethasone	Increase	Increase	Jervell and Seglen (87)
Tyrosine aminotransferase	Rat liver	In vivo	Growth hormone	Decrease	Decrease	Kenney (95)
Tyrosine aminotransferase	Rat liver	In vivo	Stress	Decrease	Decrease	Kenney and Albritton (97)
Tyrosine aminotransferase	Hepatoma cells	In culture	Dexamethasone	Increase	Increase	Granner et al. (58), Granner et al. (57)
Tyrosine aminotransferase	Rat liver	In vivo	Cyclic AMP	Increase	Increase	Wicks et al. (235)
Tyrosine aminotransferase	Hepatoma cells	In culture	Insulin	Increase	Increase	Reel et al. (171)
Tyrosine aminotransferase	Hepatoma cells	In culture	Blood serum	Increase	Increase	Gelehrter and Tomkins (50)
Tyrosine aminotransferase	Hepatoma cells	In culture	Hydrocortisone	Increase	Increase	Reel and Kenney (169), Reel et al. (171)
Tryptophan oxygenase	Rat liver	In vivo	Hydrocortisone	Increase	Increase	Schimke et al. (185)
Tryptophan oxygenase	Rat liver	In vivo	Cortisone	Increase	Increase	Berlin and Schimke (7)
Tryptophan oxygenase	Rat liver	In vivo	Hydrocortisone	Increase	Increase	Garren et al. (49)
Tryptophan oxygenase	Rat liver	Perfusion	Hydrocortisone	Increase	Increase	Seglen and Jervell (196)
α-Glycerophosphate dehydrogenase	Rat liver	In vivo	Triiodothyroxine	Increase	Increase	Lee and Miller (110)

Table VI (continued)

Enzyme	Source	Modus operandi	Stimulus	Enzyme response	Enzyme synthesis	References
Barbiturate side-chain oxidation enzyme	Rat liver	In vivo	Phenobarbital	Increase	Increase	Arias and DeLeon (3)
Arginase	Rat liver	In vivo	High protein diet	Increase	Increase	Schimke (182)
Arginase	Rat liver	In vivo	Cortisone	Increase	Increase	Berlin and Schimke (7)
Serine dehydratase	Rat liver	In vivo	Casein hydrolysate	Increase	Increase	Pitot and Peraino (149)
Serine dehydratase	Rat liver	In vivo	Glucose	Decrease	Decrease	Jost et al. (91)
Histidine decarboxylase	Rat liver	In vivo	Gastrin	Increase	Increase	Snyder and Epps (201)
Ornithine decarboxylase	Rat liver	In vivo	Partial hepatectomy	Increase	Increase	Russell and Snyder (177)
Xanthine oxidase	Rat liver	In vivo	Protein depletion	Decrease	Decrease	Rowe and Wyngaarden (175)
Catalase	Rat liver	In vivo	Protein deficiency	Decrease	Decrease	Rechcigl (159)
Catalase	Rat liver	In vivo	Severe starvation	Decrease	Decrease	Rechcigl (156)
Catalase	Rat liver	In vivo	Cycasin	Decrease	Decrease	Rechcigl and Laqueur (164)
δ-Aminolevulinate synthetase	Rat liver	In vivo	Ferric citrate	Increase	Increase	Stein et al. (204)
NADPH-cytochrome c reductase	Mouse liver	In vivo	Phenobarbital	Increase	Increase	Jick and Shuster (88)
NADPH-cytochrome c reductase	Rat liver	In vivo	Phenobarbital	Increase	Increase	Arias et al. (4)
NADPH-cytochrome c reductase	Rat liver	In vivo	Phenobarbital	Increase	Increase	Kuriyama et al. (109)
Serine dehydratase	Rat liver	In vivo	Amino acids	Increase	Increase	Jost et al. (91)
Serine dehydratase	Rat liver	In vivo	L-tryptophan	Increase	Increase	Jost et al. (91)
Serine dehydratase	Rat liver	In vivo	Glucagon	Increase	Increase	Jost et al. (91)

Table VII. Factors affecting enzyme degradation or inactivation

	Enzyme	Source	Modus operandi	Enzyme inactivation or degradation	References
Thymidine	Thymidylate kinase	Rat liver	In vivo	Decrease	Hiatt and Bojarski (77)
Tryptophan	Tryptophan oxygenase	Rat liver	In vivo	Decrease	Schimke et al. (186)
Tryptophan	Tryptophan oxygenase	Rat liver	In vitro	Decrease	Schimke et al. (186)
Arginine	Arginase	Rat liver	In vivo	Decrease	Schimke (182)
Pyridoxine	Serine dehydratase	Rat liver	In vivo	Decrease	Khairallah and Pitot (99)
Cysteine	Alkaline phosphatase	Human skin cells	In culture	Increase	DeMars (30)
Glutamine	Glutamyl transferase	HeLa cells	In culture	Increase	DeMars (30)
Manganese	Arginase	HeLa cells	In culture	Decrease	Paul and Fottrell (139)
Manganese	Arginase	Chang's liver cells	In culture	Decrease	Schimke (183)
Amethopterin	Folate reductase	S-180 cell extracts	In vitro	Decrease	Eliasson and Strecker (35)
Amethopterin	Folate reductase	Human leukemic cells	In vivo	Decrease	Hakala and Suolinna (70)
Puromycin	Tyrosine aminotransferase	Rat liver	In vivo	Decrease	Bertino et al. (8)
Puromycin	Tyrosine aminotransferase	Rat liver	In vivo	Decrease	Kenney (96)
Cycloheximide	Tyrosine aminotransferase	Rat liver	In vivo	Decrease	Schimke (184)
Cycloheximide	Tyrosine aminotransferase	Rat liver	In vivo	Decrease	Kenney (96)
Cycloheximide	Tyrosine aminotransferase	Rat liver	In vivo	Decrease	Schimke (184)
Actinomycin D	Tyrosine aminotransferase	Rat liver	In vivo	Decrease	Levitan and Webb (111, 114)
Actinomycin D	Tyrosine aminotransferase	Rat hepatoma cells	Perfusion	Decrease	Levitan and Webb (113)
Cycasin	Tyrosine aminotransferase	Rat liver	In culture	Decrease	Reel and Kenney (170)
Phenobarbital	Tyrosine aminotransferase	Rat liver	In vivo	Decrease	Grossman and Mavrides (69)
Phenobarbital	Catalase	Rat liver	In vivo	Decrease	Rechcigl and Laqueur (164)
5-Azacytidine	NADPH cytochrome c reductase	Mouse liver	In vivo	Decrease	Jick and Shuster (88)
5-Azacytidine	NADPH cytochrome c reductase	Rat liver	In vivo	Decrease	Kuriyama et al. (109)
	Tyrosine aminotransferase	Rat liver	In vivo	Decrease	Levitan and Webb (112, 114)
	Tyrosine aminotransferase	Rat liver	Perfusion	Decrease	Levitan and Webb (112)

Table VII (continued)

	Enzyme	Source	Modus operandi	Enzyme inactivation or degradation	References
8-Azaguanine	Tyrosine aminotransferase	Rat liver	In vivo	Decrease	Levitan and Webb (111, 114)
8-Azaguanine	Tyrosine aminotransferase	Rat liver	Perfusion	Decrease	Levitan and Webb (113)
Insulin	Tyrosine aminotransferase	Rat liver	Perfusion	Cessation	Seglen (195)
Nutritional deprivation	Tyrosine aminotransferase	Rat hepatoma cells	In culture	Increase	Auricchio et al. (5)
Sudden shift in diet	Arginase	Rat liver	In vivo	Increase or decrease	Schimke (182)
Fasting	Acetyl CoA carboxylase	Rat liver	In vivo	Increase	Majerus and Kilburn (120)

findings of *Knox and Piras* (106), based on the improved assay for the enzyme, it appears that the loss of the liver tryptophan pyrrolase in the absence of its protecting substrate can be reversed. The apparent stabilization of the enzyme is being attributed to the role of tryptophan in the activation process (i.e. formation of active reduced holotryptophan pyrrolase from apoenzyme or from inactive oxidized holoenzyme), and from its preserving the reduced holoenzyme in that form.

Prior to the cited studies on the tryptophan-mediated elevation of tryptophan oxygenase, *Bojarski and Hiatt* (11) had suggested that *enzyme stabilization* is involved in changes of thymidylate kinase activity of rat liver resulting from thymidylate or thymidine administration. *Kit et al.* (103) as well as *Littlefield* (118) made comparable observations using mouse fibroblast cells in culture.

Similar mechanisms were proposed for the control of glycogen metabolism (1) and for the manganese-mediated increase in arginase (183) in cultured HeLa cells. Additional factors which have been reported to affect enzyme stabilization are listed in table VII. Needless to say, direct kinetic studies of degradation rates of enzymes will have to be performed before a definite conclusion is reached.

The examination of table VII further reveals that a number of inhibitors of protein synthesis are capable of preventing the inactivation of hepatic tyrosine aminotransferase (TAT), both under basal conditions (96) and after induction of the enzyme by hydrocortisone (69, 111, 112, 170). These include puromycin (96), 8-azaguanine (111, 114), cycloheximide (96, 114), and 5-azacytidine (112, 114).

Similar effects were seen in the isolated perfused rat liver with 8-azaguanine or cycloheximide (113), showing that the prevention of enzyme degradation results from a direct action of the chemicals on the liver.

In contrast to these results, in the perfusion studies of *Jervell and Seglen* (87), the degradation of TAT remained unaffected by cycloheximide. In the extension of this work, *Seglen* (195) showed that addition of insulin prior to cycloheximide completely abolished the degradation of the enzyme 'normally' seen in the presence of the antibiotic, suggesting that insulin may be responsible for the stabilization of TAT observed after cycloheximide *in vivo*. The apparent discrepancy must await further experimentation.

A unique case of the alteration in the degradation rate of an enzyme by *genetic mutation* has been reported with catalase.

In these studies *Rechcigl and Heston* (161) described two closely related mouse strains, C57BL/6 and C57BL/He, in which a mutational event resulted in an alteration in catalase degradation in one strain (162), giving rise to a 2-fold difference in the enzyme level in the liver between the two strains. The difference in catalase activity between C57BL/6 and DBA/2, on the other hand, was ascribed to a difference in catalytic activity of the enzyme (48). In the studies of *Doyle and Schimke* (32) relating to liver δ-ALA dehydratase of C57BL/6 and

AKR mouse strains, the rate of degradation of enzyme in the two strains was the same, the differences in the enzyme level being entirely accounted for by the differential rate of enzyme synthesis.

A remarkable situation in which *both differential degradation and synthesis* play a conspicuous role is displayed by arginase (182). The rate of degradation of this enzyme in rats fed diets containing 8, 30 and 70 % protein appears to be essentially the same under all three conditions. Upon the initiation of fasting, there is a progressive rise in the total arginase content, without concomitant loss in the radioactivity from the pre-labeled enzyme, indicating that the enzyme is not degraded. On change from a high (70 %) to a low (8 %) protein diet, on the other hand, the rate of enzyme degradation is increased and the rate of synthesis is diminished, thereby decreasing the amount of the enzyme. Once the animals become adjusted to the new diets, however, the enzyme degradation proceeds at its normal pace.

The reverse situation was recently reported by *Auricchio et al.* (5) in their studies on the control of degradation and synthesis of induced tyrosine amino-transferase (TAT) in hepatoma cells in culture. When the nutritional level of the medium was suddenly reduced, a transient phase of more rapid TAT degradation appeared. This 'enhanced degradation' was inhibited by actinomycin and cyclo-heximide, in the presence of which TAT degradation approached its slower 'normal' rate. On the basis of these data the authors suggested that the earlier cited observations *in vivo* of the inhibition of rat liver TAT degradation by cycloheximide (96) or by 5-azacytidine (112) as well as those *in vitro* of the actinomycin D-mediated inhibition of induced TAT in hepatoma cells (170) might possibly be explained by a similar mechanism. They put forward an argu-ment that the animals were fasted for considerable periods of time before being killed, while in the tissue culture study of *Reel and Kenney* (170) a *nutritional step-down* was an intrinsic part of the experimental design. The reported failure of cycloheximide to inhibit TAT degradation in perfused liver (87), on the other hand, might imply, according to *Auricchio et al.* (5), that no step-down like conditions were imposed in these experiments.

These data, if confirmed with other enzyme systems, suggest that the rate of enzyme breakdown may be significantly altered during a sudden change in the external environment which, in this instance, is represented by changing dietary conditions. However, once the *internal milieu* of the organism has become ad-justed (adapted) to a given environmental change, then the breakdown will proceed at its regular pace.

If our thesis is correct, one would hardly detect this kind of shift in the enzyme degradation rate under normal circumstances since most experiments are routinely carried out under conditions in which the experimental animals have been preadjusted to a given dietary regimen or other environmental variable. Although unimportant in the afore-mentioned circumstance, a sudden, however transient, shift in the degradation rate may be of some concern in pharmacological studies, especially in acute situations where the organism has to respond rapidly to a given stimulus. A case in point are the recent experiments on the effect of phenobarbital on the turnover of microsomal NADPH-cytochrome c reductase (88). During the first 4 days of treatment, there was an almost complete cessation of enzyme breakdown, as measured by the loss of radioactivity from the pre-labeled enzyme. When phenobarbital was given longer than 4 days, however, a substantial loss of radioactivity from the enzyme followed.

The effect of phenobarbital-mediated increase in NADPH-cytochrome c reductase was independently reexamined by *Arias, Doyle and Schimke* (4) and by *Kuriyama et al.* (109). Their results, in agreement with the data of *Jick and Shuster* (88), indicate that phenobarbital brings about initially a prompt *increase in the rate of enzyme synthesis*. According to *Kuriyama et al.* (109), repeated phenobarbital doses cause a large increase in the enzyme amount resulting from a drastic *reduction in rate of enzyme degradation*. Cessation of phenobarbital treatment, on the other hand, is promptly followed by a progressive reduction in the enzyme amount brought about by a large *increase in the rate of degradation* of the enzyme in the face of continued synthesis.

Comparable regulatory situations have been encountered in several other cases. For instance, evidence of concomitant increase in synthesis and decrease in degradation of acetyl coenzyme A carboxylase in shifting rats from fasting to a fat-free diet was recently published by *Majerus and Kilburn* (120) and a reciprocal adjustment, i.e. decrease in synthesis and increase in degradation, was obtained by *Jost, Khairallah and Pitot* (91) in liver serine dehydratase of rats after administration of glucose.

Interestingly enough, in the mentioned study of *Majerus and Kilburn* (120), when the fasted animals were placed on a chow diet there was also a reduction in the enzyme degradation but without concomitant increase in the rate of its synthesis. This would seem to implicate fat as a possible inhibitor of synthesis of acetyl coenzyme A carboxylase.

The counter effects involving alteration of synthesis and degradation have also been recorded in case of extramammalian animal systems. To cite an example, the marked increase in specific activity of chick liver xanthine dehydrogenase in developing chick has been accounted for in terms of an increase in the rate of synthesis as well as a decrease in the rate of degradation (128).

The concurrent reduction of both synthesis and degradation rate constants, on the other hand, was observed in our laboratory with liver catalase following the feeding of cycasin-containing diet to rats (164). Of special interest in this study was the finding of similar effects in renal catalase, considering the fact that the treatment produced no apparent changes in the overall activity of the enzyme. The observation thus underscores the importance of kinetic studies on the relative role of enzyme synthesis and degradation under various physiological and pathological conditions, even in situations where the level of enzyme does not seem to respond to a given stimulus or treatment. Our data clearly demonstrate that the seemingly 'constant' level of enzyme activity may mask profound changes in the overall kinetics of enzyme synthesis and breakdown.

Last but not least, the heterogeneity of degradation rates of individual enzymes is of some significance in considering the specificity of response of enzymes to hormonal stimulation as has been shown by *Berlin and Schimke* (7). The differential enzyme response might be superficially interpreted as a specific effect of hormone on enzyme synthesis. In spite of the existing disparity in the time course and magnitude of responses of the enzymes studied, it was found, however, that the rates of synthesis of these enzymes were increased to essentially the same extent in the hormone-treated animals. It is evident, therefore, that the apparent enzyme specificity can be explained entirely on the basis of the inherent differences in their turnover rates.

VIII. Control Mechanisms of Enzyme Breakdown

While the understanding of the mechanisms involved in protein biosynthesis has reached a high degree of sophistication, primarily as a result of the brilliant work of *Jacob and Monod* (86), the field of protein catabolism is still in *statu nascenti*. Although the phenomenon of the protein turnover is now well established our knowledge of this process is extremely meager. We do not even know yet, with certainty, whether protein catabolism proceeds by a special process, or whether it may involve a reversal of certain steps that occur in protein biosynthesis. This is, indeed, a fertile ground for future research.

Virtually all enzymes examined to date undergo decay that conforms to *first-order kinetics* (e.g. 91, 96, 120, 124, 137, 154, 182, 185, etc.). Such observations seem to imply that enzyme degradation is a *random process* which proceeds without regard to age of individual enzyme molecules.

According to *Schimke, Sweeney and Berlin* (186), the degradation of tryptophan pyrrolase, as well as that of tyrosine transaminase, is similar to the total liver protein catabolism in that it occurs only in a metabolically and structurally intact liver tissue.

In the earlier *in vitro* studies of *Simpson* (199) and *Steinberg and Vaughan* (205) it was demonstrated that the release of radioactivity from labeled proteins in rat liver slices was inhibited by the same factors that were known then to inhibit protein synthesis, including anaerobiosis, cyanide, dinitrophenol, and amino acid analogues, implying that protein catabolism is tied to a supply of energy. *Penn* (140) confirmed the dependence of protein degradation on energy-yielding processes, and, furthermore, reported that the process is markedly stimulated by the addition of ATP and coenzyme A. In view of this evidence it would seem rather unlikely that protein catabolism proceeds by a simple proteolysis, although this possibility cannot be *a priori* discounted.

In microorganisms, inhibitors of protein synthesis are moreover found to inhibit protein breakdown, but only after a considerable lag which would suggest that the effect of inhibitors is indirect. It is conceivable, therefore, that in this system protein catabolism may be achieved by simple proteolytic action (121).

As has been discussed earlier, recently it has also been reported that inhibitors of protein synthesis may under certain conditions prevent the normally occurring breakdown of tyrosine aminotransferase in the liver of rats (69, 96, 111–114). This led *Kenney* (96) to suggest that certain protein(s) which turn over rapidly might be required for the degradative process. The earlier proposal of *Mandelstam* (121) that the cessation of protein degradation may result from the accumulation of compounds of small molecular weights may be operative just as well. According to *Schlessinger and Ben-Hamida* (187), the small molecules that accumulate may not be the amino acids themselves but rather the amino acyl sRNA complexes, or perhaps even peptide-sRNA species, while *Willets* (236, 237) proposed that the hypothetical inhibitor may be an RNA species.

The observed energy requirement for protein degradation in mammalian tissues may possibly be explained by (a) a requirement for synthesis of unknown cofactor(s), (b) a requirement for removal of degradation products, and/or (c) a requirement for maintaining the integrity of a hypothetical particle involved in protein degradation. Parenthetically, a mention should be made of lysosomes (29), sac-like structures in the liver, containing a variety of acid hydrolases, and which are implicated in cellular digestion. What role, if any, these organelles play in the turnover of enzymes remains to be elucidated.

A propos, Coffey and De Duve (27a) demonstrated a direct relationship between chemical stability of proteins and their resistance to lysosomal attack which was a basis for their conclusion that denaturation may be a prerequisite to the digestion of a protein by the lysosomal enzymes. This conclusion appears to be, however, at variance with the studies of *Segal et al.* (193) relating to aminotransferase turnover in rat muscle. The half-life of the enzyme is about 20 days as compared with that of the liver aminotransferase of about 3 days, from which it is otherwise indistinguishable using physical and immunochemical criteria. The thermal denaturation rate of the enzyme *in vitro* gave a calculated half-life of about 400 days at body temperature, this figure being 2 orders of magnitude higher than the half-life of the enzyme in liver and more than 1 order of magnitude higher than that in muscle. A liver lysosomal preparation effectively inactivated the enzyme under conditions in which it was otherwise quite stable. From these results the authors concluded that degradation of alanine aminotransferase *in vivo* is a reflection of an *active (enzymatic) process which does not depend on prior thermal denaturation.*

Another unresolved question concerns the degradation products. It is usually assumed that proteins are degraded to free amino acids although there is no concrete evidence for this. There is no medium of proof either, however, to postulate the existence of peptides as intermediates in this process. An attractive hypothesis has been formulated not too long ago (230) according to which the degradation of protein that occurs in living cells may not proceed as far as free amino acids but that the energy of the protein might be conserved by the formation of activated amino acids derivatives which could be utilized for protein synthesis.

A new light on the question of possible mechanisms involved in enzyme degradation has been shed by the yet unpublished results of *Li and Knox* (115) on rat liver tryptophan oxygenase. In their *in vivo* studies as well as those in the perfused liver the catalytic activity and antigen titer decreased in parallel. In homogenates there was an irreversible loss of catalytic activity prior to the loss of antigenecity and the latter was then degraded to amino acids. a-Methyltryptophan in concentration that combines with the enzyme specifically inhibited the degradation of tryptophan oxygenase in all three systems studied; in homogenates it inhibited the loss of antigenecity more than it did the irreversible loss of catalytic activity. If livers were perfused with whole blood without the addition of the usual amino acid mixture tryptophan oxygenase was degraded precipitously whereas four other control enzymes measured were maintained at their initial level. The results suggest that the problem of specific degradation can be best approached by resolving the process into sequential steps and the different factors that regulate each step.

IX. Summary and Conclusions

The concept of protein turnover has been traced from the early theories of protein metabolism postulated by *Carl von Voit, Pflüger, and Rubner,* through the *endogenous and exogenous protein metabolism* of *Folin,* and the *dynamic state of body constituents* as forwarded by *Schoenheimer,* to its present state. In spite of some recent attempts to reinstate *Folin*'s 'static' point of view, it may be safely concluded, on the basis of the available evidence, that intracellular turnover occurs in animal tissues, and most probably in the lower systems as well. The unequivocal evidence for the existence of protein turnover has been provided by numerous kinetic studies on enzyme synthesis and degradation *in vivo* which have been reviewed in some detail.

On the basis of the author's own work on the kinetics of synthesis and degradation of rat liver catalase it has been concluded that the *enzyme is being synthesized at a constant rate while a constant fraction of active enzyme molecules present in the tissue is being broken down per unit time.* Similar conclusions have now been reached in all of the other mammalian enzyme systems examined to date by other investigators.

A great heterogeneity exists in the turnover rates of different enzymes, ranging from as few as 11 min in terms of half-life for liver ornithine decarboxylase, to as many as 100 days for muscle glyceraldehydephosphate dehydrogenase.

Some heterogeneity seems to also occur in the degradation rates of a given enzyme of one as compared to that of another tissue origin or strain and species of an animal. Differences in half-lives have also been demonstrated between various isozymes and enzyme species localized in different components of a cell.

The *in vivo* turnover implies a *continued synthesis and degradation.* A change in the turnover of a given enzyme can thus be brought about by the alteration in the rate of enzyme synthesis or its degradation or both. The relative role of synthesis and degradation in the regulation of enzyme levels, undergoing change in response to the changes in the environment, is still little understood. The majority of the animal studies reviewed here seem to imply that the *rate of synthesis may be the preferential, although by no means the exclusive way of the mammalian organism to control its levels of enzyme activity* in response to hormonal stimulation, changes in nutritional state, and so forth.

However, a number of instances have also been noted in which the degradation rate constant of an enzyme has been changed as a result of subtle changes, such as genetic mutation, pathological state, blockage of protein synthesis, or a shift in the external environment.

Practically nothing is known about the process of enzyme degradation but the available evidence indicates that it is a complex and highly specific process that presumably takes place only in a metabolically and structurally intact tissue.

Our discussion has had to be, for the most part, limited to situations where an increase in the enzymatic activity actually represented *de novo* synthesis. This should not be in any way misconstrued as implying that a rise in the activity of any enzyme, is in fact a necessary consequence of the increased amount of newly synthesized enzyme, or that no other mechanism exists. Biologists are too well aware of the unique feature of certain proteolytic enzymes concerned in digestion to be produced as inactive proteins, which are subsequently converted into the active molecules. Conversely, one could cite examples of enzyme *inactivation,* arising from a variety of causes, many of which do not involve actual protein breakdown. Situations whereby low molecular weight compounds can either increase or decrease the rate of inactivation or activation may not, indeed, be as uncommon as one might suppose.

There are, of course, many other problems not considered or only superficially covered in our presentation. These include the question of the intracellular *versus* intramolecular turnover, the distribution and the turnover rate of isozymes, subunit construction, sites of allosteric inhibition, structural orientation, etc. Very little is known at present about these parameters, and it is to be expected that these will become fertile areas indeed for future investigation. Some of these topics are covered in other chapters of this volume and the reader should consult the appropriate section.

X. References

1   *Alpers, J.B.; Wu, R., and Racker, E.:* Regulatory mechanisms in carbohydrate metabolism. VI. Glycogen metabolism in HeLa cells. J. biol. Chem. *238:* 2274–2280 (1963).

2   *Argyris, T.S. and Magnus, D.R.:* The stimulation of liver growth and demethylase activity following phenobarbital treatment. Develop. Biol. *17:* 187–201 (1968).

3   *Arias, I.M. and DeLeon, A.:* Estimation of the turnover rate of barbiturate side-chain oxidation enzyme in rat liver. Molec. Pharmacol. *3:* 216–218 (1967).

4   *Arias, I.M.; Doyle, D., and Schimke, R.T.:* Studies on the synthesis and degradation of protein of the endoplasmic reticulum of rat liver. J. biol. Chem. *244:* 3303–3315 (1969).

5   *Auricchio, F.; Martin, D., and Tomkins, G.M.:* Control of degradation and synthesis of induced tyrosine aminotransferase studied in hepatoma cells in culture. Nature, Lond. *224:* 806–808 (1969).

6   *Berger, L.; Slein, M.W.; Colowick, S.P., and Cori, C.F.:* Isolation of hexokinase from Baker's yeast. J. gen. Physiol. *29:* 379–391 (1946).

7   *Berlin, C.M. and Schimke, R.T.:* Influence of turnover rates on the responses of enzymes to cortisone. Molec. Pharmacol. *1:* 149–156 (1965).

8   *Bertino, J.R.; Cashmore, A.; Fink, M.; Calabresi, P., and Lefkowitz, F.:* The 'induction' of leukocyte and erythrocyte dihydrofolate reductase by methotrexate. II. Clinical and pharmacologic studies. Clin. Pharmacol. Ther. *6:* 763–770 (1965).

9   *Block, R.J.:* Nitrogen requirements of animals and man: Comments on the *Folin and Schoenheimer* hypothesis. Proceedings Int. Symposium on Enzyme Chemistry, Tokyo and Kyoto 1957, pp. 444–449 (Maruzen, Tokyo 1958).

10 *Bock, K.W. and Siekevitz, P.:* Turnover studies on membrane NAD glycohydrolase. Fed. Proc. *29:* 540 (1970).

11 *Bojarski, T.B. and Hiatt, H.H.:* Stabilization of thymidylate kinase activity by thymidylate and by thymidine. Nature, Lond. *188:* 1112–1114 (1960).

12 *Borek, E.; Ponticorvo, L., and Rittenberg, D.:* Protein turnover in microorganisms. Proc. nat. Acad. Sci., Wash. *44:* 369–374 (1958).

13 *Borsook, H. and Keighley, G.L.:* The 'continuing' metabolism of nitrogen in animals. Proc. roy. Soc., London, Ser. B. *118:* 488–521 (1935).

14 *Bresnick, E. and Burleson, S.S.:* Rates of turnover of deoxythymidine kinase and of its template RNA in the Novikoff hepatoma and in neonatal liver (in preparation).

15 *Bresnick, E.; Mayfield, E.D., Jr., and Mossé, H.:* Increased activity of enzymes for *de novo* pyrimidine biosynthesis after orotic acid administration. Molec. Pharmacol. *4:* 173–180 (1968).

16 *Bresnick, E.; Williams, S.S., and Mossé, H.:* Rates of turnover of deoxythymidine kinase and of its template RNA in regenerating and control liver. Cancer Res. *27:* 469–475 (1967).

17 *Brown, E.G.:* Evidence for the involvement of ferrous iron in the biosynthesis of δ-aminolaevulic acid by chicken erythrocyte preparations. Nature, Lond. *182:* 313–315 (1958).

18 *Brown, E.G.:* Mode of action of 5,6-dimethylbenzimidazole and certain keto-acids as inhibitors of haem biosynthesis. Nature, Lond. *182:* 1091–1092 (1958).

19 *Buchanan, D.L.:* Total carbon turnover measured by feeding a uniformly labeled diet. Arch. Biochem. Biophys. *94:* 500–511 (1961).

20 *Ceriotti, G.; Spandrio, L., and Agradi, A.:* A study of the synthesis of catalase in liver of tumor-bearing mice by means of radioactive iron. Biochim. biophys. Acta *27:* 432–433 (1958).

21 *Chytil, F.:* An activator of the adaptive enzyme tryptophan pyrrolase present in fetal-rat liver. Biochim. biophys. Acta *48:* 217–218 (1961).

22 *Chytil, F.:* Activation of liver tryptophan pyrrolase. Collect. czechosl. chem. Commun. *27:* 1487–1492 (1962).

23 *Chytil, F.:* Activation of liver tryptophan oxygenase by adenosine 3′,5′-phosphate and by other purine derivatives. J. biol. Chem. *243:* 893–899 (1968).

24 *Chytil, F. and Skřivanová, J.:* Reactivation of cortisone induced liver tryptophan pyrrolase by boiled liver cell sap and by cyclic adenosine 3′,5′-phosphate. Biochim. biophys. Acta *67:* 164–166 (1963).

25 *Chytil, F. and Skřivanová, J.:* Factors influencing the conversion of the inactive form of liver tryptophan pyrrolase into the active form. Collect. czechosl. chem. Commun. *28:* 2207–2215 (1963).

26 *Chytil, F.; Skřivanová, J., and Braná, H.:* Activating effect of purines on the liver tryptophan pyrrolase system. Can. J. Biochem. *44:* 283–286 (1966).

27 *Civen, M. and Knox, W.E.:* The independence of hydrocortisone and tryptophan inductions of tryptophan pyrrolase. J. biol. Chem. *234:* 1787–1790 (1959).

27a *Coffey, J.W. and De Duve, C.:* Digestive activity of lysosomes. I. Digestion of proteins by extracts of rat liver lysosomes. J. biol. Chem. *243:* 3255–3263 (1968).

28 *Conney, A.H. and Gilman, A.G.:* Puromycin inhibition of enzyme induction by 3-methyl cholanthrene and phenobarbital. J. biol. Chem. *238:* 3682–3685 (1963).

28a *De Duve, C. and Baudhuin, P.:* Peroxisomes (microbodies and related particles). Physiol. Rev. *46:* 323–357 (1966).

29 *De Duve, C. and Wattiaux, R.:* Functions of lysosomes. Annu. Rev. Physiol. *28:* 435–492 (1966).

30 *DeMars, R.:* Some studies of enzymes in cultivated human cells. Nat. Cancer Inst. Monograph No. 13, pp. 181–195 (1964).

31 *DiPietro, D.L.; Sharma, C., and Weinhouse, S.:* Studies on glucose phosphorylation in rat liver. Biochemistry *1:* 455–462 (1962).

32 *Doyle, D. and Schimke, R.T.:* The genetic and developmental regulation of hepatic δ-aminolevulinate dehydratase in mice. J. biol. Chem. *244:* 5449–5459 (1969).

33 *Eagle, E.; Piez, K.A.; Fleischman, R., and Oyama, V.I.:* Protein turnover in mammalian cell cultures. J. biol. Chem. *234:* 592–597 (1959).

34 *Eliasson, E.E.:* Regulation of arginase activity in Chang's liver cells in tissue culture. Biochim. biophys. Acta *97:* 449–459 (1965).

35 *Eliasson, E.E. and Strecker, H.J.:* Arginase activity during the growth cycle of Chang's liver cells. J. biol. Chem. *241:* 5757–5763 (1966).

36 *Ernster, L. and Orrenius, S.:* Substrate-induced synthesis of the hydroxylating enzyme system of liver microsomes. Fed. Proc. *24:* 1190–1199 (1965).

37 *Fallon, H.J.:* Regulatory phenomena in mammalian serine metabolism. Adv. enzym. Regul. *5:* 107–120 (1967).

38 *Fallon, H.J.:* Unpublished data.

39 *Fallon, H.J.; Hackney, E.J., and Byrne, W.L.:* Serine biosynthesis in rat liver. Regulation of enzyme concentration by dietary factors. J. biol. Chem. *241:* 4157–4167 (1966).

40 *Feigelson, P.; Dashman, T., and Margolis, F.:* The half-lifetime of induced tryptophan peroxidase *in vivo.* Arch. Biochem. Biophys. *85:* 478–482 (1959).

41 *Feigelson, P.; Feigelson, M., and Greengard, O.:* Comparison of the mechanism of hormonal and substrate induction of rat liver tryptophan pyrrolase. Recent Progr. Hormone Res. *18:* 491–512 (1962).

42 *Feigelson, P. and Greengard, O.:* Immunochemical evidence for increased titers of liver tryptophan pyrrolase during substrate and hormonal enzyme induction. J. biol. Chem. *237:* 3714–3717 (1962).

43 *Folin, O.:* A theory of protein metabolism. Amer. J. Physiol. *13:* 117–138 (1905).

44 *Forssberg, A. and Révész, L.:* A study on the metabolic state of proteins in the cells of two ascites tumors. Biochim. biophys. Acta *25:* 165–171 (1957).

45 *Fox, G. and Brown, J.W.:* Protein degradation in *Escherichia coli* in the logarithmic phase of growth. Biochim. biophys. Acta *46:* 387–390 (1961).

46 *Freedland, R.A.; Cunliffe, T.L., and Zinkl, J.G.:* The effect of insulin on enzyme adaptations to diets and hormones. J. biol. Chem. *241:* 5448–5451 (1966).

47 *Fritz, P.J.; Vesell, E.S.; White, E.L., and Pruitt, K.M.:* The roles of synthesis and degradation in determining tissue concentrations of lactate dehydrogenase-5. Proc. nat. Acad. Sci., Wash. *62:* 558–565 (1969).

48 *Ganschow, R.E. and Schimke, R.T.:* Independent genetic control of the catalytic activity and the rate of degradation of catalase in mice. J. biol. Chem. *244:* 4649–4658 (1969).

49 *Garren, L.D.; Howell, R.R.; Tomkins, G.M., and Crocco, R.M.:* A paradoxical effect of actinomycin D: The mechanism of regulation of enzyme synthesis by hydrocortisone. Proc. nat. Acad. Sci., Wash. *52:* 1121–1129 (1964).

50 *Gelehrter, T.D. and Tomkins, G.M.:* Control of tyrosine aminotransferase synthesis in tissue culture by a factor in serum. Proc. nat. Acad. Sci., Wash. *64:* 723–730 (1969).

51 *Gelehrter, T.D. and Tomkins, G.M.:* Post-transcriptional control of tyrosine aminotransferase synthesis by insulin. Proc. nat. Acad. Sci., Wash. (in press).

52 *Goldstein, L.; Knox, W.E., and Behrman, E.J.:* Studies on the nature, inducibility, and assay of the threonine and serine dehydrase activities of rat liver. J. biol. Chem. *237:* 2855–2860 (1962).

53 *Goldstein, L.; Stella, E.J., and Knox, W.E.:* The effect of hydrocortisone on tyrosine-*a*-ketoglutarate transaminase and tryptophan pyrrolase activities in the isolated, perfused rat liver. J. biol. Chem. *237:* 1723–1726 (1962).

54 *Gorski, J. and Morgan, M.S.:* Estrogen effects on uterine metabolism: Reversal by inhibitors of protein synthesis. Biochim. biophys. Acta *149:* 282–287 (1967).

55 *Granick, S.:* The induction *in vitro* of the synthesis of δ-aminolevulinic acid synthetase in chemical porphyria: A response to certain drugs, sex hormones, and foreign chemicals. J. biol. Chem. *241:* 1359–1375 (1966).

56 *Granick, S. and Urata, G.:* Increase in activity of δ-aminolevulinic acid synthetase in liver mitochondria induced by feeding of 3,5-dicarbethoxy-1,4-dihydrocollidone. J. biol. Chem. *238:* 821–827 (1963).

57 *Granner, D.K.; Hayashi, S.; Thompson, E.B., and Tomkins, G.M.:* Stimulation of tyrosine aminotransferase synthesis by dexamethasone phosphate in cell culture. J. molec. Biol. *35:* 291–301 (1968).

58 *Granner, D.K.; Thompson, E.B., and Tomkins, G.M.:* Dexamethasone phosphate-induced synthesis of tyrosine aminotransferase in hepatoma tissue culture cells. J. biol. Chem. *245:* 1472–1478 (1970).

59 *Gray, G.D.:* Tryptophan pyrrolase activity: Effects of cyclic-AMP, purines, pyrimidines, nucleosides, and nucleotides. Arch. Biochem. Biophys. *113:* 502–504 (1966).

60　*Greenfield, R.E. and Price, V.E.:* Liver catalase. III. Isolation of catalase from mitochondrial fractions of polyvinylpyrrolidone-sucrose homogenates. J. biol. Chem. *220:* 607–618 (1956).

61　*Greengard, O.:* The regulation of apoenzyme levels by coenzymes and hormones. Adv. enzym. Regul. *2:* 277–288 (1964).

62　*Greengard, O. and Feigelson, O.:* The activation of the inducible enzyme, rat-liver tryptophan pyrrolase. Biochim. biophys. Acta *39:* 191–192 (1960).

63　*Greengard, O. and Gordon, M.:* The cofactor-mediated regulation of apoenzyme levels in animal tissues. I. The pyridoxine-induced rise of rat liver tyrosine-transaminase level *in vivo.* J. biol. Chem. *238:* 3708–3710 (1963).

64　*Greengard, O.; Mendelsohn, N., and Acs, G.:* Effect of cytoplasmic particles on tryptophan pyrrolase activity of rat liver. J. biol. Chem. *241:* 304–308 (1966).

65　*Greengard, O.; Smith, M.A., and Acs, G.:* Relation of cortisone and synthesis of ribonucleic acid to induced and developmental enzyme formation. J. biol. Chem. *238:* 1548–1551 (1963).

66　*Greenlees, J. and LePage, G.A.:* Protein turnover in study of host-tumor relationships. Cancer Res. *15:* 256–262 (1955).

67　*Greenstein, J.P.:* Biochemistry of cancer; 2nd ed. (Academic Press, New York 1954).

68　*Grisolia, S.:* The catalytic environment and its biological implications. Physiol. Rev. *44:* 657–712 (1964).

69　*Grossman, A. and Mavrides, C.:* Studies on the regulation of tyrosine aminotransferase in rats. J. biol. Chem. *242:* 1398–1405 (1967).

70　*Hakala, M.T. and Suolinna, E.M.:* Specific protection of folate reductase against chemical and proteolytic inactivation. Molec. Pharmacol. *2:* 465–480 (1966).

71　*Halvorson, H.:* Intracellular protein and nucleic acid turnover in resting yeast cells. Biochim. biophys. Acta *27:* 255–266 (1958).

72　*Halvorson, H.:* Studies on protein and nucleic acid turnover in growing cultures of yeast. Biochim. biophys. Acta *27:* 267–276 (1958).

73　*Harris, H. and Watts, J.W.:* Turnover of protein in a nonmultiplying animal cell. Nature, Lond. *181:* 1582–1584 (1958).

74　*Hayashi, N.; Yoda, B., and Kikuchi, G.:* Mechanism of allylisopropylacetamide-induced increase of δ-aminolevulinate synthetase in liver mitochondria. IV. Accumulation of the enzyme in the soluble fraction of rat liver. Arch. Biochem. Biophys. *131:* 83–91 (1969).

75　*Heimberg, M. and Velick, S.F.:* The synthesis of aldolase and phosphorylase in rabbits. J. biol. Chem. *208:* 725–730 (1954).

76　*Heston, W.E.; Hoffman, H.A., and Rechcigl, M., Jr.:* Genetic analysis of liver catalase activity in two substrains of C57BL mice. Genet. Res. *6:* 387–397 (1965).

77　*Hiatt, H.H. and Bojarski, T.B.:* The effects of thymidine administration on thymidylate kinase activity and on DNA synthesis in mammalian tissues. Cold Spr. Harb. Symp. quant. Biol. *26:* 367–369 (1961).

78　*Hogness, D.S.; Cohn, M., and Monod, J.:* Studies on the induced synthesis of β-galactosidase in *Escherichia coli:* The kinetics and mechanism of sulfur incorporation. Biochim. biophys. Acta *16:* 99–116 (1955).

79　*Holten, D. and Kenney, F.T.:* Regulation of tyrosine-α-ketoglutarate transaminase in rat liver. VI. Induction by pancreatic hormones. J. biol. Chem. *242:* 4372–4377 (1967).

80　*Holten, D.; Wicks, W.D., and Kenney, F.T.:* Studies on the role of vitamin B_6 derivatives in regulating tyrosine-α-ketoglutarate transaminase activity *in vitro* and *in vivo.* J. biol. Chem. *242:* 1053–1059 (1967).

81 Holtzman, J.L. and Gillette, J.R.: The effect of phenobarbital on the turnover of microsomal phospholipid in male and female rats. J. biol. Chem. 243: 3020–3028 (1968).

82 Hruban, Z. and Rechcigl, M., Jr.: Microbodies and related particles. Morphology, biochemistry and physiology (Academic Press, New York/London 1969).

83 Hruban, Z.; Swift, H., and Rechcigl, M., Jr.: Fine structure of transplantable hepatomas of the rat. J. nat. Cancer Inst. 35: 459–496 (1965).

84 Inoue, H. and Pitot, H.C.: Regulation of the rate of synthesis of two electrophoretically distinct forms of serine dehydratase. Fed. Proc. 29: 736 Abs (1970).

85 Ishikawa, E.; Ninagawa, T., and Suda, M.: Hormonal and dietary control of serine dehydratase in rat liver. J. Biochem., Tokyo 57: 506–513 (1965).

86 Jacob, F. and Monod, J.: Genetic regulatory mechanisms in the synthesis of proteins. J. molec. Biol. 3: 318–356 (1961).

87 Jervell, K.F. and Seglen, P.O.: Tyrosine transaminase degradation in perfused liver after inhibition of protein synthesis by cycloheximide. Biochim. biophys. Acta 174: 398–400 (1969).

88 Jick, H. and Shuster, L.: The turnover of microsomal reduced nicotinamide adenine dinucleotide phosphate-cytochrome c reductase in the livers of mice treated with phenobarbital. J. biol. Chem. 241: 5366–5369 (1966).

89 Jordan, H.C.; Miller, L.L., and Peters, P.A.: Constant protein catabolism of Walker carcinoma 256 and human skin epithelium in tissue culture. Cancer Res. 19: 195–200 (1959).

90 Jordan, H.C. and Schmidt, P.A.: Constant protein turnover in mammalian cells during logarithmic growth. Biochem. biophys. Res. Comm. 4: 313–316 (1961).

91 Jost, J.-P.; Khairallah, E.A., and Pitot, H.C.: Studies on the induction and repression of enzymes in rat liver. V. Regulation of the rate of synthesis and degradation of serine dehydratase by dietary amino acids and glucose. J. biol. Chem. 243: 3057–3066 (1968).

92 Julian, J.A. and Chytil, F.: A two-step mechanism for the regulation of tryptophan pyrrolase. Biochem. biophys. Res. Comm. 35: 734–740 (1969).

93 Julian, J. and Chytil, F.: Participation of xanthine oxidase in the activation of liver tryptophan pyrrolase. J. biol. Chem. 245: 1161–1168 (1970).

94 Kenney, F.T.: Induction of tyrosine-α-ketoglutarate transaminase in rat liver. IV. Evidence for an increase in the rate of enzyme synthesis. J. biol. Chem. 237: 3495–3498 (1962).

95 Kenney, F.T.: Regulation of tyrosine-α-ketoglutarate transaminase in rat liver. V. Repression in growth hormone-treated rats. J. biol. Chem. 242: 4367–4371 (1967).

96 Kenney, F.T.: Turnover of rat liver tyrosine transaminase: Stabilization after inhibition of protein synthesis. Science 156: 525–528 (1967).

97 Kenney, F.T. and Albritton, W.L.: Repression of enzyme synthesis at the translational level and its hormonal control. Proc. nat. Acad. Sci., Wash. 54: 1693–1698 (1965).

98 Kenney, F.T. and Flora, R.M.: Induction of tyrosine-α-ketoglutarate transaminase. I. Hormonal nature. J. biol. Chem. 236: 2699-2702 (1961).

99 Khairallah, E.A. and Pitot, H.C.: Studies on the turnover of serine dehydrase: Amino acid induction, glucose repression and pyridoxine stabilization: in Yamada, Katsumma and Wada Symposium on pyridoxal enzymes, pp. 159–164 (Maruzen, Tokyo 1968).

100 Kim, Y.S.: The sequential increase in the rates of synthesis of enzymes in rat liver after glucocorticoid administration. Molec. Pharmacol. 4: 168–172 (1968).

101 Kim, Y.S.: The half-life of alanine aminotransferase and of total soluble protein in livers of normal and glucocorticoid-treated rats. Molec. Pharmacol. 5: 105–108 (1969).

102 *King, D.W.; Bensch, K.G., and Hill, R.B., Jr.:* State of dynamic equilibrium in protein of mammalian cells. Science *131:* 106–107 (1960).

103 *Kit, S.; Dubbs, D.R., and Frearson, P.M.:* Decline of thymidine kinase activity in stationary phase mouse fibroblast cells. J. biol. Chem. *240:* 2565–2573 (1965).

104 *Klein, E.:* Studies on the substrate-induced arginase synthesis in animal cell strains cultured *in vitro.* Exp. Cell Res. *22:* 226–232 (1961).

105 *Knox, W.E. and Mehler, A.H.:* The adaptive increase of the tryptophan peroxidase-oxidase system of liver. Science *113:* 237–238 (1951).

106 *Knox, W.E. and Piras, M.M.:* A reinterpretation of the stabilization of tryptophan pyrrolase by its substrate. J. biol. Chem. *241:* 764–767 (1966).

107 *Knox, W.E.; Piras, M.M., and Tokuyama, K.:* Tryptophan pyrrolase of liver. I. Activation and assay in soluble extracts of rat liver. J. biol. Chem. *241:* 297–303 (1966).

108 *Koch, A.L. and Levy, H.R.:* Protein turnover in growing cultures of *Escherichia coli.* J. biol. Chem. *217:* 947–957 (1955).

109 *Kuriyama, Y.; Omura, T.; Siekevitz, P., and Palade, G.E.:* Effects of phenobarbital on the synthesis and degradation of the protein components of rat liver microsomal membranes. J. biol. Chem. *244:* 2017–2026 (1969).

110 *Lee, K.L. and Miller, O.N.:* Studies on triiodothyronine-induced synthesis of liver mito-chondrial a-glycerophosphate dehydrogenase in the thyroidectomized rat. Molec. Pharmacol. *3:* 44–51 (1967).

111 *Levitan, I.B. and Webb, T.E.:* Modification by 8-azaguanine of the effects of hydro-cortisone on the induction and inactivation of tyrosine transaminase of rat liver. J. biol. Chem. *244:* 341–347 (1969).

112 *Levitan, I.B. and Webb, T.E.:* Effects of 5-azacytidine on polyribosomes and on the control of tyrosine transaminase activity in rat liver. Biochim. biophys. Acta *182:* 491–500 (1969).

113 *Levitan, I.B. and Webb, T.E.:* Regulation of tyrosine transaminase in the isolated perfused rat liver. J. biol. Chem. *244:* 4684–4688 (1969).

114 *Levitan, I.B. and Webb, T.E.:* Hydrocortisone-mediated changes in the concentration of tyrosine transaminase in rat liver. An immunochemical study. J. molec. Biol. *48:* 339–348 (1970).

115 *Li, J.B. and Knox, W.E.:* Degradation of rat liver tryptophan oxygenase *in vivo* and *in vitro.* Fed. Proc. *29:* 736 (1970).

116 *Lin, E.C.C. and Knox, W.E.:* Adaptation of the rat liver tyrosine-a-ketoglutarate trans-aminase. Biochim. biophys. Acta *26:* 85–88 (1957).

117 *Lin, E.C.C. and Knox, W.E.:* Specificity of the adaptive response of tyrosine-a-keto-glutarate transaminase in the rat. J. biol. Chem. *233:* 1186–1189 (1958).

118 *Littlefield, J.W.:* Studies on thymidine kinase in cultured mouse fibroblasts. Biochim. biophys. Acta *95:* 14–22 (1965).

119 *Loftfield, R.B. and Bonnichsen, R.:* Incorporation of ^{59}Fe into different iron com-pounds of liver tissue. Acta chem. scand. *10:* 1547–1552 (1956).

119a*MacDonald, R.A.:* 'Lifespan' of liver cells. Arch. Intern. Med. *107:* 335–343 (1961).

120 *Majerus, P.W. and Kilburn, E.:* Acetyl coenzyme A carboxylase. The roles of synthesis and degradation in regulation of enzyme levels in rat liver. J. biol. Chem. *244:* 6254–6262 (1969).

121 *Mandelstam, J.:* Turnover of protein in growing and nongrowing populations of *Esche-richia coli.* Biochem. J. *69:* 110–119 (1958).

122 *Margoliash, E.; Novogrodsky, A., and Schejter, A.:* Irreversible reaction of 3-amino-1,2,4-triazole and related inhibitors with the protein of catalase. Biochem. J. *74:* 339–348 (1960).

123 *Markus, G.:* Protein substrate conformation and proteolysis. Proc. nat. Acad. Sci., Wash. *54:* 253–258 (1965).

124 *Marver, H.S.; Collins, A.; Tschudy, D.P., and Rechcigl, M., Jr.:* δ-Aminolevulinic acid synthetase. II. Induction in rat liver. J. biol. Chem. *241:* 4323–4329 (1966).

125 *Marver, H.S.; Tschudy, D.P.; Perlroth, M.G., and Collins, A.:* Coordinate synthesis of heme and apoenzyme in the formation of tryptophan pyrrolase. Science *154:* 501–503 (1966).

126 *Mitchell, H.:* The validity of *Folin's* concept of dichotomy in protein metabolism. J. Nutr. *55:* 193–207 (1955).

127 *Moldave, K.:* The release of labelled constituents from cellular fractions of Ehrlich ascites cells. J. biol. Chem. *225:* 709–714 (1957).

128 *Murison, G.:* Synthesis and degradation of xanthine dehydrogenase during chick liver development. Develop. Biol. *20:* 518–543 (1969).

129 *Nagabhushanam, A. and Greenberg, D.M.:* Isolation and properties of a homogeneous preparation of cystathionine synthetase-*L*-serine and *L*-threonine dehydratase. J. biol. Chem. *240:* 3002–3008 (1965).

130 *Narisawa, K. and Kikuchi, G.:* Effect of inhibitors of DNA synthesis on allylisopropyl-acetamide-induced increases of δ-aminolevulinic acid synthetase and other enzymes in rat liver. Biochim. biophys. Acta *99:* 580–583 (1965).

131 *Nebert, D.W. and Gelboin, H.V.:* Substrate-inducible microsomal aryl hydroxylase in mammalian cell culture. II. Cellular responses during enzyme induction. J. biol. Chem. *243:* 6250–6261 (1968).

132 *Nemeth, A.M.:* The effect of 5-fluorouracil on the developmental and adaptive formation of tryptophan pyrrolase. J. biol. Chem. *237:* 3703–3706 (1962).

133 *Neuberger, A. and Richards, F.F.:* Protein biosynthesis in mammalian tissues. II. Studies on turnover in the whole animal; in *Munro and Allison* Mammalian protein metabolism, vol. 1, pp. 243–296 (Academic Press, New York 1964).

134 *Niemeyer, H.:* Regulation of glucose-phosphorylating enzymes. Nat. Cancer Inst. Monograph, No. *27:* 29–40 (1967).

135 *Niemeyer, H.; Clark-Turri, L.; Pérez, N., and Rabajille, E.:* Studies on factors affecting the induction of ATP: D-hexose 6-phosphotransferase in rat liver. Arch. Biochem. Biophys. *109:* 634–645 (1965).

136 *Niemeyer, H.; Pérez, H., and Rabajille, E.:* Interrelation of actions of glucose, insulin, and glucagon on induction of adenosine triphosphate: D-hexose phosphotransferase in rat liver. J. biol. Chem. *241:* 4055–4059 (1966).

137 *Omura, T.; Siekevitz, P., and Palade, G.E.:* Turnover of constituents of the endoplasmic reticulum membranes of rat hepatocytes. J. biol. Chem. *242:* 2389–2396 (1967).

138 *Otto, K.:* Die Induktion der Glutamat-Pyruvat-Transaminase durch Corticode. Naturwissenschaften *50:* 355–356 (1963).

139 *Paul, J. and Fottrell, P.F.:* Mechanism of D-glutamyltransferase repression in mammalian cells. Biochim. biophys. Acta *67:* 334–336 (1963).

140 *Penn, N.W.:* The requirements for serum albumin metabolism in subcellular fractions of liver and brain. Biochim. biophys. Acta *37:* 55–63 (1960).

141 *Peraino, C.; Blanke, R.L., and Pitot, H.C.:* Studies on the induction and repression of enzymes in rat liver. III. Induction of ornithine-δ-transaminase and threonine dehydrase by oral intubation of free amino acids. J. biol. Chem. *240:* 3039-3043 (1965).

142 *Peraino, C.; Lamar, C., Jr., and Pitot, H.C.:* Studies on the induction and repression of enzymes in rat liver. IV. Effects of cortisone and phenobarbital. J. biol. Chem. *241:* 2944–2948 (1966).

143 *Peraino, C. and Pitot, H.C.:* Ornithine-δ-transaminase in the rat. I. Assay and some general characteristics. Biochim. biophys. Acta *73:* 222–231 (1963).

144 *Peraino, C. and Pitot, H.C.:* Studies on the induction and repression of enzymes in rat liver. II. Carbohydrate repression of dietary and hormonal induction of threonine dehydrase and ornithine-δ-transaminase. J. biol. Chem. *239:* 4308-4313 (1964).

145 *Perlroth, M.G.; Tschudy, D.P.; Marver, H.S.; Berard, C.W.; Zeigel, R.F.; Rechcigl, M., Jr., and Collins, A.:* Acute intermittent porphyria. New morphologic and biochemical findings. Amer. J. Med. *41:* 149–162 (1966).

146 *Pflüger, E.:* Über einige Gesetze des Eiweissstoffwechsels (mit besonderer Berücksichtigung der Lehre vom sogenannten „zirkulierenden Eiweiss"). Pflügers Arch. ges. Physiol. *54:* 333–419 (1893).

147 *Pine, M.J.:* Heterogeneity of protein turnover in *Escherichia coli.* Biochim. biophys. Acta *104:* 439–456 (1965).

148 *Pine, M.J.:* Metabolic control of intracellular proteolysis in growing and resting cells of *Escherichia coli.* J. Bact. *92:* 847–850 (1966).

149 *Pitot, H.C. and Peraino, C.:* Studies on the induction and repression of enzymes in rat liver. I. Induction of threonine dehydrase and ornithine-δ-transaminase by oral intubation of casein hydrolysate. J. biol. Chem. *239:* 1783–1788 (1964).

150 *Pitot, H.C.; Peraino, C.; Lamar, C., Jr., and Kennan, A.L.:* Template stability of some enzymes in rat liver and hepatoma. Proc. nat. Acad. Sci., Wash. *54:* 845–851 (1965).

151 *Pitot, H.C.; Peraino, C.; Pries, N., and Kennan, A.L.:* Template stability in liver and hepatoma. Adv. Enzym. Regul. *3:* 359–368 (1965).

152 *Poole, B.; Leighton, F., and De Duve, C.:* The synthesis and turnover of rat liver peroxisomes. II. Turnover of peroxisome proteins. J. Cell Biol. *41:* 536–546 (1969).

153 *Price, V.E.; Rechcigl, M., Jr., and Hartley, R.W., Jr.:* Methods for determining the rates of catalase synthesis and destruction *in vivo.* Nature, Lond. *189:* 62–63 (1961).

154 *Price, V.E.; Sterling, W.R.; Tarantola, V.A.; Hartley, R.W., Jr., and Rechcigl, M., Jr.:* The kinetics of catalase synthesis and destruction *in vivo.* J. biol. Chem. *237:* 3468–3475 (1962).

155 *Rechcigl, M., Jr.:* The role of synthesis and degradation in the regulation of enzyme levels. Fed. Proc. *26:* 409 (1967).

156 *Rechcigl, M., Jr.: In vivo* turnover and its role in the metabolic regulation of enzyme levels. Enzymologia, D. Haag *34:* 23–39 (1968).

157 *Rechcigl, M., Jr.:* Relative role of synthesis and degradation in the regulation of catalase activity; in *San Pietro, Lamborg and Kenney* Regulatory mechanisms for protein synthesis in mammalian cells, pp. 399–415 (Academic Press, New York 1968).

158 *Rechcigl, M., Jr.:* Studies on regulatory aspects of nutrition at molecular level. I. Enzyme turnover during fasting. Int. Arch. Physiol. Biochim. *76:* 693–706 (1968).

159 *Rechcigl, M., Jr.:* Studies on regulatory aspects of nutrition at molecular level. II. Enzyme turnover during protein deficiency. Nutr. Diet. *11:* 214–227 (1969).

160 *Rechcigl, M., Jr.; Grantham, F., and Greenfield, R.E.:* Studies on the cachexia of tumor-bearing animals. I. Body weight changes, carcass composition, and metabolic studies. Cancer Res. *21:* 238–251 (1961).

161 *Rechcigl, M., Jr. and Heston, W.E.:* Tissue catalase activity in several C57BL substrains and in other strains of inbred mice. J. nat. Cancer Inst. *30:* 855–864 (1963).

162 *Rechcigl, M., Jr. and Heston, W.E.:* Genetic regulation of enzyme activity in mammalian system by the alteration of the rates of enzyme degradation. Biochem. biophys. Res. Comm. *27:* 119–124 (1967).

163 *Rechcigl, M., Jr.; Hruban, Z., and Morris, H.P.:* The roles of synthesis and degradation of catalase levels in the neoplastic tissues. Enzym. biol. Clin. *10:* 161–180 (1969).

164 *Rechcigl, M., Jr. and Laqueur, G.L.:* Carcinogen-mediated alteration of the rates of enzyme synthesis and degradation. Enzym. biol. clin. *9:* 276–286 (1968).

165 *Rechcigl, M., Jr. and Price, V.E.:* The rates and the kinetics of enzyme formation and destruction in the living animal; in *Albanese* Newer methods of nutritional biochemistry, pp. 185–197 (Academic Press, New York 1963).

166 *Rechcigl, M., Jr. and Price, V.E.:* Studies on the turnover of catalase *in vivo.* Progr. exp. Tum. Res., vol. 10, pp. 112–132 (Karger, Basel/New York 1968).

167 *Rechcigl, M., Jr.; Price, V.E., and Morris, H.P.:* Studies on the cachexia of tumor-bearing animals. II. Catalase activity in the tissues of hepatoma-bearing animals. Cancer Res. *22:* 874–880 (1962).

168 *Rechcigl, M., Jr. and Sidransky, H.:* Isolation of two lines of transplantable, ethionine-induced rat hepatomas of high and low catalase activity from a primary tumor. J. nat. Cancer Inst. *28:* 1411–1423 (1962).

169 *Reel, J.R. and Kenney, F.T.:* Regulation of tyrosine transaminase by hydrocortisone in hepatoma cell cultures. Fed. Proc. *27:* 641 (1968).

170 *Reel, J.R. and Kenney, F.T.:* 'Superinduction' of tyrosine transaminase in hepatoma cell cultures: Differential inhibition of synthesis and turnover by actinomycin. Proc. nat. Acad. Sci., Wash. *61:* 200–206 (1968).

171 *Reel, J.R.; Lee, K.L., and Kenney, F.T.:* Regulation of tyrosine-α-ketoglutarate transaminase in rat liver. VIII. Inductions by hydrocortisone and insulin in cultured hepatoma cells (in preparation).

172 *Rosen, F. and Milholland, R.J.:* Glucocorticoids and transaminase activity. VII. Studies on the nature and specificity of substrate induction of tyrosine-α-ketoglutarate transaminase and tryptophan pyrrolase. J. biol. Chem. *238:* 3730–3735 (1963).

173 *Rosen, F.; Roberts, N.R., and Nichol, C.A.:* Glucocorticosteroids and transaminase activity. I. Increased activity of glutamic-pyruvic transaminase in four conditions associated with gluconeogenesis. J. biol. Chem. *234:* 476–480 (1959).

174 *Rotman, B. and Spiegelman, S.:* On the origin of the carbon in the induced synthesis of β-galactosidase in *Escherichia coli.* J. Bact. *68:* 419–429 (1954).

175 *Rowe, P.B. and Wyngaarden, J.B.:* The mechanism of dietary alterations in rat hepatic xanthine oxidase levels. J. biol. Chem. *241:* 5571–5576 (1966).

176 *Rubner, M.:* Über den Eiweissansatz. Arch. Physiol. 67–84 (1911).

177 *Russell, D.H. and Snyder, S.H.:* Amine synthesis in regenerating rat liver: Extremely rapid turnover of ornithine decarboxylase. Molec. Pharmacol. *5:* 253–262 (1969).

178 *Salas, M.; Viñuela, E., and Sols, A.:* Insulin-dependent synthesis of liver glucokinase in the rat. J. biol. Chem. *238:* 3535–3538 (1963).

179 *Schapira, G.; Kruh, J.; Dreyfus, J.C., and Schapira, F.:* The molecular turnover of muscle aldolase. J. biol. Chem. *235:* 1738–1741 (1960).

180 *Schimke, R.T.:* Adaptive characteristics of urea cycle enzymes in the rat. J. biol. Chem. *237:* 459–468 (1962).

181 *Schimke, R.T.:* Studies on factors affecting the levels of urea cycle enzymes in rat liver. J. biol. Chem. *238:* 1012–1018 (1963).

182 *Schimke, R.T.:* The importance of both synthesis and degradation in the control of arginase levels in rat liver. B. biol. Chem. *239:* 3808–3817 (1964).

183 *Schimke, R.T.:* Enzymes of arginine metabolism in cell culture: Studies on enzyme induction and repression. Nat. Cancer Inst. Monograph *13:* 197–217 (1964).

184 *Schimke, R.T.:* Protein turnover and the regulation of enzyme levels in rat liver. Nat. Cancer Inst. Monograph *27:* 301–314 (1967).

185 *Schimke, R.T.; Sweeney, E.W., and Berlin, C.M.:* The roles of synthesis and degradation in the control of rat liver tryptophan pyrrolase. J. biol. Chem. *240:* 322–331 (1965).

186 *Schimke, R.T.; Sweeney, E.W., and Berlin, C.M.:* Studies of the stability *in vivo* and *in vitro* of rat liver tryptophan pyrrolase. J. biol. Chem. *240:* 4609–4620 (1965).

187 *Schlessinger, D. and Ben-Hamida, F.:* Turnover of protein in *Escherichia coli* starving for nitrogen. Biochim. biophys. Acta *119:* 171–182 (1966).

188 *Schmid, R.; Figen, J.F., and Schwartz, S.:* Experimental porphyria. IV. Studies on liver catalase and other heme enzymes in sedormid porphyria. J. biol. Chem. *217:* 263–274 (1955).

189 *Schoenheimer, R.:* The dynamic state of body constituents (Harvard University Press, Cambridge 1942).

190 *Segal, H.L.; Beattie, D.S., and Hopper, S.:* Purification and properties of liver glutamic-alanine transaminase from normal and corticoid-treated rats. J. biol. Chem. *237:* 1914–1920 (1962).

191 *Segal, H.L. and Kim, Y.S.:* Glucocorticoid stimulation of the biosynthesis of glutamic-alanine transaminase. Proc. nat. Acad. Sci., Wash. *50:* 912–918 (1963).

192 *Segal, H.L. and Kim, Y.S.:* Environmental control of enzyme synthesis and degradation. J. cell. comp. Physiol. *66:* (suppl. 1): 11–22 (1965).

193 *Segal, H.L.; Matsuzawa, T.; Haider, M., and Abraham, G.J.:* What determines the half-life of proteins *in vivo?* Some experiences with alanine aminotransferase of rat tissues. Biochem. biophys. Res. Comm. *36:* 764–770 (1969).

194 *Segal, H.L.; Rosso, R.G.; Hopper, S., and Weber, M.M.:* Direct evidence for an increase in enzyme level as the basis for the glucocorticoid-induced increase in glutamic-alanine transaminase activity in rat liver. J. biol. Chem. *237:* PC3303–PC3305 (1962).

195 *Seglen, P.O.:* Insulin inhibition of tyrosine transaminase degradation in the isolated, perfused rat liver. Z. physiol. Chem. *349:* 1229–1230 (1968).

196 *Seglen, P.O. and Jervell, K.F.:* A simple perfusion technique applied to glucocorticoid regulation of tryptophan oxygenase turnover and bile production in isolated rat liver. Z. physiol. Chem. *350:* 308–316 (1969).

197 *Sharma, C.; Manjeshwar, R., and Weinhouse, S.:* Effects of diet and insulin on glucose-adenosine triphosphate phosphotransferases of rat liver. J. biol. Chem. *238:* 3840–3845 (1963).

198 *Shrago, E.; Lardy, H.A.; Nordlie, R.C., and Foster, D.O.:* Metabolic and hormonal control of phosphoenolpyruvate carboxykinase and malic enzyme in rat liver. J. biol. Chem. *238:* 3188–3192 (1963).

199 *Simpson, M.V.:* The release of labelled amino acids from the proteins of rat liver slices. J. biol. Chem. *201:* 143–154 (1953).

200 *Simpson, M.V. and Velick, S.F.:* The synthesis of aldolase and glyceraldehyde-3-phosphate dehydrogenase in the rabbit. J. biol. Chem. *208:* 61–71 (1954).

201 *Snyder, S.H. and Epps, L.:* Regulation of histidine decarboxylase in rat stomach by gastrin: The effect of inhibitors of protein synthesis. Molec. Pharmacol. *4:* 187–195 (1968).

202 *Sols, A.; Salas, M., and Viñuela, E.:* Induced biosynthesis of liver glucokinase. Adv. Enzym. Regul. *2:* 177–188 (1964).

203 *Sols, A.; Sillero, A., and Salas, J.:* Insulin-dependent synthesis of glucokinase. J. cell. comp. Physiol. *66* (suppl. 1): 23–38 (1965).

204 *Stein, J.A.; Tschudy, D.P.; Corcoran, P.L., and Collins, A.:* δ-Aminolevulinic acid synthetase. III. Synergistic effect of chelated iron on induction. J. biol. Chem. *245:* 2213–2218 (1970).

205 *Steinberg, D. and Vaughan, M.:* Observations on intracellular protein catabolism studied *in vitro.* Arch. Biochem. Biophys. *65:* 93–105 (1956).

206 *Swick, R.W.:* Measurement of protein turnover in rat liver. J. biol. Chem. *231:* 751–764 (1958).

207 *Swick, R.W. and Peraino, C.:* Unpublished data.
208 *Swick, R.W.; Rexroth, A.K., and Stange, J.L.:* The metabolism of mitochondrial proteins. III. The dynamic state of rat liver mitochondria. J. biol. Chem. *243:* 3581–3587 (1968).
209 *Szabolcsi, G.; Boross, L., and Biszku, E.:* Secondary reactions following blocking of enzyme SH groups. Acta physiol. Acad. Sci. Hung *25:* 149–159 (1964).
210 *Tarentino, A.L.; Richert, D.A., and Westerfeld, W.W.:* The concurrent induction of hepatic α-glycerophosphate dehydrogenase by thyroid hormone. Biochim. biophys. Acta *124:* 295–309 (1966).
211 *Tarver, H.:* Peptide and protein synthesis. Protein turnover; in *Neurath and Bailey* The proteins, vol. II, part B, pp. 1199–1296 (Academic Press, New York 1954).
212 *Theorell, H.; Béznak, M.; Bonnischsen, R.; Paul, K.G., and Åkerson, A.:* On the distribution of injected radioactive iron in guinea pigs and its rate of appearance of some hemoproteins and ferritins. Acta chem. scand. *5:* 445–475 (1951).
213 *Thompson, E.B.; Tomkins, G.M., and Curran, J.F.:* Induction of tyrosine-α-ketoglutarate transaminase by steroid hormones in a newly established tissue culture cell line. Proc. nat. Acad. Sci., Wash. *56:* 296–303 (1966).
214 *Tomkins, G.M.; Garren, L.D.; Howell, R.R., and Peterkofsky, B.:* The regulation of enzyme synthesis by steroid hormones: The role of translation. J. cell. comp. Physiol. *66* (suppl. 1): 137–151 (1965).
215 *Tschudy, D.P.; Perlroth, M.G.; Marver, H.S.; Collins, A.; Hunter, G., Jr., and Rechcigl, M., Jr.:* Acute intermittent porphyria: The first 'overproduction disease' localized to a specific enzyme. Proc. nat. Acad. Sci., Wash. *53:* 841–846 (1965).
216 *Tschudy, D.P.; Rose, J.; Hellman, E.; Collins, A., and Rechcigl, M., Jr.:* Biochemical studies of experimental porphyria. Metabolism *11:* 1287–1301 (1962).
217 *Tschudy, D.P.; Welland, F.H.; Collins, A., and Hunter, G., Jr.:* The effect of carbohydrate feeding on the induction of δ-aminolevulinic acid synthetase. Metabolism *13:* 396–406 (1964).
218 *Tsukuda, K. and Lieberman, I.:* Liver nuclear ribonucleic acid polymerase formed after partial hepatectomy. J. biol. Chem. *240:* 1731–1736 (1965).
219 *Turner, M.K.; Abrams, R., and Lieberman, I.:* Levels of ribonucleotide reductase activity during the division cycle of the L cell. J. biol. Chem. *243:* 3725–3728 (1968).
220 *Unger, R.H. and Eisentrant, A.M.:* Studies of the physiologic role of glucagon. Diabetes *13:* 563–568 (1964).
221 *Urbá, R.C.:* Protein breakdown in *Bacillus cereus.* Biochem. J. *71:* 513–518 (1959).
222 *Velick, S.F.:* The metabolism of myosin, the meromyosins, actin and tropomyosins in the rabbit. Biochim. biophys. Acta *20:* 228–236 (1956).
223 *Viñuela, E.; Salas, M., and Sols, A.:* Glucokinase and hexokinase in liver in relation to glycogen synthesis. J. biol. Chem. *238:* PC1175–PC1177 (1963).
224 *Voit, C.:* Physiologie des allgemeinen Stoffwechsels und der Ernährung; in *Hermanns* Handbuch der Physiologie des Gesamtstoffwechsels und der Fortpflanzung, vol. VI, part 1, pp. 301 and ff. (F.V.W. Vogel, Leipzig 1881).
225 *Walker, D.G.:* On the presence of two soluble glucosephosphorylating enzymes in adult liver and the development of one of these after birth. Biochim. biophys. Acta *77:* 209–226 (1963).
226 *Walker, D.G. and Eaton, S.W.:* Regulation of development of hepatic glucokinase in the neonatal rat by the diet. Biochem. J. *105:* 771–777 (1967).
227 *Walker, D.G. and Holland, G.:* The development of hepatic glucokinase in the neonatal rat. Biochem. J. *97:* 845–854 (1965).
228 *Walker, D.G. and Rao, S.:* The role of glucokinase in the phosphorylation of glucose by rat liver. Biochem. J. *90:* 360–368 (1964).

229 *Walker, J.B.:* End-product repression in the creatine pathway of the developing chick embryo. Adv. Enzym. Regul. *1:* 151–168 (1963).

230 *Walter, H.:* Protein catabolism. Nature, Lond. *188:* 643–645 (1960).

231 *Waxman, A.D.; Collins, A., and Tschudy, D.P.:* Oscillations of hepatic δ-aminolevulinic acid synthetase produced *in vivo* by heme. Biochem. biophys. Res. Comm. *24:* 675–683 (1966).

232 *Weber, G. and Singhal, R.L.:* Insulin: Inducer of phosphofructokinase. Life Sci. *4:* 1993–2002 (1965).

233 *Weber, G.; Singhal, R.L., and Srivastava, S.K.:* Insulin: Suppressor of biosynthesis of hepatic gluconeogenic enzymes. Proc. nat. Acad. Sci., Wash. *53:* 96–104 (1965).

234 *Weber, G.; Stamm, N.B., and Fisher, E.A.:* Insulin: Inducer of pyruvate kinase. Science *149:* 65–67 (1965).

235 *Wicks, W.D.; Kenney, F.T., and Lee, K.-L.:* Induction of hepatic enzyme synthesis *in vivo* by adenosine 3',5'-monophosphate. J. biol. Chem. *244:* 6008–6013 (1969).

236 *Willetts, N.S.:* Intracellular protein breakdown in nongrowing cells of *Escherichia coli.* Biochem. J. *103:* 453–461 (1967).

237 *Willetts, N.S.:* Intracellular protein breakdown in growing cells of *Escherichia coli.* Biochem. J. *103:* 462–466 (1967).

238 *Yagil, G. and Feldman, M.:* The stability of some enzymes in cultured cells. Exp. Cell Res. *54:* 29–36 (1969).

239 *Zito, R.; Antonini, E., and Wyman, J.:* The effect of oxygenation on the rate of digestion of human hemoglobins by carboxypeptidase. J. biol. Chem. *239:* 1804–1808 (1964).

Author's address: *Miloslav Rechcigl,* Jr., Ph.D., Research and Institutional Grants Division, Agency for International Development, US Department of State, *Washington, DC 20523* (USA)

III. Special Topics

Enzyme Synthesis and Degradation in Mammalian Systems, pp. 311–338
(Karger, Basel 1971)

Rhythmic Changes in Enzyme Activity and their Control

R. W. Fuller

Department of Metabolic Research, The Lilly Research Laboratories, Eli Lilly and
Company, Indianapolis, Ind.

Contents

I. Introduction

Somewhat contrary to a rigid concept of homeostasis is the fact that some
biological constituents, including enzymes, are not held at a constant level but
instead vary, sometimes over a fairly large magnitude, in a rhythmic manner. But
perhaps it is the process of homeostasis in a truer sense that accounts for the
rhythms. Such rhythmic variations may be of different durations. On this planet,
one of our most prominent environmental changes is the alternation of light and
dark that occurs each 24 h. Besides light, such things as temperature and
humidity also vary. Thus, a predominant rhythm in earth creatures is the
24-hour rhythm, sometimes called a diurnal rhythm or simply a daily rhythm.
Another term that has been applied to 24-hour rhythms is circadian (Latin

'circa', about and 'dies', a day). There have been suggestions that the term circadian be reserved for special kinds of rhythms, those that run freely in the absence of environmental changes as exogenous synchronizers (*Wurtman,* 1967).

Daily rhythmic changes in certain metabolic events, such as the rhythm in liver glycogen in animals and in man, have been recognized for many years (*Sollberger,* 1964). Other metabolic rhythms, notably those in enzymes, have been demonstrated only more recently. This chapter will focus on enzyme rhythms of about 24 h, and will be concerned particularly with the daily rhythm in hepatic tyrosine aminotransferase. At least some of the considerations with respect to mechanism for that enzyme rhythm will be applicable to rhythms in other enzymes.

In considering how rhythmic changes in enzymes may be produced, it is essential to review the kinds of regulation that can be imposed on enzymes. First, the *activity* of an enzyme may be regulated; it can be increased (activation or stimulation) or decreased (inhibition). In most of the examples to be discussed, there are presumably changes in the *amount* of enzyme, although that point has not been proven in all cases. The amount of enzyme present is governed by the relative rates of its synthesis and of its degradation. The rate of enzyme synthesis can be increased (induction) or decreased (repression). Regulation of enzyme synthesis can be imposed at the level of transcription (formation of messenger RNA from the DNA template of the gene) or translation (protein synthesis from the messenger RNA template).

A thorough understanding of the control responsible for rhythmic changes of enzyme activity comes only when each of these mechanisms can be specified. Such understanding is one objective of the study of enzyme rhythm control – to establish if the amount of enzyme is changing, if the change is due to altered synthesis or degradation and is brought about at the level of transcription or translation, and finally to identify the agent(s) responsible for the rhythmic change. Another objective is to learn the consequences of rhythmic enzyme changes.

II. Rhythms in Enzymes that Metabolize Aromatic Amino Acids

A. L-Tyrosine: 2-oxoglutarate Aminotransferase (E.C.2.6.1.5)

1. Discovery and Characteristics of the Rhythm

The first report of daily rhythmic changes in tyrosine aminotransferase (TAT) in rat liver seems to have been by *Boctor et al.* (1966). In an abstract concerned with the toxic effects of excess dietary tyrosine, they mentioned that diurnal changes in TAT activity were found and that rats fed a control diet (apparently *ad libitum*) had peak enzyme activity at 9 p.m.

Potter et al. (1966b) reported variations in TAT in host liver of hepatoma-bearing rats. The rats were fed on a restricted schedule from 6 p.m. to 6 a.m. (during darkness), and TAT in the host liver was found to be higher at midnight than at 6 a.m., noon, or 6 p.m. The authors mentioned unpublished studies showing a similar rhythm in liver of normal rats, presumably fed on the same schedule.

Potter et al. (1966a) also found marked fluctuation of TAT levels in the liver of rats fed 12 h during 48-hour periods. In rats fed *ad libitum,* there were smaller changes, and from the data published, it appears that the differences may not have been statistically significant. An abstract by *Watanabe and Baril* (1967) stated that TAT varied rhythmically in the liver of rats fed for 8 or 12 h in each 24-hour period. *Potter et al.* in 1967 reported that TAT in the liver of rats fed *ad libitum* varied rhythmically, with an increase during the dark period, 'presumably because (the rats) eat during the dark period'.

Wurtman and Axelrod in 1967 published data showing that an approximate four-fold rhythm of hepatic TAT occurred in rats fed *ad libitum.* Their data agree well with those reported that year by *Civen et al.* (1967b) and by *Shambaugh et al.* (1967). Subsequent to these reports, results from a number of studies aimed at elucidating the mechanism of the daily rhythm in hepatic TAT have been published.

Most of the initial reports of rhythms in hepatic TAT activity were in rats fed *ad libitum,* when most food was presumably consumed during the dark period. Although the lighting conditions varied slightly among the various laboratories from which the reports came, the dark period generally began about 5–8 p.m. In one case, food was provided only during the dark period (*Potter et al.,* 1966b).

Fuller and Snoddy (1968) showed that the pattern of food intake profoundly influenced the phase of the hepatic TAT rhythm. Rats trained to eat a single daily meal (8 a.m. until noon) had a TAT rhythm much like that of rats fed *ad libitum* (presumably eating mainly at night) except that the rhythms were out of phase by 12 h. Highest TAT activity in the liver of rats fed *ad libitum* was at 11 p.m., whereas highest TAT activity in the rats fed a single daily meal was at 11 a.m. Figure 1 shows results of another experiment in which two groups of rats were allowed access to food only during the light period or only during the dark period. Lights were on from 7 a.m. until 7 p.m. The TAT rhythm was similar, but opposite in phase, in the 2 groups of rats. The phase of the rhythm thus is determined by the pattern of food intake. Light-dark cycles are not of primary importance, except that the light-dark cycle usually sets the feeding patterns of rats that are allowed free access to food.

Black and Axelrod (1968b) found that the phase of the TAT rhythm could be reversed by reversing the light-dark cycles. That effect was probably mediated by a shift in feeding habits, since the rats were fed *ad libitum.* The conclusion

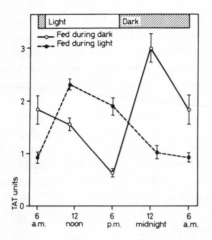

Fig. 1. Hepatic TAT rhythm in rats fed during light or dark periods only. Male Wistar rats weighing about 150 g each were kept in a room with lights on from 7 a.m. to 7 p.m. For 2 weeks before the rats were killed, food was available for only 12 h of each day as shown; water was available at all times. Hepatic TAT activity in 105,000 x g supernatant fractions was assayed spectrophotometrically (*Lin et al.*, 1958) in this experiment and in those shown in subsequent figures. A unit of enzyme activity is that amount producing 1 μmole of p-hydroxyphenylpyruvate/min/g of liver at 37°. Means and SEM for 5 rats per group are shown.

that 'environmental lighting is capable of entraining the enzyme rhythm' is true probably in an indirect sense, i.e. the lighting cycle entrains the feeding cycle which in turn entrains the enzyme rhythm. The situation with TAT seems to be quite different from that in the case of enzyme rhythms in the pineal. *Wurtman et al.* (1968a) have reviewed the data showing that the hydroxyindole O-methyl transferase rhythm in pineal is totally dependent upon environmental light.

Honova et al. (1968) studied the development of the TAT rhythm in the liver of neonatal rats. They found a recognizable rhythm within 48 h after birth. Enzyme activity in the suckling rats was higher during the light period (10 a.m.) than during the dark period (10 p.m.). After the rats were weaned (21–23 days), the phase of the rhythm was shifted so that the enzyme was higher in the dark period as is characteristic of the adult rat. The results most likely represent another demonstration that the enzyme rhythm is related to the time of food intake. Presumably the suckling rats consumed more food during the daylight hours when the mother was less active. After weaning, the tendency of rats to eat at night resulted in the shift in enzyme rhythm.

Still further indication of an association between food intake and the TAT rhythm comes from some work, not yet published in detail, by *Cohn and Wurtman* (*Zigmond et al.*, 1969). They studied rats trained to eat a fixed amount of food each hour throughout the 24-hour day. These rats without rhythmic food intake had no TAT rhythm. It seems that the hepatic TAT rhythm in rats with free access to food is generated by the rhythmic consumption of food.

The tendency for the laboratory rat to consume most of its food at night is well known. *Suttie* (1968), with an elaborate system for measuring consumption of food available continuously to rats, found that greater than 90 % of the food consumption was during the dark period of a 12-hour light-dark cycle. Most other workers have reported similar results (*Suttie*, 1968; *Zigmond et al.*, 1969). In one case, however, rats were reported to have eaten almost as many meals in daylight as in the dark (*LeMagnen and Tallon*, 1966).

The governing influence of feeding pattern on the TAT rhythm and the apparent variation in feeding habits of rats in different laboratories make it desirable to have information on the pattern of food intake when food is freely accessible. Otherwise results may be misinterpreted. Even subtle differences in the *pattern* of food consumption can cause differences in some enzyme levels between groups of rats consuming the same *amount* of food (*Suttie*, 1968).

2. Mechanism of the Rhythm

Although the experiments cited so far revealed that the TAT rhythm was related to food intake, they provided little information on the nature of that relationship. Apparently some biochemical changes that occur in the animal as a response to the rhythmic intake of food cause a signal to be received by the liver that results in an increase of TAT levels. Several questions are immediately raised. What are those biochemical changes that generate the signal, what is the signal, and by what pathway does it reach the liver? These questions are not yet satisfactorily answered, but some possibilities are discussed below.

a) Hormonal Control

The most obvious mechanism to be considered when the TAT rhythm was recognized was hormonal control, for many hormones were known to influence hepatic TAT levels.

Glucocorticoids were shown to elevate TAT in rat liver by *Lin et al.* in 1958. *Kenney* (1962) showed by immunochemical analysis that an increased amount of enzyme protein was present. The induction by glucocorticoids was apparently produced at the transcriptional level, inasmuch as blockade of RNA synthesis with actinomycin D prevented the rise in TAT (*Csányi et al.*, 1967). The levels of corticosterone in the plasma of rats vary diurnally and are highest a few hours before the peak TAT levels in rats fed *ad libitum* (*Civen et al.*, 1967b; *Wurtman and Axelrod*, 1967). A logical consideration was that corticosterone generated the TAT rhythm. However, *Civen et al.* (1967b) showed that the TAT rhythm persisted essentially unaltered in adrenalectomized rats. Others have found only slight changes in the TAT rhythm in adrenalectomized rats (*Sham-*

baugh et al., 1967; *Wurtman and Axelrod,* 1967). The presence of the TAT rhythm in the absence of the adrenals indicates that the glucocorticoid rhythm is not essential to the enzyme rhythm.

The adrenomedullary hormone, epinephrine, can also induce hepatic TAT in rats (*Fuller and Snoddy,* in press; *Reshef and Greengard,* 1969). Epinephrine probably exerts a direct effect on the liver, for it increases the enzyme in cultures of fetal rat liver as well (*Wicks,* 1968c). But, because the hepatic TAT rhythm in rats persists in the absence of the adrenal glands, rhythmic epinephrine secretion must not produce the TAT rhythm.

Growth hormone can alter hepatic TAT levels. *Harding and Rosen* (1963) found that growth hormone lowered TAT activity and inhibited its induction by glucocorticoids in the liver of hypophysectomized rats. *Kenney* (1967) showed that growth hormone repressed TAT in liver of hypophysectomized or adrenalectomized rats. The repression was blocked by actinomycin. An action of growth hormone does not, however, appear to cause the daily TAT rhythm, since hypophysectomy does not abolish the enzyme rhythm (*Shambaugh et al.,* 1967; *Wurtman and Axelrod,* 1967).

Pancreatic hormones (both glucagon and insulin) induce TAT. *Greengard and Baker* (1966) first reported that glucagon induced TAT and potentiated the induction by hydrocortisone in adrenalectomized rats, and others have also studied TAT induction by glucagon (*Brown and Civen,* 1969; *Civen et al.,* 1967a; *Labrie and Korner,* 1969). *Holten and Kenney* (1967) showed by immunochemical analysis that glucagon stimulated TAT synthesis and they, as well as *Csányi and Greengard* (1968) found that the induction was blocked by actinomycin D. Insulin induces TAT in a similar way (*Brown and Civen,* 1969; *Holten and Kenney,* 1967; *Labrie and Korner,* 1969; *Reshef and Greengard,* 1969; *Staib et al.,* 1969). Although one of these pancreatic hormones, when administered to animals, can cause release of the other, the induction of TAT seems to be a direct effect of each hormone. Both glucagon and insulin can induce TAT in the isolated, perfused liver (*Hager and Kenney,* 1968; *Knox and Sharma,* 1968; *Staib et al.,* 1969) and in cultures of rat liver (*Wicks,* 1968a). Rhythmic secretion of the pancreatic hormones in response to periodic ingestion of food might be expected. The hepatic TAT rhythm persists in pancreatectomized rats (*Fuller et al.,* 1969), however, apparently ruling out the possibility that pancreatic hormones generate the TAT rhythm.

From all presently available evidence, it appears that control by any single hormone does not account for the TAT rhythm.

b)   Neural Control

The possibility of a neural involvement in the TAT rhythm was considered by *Wurtman and Axelrod* (1967). They pointed out that the amount of the sympathetic neurotransmitter, norepinephrine, varied rhythmically in pineal, salivary glands, and urine and considered that rhythmic variation in sympathetic nervous system activity might be related to the TAT rhythm.

Fuller and Slater (1968) proposed a neural mechanism in the TAT rhythm with a central nervous system component on the basis of experiments in mice made obese by gold thioglucose treatment. Gold thioglucose, when injected into mice, localizes in the ventromedial nucleus of the hypothalamus and produces irreversible lesions there (*Deter and Liebelt*, 1964). Mice so affected become hyperphagic, then obese. TAT in the liver of such mice did not vary significantly during a twenty-four period as it did in untreated mice (figure 2). The differences between the 2 groups of mice were statistically significant (P < 0.05) at 5 and 11 a.m. and at 2, 8, and 11 p.m., when the peak occurred in normal mice but not in gold thioglucose-treated mice. Thus some factors determining the daily variation of the enzyme do not seem to be functional in the gold thioglucose-treated mice. Several possible explanations were considered. The lack of a normal TAT rhythm in the gold thioglucose-treated mice might be due to (1) the gross obesity, (2) lack of normal protein synthesis in the liver, i.e. lack of response to a signal calling for increased TAT synthesis, (3) an altered feeding pattern due to hyperphagia, possibly resulting in the mice eating at a nearly constant rate throughout the 24-hour period, or (4) a hypothalamic control as

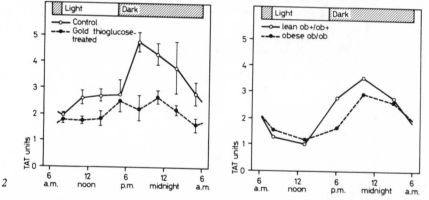

Fig. 2. Hepatic TAT rhythm in mice fed *ad libitum*. Male albino mice were given gold thioglucose (700 mg/kg, i.p.) when they were 6 weeks old. Those animals that developed symptoms indicative of hypothalamic damage — hyperphagia and obesity — were selected and used in studies described in this chapter 3 to 8 months after treatment. The control mice were of the same age; body weights were 39.9 ± 0.7 g for all controls and 66.2 ± 1.3 g for all treated mice in this experiment. Means and SEM for 5 mice per group are shown.

Fig. 3. Hepatic TAT rhythm in mice fed *ad libitum*. Male C57BL/6J mice were used; the ob+/ob+ mice had an average body weight of 24.0 ± 1.0 g, whereas the ob/ob mice has an average body weight of 38.7 ± 1.9 g. Mean values of groups of two mice are shown at each time.

the mechanism for the TAT rhythm and the impairment of the control as a result of the hypothalamic damage.

Evidence against the obesity itself accounting for the lack of TAT rhythm was obtained when genetic obese mice were shown to have an essentially normal rhythm in hepatic TAT (figure 3).

A normal increase in TAT in the liver of gold thioglucose-treated mice was found to occur in response to a hormonal stimulus. Zinc glucagon (3.3 mg/kg) injected i.p. caused a 6.8-fold increase in hepatic TAT of control mice and an 8.3-fold increase in gold thioglucose-treated mice. Gold thioglucose-treated mice were equally responsive at 11 a.m. and at 11 p.m. Hydrocortisone (25 mg/kg, i.p.) increased hepatic TAT 2.6-fold in control mice and 3.4-fold in gold thioglucose-treated mice. Thus, the livers of gold thioglucose-treated mice were capable of responding in these conditions with an increased TAT level when stimulated to do so. Interestingly, hepatic TAT did not increase in gold thioglucose-treated mice fasted for 48 h, although it did increase in control mice.

Gold thioglucose treatment initially abolishes the feeding rhythm in mice, but the rhythm returns after the mice have become obese (*Anliker and Mayer,* 1955; *Wiepkema et al.,* 1966). One explanation for the lack of TAT rhythm (figure 2) could be the lack of a feeding rhythm, although these mice had entered the obese phase when the feeding rhythm should be normal. Furthermore, the presence of a liver glycogen rhythm in the gold thioglucose-treated mice

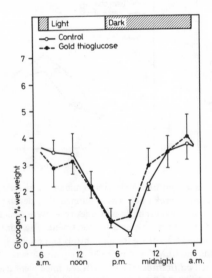

Fig. 4. Hepatic glycogen rhythm in mice fed *ad libitum.* Glycogen in liver samples from the mice in figure 2 was extracted, precipitated and hydrolyzed; the glucose so formed was determined colorimetrically with glucose oxidase.

identical to that in untreated control mice (figure 4) was considered as evidence that the feeding rythm was unchanged. This view rested on the assumption that the glycogen rhythm was simply a manifestation of rhythmic food intake.

Seemingly the most plausible explanation remaining was that the ventromedial portion of the hypothalamus destroyed by gold thioglucose controlled in some way the hepatic TAT rhythm. Consideration of neural control of enzyme activity has precedence in the results reported by *Shimazu* and his collaborators (1965, 1966, 1967, 1968a, 1968b).

They initially found that electrical stimulation of the ventromedial hypothalamus (a sympathetic region) led to increased blood glucose and decreased liver glycogen in rabbits, whereas stimulation in the lateral hypothalamus (a parasympathetic area) caused a decrease in blood glucose (*Shimazu et al.,* 1966). The changes in blood glucose were also observed in adrenalectomized, in thyroidectomized, and in hypophysectomized rabbits. In further experiments, *Shimazu* (1967) reported that electrical stimulation of the vagus nerve in intact or pancreatectomized rabbits resulted in a rapid (within 5 min) increase in the activity of glycogen synthetase in liver. The effect was completely counteracted by simultaneous stimulation of the splanchnic nerve.

Shimazu and Amakawa (1968b) showed that stimulation of the splanchnic nerve (sympathetic) caused a rapid increase in the activity of hepatic glucose-6-phosphatase and glycogen phosphorylase in rabbits (*Shimazu and Fukuda,* 1965). They proposed a direct neural control of phosphorylase activity in addition to a slower hormonal control (*Shimazu and Amakawa,* 1968a and 1968b). This work lends credence to the idea of neural regulation of enzyme activity.

If the lack of hepatic TAT rhythm in gold thioglucose-treated mice were indeed not due to the lack of a feeding rhythm in the mice, then a TAT rhythm ought not to appear even if a feeding rhythm were imposed. We therefore gave single daily meals to control and to gold thioglucose-treated mice in the same way that we had earlier done with rats (*Fuller and Snoddy,* 1968). The data in figure 5 show that a rhythm of hepatic TAT in gold thioglucose-treated mice was generated by imposing the feeding rhythm, although the rhythm was still less than in control mice. Results from glycogen analysis revealed that the glycogen rhythm (normal in gold thioglucose-treated mice fed *ad libitum*) was nearly completely suppressed in gold thioglucose-treated mice fed the single daily meal (figure 6). That finding makes it doubtful that the normal glycogen rhythm in the mice fed *ad libitum* should be considered as evidence supporting a normal feeding rhythm. Hence, the interpretation of the lack of TAT rhythm in mice fed *ad libitum* after gold thioglucose treatment becomes open to qestion. The results in figure 5 show that some rhythmic changes in hepatic TAT can occur in mice with hypothalamic damage.

However, other results consistent with neural regulation of TAT have been reported by *Black and Axelrod* (1968a). They showed that depletion of brain

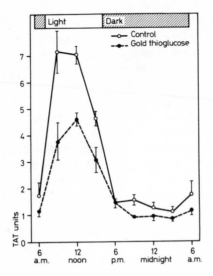

Fig. 5. Hepatic TAT rhythm in mice fed a single daily meal. Male albino mice treated with gold thioglucose as described in figure 2 were used. They were given free access to drinking water but were provided food only from 8 a.m. until noon for the two weeks before they were killed. The control mice weighed 37.6 ± 0.3 g each; treated mice weighed 47.8 ± 0.5 g each. Means and SEM for 3 mice per control group or 5 mice per treated group are shown.

norepinephrine by reserpine or a-methyltyrosine was associated with increased hepatic TAT levels in adrenalectomized rats. Repletion of norepinephrine in the a-methyltyrosine-treated rats by administration of dihydroxyphenylalanine reversed the elevation of TAT. *Black and Axelrod* called attention to the daily change in brain stem norepinephrine (*Manshardt and Wurtman,* 1968; *Reis and Wurtman,* 1968; *Walker and Friedman,* 1967) and suggested that the TAT rhythm might 'be dependent upon such daily variations in the activity of central adrenergic neural pathways'. *Margules* (1969) has suggested that noradrenergic synapses are involved in the regulation of food intake by cells in the ventromedial hypothalamus.

Westermann (1963) had previously reported that hepatic TAT was elevated by reserpine in intact rats, in which case plasma corticoid levels were also increased. However, the induction of the enzyme lasted much longer than the elevated corticoids. At 24 h, when plasma corticosterone levels had been back to normal for 8 h, hepatic TAT was still markedly increased.

In apparent contradiction to the involvement of norepinephrine in the TAT rhythm, though, is the finding of *Wurtman et al.* (1968c) that reserpine did not block the daily TAT rhythm. Their treatment led to 77 % reduction of brain

catecholamine, whereas the treatment used by *Black and Axelrod* (1968a) caused a slightly greater reduction.

Black and Axelrod (1969) have further shown that a ganglionic blocking
agent, chlorisondamine, suppressed the daily rhythm in TAT at high enzyme
levels. On the other hand, treatments that maintained tissue norepinephrine at
high levels suppressed the daily rhythm in TAT at low enzyme levels (*Axelrod
and Black,* 1968). *Govier et al.* (1969) did not find changes in hepatic TAT in
rats treated with other depleters and blockers of norepinephrine.

Thus, although it seems possible that a direct or indirect neural control
drives the rhythm in hepatic TAT, evidence to support that possibility is at the
present mainly circumstantial.

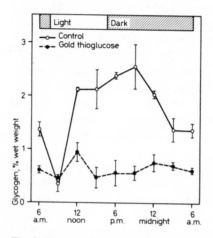

Fig. 6. Hepatic glycogen rhythm in mice fed a single daily meal. Glycogen in liver
samples from the mice in figure 5 was extracted, precipitated and hydrolyzed; the glucose so
formed was determined colorimetrically with glucose oxidase.

c)   Metabolite Level Control

Among the agents that induce TAT are mixtures of amino acids (*Korner and
Labrie,* 1967; *Reshef and Greengard,* 1969; *Rosen and Milholland,* 1968) and
specific amino acids such as tyrosine (*Knox et al.,* 1964) and tryptophan
(*Korner and Labrie,* 1967; *Kroger et al.,* 1968).

Potter et al. (1966b) showed that the height of the midnight peak in host
liver TAT of hepatoma-bearing rats was directly dependent upon the protein
content of the diet. Little or no rhythm occurred when the rats were fed a
protein-free diet.

Wurtman et al. (1968b, 1968d) considered the possible role of two amino acids, tyrosine and tryptophan, in generating the TAT rhythm. They found that the rise in hepatic TAT between 5 p.m. and 8 p.m. was associated with a fall in hepatic tyrosine levels, but that hepatic tryptophan increased several hours before the TAT rise. Almost no TAT rhythm was seen if the diet contained no protein, but if a mixture of amino acids was substituted for protein, the TAT rhythm occurred provided the mixture contained tryptophan. If diets containing no protein and supplemented only with $1/2$% tryptophan or 1% tyrosine were fed, no TAT rhythm was seen, but rats fed a protein-free diet with 6% tryptophan added did have a TAT rhythm. On the basis of these results, *Wurtman et al.* (1968b) suggested that 'dietary tryptophan alone may initiate the processes responsible for the afternoon rise in tyrosine transaminase activity'.

Indeed, it seems to be true that there is 'a special relationship between tryptophan and the transaminase rhythm' (*Wurtman et al.*, 1968b). Such a relationship may exist because tryptophan has a unique action on the aggregation of hepatic ribosomes in the rat. *Munro*'s group (1968), *Wunner et al.* (1966) and *Sidransky et al.* (1967) found that injection of tryptophan into rats led to a shift in polyribosomes from lighter to heavier aggregates, thereby enhancing protein synthesis. Injection of L-tryptophan also increased hepatic TAT levels. Different mechanisms seem to be involved in these two effects of tryptophan, for although actinomycin D blocked the induction of TAT by tryptophan (*Labrie and Korner,* 1968), it did not block tryptophan's effect on polysome profiles (*Munro,* 1968).

All of the currently available data are consistent with the idea that the hepatic TAT rhythm cannot occur without adequate dietary tryptophan. The data do not seem to prove that tryptophan generates the TAT rhythm; tryptophan availability may be necessary but not sufficient.

Although the data showing that hepatic tryptophan in rats fed *ad libitum* increased prior to the rise in TAT (*Wurtman et al.,* 1968b; *Fuller,* 1970) might suggest that tryptophan caused the rhythmic increase in TAT, the association may be fortuitous. In rats fed a single meal during the day, the TAT rhythm was shifted by 12 h, whereas the hepatic tryptophan rhythm was changed only slightly (*Fuller,* 1970). Apparently the rhythm in hepatic tryptophan is dependent primarily upon the light-dark cycle and may be caused by rhythmic changes in amino acid uptake into tissues (*Baril and Potter,* 1968). In any case, there was no direct relationship between hepatic tryptophan content and TAT in rats fed a single daily meal. It is doubtful that the TAT rhythm, at least in those rats, was generated by a rise in liver tryptophan.

Other data also are not consistent with the TAT rise being caused directly by the ingestion of tryptophan or other dietary constituents. In rats fed only eight hours during a 48-hour period (*Potter et al.,* 1968; *Watanabe et al.,* 1968), there was a 'secondary rise' in TAT during the fasting period at about the same time in the light-dark cycle as the rise had occurred when the rats were fed. The peak TAT activity, even during the fasting period, was directly related to the protein content of the diet. Clearly, the TAT rhythm on the day the rats were not fed was not generated by absorption of a metabolite from the diet. Finally, our data (figure 7) show that hepatic TAT in mice and rats fed a single daily meal increased at the time the meal was normally fed even though food was not presented.

To summarize the discussion of metabolite level control of the TAT rhythm, tryptophan may be a metabolite that exerts a regulatory effect on TAT, but not all data support the contention that the hepatic TAT rhythm is primarily a function of tryptophan levels in the liver. Perhaps some metabolite other than tryptophan generates the daily rhythm. Such a metabolite might or might not be dietary in origin, though it must be related in some way to food intake.

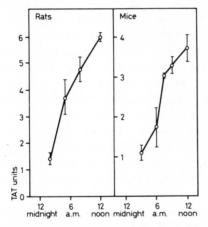

Fig. 7. Increase of hepatic TAT prior to ingestion of food in rats and mice fed a single daily meal. Rats and mice were allowed access to food only from 8 a.m. until noon for 2 weeks prior to the experimental day, at which time they were not given food. Water was freely available at all times. Means and SEM for 5 animals per group are shown.

d) Location of the Control

No direct experiments have yet been reported that would locate the control of the TAT rhythm exclusively at the level of transcription or of translation. Studies with inhibitors such as actinomycin may be helpful, though the use of such agents must be with caution, for disruption of the rhythm of food consumption could lead to spurious changes in the TAT rhythm. Establishing the location of the control may be difficult until the signal or signals responsible for the TAT rhythm have been unequivocally identified.

Most of the discussion has implied that a single signal produces the hepatic TAT rhythm. Perhaps that is not so. The physiological control of hepatic TAT may be complex, and several factors rather than a single signal may impinge with the result being a daily rhythm of enzyme activity. Elimination of a single factor might not abolish the rhythm, yet that factor may have a distinct physiological influence on TAT.

Recently, *Szepesi and Freedland* (1968) have presented evidence for both translational and transcriptional control of enzyme synthesis in rat liver as a result of dietary alterations. They studied pyruvate kinase. When the diet contained no protein, the limiting factor controlling synthesis of that enzyme was at the level of translation, whereas on a high protein diet the limiting factor involved transcription. It is possible that a similar situation exists with TAT. When the diet contains protein, the initiation of the daily rhythm might be at the level of transcription – such initiation might involve hormonal, neural, or metabolite level regulation. If, when the diet contains no protein (or perhaps even no tryptophan), translation limited TAT synthesis, the daily rhythm might not be manifest even if rhythmic changes in transcriptional control persisted.

The question of whether transcriptional or translational control accounts for the TAT rhythm rests in part on the question of stability of the messenger RNA template for TAT. If the amount of template remains constant throughout a 24-hour period, translational control probably accounts for the TAT rhythm. *Kenney et al.* (1968) have proposed that the messenger RNA coding for TAT is labile, i.e. has a short half-life, in rat liver. *Pitot et al.* (1965) proposed a 3-hour lifetime for the messenger RNA template for TAT in rat liver. If there is a transcriptional basis for the TAT rhythm, the inference would be that a rhythm of the messenger RNA coding for TAT would occur. *Lang et al.* (1968) have reported that a messenger RNA fraction from the liver of cortisol-treated rats stimulated the synthesis of TAT in an *in vitro* system. Such techniques might permit one to determine if daily variation of the messenger RNA occurs in liver.

Another question concerning the TAT rhythm can be raised, namely if there is any 'second messenger' interposed between the primary signal and the expression of the enzyme rhythm. An obvious candidate is adenosine 3',5'-monophosphate (cyclic AMP), which is thought to function as a second messenger in a variety of regulatory systems. Some of the agents that induce TAT are known to elevate cyclic AMP levels. Such agents include hormones (glucocorticoids, glucagon, epinephrine) and theophylline (*Fuller and Snoddy,* in press), which inhibits cyclic AMP hydrolysis. *Wicks* (1968b) has shown that cyclic AMP and its dibutyryl derivative both induce TAT in cultures of fetal rat liver. He has suggested that the induction by glucagon and by epinephrine is mediated by cyclic AMP and differs in mechanism from the induction by hydrocortisone. There seems to be no information in the literature on whether a daily rhythm of cyclic AMP occurs in liver.

3. Significance of the Rhythm

The consequences of the rhythm in hepatic TAT remain to be established, though some data suggest that the amount of tyrosine transaminated does indeed vary diurnally. *Wurtman et al.* (1967, 1968b) found that plasma and liver tyrosine levels fell after the normal daily rise in TAT in rats fed *ad libitum*. There is a daily rhythm in plasma tyrosine concentration in rats (*Coburn et al.*, 1968), and the liver enzyme rhythm may be at least partly responsible for the rhythm in its circulating substrate. Some evidence consistent with this possibility has been published. *Coburn et al.* (1968) found, in rats on a restricted feeding schedule, that the period during which hepatic TAT increased (2 a.m.–8 a.m.) was one in which plasma tyrosine levels decreased. Their results also showed that ingestion of food could significantly influence plasma tyrosine levels, at least in rats fed only a single daily meal.

It is likely that multiple factors account for the plasma tyrosine rhythm and that hepatic TAT variation is just one of them. The levels of other amino acids besides tyrosine vary diurnally (*Feigin et al.*, 1968). *Baril et al.* (1968) showed diurnal variation in tissue/plasma ratios of a non-metabolizable amino acid, cycloleucine. So there seems to be rhythmic variation of amino acid uptake into tissues. Hormonal influences may be important here, for the levels of some hormones – glucocorticoids, glucagon, and insulin – known to affect amino acid uptake into tissues (*Chambers et al.*, 1965) may vary diurnally (*Civen et al.*, 1967b; *Freinkel et al.*, 1968; *Hellman and Hellerstrom*, 1959).

There is a daily rhythm in plasma tyrosine in humans (*Wurtman et al.*, 1967). A hepatic TAT rhythm has not been measured directly in humans, though some published results suggest the amount of tyrosine transaminated may vary diurnally. *Seegmiller et al.* (1961) studied plasma levels of homogentisic acid in patients with alcaptonuria, in whom further metabolism of homogentisic acid does not occur because of the genetic lack of homogentisic acid oxidase. Homogentisic acid is formed from tyrosine by the combined action of TAT and p-hydroxy-phenylpyruvic acid oxidase. In normal subjects, homogentisic acid levels in plasma were too low to be detected. But, in alcaptonuric patients, homogentisic acid was present in plasma, and the amount varied diurnally. Such variation may have been due to variations in hepatic TAT.

In spite of this suggestive evidence, one cannot say with certainty that there is more tyrosine actually transaminated when TAT activity is high. *Kim and Miller* (1969) have shown that 7 to 14-fold increases in TAT induced by hydrocortisone injection into adrenalectomized rats did not substantially increase the amount of radioactive tyrosine converted to radioactive carbon dioxide. Perhaps other factors, such as amino acid transport into liver, have greater influence on the amount of tyrosine transaminated than does the actual TAT level.

In considering the biological significance of the TAT rhythm, one must ask: what is the physiological function of TAT? Is it to detoxify tyrosine by conversion to a more easily excretable metabolite? Or is its function to contribute to gluconeogenesis by starting tyrosine along a pathway that can feed into the gluconeogenic pathway? It is probably not possible at present to distinguish between these alternatives, and perhaps TAT serves both purposes.

B.   L-Tryptophan: Oxygen Oxidoreductase (E.C.1.13.1.12)

Tryptophan oxygenase (TPO), like TAT, is an enzyme in liver that metabolizes an aromatic amino acid and is induced by glucocorticoid hormones.

Rapoport et al. (1966) reported that there was a daily rhythm in hepatic TPO in mice fed *ad libitum* with lights on from 6 a.m. to 6 p.m. The peak in enzyme activity was soon after the onset of darkness. *Hardeland and Resning* (1968) obtained similar results in rats and concluded that activity increased by *de novo* synthesis of enzyme throughout the light period (9 a.m. to 6 p.m.).

The possible control mechanisms (hormonal, neural, metabolite) considered for TAT can also be considered for the TPO rhythm. Glucocorticoids induce TPO (*Knox and Auerbach,* 1955). *Rapoport et al.* (1966) found that adrenalectomized mice 'had a lower, less sharpened, and reversed rhythmicity' of TPO. They proposed that adrenal steroids played at least a permissive role in maintain-

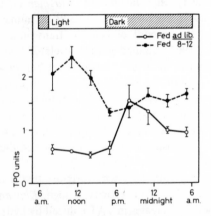

Fig. 8. Hepatic TPO rhythm in rats fed *ad libitum* or a single daily meal. Male Wistar rats weighing about 150 g each had access to food continually or only from 8 a.m. until noon for 2 weeks before the experiment. TPO, in the 105,000 x g supernatant fraction, was measured by the method of *Knox and Auerbach* (1955). A unit of enzyme activity is that amount oxidizing 1 μmole of tryptophan/min/g of liver. Means and SEM for 5 rats per group are shown.

ing rhythmicity of TPO and suggested that TPO might represent only part of an overall rhythmic pattern in hepatic protein synthesis. The TPO rhythm is related to food intake in much the same way as the TAT rhythm (figure 8). In rats fed a single daily meal, the phase of the TPO rhythm was shifted by 12 h. In addition, there was a general increase in TPO of such meal-fed rats compared to rats fed *ad libitum*. There was no appreciable shift of plasma corticosterone rhythm in the rats fed the single daily meal (*Fuller,* 1970), so that in those rats the rhythm in TPO seemed to be dissociated from the plasma corticoid rhythm. Such a dissociation would be consistent with the suggestion by *Rapoport et al.* (1966) of a permissive role of glucocorticoids — so long as adequate glucocorticoid levels are maintained, the TPO rhythm need not parallel the corticosterone rhythm.

Studies on the precise mechanism by which the TPO rhythm is related to the feeding rhythm are fewer than with TAT. There is a rhythm in tryptophan in plasma (*Rapoport et al.,* 1966) and in liver (*Fuller,* 1970), and tryptophan can lead to increased TPO activity (*Labrie and Korner,* 1968). The TPO rhythm apparently does not result from the tryptophan rhythm, though, for in meal-fed rats, in which the TPO rhythm was shifted in phase, the hepatic tryptophan rhythm was not shifted (*Fuller,* 1970). *Hardeland* (1969) has suggested that the rate of synthesis of hepatic TPO varies in a daily rhythm due to varying repressor concentrations.

A neural control of the TPO rhythm can also be considered. *Shimazu* (1962) has shown that electrical stimulation in the hypothalamus of rabbits caused an increase of hepatic TPO, although secondary influences via adrenal corticoids or other hormones may have accounted for the enzyme change. Additional evidence suggestive of neural control of TPO has been published by *Vaptzarova et al.* (1969) who found that transection of the spinal cord at C_7 in rats was followed by a two- to three-fold increase in hepatic TPO, even in adrenalectomized rats. The same operation at the level of L_3 did not change TPO activity.

The functional consequences of the rhythm in hepatic TPO are not yet clear. *Kim and Miller* (1969) showed that several-fold increases in TPO activity in rats were associated with at most a 20 % increase in the proportion of radioactive tryptophan converted to radioactive carbon dioxide. However, *Moran and Sourkes* (1963) have shown that induction of TPO by α-methyltryptophan resulted in increased oxidation of tryptophan and concluded that TPO plays a major physiological role in the control of catabolism of tryptophan in the rat. Subsequently, *Oravec and Sourkes* (1969) have shown that α-methyltryptophan stimulated the incorporation of labelled tryptophan and other precursors into hepatic glycogen, perhaps because of induction of TPO.

There are other indications that TPO levels influence the amount of tryptophan oxidized. For example, *Altman and Greengard* (1966) found that high TPO activity in liver was correlated with high levels of urinary excretion of kynurenine in humans given hydrocortisone. *Mandell and Rubin* (1966) found that ACTH, possibly by inducing hepatic TPO, increased the conversion of radio-actively-labeled tryptophan to kynurenine in man. *Rapoport and Beisel* (1968) demonstrated a 24-hour rhythm in the amount of tryptophan metabolized to kynurenine and suggested that a rhythm in hepatic TPO probably occurred in man.

Green and Curzon (1968) have speculated on one possible physiological consequence of increased TPO activity. Tryptophan is the precursor of sero-tonin, which is thought to function as a neurohormone in brain. Changes in TPO, if they significantly altered the amount of tryptophan metabolized by it, might alter the rate of serotonin formation. In mental depression, there is an abnormal rhythm in plasma steroids, with abnormally high values occurring in the early morning (*Fullerton et al.,* 1968). If the increased adrenal corticoid activity leads to increased liver TPO, as *Rubin* (1967) has suggested, serotonin might be abnormally low at this time. Indeed, *Sarai and Kayano* (1968) have reported an abnormal rhythm of blood serotonin in depressed patients, with low levels in the early morning. If brain serotonin followed a pattern similar to that in blood, low brain serotonin might be related to the characteristic early morn-ing wakefulness and depression in such patients.

III. Rhythms in Other Enzymes in Intermediary Metabolism

In rats that have free access to food, 24-hour rhythms occur in liver glyco-gen (*Sollberger,* 1964) and in plasma free fatty acids (*Barrett,* 1964; *Fuller and Diller,* 1970). Presumably such rhythms occur in other species, but the phase and perhaps amplitude of the rhythms may depend on the exact feeding habits of the animals. There seem to be rhythms in the overall metabolic economy, i.e. of the substrates serving as energy sources, over the 24-hour period. Such rhythms probably come about in part by rhythmic changes in enzymes, in part by rhythmic ingestion of nutrients, and in part by other factors such as rhythmic fluctuation in hormones that affect transport of metabolites across cell mem-branes.

Glick and Cohen (1964) showed that the rate of succinate oxidation in hepatic mitochondria from rats with free access to food was greater at night than during the day and that there were higher phosphate to oxygen ratios resulting from malate, citrate, and α-ketoglutarate oxidation at night.

Krebs et al. (1966) have published data showing that gluconeogenesis from several substrates apparently varies during the 24-hour period in mice with free access to food. Rates of gluconeogenesis in mice killed at 2 p.m. were higher than in mice killed at 9 a.m. though not as high as in starved mice. Perhaps several enzymes associated with glycolysis and gluconeogenesis vary rhythmically. *Potter et al.* (1966a) found that, in rats fed *ad libitum,* citrate cleavage enzyme and glucose-6-phosphate dehydrogenase levels had a reciprocal relationship to liver glycogen. Cyclic changes in liver of the incorporation of acetate into fatty acids and sterols and in hydroxymethylglutaryl coenzyme A reductase have been found in mice (*Kandutsch and Saucier,* 1969). The rate of sterol synthesis was inversely related to the rate of fatty acid synthesis.

Phillips and Berry (1969) have recently studied a daily rhythm in phosphoenolpyruvate carboxykinase (PEPCK), a key enzyme in gluconeogenesis. They found that PEPCK in the liver of mice fed *ad libitum* varied over a two- to three-fold range during a 24-hour period. The levels of PEPCK were inverse to those of liver glycogen. In terms of the phase of the rhythm (peak at 8 p.m.), the lack of abolition of the rhythm by adrenalectomy, and the flattening but not abolition of the rhythm by fasting, the PEPCK rhythm resembled the rhythm in hepatic TAT. The rhythmic variation in PEPCK as found by *Phillips and Berry*

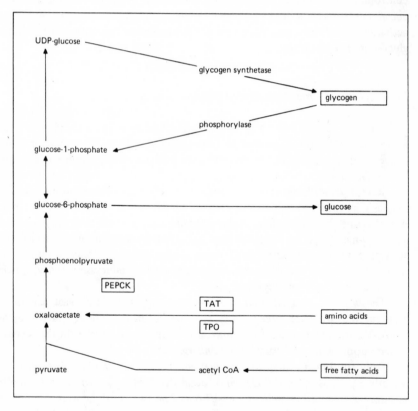

Fig. 9. Simplified pathway for glucose metabolism. Circled components have been shown to vary in a daily rhythm in rats fed *ad libitum.*

(1969) may well have partially accounted for the variation in gluconeogenesis observed earlier by *Krebs et al.* (1966).

Some of the enzymes discussed so far in this chapter can be related to each other as shown in figure 9. Although glycogen stores in liver vary rhythmically, daily rhythmic changes in glycogen synthetase and in glycogen phosphorylase have not been reported. Daily rhythmic variation in glucose tolerance tests (*Bowen and Reeves,* 1967) and in insulin secretion (*Freinkel et al.,* 1968) have been found. When glycogen stores are low in rats fed *ad libitum,* plasma free fatty acids and liver PEPCK, TAT, and TPO levels are high. If the glycogen rhythm is reversed in phase by feeding rats only during the day, the rhythms in plasma free fatty acids (*Fuller and Diller,* 1970) and in hepatic TAT (*Fuller and Snoddy,* 1968) and TPO (*Fuller,* 1970) are also reversed so that their opposite relationship to glycogen level is maintained. That fact implies either some connection between glycogen and the other parameters or a common mechanism controlling all of them. That common mechanism might involve the central or autonomic nervous system. To some extent, daily rhythms in the amounts or activities of the enzymes shown may act to maintain a rather constant supply of glucose in the circulation.

IV. Rhythms in Drug Metabolizing Enzymes

Twenty-four-hour rhythms in the susceptibility of animals to drugs have been recognized for a number of years (see, for example, *Reinberg,* 1967; *Scheving et al.,* 1968). In few cases has the basis for the rhythmic susceptibility been established.

Recently, *Radzialowski and Bousquet* (1967 and 1968) found 24-hour rhythms in the level of hepatic drug-metabolizing enzymes in rats and mice. Aminopyrine N-demethylase, 4-dimethylaminoazobenzene reductase, p-nitro-anisole 9-methylase, and the hexobarbital oxidizing enzyme in the liver of rats all varied with peak activity at about 2 a.m. Adrenalectomy abolished the rhythms in all but the 4-dimethylaminoazobenzene reductase activity. Fasting did not alter the enzyme rhythms.

The work of *Radzialowski and Bousquet* suggests that at least part of the basis for the rhythmic susceptibility to drugs lies in the rate at which they are detoxified by metabolism. There may also be rhythmic changes in the receptor systems upon which the drugs act (*Reinberg,* 1967).

Daily rhythms in other microsomal enzymes of rat liver have been reported recently by *Colas et al.* (1969), who found rhythms in the 7a- and 16a-hydroxylation of 3β-hydroxyandrost-5-en-17-one. The two rhythms were opposite in phase, i.e. 16a-hydroxylase was highest at 10 a.m., at which time 7a-hydroxylase activity was lowest.

V. Summary and Comment

The rhythmic changes that have been considered have been of 24-hour duration. Enzyme rhythms of longer duration also occur. When a physiological function changes in a rhythmic cycle, it is not surprising that enzymes involved in that function would change. Of greatest interest are enzymes possibly related in a causal way to the cycle.

Sexual cycles represent an example of a physiological function that varies in a rhythm other than 24 h in length. A possible role of brain monoamines like serotonin or norepinephrine in regulating sexual cycles has been considered. In this regard, rhythmic changes that have been reported in monoamine oxidase (MAO) may be important. Such rhythms have been reported both in species having an estrous cycle (female rats) and a menstrual cycle (women).

Zolovick et al. (1966) reported that MAO in amygdala, frontal cortex, and hypothalamus was higher during proestrus and estrus than in diestrus in rats. In our laboratory, we were unable to confirm those results. *Salseduc et al.* (1966) found MAO activity in cerebellum, diencephalon, and rhombencephalon higher in estrus than in diestrus or proestrus. Further work may substantiate these reports and reveal the significance of the rhythmic changes that may occur in brain MAO. There are cyclic variations in catecholamines, which are substrates for MAO, in hypothalamic nerve cells of the rat during the estrous cycle (*Lichtensteiger,* 1969). *Salseduc et al.* (1966) have also reported changes in catechol 0-methyl transferase, another enzyme that metabolizes catecholamines.

Southgate et al. (1968) found, by histochemical and biochemical assay, large increases of MAO in the endometrium during the late secretory phase of the menstrual cycle in humans. A possible physiological consequence of such changes can be deduced from the work of *Grant and Pryse-Davies* (1968) who studied the effect of oral contraceptives on MAO and other enzymes in the endometrium. High levels of MAO during most of the cycle were found in patients receiving strongly progestogenic compounds, who also had the highest incidence of depression and loss of libido. It is conceivable that changes in the metabolism of amines like norepinephrine and serotonin were related to the mental changes.

Rhythmic changes of enzyme activity of still a longer duration may also occur. Seasonal changes in metabolism as a consequence of changes in day length or other environmental influences are likely associated with changes in the activity of some enzymes.

Rhythmic changes in enzymes are of interest for several reasons. Such changes can be useful to the biochemist in studying mechanisms of enzyme regulation. A common technique for studying the properties of a system is to determine its response to some perturbation from a steady state. Enzymes that vary in a rhythm have a variable rather than a constant system of control, i.e. are undergoing continued perturbation. Enzyme rhythms are a reality of living organisms that ought not to be ignored by the biologist in studies of those organisms. It is difficult if not impossible and undesirable to circumvent such rhythmic changes in studies of enzymes or their metabolites. A 'normal' level of a biological constituent that varies rhythmically cannot be defined in the same way as a 'normal' level for a constituent that does not so vary. The aim of drug therapy in the treatment of many disease states is to impose pharmacologically a control that will lead to therapeutic benefit. It is crucial to recognize that such control is being imposed on a system that already is regulated and to understand as well as possible the kinds of regulatory mechanisms that are operating. Rhythmic changes in enzymes involved in the disposition of drugs and in the functioning of biological systems upon which drugs act are of importance to the pharmacologist. In addition, the toxicologist may find that rhythmic changes of drug-metabolizing enzymes result in variable toxicity related to the time of drug administration. To the endocrinologist, rhythmic changes in enzymes may be seen as the cause or an effect of rhythmic hormone release. Studies of enzyme rhythms may ultimately guide medical practitioners in understanding disease states associated with disturbances in daily rhythms.

One of the challenges in studies of enzyme or metabolic rhythms is that adventitious factors may affect them. Cause and effect are hard to separate, and adequate experimental control is not easy because of the complex regulatory processes that probably are inherent in the animal. In spite of these difficulties, further studies of rhythmic changes in enzymes seem essential.

Acknowledgments

I am grateful to *Harold D. Snoddy* for technical assistance in the experiments; to *Betty J. Warren* for technical assistance and preparation of the illustrations; to Dr. *I.H. Slater* for his continual advice and assistance during the experimental studies and in the preparation of this chapter; and to Mrs. *Carole Burgess* for her typing and other assistance in preparing the manuscript.

VI. References

Altman, K. and Greengard, O.: Tryptophan pyrrolase induced in human liver by hydro-cortisone: Effect on excretion of kynurenine. Science *151:* 332–333 (1966).

Anliker, J. and Mayer, J.: An operant conditioning technique for studying feeding-fasting patterns in normal and obese mice. J. appl. Physiol. *8:* 667–670 (1955).

Axelrod, J. and Black, I.B.: Suppression of the daily rhythm in tyrosine transaminas activity by acute elevation of norepinephrine. Nature, Lond. *220:* 161–162 (1968).

Baril, E.F. and Potter, V.R.: Systematic oscillations of amino acid transport in liver from rats adapted to controlled feeding schedules. J. Nutr. *95:* 228–237 (1968).

Barrett, A.M.: Adventitious factors affecting the concentration of free fatty acids in the plasma of rats. Brit. J. Pharmacol. *22:* 577–584 (1964).

Black, I.B. and Axelrod, J.: Elevation and depression of hepatic tyrosine transaminase activity by depletion and repletion of norepinephrine. Proc. nat. Acad. Sci., Wash. *59:* 1231–1234 (1968a).

Black, I.B. and Axelrod, J.: Regulation of the daily rhythm in tyrosine transaminase activity by environmental factors. Proc. nat. Acad. Sci., Wash. *61:* 1287–1291 (1968b).

Black, I.B. and Axelrod, J.: Neural regulation of rat hepatic tyrosine transaminase (TT) activity. Fed. Proc. *28:* 729 (1969).

Boctor, A.M.; Rogers, Q.R., and Harper, A.E.: Some metabolic aspects of tyrosine toxicity. Fed. Proc. *25:* 364 (1966).

Bowen, A.J. and Reeves, R.L.: Diurnal variation in glucose tolerance. Arch. intern. Med. *119:* 261–264 (1967).

Brown, C.B. and Civen, M.: Control of rat liver aromatic amino acid transaminases by glucagon and insulin. Endocrinology *84:* 381–385 (1969).

Chambers, J.W.; Georg, R.H., and Bass, A.D.: Effect of hydrocortisone and insulin on uptake of α-aminoisobutyric acid by isolated perfused rat liver. Mol. Pharmacol. *1:* 66–76 (1965).

Civen, M.; Trimmer, B.M., and Brown, C.B.: The induction of hepatic tyrosine α-ketoglutarate and phenylalanine pyruvate transaminases by glucagon. Life Sci. *6:* 1331–1338 (1967a).

Civen, M.; Ulrich, R.; Trimmer, B.M., and Brown, C.B.: Circadian rhythms of liver enzymes and their relationship to enzyme induction. Science *157:* 1563–1564 (1967b).

Coburn, S.P.; Seidenberg, M., and Fuller, R.W.: Daily rhythm in plasma tyrosine and phenylalanine. Proc. Soc. exp. Biol., NY *129:* 338–343 (1968).

Colas, A.; Gregonis, D., and Moir, N.: Daily rhythms in the hydroxylation of 3β-hydroxy-androst-5-en-17-one by rat liver microsomes. Endocrinology *84:* 165–167 (1969).

Csányi, V.; Greengard, O., and Knox, W.E.: The inductions of tyrosine aminotransferase by glucagon and hydrocortisone. J. biol. Chem. *242:* 2688–2692 (1967).

Csányi, V. and Greengard, O.: Effect of hypophysectomy and growth hormones on the inductions of rat liver tyrosine aminotransferase and tryptophan oxygenase by hydrocortisone. Arch. Biochem. *125:* 824–828 (1968).

Deter, R.L. and Liebelt, R.A.: Gold thioglucose as an experimental tool. Texas Rep. Biol. Med. *22:* 229–243 (1964).

Feigin, R.D.; Klainer, A.S., and Beisel, W.R.: Factors affecting circadian periodicity of blood amino acids in man. Metabolism *17:* 764–775 (1968).

Freinkel, N.; Mager, M., and Vinnick, L.: Cyclicity in the interrelationships between plasma insulin and glucose during starvation in normal young men. J. Lab. clin. Med. *71:* 171–178 (1968).

Fuller, R.W.: Daily variation in liver tryptophan, tryptophan pyrrolase, and tyrosine trans-aminase in rats fed *ad libitum* or single daily meals. Proc. Soc. exp. Biol., NY *133:* 620–622 (1970).

Fuller, R.W. and Diller, E.R.: Diurnal variation of liver glycogen and plasma free fatty acids in rats fed *ad libitum* or a single daily meal. Metabolism *19:* 226–229 (1970).

Fuller, R.W.; Jones, G.T.; Snoddy, H.D., and Slater, I.H.: Daily rhythm of liver tyrosine transaminase and of plasma tyrosine and glucose after removal of the pancreas of rats. Life Sci. *8:* 685–691 (1969).

Fuller, R.W. and Slater, I.H.: Daily cyclic variation in liver tyrosine α-ketoglutarate trans-aminase (TKT) in rats and mice. Fed. Proc. *27:* 403 (1968).

Fuller, R.W. and Snoddy, H.D.: Feeding schedule alteration of daily rhythm in tyrosine α-ketoglutarate transaminase of rat liver. Science *159:* 738 (1968).

Fuller, R.W. and Snoddy, H.D.: Induction of tyrosine aminotransferase in rat liver by epinephrine and theophylline. Biochem. Pharmacol. (in press).

Fullerton, D.T.; Wenzel, F.J.; Lohrenz, F.N., and Fahs, H.: Circadian rhythm of adrenal cortical activity in depression. Arch. gen. Psychiat. *19:* 674–681 (1968).

Glick, J.L. and Cohen, W.D.: Nocturnal changes in oxidative activities of rat liver mito-chondria. Science *143:* 1184–1185 (1964).

Govier, W.C.; Lovenberg, W., and Sjoerdsma, A.: Catecholamines and the control of tyrosine transaminase. Fed. Proc. *28:* 741 (1969).

Grant, E.C.G. and Pryse-Davies, J.: Effect of oral contraceptives on depressive mood changes and on endometrial monoamine oxidase and phosphatases. Brit. med. J. *3:* 777–780 (1968).

Green, A.R. and Curzon, G.: Decrease of 5-hydroxytryptamine in the brain provoked by hydrocortisone and its prevention by allopurinol. Nature, Lond. *200:* 1095–1097 (1968).

Greengard, O. and Baker, G.T.: Glucagon, starvation, and the induction of liver enzymes by hydrocortisone. Science *154:* 1461–1462 (1966).

Hager, C.B. and Kenney, F.T.: Regulation of tyrosine-α-ketoglutarate transaminase in rat liver. J. biol. Chem. *243:* 3296–3300 (1968).

Hardeland, R.: Circadian rhythm and regulation of enzymes of tryptophan metabolism in rat liver and kidney. Z. vergl. Physiol. *63:* 119–136 (1969).

Hardeland, R. and Rensing, L.: Circadian oscillation in rat liver tryptophan pyrrolase and its analysis by substrate and hormone induction. Nature, Lond. *219:* 619–621 (1968).

Harding, H.R. and Rosen, F.: Effects of hypophysectomy and growth hormone (GH) on the endogenous levels and induction of two adaptive enzymes. Fed. Proc. *22:* 409 (1963).

Hellman, B. and Hellerstrom, C.: Diurnal changes in the function of the pancreatic islets of rats as indicated by nuclear size in the islet cells. Acta endocrinol. *31:* 267–281 (1959).

Holten, D. and Kenney, F.T.: Regulation of tyrosine-α-ketoglutarate transaminase in rat liver. VI. Induction by pancreatic hormones. J. biol. Chem. *242:* 4372–4377 (1967).

Honova, E.; Miller, S.A.; Ehrenkranz, R.A., and Woo, A.: Tyrosine transaminase: Develop-ment of daily rhythm in liver of neonatal rat. Science *162:* 999–1001 (1968).

Kandutsch, A.A. and Saucier, S.E.: Prevention of cyclic and triton-induced increases in hydroxymethylglutaryl coenzyme A reductase and sterol synthesis by puromycin. J. biol. Chem. *244:* 2299–2305 (1969).

Kenney, F.T.: Induction of tyrosine-α-ketoglutarate transaminase in rat liver. III. Immuno-chemical analysis. J. biol. Chem. *237:* 1610–1614 (1962).

Kenney, F.T.: Regulation of tyrosine-α-ketoglutarate transaminase in rat liver. V. Repres-sion in growth hormone-treated rats. J. biol. Chem. *242:* 4367–4371 (1967).

Kenney, F.T.; Reel, J.R.; Hager, C.B., and Wittliff, J.L.: Hormonal induction and repression; in *San Pietro, Lamborg and Kenney* Regulatory mechanisms for protein synthesis in mammalian cells, pp. 119–142 (Academic Press, New York 1968).

Kim, J.H. and Miller, L.L.: The functional significance of changes in activity of the en-zymes, tryptophan pyrrolase and tyrosine transaminase, after induction in intact rats and in the isolated, perfused rat liver. J. biol. Chem. *244:* 1410–1416 (1969).

Knox, W.E. and Auerbach, V.H.: The hormonal control of tryptophan peroxidase in the rat. J. biol. Chem. *214:* 307–313 (1955).

Knox, W.E.; Linder, M.C.; Lynch, R.D., and Moore, C.L.: The enzymatic basis of tyrosyluria in rats fed tyrosine. J. biol. Chem. *239:* 3821–3825 (1964).

Knox, W.E. and Sharma, C.: Enzyme induction in perfused rat liver by glucagon and other agents. Enzym. biol. clin. *9:* 21–30 (1968).

Korner, A. and Labrie, F.: Induction of tyrosine transaminase and tryptophan pyrrolase by amino acids by a process requiring ribonucleic acid synthesis. Biochem. J. *105:* 49P (1967).

Krebs, H.A.; Notton, B.M., and Hems, R.: Gluconeogenesis in mouse-liver slices. Biochem. J. *101:* 607–617 (1966).

Kroger, H.; Lowel, M., and Kessel, H.: Induction of tyrosine transaminase in the presence of L-tryptophan. Z. Physiol. Chem. *349:* 1221–1224 (1968).

Labrie, F. and Korner, A.: Actinomycin-sensitive induction of tyrosine transaminase and tryptophan pyrrolase by amino acids and tryptophan. J. biol. Chem. *243:* 1116–1119 (1968).

Labrie, F. and Korner, A.: Effect of glucagon, insulin, and thyroxine on tyrosine transaminase and tryptophan pyrrolase of rat liver. Arch. Biochem. *129:* 75–78 (1969).

Lang, N.; Herrlich, P., and Sekeris, C.E.: On the mechanism of hormone action. VII. Induction by cortisol of a messenger RNA coding for tyrosine-a-ketoglutarate transaminase in an *in vitro* system. Acta endocrin., Kbh. *57:* 33–44 (1968).

LeMagnen, J. and Tallon, S.: The spontaneous periodicity of *ad libitum* food intake in rats. J. Physiol., Paris *58:* 323 (1966).

Lichtensteiger, W.: Cyclic variations of catecholamine content in hypothalamic nerve cells during the estrous cycle of the rat, with a concomitant study of the substantia nigra. J. Pharmacol. exp. Ther. *165:* 204–215 (1969).

Lin, E.C.C.; Pitt, B.M.; Civen, M., and Knox, W.E.: The assay of aromatic amino acid transaminations and keto acid oxidation by the enol borate-tautomerase method. J. biol. Chem. *233:* 668–673 (1958).

Mandell, A.J. and Rubin, R.T.: ACTH-induced changes in tryptophan turnover along induceable pathways in man. Life Sci. *5:* 1153–1161 (1966).

Manshardt, J. and Wurtman, R.J.: Daily rhythm in the noradrenaline content of rat hypothalamus. Nature, Lond. *217:* 574–575 (1968).

Margules, D.L.: Noradrenergic synapses for the suppression of feeding behavior. Life Sci. *8:* 693–704 (1969).

Moran, J.F. and Sourkes, T.L.: Induction of tryptophan pyrrolase by a-methyltryptophan and its metabolic significance *in vivo*. J. biol. Chem. *238:* 3006–3008 (1963).

Munro, H.N.: Role of amino acid supply in regulating ribosome function. Fed. Proc. *27:* 1231–1237 (1968).

Oravec, M. and Sourkes, T.L.: Action of a-methyl-DL-tryptophan *in vivo* on catabolism of amino acids and their conversion to liver glycogen. Canad. J. Biochem. *47:* 179–184 (1969).

Phillips, L.J. and Berry, L.J.: Regulation of the circadian rhythm of mouse liver phosphoenol pyruvate carboxykinase (PEPCK). Fed. Proc. *28:* 888 (1969).

Pitot, H.C.; Peraino, C.; Lamar, C., Jr., and Kennan, A.L.: Template stability of some enzymes in rat liver and hepatoma. Proc. nat. Acad. Sci., Wash. *54:* 845–851 (1965).

Potter, V.R.; Baril, E.F.; Watanabe, M., and Whittle, E.D.: Systematic oscillations in metabolic functions in liver from rats adapted to controlled feeding schedules. Fed. Proc. *27:* 1238–1245 (1968).

Potter, V.R.; Gebert, R.A., and Pitot, H.C.: Enzyme levels in rats adapted to 36-hour fasting. Adv. Enzymol. *4:* 247–265 (1966a).

Potter, V.R.; Gebert, R.A.; Pitot, H.C.; Peraino, C.; Lamar, C., Jr.; Lesher, S., and Morris, H.P.: Systematic oscillations in metabolic activity in rat liver and in hepatomas. I. Morris hepatoma No. 7793. Cancer Res. *26:* 1547–1560 (1966b).

Potter, V.R.; Watanabe, M.; Becker, J.E., and Pitot, H.C.: Hormonal effects on enzyme activities in tissue culture and in whole animals. Adv. Enzymol. 5: 303–316 (1967).

Radzialowski, F.M. and Bousquet, W.F.: Circadian rhythm in hepatic drug metabolizing activity in the rat. Life Sci. 6: 2545–2548 (1967).

Radzialowski, F.M. and Bousquet, W.F.: Daily rhythmic variation in hepatic drug metabolism in the rat and mouse. J. Pharmacol. exp. Ther. 163: 229–238 (1968).

Rapoport, M.I. and Beisel, W.R.: Circadian periodicity of tryptophan metabolism. J. clin. Invest. 47: 934–939 (1968).

Rapoport, M.I.; Feigin, R.D.; Bruton, J., and Beisel, W.R.: Circadian rhythm for tryptophan pyrrolase activity and its circulating substrate. Science 153: 1642–1644 (1966).

Reinberg, A.: The hours of changing responsiveness or susceptibility. Perspect. Biol. méd. 11: 111–128 (1967).

Reis, D.J. and Wurtman, R.J.: Diurnal changes in brain noradrenalin. Life Sci. 7: 91–98 (1968).

Reshef, L. and Greengard, O.: The effect of amino acid mixtures, insulin, epinephrine and glucagon in vivo on the levels of rat liver tyrosine aminotransferase. Enzym. biol. clin. 10: 113–121 (1969).

Rosen, F. and Milholland, R.J.: Effects of casein hydrolysate on the induction and regulation of tyrosine-α-ketoglutarate transaminase in rat liver. J. biol. Chem. 243: 1900–1907 (1968).

Rubin, R.T.: Adrenal cortical activity changes in manic-depressive illness. Influence on intermediary metabolism of tryptophan. Arch. gen. Psychiat. 17: 671–679 (1967).

Salseduc, M.M.; Jofre, I.J., and Isquierdo, J.A.: Monoamine oxidase (E.C.1.4.3.4) and catechol-0-methyltransferase (E.C.2.1.1.a) activity in cerebral structures and sexual organs of rats during their sexual cycle. Med. Pharmacol. exp. 14: 113–119 (1966).

Sarai, K. and Kayano, M.: The level and diurnal rhythm of serum serotonin in manic-depressive patients. Folia psychiat. neurol. jap. 22: 271–281 (1968).

Scheving, L.E.; Vedral, D.F., and Pauly, J.E.: Daily circadian rhythm in rats to d-amphetamine sulfate: Effect of blinding and continuous illumination on the rhythm. Nature, Lond. 219: 621–622 (1968).

Seegmiller, J.E.; Zannoni, V.G.; Laster, L., and LaDu, B.N.: An enzymatic spectrophotometric method for the determination of homogentisic acid in plasma and urine. J. biol. Chem. 236: 774–777 (1961).

Shambaugh, III, G.E.; Warner, D.A., and Beisel, W.R.: Hormonal factors altering rhythmicity of tyrosine-α-ketoglutarate transaminase in rat liver. Endocrinology 81: 811–818 (1967).

Shimazu, T.: The effect of electric stimulation of hypothalamus on rabbit liver tryptophan pyrrolase. Biochim. biophys. Acta 65: 373–375 (1962).

Shimazu, T.: Glycogen synthetase activity in liver: Regulation by the autonomic nerves. Science 156: 1256–1257 (1967).

Shimazu, T. and Amakawa, A.: Regulation of glycogen metabolism in liver by the autonomic nervous system. II. Neural control of glycogenolytic enzymes. Biochim. biophys. Acta 165: 335–348 (1968a).

Shimazu, T. and Amakawa, A.: Regulation of glycogen metabolism in liver by the autonomic nervous system. III. Differential effects of sympathetic-nerve stimulation and of catecholamines on liver phosphorylase. Biochim. biophys. Acta 165: 349–356 (1968b).

Shimazu, T. and Fukuda, A.: Increased activities of glycogenolytic enzymes in liver after splanchnic-nerve stimulation. Science 156: 1607–1608 (1965).

Shimazu, T.; Fukuda, A., and Ban, T.: Reciprocal influences of the ventromedial and lateral hypothalamic nuclei on blood glucose level and liver glycogen content. Nature, Lond. *210:* 1178–1179 (1966).

Sidransky, H.; Bongiorno, M.; Sarma, D.S.R., and Verney, E.: The influence of tryptophan on hepatic polyribosomes and protein synthesis in fasted mice. Biochem. biophys. Res. Commun. *27:* 242–248 (1967).

Sollberger, A.: The control of circadian glycogen rhythms. Ann. NY Acad. Sci. *117:* 519–554 (1964).

Southgate, J.; Grant, E.C.G.; Pollard, W.; Pryce-Davies, J., and Sandler, M.: Cyclical variation in endometrial monoamine oxidase: Correlation of histochemical and quantitative biochemical assays. Biochem. Pharmacol. *17:* 721–726 (1968).

Staib, R.; Thienhaus, R.; Ammedick, U., and Staib, W.: Direct influence of insulin and glucagon on the activity of tyrosine-a-ketoglutarate transaminase. Europ. J. Biochem. *8:* 23–25 (1969).

Suttie, J.W.: Effect of dietary fluoride on the pattern of food intake in the rat and the development of a programmed pellet dispenser. J. Nutr. *96:* 529–536 (1968).

Szepesi, B. and Freedland, R.A.: Evidence for both translational and transcriptional control of enzyme synthesis in rat liver by dietary alterations. Fed. Proc. *27:* 257 (1968).

Vaptzarova, K.I.; Davidov, M.S.; Markov, D.V.; Popov, P.G., and Galabov, G.P.: The effect of spinal cord section on rat liver tryptophan pyrrolase activity. Life Sci. *8:* 905–909 (1969).

Walker, C.A. and Friedman, A.H.: Circadian rhythms of biogenic amine levels in brain stem and caudate nucleus of rat. Pharmacologist *9:* 239 (1967).

Watanabe, M. and Baril, E.F.: Oscillations in tyrosine transaminase activity and amino acid transport in liver of rats adapted to 24- or 48-hour cycles of feeding and fasting. J. Cell Biol. *35:* A139 (1967).

Watanabe, M.; Potter, V.R., and Pitot, H.C.: Systematic oscillations in tyrosine transaminase and other metabolic functions in liver of normal and adrenalectomized rats on controlled feeding schedules. J. Nutr. *95:* 207–227 (1968).

Westermann, E.O.: Cumulative effects of reserpine on the pituitary-adrenocortical and sympathetic nervous system. Proc. 2nd Internat. Pharmacol. Meeting, vol. 4, pp. 381–392, 1963.

Wicks, W.D.: Induction of tyrosine-a-ketoglutarate transaminase in fetal rat liver. J. biol. Chem. *243:* 900–906 (1968a).

Wicks, W.D.: The possible role of cyclic adenylic acid in regulation of enzyme synthesis; in *San Pietro, Lamborg and Kenney* Regulatory mechanisms for protein synthesis in mammalian cells, pp. 143–155 (Academic Press, New York 1968b).

Wicks, W.D.: Tyrosine-a-ketoglutarate transaminase: Induction by epinephrine and adenosine-3',5'-cyclic phosphate. Science *160:* 997–998 (1968c).

Wiepkema, P.R.; DeRuiter, L., and Reddinguis, J.: Circadian rhythms in the feeding behavior of CBA mice. Nature, Lond. *209:* 935–936 (1966).

Wunner, W.H.; Bell, J., and Munro, H.N.: The effect of feeding with a tryptophan-free amino acid mixture on rat-liver polysomes and ribosomal ribonucleic acid. Biochem. J. *101:* 417–428 (1966).

Wurtman, R.J.: Ambiguities in the use of the term circadian. Science *156:* 104 (1967).

Wurtman, R.J. and Axelrod, J.: Daily rhythmic changes in tyrosine transaminase activity of the rat liver. Proc. nat. Acad. Sci., Wash. *57:* 1594–1598 (1967).

Wurtman, R.J.; Axelrod, J., and Kelly, D.E.: The pineal (Academic Press, New York 1968a).

Wurtman, R.J.; Chou, C., and Rose, C.M.: Daily rhythm in tyrosine concentration in human plasma: Persistence on low-protein diets. Science *158:* 660–662 (1967).

Wurtman, R.J.; Shoemaker, W.J., and Larin, F.: Mechanism of the daily rhythm in hepatic tyrosine transaminase activity: Role of dietary tryptophan. Proc. nat. Acad. Sci., Wash. *59:* 800–807 (1968b).

Wurtman, R.J.; Shoemaker, W.J.; Larin, F., and Zigmond, M.: Failure of brain norepinephrine depletion to extinguish the daily rhythm in hepatic tyrosine transaminase activity. Nature, Lond. *219:* 1049–1050 (1968c).

Wurtman, R.J.; Shoemaker, W.J.; Larin, F., and Zigmond, M.J.: Mechanism of the daily rhythm in hepatic tyrosine transaminase activity in the rat. Fed. Proc. *27:* 420 (1968d).

Zigmond, M.J.; Shoemaker, W.J.; Larin, F., and Wurtman, R.J.: Hepatic tyrosine transaminase rhythm: Interaction of environmental lighting, food consumption, and dietary protein content. J. Nutr. *98:* 71–75 (1969).

Zolovick, A.J.; Pearse, R.; Boehlke, K.W., and Eleftheriou, B.E.: Monoamine oxidase activity in various parts of the rat brain during the estrous cycle. Science *154:* 649 (1966).

Author's address: *Ray W. Fuller,* Dept. of Metabolic Research, The Lilly Research Laboratories, Eli Lilly and Company, *Indianapolis, IN 46227* (USA)

Enzyme Synthesis and Degradation in Mammalian Systems, pp. 339–374
(Karger, Basel 1971)

Factors Affecting the Activity,
Tissue Distribution, Synthesis and Degradation of Isozymes [1]

E.S. Vesell and P.J. Fritz

Department of Pharmacology, The Milton S. Hershey Medical Center,
The Pennsylvania State University, College of Medicine, Hershey, Pa.

Contents

I. Introduction

Inclusion of isozymes in a book devoted to factors influencing enzyme activity and turnover provides another illustration of how the phenomenon of multiple molecular forms of enzymes has changed over the past decade from a laboratory accident or curiosity to a general biological principle. Early examples of multiple forms of an enzyme, that should by all previously accepted criteria have been homogeneous, were dismissed as artifacts. However, as more instances developed and as certain examples of enzyme heterogeneity seemed to reflect biological alterations that were occurring *in vivo* (*Vesell and Bearn,* 1957), multiple molecular forms of enzymes had to be reconsidered and reevaluated. This reassessment is still in progress, and it is the purpose of our chapter to emphasize not only recent acquisitions of knowledge but also challenging problems in the field that still remain to be solved. Furthermore, we anticipate that in the future the application of isozymes as a laboratory tool will result in better understand-

1   This review was aided by grant No. P-535 from the American Cancer Society.

ing of a variety of biological problems. Isozymes themselves have already become a general biological principle. Thus multiple molecular forms of an enzyme have become the rule, rather than the exception. Of the many enzymes that exist in multiple molecular forms, the most intensively investigated is that of lactate dehydrogenase (LDH). Extensive research on LDH isozymes has established them as a model to which other systems can be conveniently compared. Therefore, this chapter considers primarily LDH isozymes. Furthermore, for the primary purpose of this book, the multiple forms of LDH merit special consideration since they represent the only system in which measurement of rates of synthesis and degradation of the individual molecular forms has thus far been accomplished (*Fritz et al.,* 1969).

Markert and Møller (1959) first introduced the term isozyme and developed the conceptual framework from which it emerged. They defined isozymes as multiple molecular forms of an enzyme with similar substrate specificity. The only restrictions placed on these multiple forms were that they had to exist within a single organism, either within one cell or in different tissues, and they had to share all or most of their substrates. Enzymes with very broad substrate specificity that shared only a few of their numerous substrates were initially excluded, although recently many reports have been published on esterase and phosphatase isozymes whose substrate specificities have not been explored. Problems arising from limitations imposed by this operational definition, as well as the desirability for including in the definition information concerning the structural and genetic basis of different isozymic systems, have been discussed by *Vesell* (1968), *Markert and Whitt* (1969) and *Shaw* (1969).

LDH (L-lactate: NAD + oxidoreductase, EC 1.1.1.27), catalyzes the interconversion of pyruvate and lactate and the simultaneous interconversion of reduced and oxidized nicotinamide adenine dinucleotide (NAD). The enzyme is ubiquitous in vertebral cells and is generally present in 5 electrophoretically and chromatographically separable forms. However, there have been abundant reports of tissues in which, under basal conditions, either more (testes and sperm) or less (liver) than 5 LDH isozymes occur. In some species (*Salmonidae*) more than 5 forms are regularly encountered (*Morrison and Wright,* 1967). Its availability in high concentrations, stability, specificity, ease of purification and convenient assay, have made LDH an attractive object of biochemical investigation for the past 20 years.

Meister (1950) and *Neilands* (1952) first described two forms of bovine LDH. When human erythrocytes and serum were demonstrated to contain several electrophoretically distinguishable LDHs changing independently in various disease states, multiple molecular forms of an enzyme were recognized as reflections of biological events occurring *in vivo* (*Vesell and Bearn,* 1957). Early studies on the LDH isozymes in tissues of various vertebrates include those of *Wieland and Pfleiderer* (1957), *Sayre and Hill* (1957) and *Hess* (1958). The introduction of rapid and convenient techniques for analysis of isozymes by histochemical staining after starch gel electrophoresis greatly facilitated studies on tissue, ontogenetic and species specificity of various isozymes (*Hunter and Markert* 1957; *Markert and Møller*, 1959), and consequently led to a vast increase in the number of investigations on isozymes.

Before 1957 enzyme heterogeneity was regarded with suspicion and displeasure by enzymologists who generally interpreted it as an artifact introduced by the laboratory procedures commonly employed for protein purification. Enzymologists attempted to make their proteins as homogeneous as possible by all physicochemical criteria. *Colvin, Smith and Cook* (1954) were exceptions in their recognition that enzyme heterogeneity reflected biological realities rather than laboratory artifacts. *Colvin, Smith and Cook* (1954) maintained that in most organisms proteins actually existed within cells in multiple forms; they believed that this heterogeneity arose from occasional lapses in the protein synthesizing machinery. Although this concept was far in advance of its time and although it has been shown that ambiguity in the genetic code does indeed lead to heterogeneity of the hemoglobin molecule in rabbit reticulocytes (*Rifkin et al.*, 1966; *von Ehrenstein*, 1966), recent work on many diverse isozymic systems reveals that errors at the level of protein synthesis are an exceedingly rare cause of protein heterogeneity.

Regarding the genetic origin of isozymes, the current view is that LDH isozymes, and other isozymic systems based on two or more different subunits, arose by chromosomal duplication with subsequent mutations at daughter and parental loci (*Vesell and Bearn*, 1962a). Chromosomal duplication could result in the production of two dissimilar subunits where a single subunit previously existed (*Vesell and Bearn*, 1962a). At genetic loci controlling vital proteins, multiple mutations might prove lethal, whereas duplicated genes performing initially identical functions permit a greater latitude for mutations at each locus. Duplication thus helps to assure retention of critical enzymatic function while simultaneously facilitating development of new, potentially advantageous, enzymatic characteristics. Selective advantages might operate to preserve these mutations during evolution and make them a generalized feature of intracellular metabolism. There is evidence for the subsequent occurrence of divergent evolution of the LDH isozymes through independent mutations of the 2 LDH loci in several classes of vertebrates (*Whitt*, 1969). *Shaw and Barto* (1963), working with mutants at the A and B structural loci of the deer mouse *Peromyscus maniculatis*, provided genetic evidence for the hypothesis of *Appella and Markert* (1961) that the 5 commonly encountered forms of LDH were tetramers of 140,000 mol. wt. and were composed of A and B subunits in all possible combinations of 4. *Markert* (1963) obtained biochemical proof of this thesis. He froze LDH-1 and LDH-5 in sodium chloride; through dissociation of the active tetramers and recombination of the A and B subunits in all possible combinations of

4, the intermediate forms were produced. The subunits of 35,000 mol. wt. can be further reduced to subunits of 19,000 mol. wt. so that the LDH molecule may be an octamer rather than a tetramer (*Appella*, 1964; *Appella and Zito*, 1968; *Millar et al.*, 1969).

If chromosomal duplication represents one possible genetic mechanism whereby isozymes came into being, there are numerous chemical ways in which multiple forms of an enzyme can arise. These have recently been considered by *Markert* (1968), *Epstein and Schechter* (1968), *Kaplan* (1968), *Vesell* (1968) and *Markert and Whitt* (1969) and therefore merit only a brief review here. The most common mechanism thus far described involves the combination of two or more dissimilar subunits, each subunit being controlled by separate genetic loci or by alleles at a single locus. One unsettled problem concerns whether or not this combination of subunits occurs entirely at random *in vivo*. Evidence has been presented to suggest that the association of subunits *in vitro* is not invariably random (*Anderson and Weber*, 1966; *Khan and Südi*, 1968). The mechanism for this control and possible sites of isozyme assembly from constituent subunits will be discussed in subsequent sections of this chapter. However, it

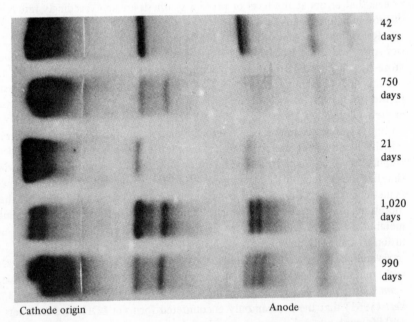

Cathode origin   Anode

Fig. 1. Photograph of a starch gel showing LDH isozymes in human skeletal muscles from 5 individuals. These muscles were treated identically except for length of storage at −25 °C. Note that splitting of bands increases with length of storage of the whole muscle (from *Vesell and Brody*, 1964).

should be mentioned here that isozyme patterns such as those of rat heart muscle are difficult to reconcile with the concept of random subunit association.

Multiple forms of some enzymes arise from different polymeric states of a single subunit (*Davis et al.*, 1967; *Yielding and Tomkins*, 1962) or from a single polymeric form to which varying amounts of a prosthetic group are attached (*Jacobson*, 1968). Divalent metals, sialic acid, AMP and various coenzymes such as NAD have been implicated as prosthetic groups in certain isozymic systems (*Cabello et al.*, 1966; *Butterworth and Moss*, 1966; *Beckman, L. and Beckman, G.*, 1967; *Shapiro and Stadtman*, 1968). The electrophoretic mobility of LDH has been altered by acetylation (*Rajewski*, 1966). Enzymatic cleavage of proteins, including removal of whole residues or of amino, carboxyl, or hydroxyl groups, is part of the large category of catabolic processes that theoretically could produce multiple forms of enzymes such as pepsinogen, trypsinogen and chymotrypsinogen. Proteolytic cleavage of part of a polypeptide chain may produce isozymes, as it is known to be essential to confer activity on many proteins. Multiple forms of certain enzymes may represent what *Hotchkiss* (1964) has termed 'configurational isomers' and what *Kitto et al.* (1966) recently termed 'conformers', signifying identical primary structure but different tertiary or quarternary structure. Finally, multiple forms of some enzymes may arise *in vitro* during storage of tissues (fig. 1) or during preparative procedures. They may be produced by homogenization of tissues, during purification of enzymes, or while homogenates are being analyzed for isozymes by electrophoresis, chromatography, ultracentrifugation, or immunochemical techniques; and they may also be formed during histochemical treatment.

II. Factors Affecting the Activity and Measurement of Lactate Dehydrogenase Isozymes

A. Methodology

Assays of total LDH activity as well as identification and quantitation of LDH isozymes are generally regarded as comparatively simple, accurate, inexpensive and convenient procedures. These methods have become standardized and in laboratories of clinical chemistry often are automated; however, like many other enzymes, LDH activity is markedly affected by a variety of conditions. Therefore, interpretations of the biological significance of changes in isozyme activities or patterns should be approached cautiously until the methods for identifying isozymes have been sufficiently studied and standardized to insure that alterations observed do not arise from manipulations *in vitro*. For this reason we feel it important to consider some of the commonly employed methods used in estimating LDH isozymes.

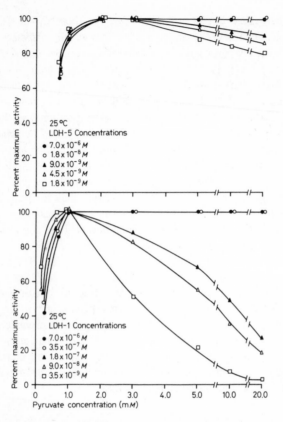

Fig. 2. Effect of increasing pyruvate concentration on the activity of several concentrations of LDH-1 and LDH-5 purified from rat kidney. All reagents were prepared in 0.1 M sodium phosphate buffer, pH 7.0, and the final NADH concentration was 0.56 mM. Molar concentrations of partially purified LDH-1 and LDH-5 were calculated from turnover numbers (*Pesce et al.*, 1964). LDH-1 was purified 130-fold, and LDH-5 purified 98-fold. The maximum specific activity remained constant over the range of enzyme concentrations examined. With 1.0 mM pyruvate, approximately 3.2 μM of NADH were oxidized/sec/mg of LDH-1, and approximately 4.5 μM of NADH were oxidized/sec/mg of LDH-5 with 2.0 mM pyruvate (from *Wuntch et al.*, 1970).

The spectrophotometric technique has been employed extensively to measure LDH activity. The method is based on the absorption peak at 340 mμ exhibited by NADH, but not NAD. Thus changes in absorption that occur at this wavelength with oxidation of NADH or reduction of NAD form a convenient, sensitive and accurate assay for LDH activity, provided that changes in absorbance are of the order of 0.100/min. To maintain absorbance changes that are linear, highly dilute LDH concentrations must be used, and these LDH activities

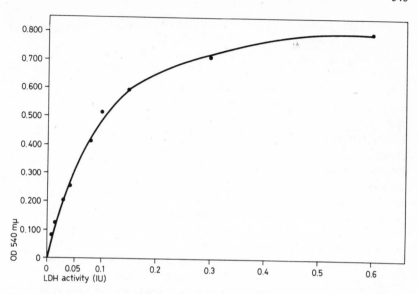

Fig. 3. Nonlinearity of the colorimetric method for estimating LDH activity (*Babson and Phillips,* 1965) when the activity is greater than that customarily observed in normal and pathological serum, but less than the LDH activity present in most mammalian tissues. The source of LDH was partially purified rat heart LDH-1 and the conditions of the assay were identical to those described by *Babson and Phillips* (1965).

are many orders of magnitude below the activities that occur in mammalian tissues (*Wuntch et al.,* 1969, 1970). One consequence of this requirement for high dilution can be seen in figure 2. The figure shows the marked effect of changing enzyme concentrations on the kinetics of LDH isozymes, measured by a stopped flow spectrophotofluorometric technique capable of measuring LDH activity with sensitivity at high enzyme concentrations (*Wuntch et al.,* 1969, 1970). Thus the kinetic properties of LDH isozymes that have been published prior to 1969 are applicable only under unphysiologic conditions. The kinetic properties observed under physiologic conditions are quite different, particularly as regards substrate inhibition. The substrate inhibition so marked with dilute enzyme is not observed at physiologic enzyme concentrations (fig. 2).

The zymogram technique (*Hunter and Markert,* 1957; *Markert and Møller,* 1959) in which various histochemical stains are used to visualize isozymes after their electrophoretic separation has proved unequaled for rapid, qualitative identification of the multiple forms of many different enzymes (*Shaw,* 1965). It has been employed with particular success in various aspects of genetic research and in clinical chemistry. The tetrazolium salts used for enzyme staining after electrophoresis, of which nitroblue tetrazolium is probably the most popular, give insoluble formazan precipitates when reduced. However, not all formazans are insoluble in water.

A technique using the tetrazolium salt, 2-p-iodophenyl-3-p-nitrophenyl-5-phenyl tetrazolium chloride (INT), which yields a water soluble formazan on reduction, has been described (*Babson and Phillips*, 1965; *Briere et al.*, 1966; *Capps et al.*, 1966). When tetrazolium salts are used to identify isozymes, phenazine methosulfate serves as an intermediate electron carrier. Quantitation of LDH activity by densitometric determination of the amount of insoluble colored formazan precipitated at various locations on starch or polyacrylamide gels, however, yields linear results only over a small range of LDH activity; at higher activities the quantity of formazan precipitated on the gels is not proportional to the LDH activity present (*Fritz et al.*, 1970a). Even with the use of INT, which is reduced to a soluble colored formazan, the method gives linear results only for low LDH activity such as that occurring in serum; at higher LDH activities, such as those encountered in most mammalian tissues, the reduction of tetrazolium salts to colored formazans is nonlinear (fig. 3). Therefore, the tetrazolium technique yields inaccurate results when, without sufficient enzyme dilution, it is employed to obtain quantitative data on LDH activities in various tissues. This method has been frequently utilized in the past for such purposes. When high LDH activities are being determined, results obtained with this technique may be misleading, particularly when they form the basis for conclusions concerning the biological significance of small differences in intensity of staining of LDH isozymes in different tissues or in a single tissue under various conditions (*Fritz et al.*, 1970a).

A convenient technique that obviates the necessity for electrophoretic separation of LDH isozymes has been described for the determination of the percentage of A and B subunits in a mixture of LDH isozymes. Measurement of the total LDH activity of the mixture is performed with high and then with low concentrations of pyruvate or lactate (*Plagemann et al.*, 1960; *Kaplan and Cahn*, 1963; *Stambaugh and Post*, 1966a). This method is based on LDH-1 inhibition at 25 °C by substrate concentrations to which LDH-5 is resistant. In this technique it is assumed that the active LDH tetramer is composed of catalytically independent subunits. This assumption is supported by the stepwise decrease in molecular activity from LDH-5 to LDH-1 (*Pesce et al.*, 1967). Recently, however, certain theoretical limitations of this approach have been emphasized (*Rouslin and Braswell*, 1968); and the catalytic interdependence of the 4 A subunits in LDH-5 has been demonstrated (*Anderson and Weber*, 1966).

A highly reproducible method for the quantitative determination of LDH isozymes is anion exchange chromatography on either DEAE Sephadex or DEAE cellulose. Although requiring more time than other techniques, this method, first used by *Hess and Walter* in 1960, is the one of choice for preparation of large amounts of purified LDH isozymes.

B. Factors Influencing the Activity of LDH Isozymes

It is known that LDH isozymes differ markedly from each other with respect to their physicochemical properties. Differences among the LDH isozymes
have been described in their pH optima (*Vesell and Bearn*, 1958; *Fritz*, 1967;
Vesell et al., 1968), thermal stability (*Plagemann et al.*, 1961; *Zondag*, 1963;
Vesell and Yielding, 1966), molecular activities (*Pesce et al.*, 1967), kinetic
properties including V_{max}, K_m (*Pesce et al.*, 1967; *Plagemann et al.*, 1960),
degree of inhibition by substrate and other compounds such as oxalate, urea and
sulfite (*Wieland and Pfleiderer*, 1957; *Emerson et al.*, 1964; *Schoenenberger and
Wacker*, 1966; *Brody*, 1968). As shown in figures 4 to 7, LDH isozymes differ in
properties under a given set of conditions and are altered to different extents
when the pH or the buffer is changed (*Vesell et al.*, 1968). Therefore, if a given
set of conditions is employed to estimate the relative concentrations of LDH-1
and LDH-5, the assay system may be optimum for one isozyme but not for
another. Such a situation may cause underestimation of the activity of one of
the isozymes. For example, the relative concentrations of LDH-1 and LDH-5 in a
tissue homogenate can change markedly on dilution in distilled water because

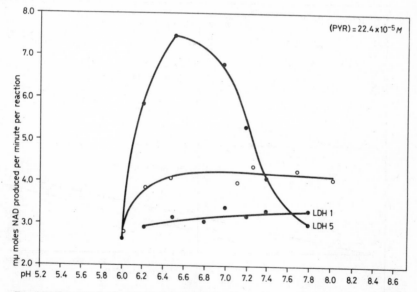

Fig. 4. Effect of pH on LDH-5 and LDH-1 activity. In the curve delineated o all assays
were performed with the same concentration of LDH-5 as in the curve delineated ●. However, instead of direct assay at the pH on the abscissa as in the curve marked ●, LDH-5 was
assayed at pH 7.4 after 3 min of prior incubation at the pH indicated on the abscissa (from
Fritz, 1967).

LDH-5 is more sensitive than LDH-1 to dilution in distilled water (*Vesell*, 1962). These problems associated with the determination of the relative levels of the various isozymes in a mixture will be augmented if the assay system depends on the coupling of several enzymatically independent reactions. Such coupled reactions are used for the spectrophotometric determination of a number of enzymes and for indentification after electrophoresis of isozymes of creatinine phosphokinase (CPK) (*Kar and Pearson*, 1965), hexokinase (*Katzen et al.*, 1968; *Schimke and Grossbard*, 1968), pyruvate kinase (*Schloen et al.*, 1969), phosphoglucomutase (*Spencer et al.*, 1964), adenylate kinase (*Fildes and Harris*, 1966) and acid phosphatase (*Hopkinson et al.*, 1963). It is not yet clear what relationship the relative intensity of formazan staining on starch gels of the isozymes of any one of these systems bears to their actual concentration within tissues and indeed whether all the bands on the gel do in fact represent isozymes of the enzyme being investigated. With coupled reactions slight differences in conditions would be anticipated to produce major differences in results, both with respect to relative intensity of the bands observed and also the numbers of bands appearing on the gel. In accord with this expectation, different labora-

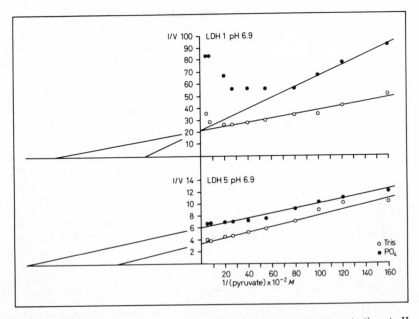

Fig. 5. Activities of rat LDH-1 and LDH-5 as affected by pyruvate concentration at pH 6.9 and 25 °C plotted according to the method of *Lineweaver and Burk* (1934). The concentration of NADH was $12.5 \times 10^{-5}\ M$; the concentration of LDH-1 was $1.8 \times 10^{-9}\ M$; and the concentration of LDH-5 was $1.0 \times 10^{-9}\ M$ (from *Vesell et al.*, 1968).

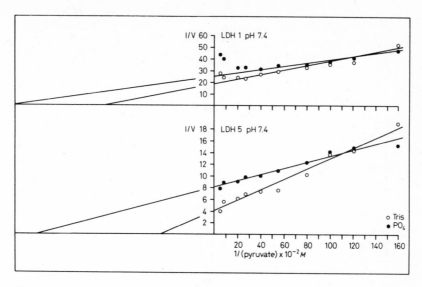

Fig. 6. Activities of rat LDH-1 and LDH-5 as affected by pyruvate concentration at pH 7.4 and 25 °C plotted according to the method of *Lineweaver and Burk* (1934). The concentration of NADH was 12.5 x 10^{-5} *M;* the concentration of LDH-1 was 1.8 x 10^{-9} *M;* and the concentration of LDH-5 was 1.0 x 10^{-9} *M* (from *Vesell et al.,* 1968).

tories using slight variations in the conditions under which reactions were coupled to identify the serum CPK isozymes reported variable numbers of CPK isozymes in serum (*Kar and Pearson,* 1965). These differences have not yet been resolved. Similar problems have arisen with hexokinase isozymes (*Holmes et al.,* 1967; *Brown et al.,* 1967; *Kaplan and Beutler,* 1968).

Even with isozymes identified by a single uncoupled reaction numerous factors other than those previously mentioned may significantly affect the pattern observed on the starch gel. *Vesell and Brody* (1963) have enumerated several such conditions with particular reference to LDH isozymes. Some of these factors probably also influence other isozymic systems. It was demonstrated that the LDH isozyme pattern was significantly influenced by the length of storage of the tissue before it was subjected to homogenization (fig. 1), the type of homogenization process employed, the buffer used to homogenize the tissue, the conditions of electrophoresis, especially the pH, the buffers used, the time, voltage and amperage of electrophoresis including the temperature at which electrophoresis was performed. Finally the LDH isozyme pattern was significantly influenced by all the conditions of staining including the compounds used, their relative concentrations, the pH and ionic strength of the buffers and the length of time of staining (*Vesell and Brody,* 1963).

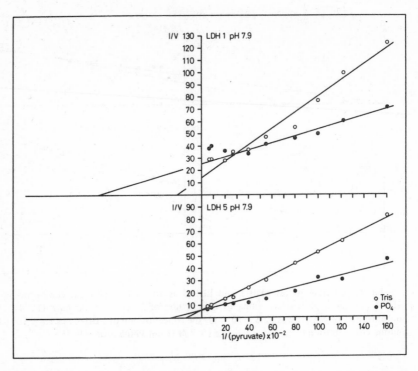

Fig. 7. Activities of rat LDH-1 and LDH-5 as affected by pyruvate concentration at pH 7.9 and 25 °C plotted according to *Lineweaver and Burk* (1934). The concentration of NADH was 12.5 x 10^{-5} *M;* the concentration of LDH-1 was 1.8 x 10^{-9} *M;* and the concentration of LDH-5 was 1.0 x 10^{-9} *M* (from *Vesell et al.,* 1968).

III. *Tissue Distribution and Physiologic Function of Isozymes*

Once methodological problems have been settled and the assay system standardized, relative differences in isozyme activity or mobility can generally, but not invariably, be equated with actual differences in intracellular isozyme concentration or chemical structure, respectively. Such differences among isozymes have been interpreted in various ways in attempts to provide new insights into isozyme function and into numerous biological problems. However, an interesting example of how such interpretations may occasionally be spurious is provided by apparent differences between the glucose-6-phosphate dehydrogenase (G-6-PD) isozymes of mouse kidney and liver (*Shaw*, 1969). These differences in G-6-PD isozyme pattern of mouse kidney and liver disappeared when the tissue homogenates were dialyzed prior to electrophoresis (*Shaw*, 1969). The liver pattern changed on dialysis to become identical to that in kidney. Some dialyzable substance in liver altered the G-6-PD isozyme pattern to suggest erroneously the existence in liver of a biochemically and genetically distinct isozyme from that in kidney. Another interesting example of an environmentally induced alteration in isozyme pattern is that of alcohol dehydrogenase (ADH) from *Drosophila melanogaster* (*Grell et al.*, 1968; *Jacobson*, 1968); ADH isozymes are changed in mobility by addition of NAD to the acrylamide gel and increased in number by passage through a DEAE Sephadex column.

Additions to biological knowledge through use of isozymes as a laboratory tool, although too diversified to enumerate in detail here, should be identified in broad outline. Isozymes provide the clinical pathologist with a sensitive diagnostic tool, the geneticist with a fruitful method for discovering and investigating polymorphic systems, marking chromosomes and following changing gene action, the biochemist with a new approach to understanding the control and regulation of intermediary metabolism and the student of growth and development with precise markers along the stages of ontogeny and phylogeny.

The presence of isozymes of LDH in most vertebral cells has been taken as evidence that multiple molecular forms of LDH offer the organism distinct adaptive advantages. Otherwise, it is argued that LDH heterogeneity would have disappeared long ago in evolution (*Markert and Møller*, 1959; *Markert*, 1965, 1968). Similarly the occurrence of other enzymes in multiple forms throughout the biosphere suggests that each isozyme performs a unique function. That function may vary with the organism, the tissue, the cell and the environmental conditions. Thus in cells of different tissues and under altering environmental conditions, isozyme patterns can change. What physiological role these isozymes subserve in a given cell and why isozyme patterns change in different tissues and under varying environmental circumstances are understood in some isozymic systems, not understood at all in most cases, and a matter of controversy in still others. Nevertheless, much support for the conception that each isozyme serves a specific and distinct function within the cell is derived from numerous studies establishing physicochemical differences among the multiple molecular forms of an enzyme.

The role of isozymes in regulating rates of metabolite flow at branched metabolic pathways in bacteria is one of most elegant and biologically significant functions yet ascribed to them (*Stadtman*, 1968). The situation where several end products are derived from a common precursor controlled by a single enzyme is potentially dangerous for the cell. If the concentration of precursor is regulated by a feedback mechanism, an accumulation of one of the end products could lead to a deficiency of the others. In *E. coli* aspartyl phosphate is an intermediate in the biosynthesis of methionine, threonine, and lysine. Aspartokinase converts aspartate to aspartyl phosphate in this organism; and the aspartokinase activity in crude extracts is inhibited by threonine, lysine, and methionine. Thus, by feedback inhibition, excessive accumulation of threonine, for example, would be expected to lead to a deficiency of lysine and methionine. However, such a deficiency fails to occur because the organism possesses 3 aspartokinase isozymes, one sensitive only to threonine, another only to lysine, and a third only to methionine. These observations with aspartokinase were extended to other bacterial enzymes; and it was concluded that many isozymic systems in bacteria function in a similar fashion at branched pathways to regulate the relative flow of metabolites over each pathway (*Stadtman,* 1968).

Another example of an isozymic system for which a regulatory function has been adduced is that of aldolase. Class A aldolases favor cleavage of fructose-1,6-diphosphate to dihydroxyacetone phosphate and glyceraldehyde-3-phosphate and therefore glycolysis; whereas Class B aldolases are more suited for fructose-1,6-diphosphate synthesis and thus gluconeogenesis (*Rutter et al.,* 1968). No discrete physiological function has yet been associated with Class C aldolases (*Rutter et al.,* 1968).

In still other isozymic systems, such as LDH, several distinct, but not necessarily mutually exclusive, functions have been attributed to the individual forms. Although LDH is known to be a soluble or cytoplasmic enzyme, early work both on intact and disrupted cells established that LDH activity and even specific LDH isozymes are associated with different subcellular organelles (*Vesell,* 1966). These conclusions are derived from two separate techniques. The first is histochemical staining of intact cells (*Allen,* 1961; *Brody and Engel,* 1964; *Brody,* 1968). The second involves isolation of subcellular particles by centrifugation in sucrose gradients followed by ultrasonic disruption of the isolated organelles and identification of their LDH isozymes patterns on starch gels after electrophoresis (*Vesell,* 1966). According to *Keck and Choules* (1962), conclusions based on the latter technique alone may be erroneous because at physiological pH, the LDH isozymes in the cytoplasm are charged and may become attached to differently charged organelles during homogenization and centrifugal partitioning. Once attached to these particles, they may be difficult to release.

Other examples of isozymes located in different subcellular fractions are NAD-malate dehydrogenase (*Delbrück et al.,* 1959; *Siegel and Englard,* 1962; *Englard et al.,* 1960; *Thorne,* 1960; *Thorne and Cooper,* 1964; *Grimm and Doherty,* 1961), aspartate amino transferase (*Borst and Peeters,* 1961; *Eichel and Bukovsky,* 1961), glutaminase (*Shepherd and Kalnitsky,* 1951; *Katunuma et al.,* 1966), alanine amino transferase (*Swick et al.,* 1965), and mitochondrial and supernatant isocitrate dehydrogenase (*Henderson,* 1965, 1968).

An association of LDH-5 with the cell nucleus rests on a series of observations with platelets, bovine lens fibers and nucleated and anucleated erythrocytes (*Vesell and Bearn,* 1962b; *Vesell,* 1965). These studies suggested that a nucleus or active protein synthesis is required to maintain the stability of LDH-5 more than that of other isozymes. Evidence for nuclear synthesis of LDH-5 in rat liver has been published (*Kuehl*, 1967).

Certain isozymes are confined to specific tissues, as in the case of the LDH-X, a tetramer composed of 4 C subunits. LDH-X is present in the testis and the sperm of various mammals (*Blanco and Zinkham*, 1963; *Zinkham et al.*, 1964; *Hawtrey and Goldberg*, 1968). Other examples of specific localization of LDH isozymes in certain tissues are available; teleost fish possess, in their eye and brain, isozymes in addition to the customarily encountered 5 (*Morrison and Wright*, 1966).

These differences among LDH isozymes, not only in distribution between different tissues of an organism but even in subcellular localization, suggest that certain isozymes might meet the specific metabolic requirements of various subcellular organelles better than others (*Vesell and Bearn*, 1961; *Allen*, 1961; *Vesell*, 1966). *Munkres* (1968) has emphasized in his work on malic dehydrogenase isozymes that metabolically related enzymes may function better if they are located in the same subcellular organelles in what he terms multienzyme complexes. He suggests that this aggregation may produce, enhance or modify the activity of several enzymes in this group or that such spatial proximity of sequentially operating enzymes may improve the efficiency of the entire pathway even if aggregation failed to alter the catalytic properties of individual enzymes (*Munkres*, 1968). Multiple molecular forms of an enzyme clearly would offer greater opportunities for interaction at differently charged sites within a cell as well as at different reaction sites in different metabolic pathways.

LDH is critical in intermediary metabolism for several reasons. It is the only enzyme in mammals capable of catalyzing the formation and degradation of lactate. LDH catalyzes a reaction which stands at a branch in a pathway, thereby influencing the fate of pyruvate and being available when shortage of oxygen prevents complete breakdown of pyruvate to CO_2 and water by the Krebs tricarboxylic acid cycle. Finally, by catalyzing the reversible reaction between lactate and pyruvate, LDH regulates the cellular relationship between NAD and NADH and thereby primes glycolysis by providing one source of NAD.

Control mechanisms for this critical enzymatic reaction involving the participation of different LDH isozymes, each subject to a different type of regulation, have been proposed. For example, in *Butyribacterium rettgeri* two LDH isozymes exist (*Wittenberger and Haaf*, 1964). In this organism fermentation of lactate is inducible with lipoic acid as a requirement. However, lipoic acid is not necessary for the fermentation of either glucose or pyruvate (*Kline and Barker*, 1950). One LDH isozyme is present in high concentrations in cells grown on glucose; this LDH requires NADH, catalyzes the reduction of pyruvate by NADH, but not the oxidation of lactate by NAD; hence this isozyme probably functions in lactate synthesis (*Wittenberger*, 1966). The second LDH isozyme is induced by growth on lactate, catalyzes only the oxidation of lactate to pyruvate, uses ferricyanide, but not NAD, as the electron carrier and is repressed by growth on glucose; hence this isozyme probably functions in lactate degradation (*Wittenberger*, 1966).

Another type of regulatory function has been attributed to a specific LDH isozyme. *Fritz* (1965, 1967) proposed that LDH-5, but not LDH-1, is a regulatory protein. He demonstrated that the Krebs Cycle intermediate oxalacetate activates LDH-5 at low pyruvate concentrations but competitively inhibits LDH-5 at high pyruvate concentrations. Such a regulatory effect on LDH-5 by a compound in the adjacent, interrelated Krebs tricarboxylic acid cycle suggests the possible coordination of reactions in the Embden-Myerhoff pathway by levels of intermediates in other pathways. Feedback control of other glycolytic enzymes by Krebs Cycle intermediates and by adenine necleotides is another example of this type of coordination (*Atkinson*, 1966).

Several additional theories to account for the multiple forms of LDH and for their differential tissue distribution have been proposed. Attention has been called to the relationship between the rate of mitosis in a tissue and the prevalence of LDH-5 (*Papaconstantinou*, 1967). LDH-5 is frequently observed to predominate in cells capable of rapid division such as liver and leukocytes, whereas tissues such as brain and heart in which capacity for cell division is low are characterized mainly by LDH-1. Further evidence in support of this association between rate of mitosis and predominance of LDH-5 is derived from conditions where the environment changes rapidly as, for example, when cells are grown in culture (*Philip and Vesell*, 1962; *Vesell et al.*, 1962) or when malignancy develops (*Goldman et al.*, 1964). In these situations, shifts from LDH-1 to predominance of LDH-5 have been reported. However, the isozyme pattern of skeletal muscle, a tissue incapable of regeneration and characterized by predominance of LDH-5, cannot be readily reconciled with this hypothesis.

Still another possible function of multiple LDHs emerged from the data presented in the final section of this chapter; namely, markedly different rates of degradation of a given isozyme in various tissues (*Fritz et al.*, 1969). Certain LDH isozymes might serve a conservative metabolic role in which one LDH isozyme would be required to maintain critical enzymatic function in a tissue where another LDH isozyme was rapidly degraded (*Fritz et al.*, 1969). Previously, differences in LDH isozyme pattern were attributed entirely to differences in rates of synthesis of the A and B subunits (*Markert and Ursprung*, 1962; *Dawson et al.*, 1964; *Goodfriend and Kaplan*, 1964; *Goodfriend et al.*, 1966; *Goldberg et al.*, 1969). The demonstration that degradation plays as significant a role as synthesis in establishing tissue LDH isozyme patterns raises the possibility that differential rates of isozyme degradation not only in various tissues but also within different subcellular particles may exert a significant regulatory effect.

Finally, the 'aerobic-anaerobic' theory developed to account for tissue specific isozyme patterns must be considered. This theory received early favor and widespread attention. Consequently the theory is now almost uniformly accepted and accorded recognition in many textbooks as the true explanation for the tissue distribution of LDH isozymes. Exclusive dependence on the 'aerobic-anaerobic' explanation of data has substantially reduced the power of LDH isozymes as a research tool.

The theory states that LDH-1 is the main isozyme in so-called 'aerobic' tissues, whereas LDH-5 predominates in so-called 'anaerobic' tissues (*Cahn et al.,* 1962; *Dawson et al.,* 1964; *Kaplan,* 1964, 1968). Quantitative oxygen tensions that would be required to place a tissue either in the 'aerobic' or 'anaerobic' category have not been established. Thus tissues are considered either 'aerobic' or 'anaerobic' to suit their LDH isozyme patterns which are then interpreted as conforming to the predictions of the theory. For example, a recent textbook of genetics stated that liver was an anaerobic tissue and contained almost exclusively LDH-5 (*Porter,* 1968). Liver does contain almost exclusively LDH-5, but it is one of the most aerobic tissues. The fact that its isozyme pattern is contrary to the predictions of the theory is evidence against the theory. Other tissues whose isozyme patterns do not conform to the predictions of the theory are mature erythrocytes, platelets and lens fibers (*Vesell,* 1965).

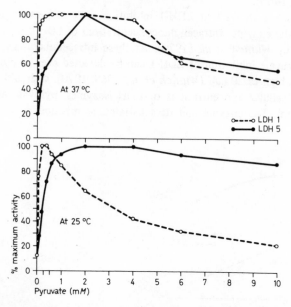

Fig. 8. Effect of temperature on minimizing differences in pyruvate inhibition between electrophoretically purified human LDH-1 and LDH-5 from heart and liver respectively. Reagents and enzyme were heated separately for one-half hour at 37°C prior to assay. Enzyme, coenzyme, and substrate were mixed immediately before assay (from *Vesell,* 1968).

The theory has been supported by the kinetic difference between LDH-1 and LDH-5. Figure 8 shows that at 25 °C LDH-1 is inhibited by pyruvate concentrations to which LDH-5 is resistant but that this great difference between LDH-1 and LDH-5 in capacity to withstand substrate inhibition at 25 °C is substantially reduced at temperatures more physiological for mammals. This fact has been known since 1961 when *Plagemann et al.* published the curve reproduced in figure 9. Figure 9 shows that although LDH-1 is very sensitive to substrate inhibition at 6 °C, at the more physiologic temperature of 40 °C its sensitivity to the concentrations of pyruvate used in the experiment is negligible.

Another question relating to the 'aerobic-anaerobic' theory is whether, under even the most anaerobic circumstances, concentrations of pyruvate sufficiently high to inhibit LDH-1 are ever attained *in vivo*. Concentrations of pyruvate and lactate in various tissues have been available for many years (*Himwich et al.,* 1924; *Johnson and Edwards,* 1937; *Newman,* 1938; *Vesell and Pool,* 1966). Concentrations of pyruvate and lactate in canine skeletal muscle under the most anaerobic conditions never exceed 2.0 m*M* and 25 m*M* respectively (*Vesell and Pool,* 1965). Even under unphysiologic conditions of temperature (25 °C) LDH-1 is not significantly inhibited by these substrate concentrations (*Vesell and Pool,* 1966).

Furthermore, figure 2 shows that LDH-1 inhibition is dependent on the degree of dilution of the enzyme. Intracellular concentrations of LDH isozymes have been estimated by *Wuntch et al.* (1970). At these intracellular isozyme concentrations no pyruvate inhibition of LDH-1 can be detected even at pyruvate concentrations as high as 20 m*M* (*Wuntch et al.,* 1969, 1970). It should be emphasized that intracellular concentrations of LDH isozymes, pyruvate and lactate are not uniform. The enzyme and its substrates, as mentioned in the

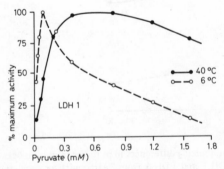

Fig. 9. Effect of temperature on pyruvate inhibition of rabbit LDH-1. Note that at physiological temperature for the rabbit the substrate inhibition is negligible compared to that observed at 6 °C (from *Plagemann et al.,* 1961).

section on subcellular localization, exhibit specific subcellular distributions; and hence estimations of concentrations based on tissue homogenates are probably at best only approximations. Because the concentrations of LDH isozymes at certain regions within the cell are much higher than at other regions, the concentrations of LDH isozymes in certain subcellular sites are higher than those calculated from tissue homogenates (*Wuntch et al.*, 1970). Similarly, substrate concentrations are also probably higher at certain sites within the cell than calculations made from tissue homogenates would indicate. However, such differences that may occur between estimations based on tissue homogenates and the local concentrations actually existing within cells do not alter the conclusion that substrate inhibition of LDH-1 as measured by the usual spectrophotometric methods is probably an artifact of dilution. With stopped-flow spectrophotofluorometric techniques (*Chen et al.*, 1969) no LDH-1 inhibition could be detected when the isozyme was present at physiologic concentrations (*Wuntch et al.*, 1969, 1970).

Recognizing that at physiological concentrations of pyruvate and lactate substrate inhibition of neither LDH-1 nor LDH-5 occurs, *Stambaugh and Post* (1966b) demonstrated a difference between LDH-1 and LDH-5 in degree of inhibition by the end product lactate. They suggested that LDH isozymes might be distributed in tissues not according to inhibition by substrate, but rather by the end product lactate. With pyruvate as substrate, LDH-1, when assayed in dilute concentrations and at pH 7.4, was inhibited significantly more by physiologic concentrations of lactate placed in the reaction mixture than was LDH-5 (*Stambaugh and Post*, 1966b). *Vesell* (1968) confirmed this observation but showed that at 37°C this difference between LDH-1 and LDH-5 in end product inhibition diminished substantially (fig. 10). Furthermore, as the pH was lowered from 7.4 to 6.8, no significant difference between LDH-1 and LDH-5 in end product inhibition occurred (fig. 10) (*Vesell*, 1968). Under anaerobic conditions, lactate accumulates in skeletal muscle and pH's of 6.8 and 6.4 develop within the muscle fiber. Thus in skeletal muscle under anaerobic conditions, pH 6.8 rather than pH 7.4 may be considered physiological.

Concentrations of NADH, rather than of pyruvate, are rate-limiting for the LDH reaction because under basal and particularly under anaerobic conditions NADH levels within the cell are appreciably lower than pyruvate concentrations (*Kaplan*, 1966).

The 'aerobic-anaerobic' theory was utilized early as a basis for several imaginative studies on biochemical evolution in which a relationship between biochemical constitution and anatomical structure was claimed (*Wilson et al.*, 1963; *Wilson et al.*, 1964; *Salthe*, 1965). A correlation was reported between the flight habits of various birds and the amount of A subunits of LDH in their breast muscles (*Wilson et al.*, 1963). Birds capable of sustained flight were observed to possess a higher percentage of subunits in their breast muscles than birds that flew less. This pattern was used to support the 'aerobic-anaerobic' theory that LDH-5 and not LDH-1 was adaptively suited to function where sporadic, sudden releases of energy were required in a relative absence of oxygen. Relative oxygen tension in tissues may indeed influence isozyme pattern, although this seems not to be accomplished by direct substrate inhibition of isozymes *in vivo* because physiological levels of lactate and pyruvate at physiological temperatures are incapable of inhibiting physiological concentrations of LDH-1.

In the next section evidence is presented to show that the tissue LDH isozyme pattern is determined not only by rates of synthesis of the 2 subunits

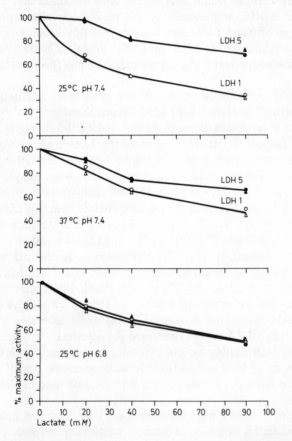

Fig. 10. Effect of increasing concentrations of lactate on pyruvate inhibition of electrophoretically purified human LDH-1 and LDH-5 at 25 °C and pH 7.4 (top), at 37 °C and pH 7.4 (middle) and at 25 °C and pH 6.8 (bottom). Two concentrations of the substrate pyruvate (circles for 0.1 mM and triangles for 0.2 mM) were employed to assay LDH activity spectrophotometrically in the absence and presence of 20, 40 and 90 mM of the product lactate (from *Vesell et al.*, 1968).

but also by rates of isozyme degradation (*Fritz et al.,* 1969). The mechanisms by which isozymes are degraded remain obscure; and the factors capable of influencing these mechanisms are not well defined. However, relative oxygen tension by virtue of its role in determining the concentration of substrate within a tissue may differentially affect the stability and rate of degradation of LDH-1 and LDH-5. Thus tissue isozyme patterns may be substantially affected by the action of different cellular oxygen tensions in determining lactate, pyruvate, NAD and NADH concentrations in various tissues. Whether oxygen tension and substrate concentrations can differentially modify the rate at which each isozyme is degraded remains to be determined. Such a mechanism is attractive because it could account for several observations, including the decrease in A subunits of skeletal muscle, brain, heart and liver with increasing age (*Singh and Kanungo,* 1968) and the increased proportion of A subunits in cells grown in culture with reduced oxygen tension (*Dawson et al.,* 1964; *Goodfriend and Kaplan,* 1966). However, in other circumstances, different mechanisms, opposite in effect to those apparently dependent on cellular oxygen tension, determine rates of isozyme degradation. For example, LDH-5 predominates in the highly aerobic nucleated erythrocytes (*Starkweather et al.,* 1965); with age the nucleus is lost and the cell becomes more dependent on anaerobic metabolism, as indicated by the disappearance of Krebs cycle enzymes from the anucleated erythrocyte. However, with this shift toward anaerobic metabolism, LDH-5, rather than LDH-1, vanishes from the cell (*Starkweather et al.,* 1965). Under conditions prevailing in the anaerobic erythrocyte, stabilization of LDH-5 rather than of LDH-1 is impaired. Were LDH-5 better suited than LDH-1 for function in environments with decreased oxygen tension, and were this the only factor that determined the distribution of LDH in various tissues, as claimed by the 'aerobic-anaerobic' theory, the opposite result should occur.

Interesting investigations by *Coulson and Rabin* (1969) revealed that substrate inhibition of LDH-1 is produced by an inhibitor present in commercial sources of pyruvate. When this inhibitor was chromatographically removed, even high concentrations of pyruvate failed to inhibit LDH-1. This inhibitor was identified as the enol form of pyruvate; thus the actual extent of intracellular LDH inhibition would be restricted by the enol-keto tautomerization rate of pyruvate.

IV. Synthesis and Degradation of LDH Isozymes

Of the many unexpected results obtained during the last 14 years of intensive investigations on multiple molecular forms of enzymes, one of the earliest was the tissue specific patterns of LDH (fig. 11). Different concentrations of a protein in the tissues of an organism, all of whose cells presumably possess the same genome, constitute a major problem and paradox in biology. Since concentrations of various intracellular proteins have been interpreted as indications of gene expression, estimations of rate constants for synthesis and degradation of the same protein in different cells of an animal offer a better approach to the resolution of this problem of gene action than do measurements of intracellular protein concentration alone.

In view of the extensive work on protein turnover, well documented in other chapters of this book, it is clear that the tissue concentration of any protein depends both on its rate constant for synthesis and its rate constant for degradation. The relationship between rates and rate constants for synthesis and degradation and the role of these values in determining tissue concentrations of the various isozymes are illustrated in the experimental data of table I and also in the theoretical examples presented in table II. Table II is designed to clarify the complex relationship between these parameters and to show how changes in protein concentration that occur during development result in attainment of a steady state in adulthood. With reference to table II, consider a hypothetical cell which contains at time zero none of a particular protein P. At this time the cell begins to produce P at a rate determined by the relationship $dP/dt = V_{AP} - V_{PA}$. During the time that the concentration of P is increasing in the cell, V_{AP} is greater than V_{PA}. If we assume that the rate constants for synthesis and degradation do not change, then as the value of V_{PA} approaches the value of V_{AP} the concentration of P in the cell will reach a maximum value. At this time V_{AP} equals V_{PA}, and the protein concentration ceases to change. With similar considerations for the same protein in another tissue B (table II) it can be seen how

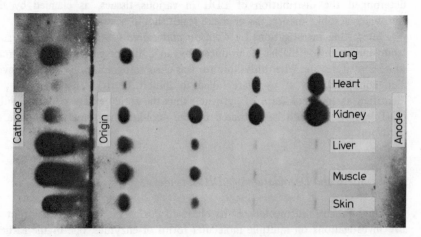

Fig. 11. Photograph of a starch gel stained to reveal the LDH isozymes in homogenates of 6 human tissues obtained at autopsy (from *Vesell*, 1966).

Table I. Turnover of LDH isozymes in rat tissues

Isozyme	Subunit composition	Rate constants		pmoles/g
		Synthesis (pmoles/day/g)	Degradation (day^{-1})	
Cardiac muscle				
1	B$_4$	98.2	0.042	2,340
5	A$_4$	12.4	0.316	39
Skeletal muscle				
1	B$_4$	5.8	0.112	52
5	A$_4$	8.6	0.005	1,720
Liver				
5	A$_4$	74.8	0.047	1,600

From *Fritz et al.*, 1970b.

Table II. Theoretical role of protein turnover in establishing tissue protein concentrations

Tissue	Day	$V_{AP} = k_{AP}$ (pmoles/day/g)	V_{PA} (pmoles/day/g)	k_{PA} (day^{-1})	P pmoles/g
A	1	10	5.00	0.5	8.02
	2	10	7.50	0.5	12.83
	3	10	8.75	0.5	15.71
	4	10	9.38	0.5	17.43
	5	10	9.68	0.5	18.46
	10	10	9.99	0.5	20.00
B	1	10	0.50	0.05	9.78
	2	10	1.00	0.05	19.08
	3	10	1.40	0.05	27.92
	4	10	1.85	0.05	36.34
	5	10	2.26	0.05	44.34
	50	10	9.22	0.05	183.70
C	1	1	0.05	0.05	0.97
	2	1	0.09	0.05	1.91
	3	1	0.14	0.05	2.79
	4	1	0.18	0.05	3.63
	5	1	0.23	0.05	4.43
	50	1	0.92	0.05	18.37

its final concentration would differ if the rate constant for synthesis is identical but the rate constant for degradation is changed from its value in tissue A (table II). Table II also illustrates the hypothetical situation where the rate constants for degradation are identical in two different tissues B and C but the rate constants for synthesis differ.

In the liver of a substrain of C57B1 mice the genetically determined trait of low catalase activity arises from accelerated catalase destruction rather than from any change in synthetic rates (*Rechcigl and Heston*, 1967; *Ganschow and Schimke*, 1969), whereas decreased hepatic activities of δ-aminolevulinate dehydratase in another substrain of C57B1 mice arise from decreased rates of synthesis (*Doyle and Schimke*, 1969). Differences in catalase levels of rat kidney and liver result from a four-fold higher rate of synthesis in liver, whereas rate constants of catalase degradation in these tissues are identical (*Price et al.*, 1962). To answer the question of why different tissues have different levels of LDH isozymes, it is therefore necessary to measure the rate constants for synthesis and the rate constants for degradation of the isozyme in each tissue. Methods were developed for accomplishing these measurements based on techniques described by *Schimke* (1964) and may be summarized as follows: (1) Antibodies to pure rat LDH-1 and LDH-5 were prepared in rabbits. Both of these antibodies will cross react with the heteropolymers LDH-4, LDH-3, and LDH-2 (*Markert and Appella*, 1963). (2) Rats were continuously fed a diet containing C^{14} amino acids. The animals were killed at intervals after starting the diet. LDH isozymes from their heart muscle, skeletal muscle, and liver were separated on anion exchange columns. (3) The isozymes thus separated were precipitated with the appropriate antibody, and incorporation of C^{14} amino acids into each isozyme was determined. (4) The amount of radioisotope incorporated into each isozyme was plotted against the number of days on the special diet. Rate constants for synthesis and degradation were calculated from these data. Details of the methods used for the collection and treatment of data are discussed by *Fritz et al.* (1969). Figure 12 shows the results for LDH-5 from three rat tissues. From these and similar data for the other isozymes rate constants for synthesis and degradation of the 5 LDH isozymes from cardiac and skeletal muscle and of liver LDH-5 and LDH-4 were estimated (*Fritz et al.*, 1970b) using the model for protein synthesis described below.

The problems involved in studying protein turnover by means of isotopic tracers have been considered extensively by *Reiner* (1953) and by *Buchanan* (1961). In the interpretation of our data, we have used a simplified model that is conceptually similar to one proposed in general form by *Reiner* (1953). We assume that there is an intracellular amino acid pool and an intracellular protein pool. For any protein in the pool, V_{AP} is the rate at which it is synthesized, and V_{PA} is the rate at which it is degraded. The following additional assumptions are also made: (1) A steady state in which $V_{AP} = V_{PA}$ exists in the adult healthy

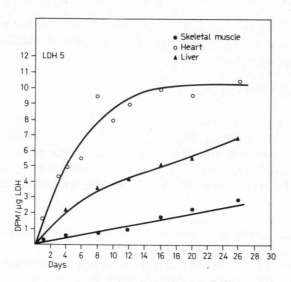

Fig. 12. Time course of C^{14} amino acid incorporation into the LDH-5 of rat heart muscle, skeletal muscle, and liver (from *Fritz et al.,* 1969).

animal. Concentration and composition of the protein and amino acid pools are constant. (2) Only amino acids from the amino acid pool are incorporated into the proteins in the protein pool. (3) When protein is degraded, the products are always amino acids and these are always returned to the amino acid pool. (4) The concentrations of protein, enzyme, and amino acids measured in the tissue extracts are a true measure of the intracellular concentration of these substances. According to this assumption, rates of intracellular protein and amino acid turnover are much greater than rates of cell turnover.

These assumptions permit a simplified treatment of what is an otherwise very difficult kinetic problem. With these assumptions, the total rate of change of any protein P in the protein pool is:

$$dP/dt = V_{AP} - V_{PA} = 0,$$

where P = moles of protein/unit weight of tissue.

Now let $A^* = $ dpm/mole amino acid, and $P^* = $ dpm/mole protein.

Then the rate of labeling is $dP^*/dt = V^*_{AP} - V^*_{PA}$.

$V^*_{AP} = $ rate of incorporating label into protein $= (V_{AP} \Sigma n_i A_i^*)/P$,

where $n_i = $ number of moles of amino acid i in protein

and $A^*_i = $ dpm/mole amino acid i.

Also, $V^*_{PA} = (V_{PA} P^*)/P$.

Therefore, $dP^*/dt = (V_{AP} \Sigma n_i A^* - V_{PA} P^*) \, 1/P$.

V_{AP} and V_{PA} are constant.

However, A_i^* and P^* vary with time.

If we continue feeding the animal a constant amount of randomly and uniformly labeled amino acids, A_i^* and P^* eventually reach constant maximum values A^*_{max} and P^*_{max}, respectively.

At this point $dP^*/dt = 0$;

so, $(V_{AP} \Sigma n_i A^*_{imax})/P = (V_{PA} P^*_{max})/P$.

But $V_{AP} = V_{PA}$;

thus, $\Sigma n_i A^*_{imax} = P^*_{max}$ = constant.

Therefore, $dP^*/dt = V_{AP}/P (P^*_{max} - P^*)$.

Integration of this equation between the limits of $t = 0$, taken as the time at which the amino acid pool has reached constant specific activity, and any arbitrary time, t, prior to the time at which the protein pool has reached constant specific activity yields:

$$\ln (P^*_{max} - P^*) = -V_{AP}t/P - \ln (P^*_{max} - P^*_o),$$

where P^*_o is the specific activity of the protein at $t = 0$ and $-V_{AP}/P$ is the slope of the straight line resulting from a plot of $\ln(P^*_{max} - P^*)$ vs. t.

Assuming that synthesis is zero-order and degradation is first-order (*Price et al.*, 1962; *Segal and Kim*, 1963; *Berlin and Schimke*, 1965; *Rechcigl*, 1968), V_{AP} must be equal to the rate constant for synthesis (k_{AP}), and V_{PA} must be equal to the product of the steady-state concentration of protein and the first-order rate constant for degradation (k_{PA}).

In the adult animal LDH isozyme levels do not change, and therefore rates of isozyme synthesis are equal to rates of isozyme degradation. Rate constants for isozyme synthesis that are zero order have physical significance because they equal rates of isozyme synthesis. However, rate constants for degradation that are first order do not directly indicate how rapidly the enzyme is degraded in the adult healthy rat because they are not equal to the rates of degradation. During ontogeny the concentrations of the LDH isozymes in most mammalian tissues are in a continuous process of change (*Philip and Vesell*, 1962; *Vesell et al.*, 1962; *Cahn et al.*, 1962). During this period, if the concentration of an isozyme is increasing in the cell, then the rate of synthesis of this isozyme is greater than its rate of degradation.

If it is true that the rate constants for synthesis and degradation do not change during the life of the animal, then according to this model the steady-state concentrations of isozymes in various tissues would be determined entirely by the values of the enzyme rate constants. Changes of isozyme concentrations during development would simply be a result of an approach to steady state from an initial low level of isozyme. It has been previously reported that the LDH isozyme concentrations change in the mouse and rat during postnatal development and that the distribution of isozymes is dramatically altered in the developing mouse and rat heart (*Markert and Ursprung*, 1962, and *Blatt et al.*, 1966). For example, LDH-1, the isozyme present in highest concentration in the adult rat heart, is present in lowest concentration in the postpartum rat heart.

The enzyme levels within a cell will change as long as the rates of synthesis and degradation are not equal. If V_{AP} is greater than V_{PA} they will increase. If V_{AP} is less than V_{PA} they will decrease.

It is important to emphasize that this method which involves the continuous administration of C^{14} amino acids in the rat diet is valid only for measuring rate constants for synthesis and degradation of relatively slowly turning over proteins. This is so because the time required for the free amino acid pools in the various tissues to become saturated with radioisotope may be rate limiting. If a protein becomes completely labeled simultaneously with full labeling of the amino acid pool, the rate constants for synthesis and degradation of that protein will be only minimum values. Additional problems in determining turnover rates of proteins are discussed by *Poole et al.* (1969) in their study of catalase turnover employing various techniques.

The mechanisms responsible for maintaining different rate constants for synthesis and degradation of the LDH isozymes within rat tissues remain to be elucidated. The chemical steps involved in protein biosynthesis are known in outline, but the regulation of these processes is poorly defined. Thus the observation that rat LDH-5 is made 6 times faster in liver than in cardiac muscle could be explained in several ways: (1) messenger RNA for the A subunit could be made 6 times faster in rat liver than in rat heart. This assumes that the messenger is used only once; (2) the messenger could bind to ribosomes 6 times faster in the liver; (3) a combination of (1) and (2) e.g., the messenger could be made 3 times faster in the liver and could bind to ribosomes twice as fast. Additional possibilities could be involved, including availability of transfer RNA, binding of transfer RNA to ribosomes, activation of amino acids, etc.

Recent evidence indicates that mammalian messenger RNA is more stable than the bacterial form and that translation may be more important than transcription in regulating protein levels in mammalian cells (*Stent,* 1964; *Tomkins et al.,* 1969). In the case of subunit enzymes such as LDH, the assembly of the subunits is an additional site of possible regulation which should be considered.

Although biodegradation of LDH isozymes in rat tissues appears from our work to proceed with great specificity, the exact nature of the process is entirely unknown. It is difficult to imagine that all the intracellular proteins have their own protease. With the exception of lysosomes, even nonspecific proteases have thus far not been detected in most mammalian cells. When tyrosine transaminase synthesis is blocked by cycloheximide or puromycin, its degradation is also prevented (*Kenney,* 1967). This observation could be construed as evidence in favor of a rapidly turning over protease which digests itself as well as other proteins. Such a protein would be difficult to isolate because of its capacity for self-destruction. The activity of such a nonspecific protease toward its enzyme substrates could be regulated by numerous factors including intracellular metabolite concentration. The recent report that ornithine decarboxylase in rat liver has a half-life of 11 minutes demonstrates the existence of very rapidly turning over proteins in mammals (*Russell and Snyder,* 1969).

It may also be significant that protein degradation depends upon an intra-cellular source of energy (*Simpson,* 1953; *Steinberg and Vaughan,* 1956; *Penn,* 1960). Protein biosynthesis was once considered to proceed by a reversal of proteolysis and perhaps, as *Kenney* (1967) suggested, the ribosomal mechanism for enzyme synthesis functions reversibly.

The example of the numerous proteolytic enzymes required for blood clotting may also be relevant; these enzymes are present in an inactive form and require stepwise conversion through cleavage of certain residues by a series of enzymes and environmental changes triggered by bleeding (*Ratnoff,* 1966). Degradation of LDH isozymes could also proceed by dissociation of the active tetramer into monomers which may then be available for re-utilization. The mechanisms responsible for regulating the assembly of subunits and the intra-cellular sites where this assembly occurs remain unidentified. With respect to the possibility that thermal denaturations play a role in LDH degradation, it is known that LDH-1 in certain animals is more heat stable than LDH-5 (*Zondag,* 1962; *Vesell and Yielding,* 1966). Table I illustrates that the former is degraded 22 times faster than the latter in rat skeletal muscle, although LDH-5 of cardiac muscle is degraded almost 8 times faster than cardiac muscle LDH-1. Similarly, factors other than thermal denaturation appear to be involved in the decay of rat alanine aminotransferase which has a half-life of 400 days at 37.5 °C *in vitro; in vivo* this enzyme exhibits a half-life of 3 days in rat liver and 20 days in rat skeletal muscle (*Segal et al.,* 1969). Common intracellular metabolites have been reported to protect specific LDH isozymes against thermal and proteolytic digestion *in vitro* (*Vesell and Yielding,* 1966). Large differences among tissues in rate constants for LDH degradation suggest that certain intracellular physico-chemical conditions and concentrations of various metabolites shown to in-fluence isozyme decay *in vitro* might play a role in regulating their catabolism *in vivo*.

Finally, enzyme turnover appears to offer a means of regulating various metabolic pathways. Modification in rates of enzyme degradation could permit the cell an additional control of the relative speed at which compounds flowed over different metabolic pathways. Changes in rates of degradation might be accomplished by alterations in any one of numerous intracellular physicochemi-cal conditions, or in the relative concentrations of cofactors, vital nutrients, activators, inhibitors or hormones. *Schimke et al.* (1965a, b) demonstrated that increased tryptophan pyrrolase activity in rat liver after tryptophan injection was due to decreased intracellular degradation of the enzyme.

Changing rates of isozyme degradation have not previously been considered as a possible mechanism for regulating metabolic pathways. As data accumulate on turnover of critical rate limiting or regulatory enzymes, it is anticipated that differential rates of enzyme degradation will be recognized as being of biological significance in the regulation of metabolic pathways.

Acknowledgment

The model discussed on page 363 was largely developed by Dr. *Kenneth M. Pruitt*, Laboratory of Molecular Biology, University of Alabama Medical Center.

V. References

Allen, J.M.: Multiple forms of lactic dehydrogenase in tissues of the mouse: their specificity, cellular localization, and response to altered physiological conditions. Ann. N.Y. Acad. Sci. *94:* 937–951 (1961).

Amador, E.; Dorfman, L.E., and Wacker, W.E.C.: Serum lactic dehydrogenase: an analytical assessment of current assays. Clin. Chem. *9:* 391–399 (1963).

Anderson, S.R. and Weber, G.: Multiplicity of binding by lactate dehydrogenases. Biochemistry *4:* 1948–1957 (1965).

Anderson, S.R. and Weber, G.: The reversible acid dissociation and hybridization of lactic dehydrogenase. Arch. Biochem. Biophys. *116:* 207–223 (1966).

Appella, E.: Subunit structure of proteins biochemical and genetic aspects. Brookhaven Symp. Biol. *17:* 151–152 (1964).

Appella, E. and Markert, C.L.: Dissociation of lactate dehydrogenase into subunits with guanidine hydrochloride. Biochem. biophys. res. Commun. *6:* 171–176 (1961).

Appella, E. and Zito, R.: A chemical study of lactate dehydrogenase isozyme B. Ann. N.Y. Acad. Sci. *75:* 568–577 (1968).

Atkinson, D.E.: Regulation of enzyme activity. Annu. Rev. Biochem. *35:* 85–124 (1966).

Babson, A.L. and Phillips, G.E.: A rapid colorimetric assay for serum lactic dehydrogenase. Clin. chim. Acta *12:* 210–215 (1965).

Beckman, L. and Beckman, G.: Individual and organ-specific variations of human acid phosphatase. Biochem. Genetics *1:* 145–153 (1967).

Berlin, C.M. and Schimke, R.T.: Influence of turnover rates on the responses of enzymes to cortisone. Mol. Pharmacol. *1:* 149–156 (1965).

Blanco, A. and Zinkham, W.H.: Lactate dehydrogenase in human testes. Science *139:* 601–602 (1963).

Blatt, W.F.; Blatteis, C.M., and Mager, M.: Tissue lactate dehydrogenase isozymes. Developmental patterns in the neonatal rat. Canad. J. Biochem. *44:* 537–543 (1966).

Borst, P. and Peeters, E.: Intracellular localization of glutamate-oxalacetic transaminase in heart. Biochem. biophys. Acta *54:* 188–189 (1961).

Briere, R.O.; Preston, J.A., and Batsakis, J.G.: Rapid colorimetric (tetrazolium salt) assay for lactate dehydrogenase. Amer. J. clin. Path. *45:* 544–547 (1966).

Brody, I.A.: Histochemistry of lactate dehydrogenase isozymes in intact muscle. Ann. N.Y. Acad. Sci. *151:* 587–593 (1968).

Brody, I.A. and Engel, W.K.: Isozyme histochemistry: a new method for the display of selective lactic dehydrogenase isozymes on an electrophoretic pattern. Nature, Lond. *201:* 685–687 (1964).

Brown, J.; Miller, D.M.; Holloway, M.T., and Leve, G.D.: Hexokinase isoenzymes in liver and adipose tissue of man and dog. Science *155:* 205–207 (1967).

Buchanan, D.L.: Analysis of continuous dosage isotope experiments. Arch. Biochem. Biophys. *94:* 489–499 (1961).

Butterworth, P.J. and Moss, D.W.: Action of neuraminidase on human kidney alkaline phosphatase. Nature, Lond. *209:* 805–806 (1966).

Cabello, J.; Prajoux, V., and Plaza, M.: Separation of 2 enzyme proteins in purified preparations of human liver and erythrocyte arginases by zone electrophoresis – mobility of fast components in both preparations. Arch. Biol. *3:* 7–13 (1966).

Cahn, R.D.; Kaplan, N.O.; Levine, L., and Zwilling, E.: Nature and development of lactic dehydrogenases. Science *136:* 962–969 (1962).

Capps, R.D.; Batsakis, J.G.; Briere, R.O., and Calam, R.R.L.: An automated (tetrazolium salt) assay for serum lactic dehydrogenase. Clin. Chem. *12:* 406–413 (1966).

Chen, R.F.; Schechter, A.N., and Berger, R.L.: Stopped-flow fluorometry with available instrumentation. Anal. Biochem. *29:* 68–75 (1969).

Colvin, J.R.; Smith, D.B., and Cook, W.H.: The microheterogeneity of proteins. Chem. Rev. *54:* 687–714 (1954).

Coulson, C.J. and Rabin, B.R.: Inhibition of lactate dehydrogenase by high concentrations of pyruvate: the nature and removal of the inhibitor. FEBS Letters *3:* 333–337 (1969).

Davis, C.H.; Schliselfeld, L.H.; Wolf, D.P.; Leavitt, C.A., and Krebs, E.G.: Interrelationships among glycogen phosphorylase isozymes. J. biol. Chem. *242:* 4824–4833 (1967).

Dawson, D.M.; Goodfriend, T.L., and Kaplan, N.O.: Lactic dehydrogenases: functions of the two types. Science *143:* 929–933 (1964).

Delbrück, A.H.; Schimassek, H.; Bartsch, K., und Bücher, Th.: Enzym-Verteilungsmuster in einigen Organen und in experimentellen Tumoren der Ratte und der Maus. Biochem. Z. *331:* 297–311 (1959).

Doyle, D. and Schimke, R.T.: The genetic and developmental regulation of hepatic δ-aminolevulinate dehydratase in mice. J. biol. Chem. *244:* 5449–5459 (1969).

Ehrenstein, G. von: Translational variations in the amino acid sequence of the α-chain of rabbit hemoglobin. Cold Spr. Harb. Symp. quant. Biol. *31:* 705–714 (1966).

Eichel, H. and Bukovsky, J.: Intracellular distribution pattern of rat liver glutamic-oxalacetic transaminase. Nature, Lond. *191:* 243–245 (1961).

Emerson, P.M., Wilkinson, J.H., and Withycombe, W.A.: Effect of oxalate on the activity of lactate dehydrogenase isoenzymes. Nature, Lond. *202:* 1337–1338 (1964).

Englard, S.; Siegel, L., and Breiger, H.H.: Purification and properties of beef heart muscle 'cytoplasmic' malate dehydrogenase. Biochem. biophys. res. Commun. *3:* 323–327 (1960).

Epstein, C.J. and Schechter, A.N.: An approach to the problem of conformational isozymes. Ann. N.Y. Acad. Sci. *151:* 85–101 (1968).

Fildes, R.A. and Harris, H.: Genetically determined variation of adenylate kinase in man. Nature, Lond. *209:* 261–263 (1966).

Fritz, P. J.: Rabbit muscle lactate dehydrogenase-5: a regulatory enzyme. Science *150:* 364–366 (1965).

Fritz, P.J.: Rabbit lactate dehydrogenase isozymes: effect of pH on activity. Science *156:* 82–83 (1967).

Fritz, P.J.; Morrison, W.J.; White, E.L., and Vesell, E.S.: Comparative study of methods for quantitative measurement of lactate dehydrogenase isozymes. Analyt. Biochem. (1970a).

Fritz, P.J.; Vesell, E.S.; White, E.L., and Pruitt, K.M.: The roles of synthesis and degradation in determining tissue concentrations of lactate dehydrogenase-5. Proc. nat. Acad. Sci., Wash. *62:* 558–565 (1969).

Fritz, P.J.; White, E.L., and Vesell, E.S.: Biosynthesis and biodegradation of lactate dehydrogenases in rats. Fed. Proc. *29:* 735 (1970b).

Ganschow, R.E. and Schimke, R.T.: Independent genetic control of the catalytic activity and the rate of degradation of catalase in mice. J. biol. Chem. *244:* 4649–4658 (1969).

Goldberg, E.; Cuerrier, J.P., and Ward, J.C.: Lactate dehydrogenase ontogeny, paternal gene activation, and tetramer assembly in embryos of brook trout, lake trout and their hybrids. Biochem. Genet. *2:* 335–350 (1969).

Goldman, R.D.; Kaplan, N.O., and Hall, T.C.: Lactic dehydrogenase in human neoplastic tissues. Cancer Res. *24:* 389–399 (1964).

Goodfriend, T.L. and Kaplan, N.O.: Effects of hormone administration on lactic dehydrogenase. J. biol. Chem. *239:* 130–135 (1964).

Goodfriend, T.L.; Sokol, D.M., and Kaplan, N.O.: Control of synthesis of lactic acid dehydrogenases. J. molec. Biol. *15:* 18–31 (1966).

Grell, E.H.; Jacobson, K.B., and Murphy, J.B.: Alterations of genetic material for analysis of alcohol dehydrogenase isozymes of *Drosophila melanogaster.* Ann. N.Y. Acad. Sci. *151:* 441–455 (1968).

Grimm, F.C. and Doherty, D.G.: Properties of the two forms of malate dehydrogenase from beef heart. J. biol. Chem. *236:* 1980–1985 (1961).

Hawtrey, C. and Goldberg, E.: Differential synthesis of LDH in mouse testes. Ann. N.Y. Acad. Sci. *151:* 611–615 (1968).

Henderson, N.S.: Isozymes of isocitrate dehydrogenase: subunit structure and intracellular location. J. exp. Zool. *158:* 263–273 (1965).

Henderson, N.S.: Intracellular location and genetic control of isozymes of NADP – dependent isocitrate dehydrogenase and malate dehydrogenase. Ann. N.Y. Acad. Sci. *151:* 429–440 (1968).

Hess, B.: DPN-dependent enzymes in serum. Ann. N.Y. Acad. Sci. *75:* 292–303 (1958).

Hess, B. und Walter, S.I.: Über das Protein der Laktatdehydrogenase im menschlichen Serum und Geweben. Klin. Wschr. *38:* 1080–1088 (1960).

Himwich, H.E.; Loebel, R.O., and Barr, D.P.: Studies of the effect of exercise in diabetes. 1. Changes in acid-base equilibrium and their relation to the accumulation of lactic acid and acetone. J. biol. Chem. *59:* 265–293 (1924).

Holmes, E.W.; Malone, J.I.; Winegrad, A.I., and Oski, F.A.: Hexokinase isoenzymes in human erythrocytes: association of type II with fetal hemoglobin. Science *156:* 646–648 (1967).

Hopkinson, D.A.; Spencer, N., and Harris, H.: Red cell acid phosphatase variance: A new human polymorphism. Nature, Lond. *199:* 969–971 (1963).

Hotchkiss, R.D.: Subunit structure of proteins biochemical and genetic aspects. Brookhaven Symp. Biol. *17:* 129–130 (1964).

Hunter, R.L. and Markert, C.L.: Histochemical demonstration of enzymes separated by zone electrophoresis in starch gels. Science *125:* 1294–1295 (1957).

Jacobson, K.B.: Alcohol dehydrogenase of *Drosophila:* interconversion of isoenzymes. Science *159:* 324–325 (1968).

Johnson, R.E. and Edwards, H.T.: Lactate and pyruvate in blood and urine after exercise. J. biol. Chem. *118:* 427–432 (1937).

Kaplan, J.C. and Beutler, E.: Hexokinase isoenzymes in human erythrocytes. Science *159:* 215–216 (1968).

Kaplan, N.O.: Lactate dehydrogenase – structure and function. Brookhaven Symp. Biol. *17:* 131–149 (1964).

Kaplan, N.O.: Current aspects of biochemical energetics; pp. 447–458 (Academic Press, New York 1966).

Kaplan, N.O.: Nature of multiple molecular forms of enzymes. Ann. N.Y. Acad. Sci. *151:* 382–399 (1968).

Kaplan, N.O. and Cahn, R.D.: Lactic dehydrogenases and muscular dystrophy in the chicken. Proc. nat. Acad. Sci., Wash. *48:* 2123–2130 (1962).

Kaplan, N.O.; Everse, J., and Admiraal, J.: Significance of substrate inhibition of dehydrogenases. Ann. N.Y. Acad. Sci. *75:* 400–412 (1968).

Kar, N.C. and Pearson, C.M.: Creatine phosphokinase isoenzymes in muscle in human myopathies. Amer. J. clin. Path. *43:* 207–209 (1965).

Katunuma, N.; Tomino, I., and Nishino, H.: Glutaminase isozymes in rat kidney. Biochem. biophys. res. Commun. *22:* 321–328 (1966).

Katzen, H.M.; Soderman, D.D., and Cirillo, V.J.: Tissue distribution and physiological significance of multiple forms of hexokinase. Ann. N.Y. Acad. Sci. *75:* 351–358 (1968).

Keck, K. and Choules, E.A.: The differential binding of lactate dehydrogenase isozymes to ribonucleoprotein particles. Arch. Biochem. *99:* 205–209 (1962).

Kenney, F.T.: Turnover of rat liver tyrosine transaminase: stabilization after inhibition of protein synthesis. Science *156:* 525–527 (1967).

Khan, M.G. and Südi, J.: Temperature and concentration dependence of the stability of pig lactate dehydrogenase isoenzymes. H_4, H_2M_2 and M_4. Acta biochim. biophys. Acad. Sci. Hung. *3:* 409–420 (1968).

Kitto, G.B.; Wassarman, P.M., and Kaplan, N.O.: Enzymatically active conformers of mitochondrial malate dehydrogenase. Proc. nat. Acad. Sci., Wash. *56:* 578–585 (1965).

Kline, L. and Barker, H.A.: A new growth factor required by *Butyribacterium rettgeri.* J. Bact. *60:* 349–363 (1950).

Kuehl, L.: Evidence for nuclear synthesis of lactic dehydrogenase in rat liver. J. biol. Chem. *242:* 2199–2206 (1967).

Lineweaver, H. and Burk, D.: The determination of enzyme dissociation constants. J. amer. chem. Soc. *56:* 658–666 (1934).

Markert, C.L.: Lactate dehydrogenase isozymes: dissociation and recombination of subunits. Science *140:* 1329–1330 (1963).

Markert, C.L.: Developmental genetics. Harvey Lect. *59:* 187–218 (1965).

Markert, C.L.: The molecular basis for isozymes. Ann. N.Y. Acad. Sci. *151:* 14–40 (1968).

Markert, C.L. and Appella, E.: Immunological properties of lactate dehydrogenase isozymes. Ann. N.Y. Acad. Sci. *103:* 915–929 (1963).

Markert, C.L. and Møller, F.: Multiple forms of enzymes: tissue, ontogenetic, and species specific patterns. Proc. nat. Acad. Sci., Wash. *45:* 753–763 (1959).

Markert, C.L. and Ursprung, H.: The ontogeny of isozyme patterns of lactate dehydrogenase in the mouse. Develop. Biol. *5:* 363–381 (1962).

Markert, C.L. and Whitt, G.S.: Molecular varieties of isozymes. Experientia *24:* 977–991 (1968).

Meister, A.: Reduction of a, γ-diketo and a-keto acids catalyzed by muscle preparations and by crystalline lactic dehydrogenase. J. biol. Chem. *184:* 117–129 (1950).

Millar, D.B.; Fratelli, V., and Willick, G.E.: The quarternary structure of lactate dehydrogenase. 1. The subunit molecular weight and the reversible association at acid pH. Biochemistry *8:* 2416–2421 (1969).

Morrison, W.J. and Wright, J.E.: Genetic analysis of three lactate dehydrogenase isozyme systems in trout: evidence for linkage of genes coding subunits A and B. J. exp. Zool. *163:* 259–270 (1966).

Munkres, K.D.: Genetic and epigenetic forms of malate dehydrogenase in *Neurospora.* Ann. N.Y. Acad. Sci. *151:* 294–306 (1968).

Neilands, J.B.: Studies on lactic dehydrogenase of heart. J. biol. Chem. *199:* 373–381 (1952).

Newman, E.V.: Distribution of lactic acid between blood and muscle of rats. Amer. J. Physiol. *122:* 359–366 (1938).

Papaconstantinou, J.: Molecular aspects of lens cell differentiation. Science *156:* 338–346 (1967).

Penn, N.W.: The requirements for serum albumin metabolism in subcellular fractions of liver and brain. Biochim. biophys. Acta *37:* 55–63 (1960).

Pesce, A.; Fondy, T.P.; Stolzenbach, F.; Castillo, F., and Kaplan, N.O.: The comparative enzymology of lactic dehydrogenases. III. Properties of the H_4 and M_4 enzymes from a number of vertebrates. J. biol. Chem. *239:* 2151–2167 (1967).

Philip, J. and Vesell, E.S.: Sequential alterations of lactic dehydrogenase isozymes during embryonic development and in tissue culture. Proc. Soc. exp. Biol., N.Y. *110:* 582–585 (1962).

Plagemann, P.G.W.; Gregory, K.F., and Wróblewski, F.: The electrophoretically distinct forms of mammalian lactic dehydrogenase. II. Properties and interrelationships of rabbit and human lactic dehydrogenase isozymes. J. biol. Chem. *235:* 2288–2293 (1960).

Plagemann, P.G.W.; Gregory, K.F., and Wróblewski, F.: Die elektrophoretisch trennbaren Lactat-dehydrogenasen des Säugetieres. III. Einfluss der Temperatur auf die Kaninchen. Biochem. Z. *334:* 37–48 (1961).

Poole, B.; Leighton, F., and de Duve, C.: The synthesis and turnover of rat liver peroxisomes. II. Turnover of peroxisome proteins. J. cell. Biol. *41:* 536–546 (1969).

Porter, I.H.: Heredity and disease; p. 222 (McGraw-Hill, New York 1968).

Price, V.E.; Sterling, W.R.; Tarantola, V.A.; Hartley, R.W., Jr., and Rechcigl, M., Jr.: The kinetics of catalase synthesis and destruction *in vivo.* J. biol. Chem. *237:* 3468–3475 (1962).

Rajewski, K.: Kreuzreagierende antigene Determinanten auf Lactatdehydrogenasen I und V durch Acetylierung. Biochim. biophys. Acta *121:* 51–68 (1966).

Ratnoff, O.D.: The biology and pathology of the initial stages of blood coagulation. Progr. Hemat., vol. 5, pp. 204–245 (Grune and Stratton, New York 1966).

Rechcigl, M., Jr.: In vivo turnover and its role in the metabolic regulation of enzyme levels. Enzymologia, D. Haag *34:* 23–39 (1968).

Rechcigl, M., Jr. and Heston, W.E.: Genetic regulation of enzyme activity in mammalian system by the alteration of the rates of enzyme degradation. Biochem. biophys. res. Commun. *27:* 119–124 (1967).

Reiner, J.M.: The study of metabolic turnover rates by means of isotopic tracers. I. Fundamental relations. Arch. Biochem. Biophys. *46:* 53–79 (1953).

Rifkin, D.B.; Hirsch, D.I.; Rifkin, M.R., and Konigsberg, W.: A possible ambiguity in the coding of mouse hemoglobin. Cold Spr. Harb. Symp. quant. Biol. *31:* 715–718 (1966).

Rouslin, W. and Braswell, E.: Analysis of factors affecting lactic dehydrogenase subunit composition determinations. J. theor. Biol. *19:* 169–182 (1968).

Russell, D.H. and Snyder, S.H.: Amine synthesis in regenerating rat liver: extremely rapid turnover of ornithine decarboxylase. Molec. Pharmacol. *5:* 253–262 (1969).

Rutter, W.J.; Rajkumar, T.; Penhoet, E.; Kochman, M., and Valentine, R.: Aldolase variants: structure and physiological significance. Ann. N.Y. Acad. Sci. *151:* 102–117 (1968).

Salthe, S.N.: Comparative catalytic studies of lactic dehydrogenases in the amphibia: environmental and physiological correlations. Comp. Biochem. Physiol. *16:* 393–408 (1965).

Sayre, F.W. and Hill, B.R.: Fractionation of serum lactic dehydrogenase by salt concentration gradient elution and paper electrophoresis. Proc. Soc. exp. Biol., N.Y. *96:* 695–697 (1957).

Schimke, R.T.: The importance of both synthesis and degradation in the control of arginase levels in rat liver. J. biol. Chem. *239:* 3808–3817 (1964).

Schimke, R.T. and Grossbard, L.: Studies on isozymes of hexokinase in animal tissues. Ann. N.Y. Acad. Sci. *75:* 332–350 (1968).

Schimke, R.T.; Sweeney, E.W., and Berlin, C.M.: The roles of synthesis and degradation in the control of rat liver tryptophan pyrrolase. J. biol. Chem. *240:* 322–331 (1965a).

Schimke, R.T.; Sweeney, E.W., and Berlin, C.M.: Studies on the stability *in vivo* and *in vitro* of rat liver tryptophan pyrrolase. J. biol. Chem. *240:* 4609–4620 (1965b).

Schloen, L.H.; Bamburg, J.R.; and Sallach, H.J.: Isozymes of pyruvate kinase in tissues and eggs of *Rana pipiens.* Biochem. biophys. res. Commun. *36:* 823–829 (1969).

Schoenenberger, G. and Wacker, W.E.C.: Peptide inhibitors of lactic dehydrogenase (LDH). II. Isolation and characterization of peptides I and II. Biochemistry *5:* 1375–1379 (1966).

Segal, H.L. and Kim, Y.S.: Glucocorticoid stimulation of the biosynthesis of glutamic-alanine transaminase. Proc. nat. Acad. Sci., Wash. *50:* 912–918 (1963).

Segal, H.L.; Matsuzawa, T.; Haider, M., and Abraham, G.J.: What determines the half-life of proteins *in vivo*? Some experiences with alanine aminotransferase of rat tissues. Biochem. biophys. res. Commun. *36:* 764–770 (1969).

Shapiro, B.M. and Stadtman, E.R.: Glutamine synthetase deadenylylating enzyme. Biochem. biophys. res. Commun. *30:* 32–37 (1968).

Shaw, C.R.: Electrophoretic variation in enzymes. Science *149:* 936–943 (1965).

Shaw, C.R.: Isozymes: classification, frequency and significance. Int. Rev. Cytol. *25:* 297–332 (1969).

Shaw, C.R. and Barto, E.: Genetic evidence for the subunit structure of lactate dehydrogenase isozymes. Proc. nat. Acad. Sci., Wash. *50:* 211–214 (1963).

Shepherd, J.A. and Kalnitsky, G.: Intracellular distribution of the phosphate – activated glutaminase of rat liver. J. biol. Chem. *192:* 1–7 (1951).

Siegel, L. and Englard, S.: Beef heart malate dehydrogenases. III. Comparative studies of some properties of M-malic dehydrogenase. Biochem. biophys. Acta *64:* 101–110 (1962).

Simpson, M.V.: The release of labeled amino acids from the proteins of rat liver slices. J. biol. Chem. *201:* 143–154 (1953).

Singh, S.N. and Kanungo, N.S.: Alterations in lactate dehydrogenase of the brain, heart, skeletal muscle, and liver of rats of various ages. J. biol. Chem. *243:* 4526–4529 (1968).

Spencer, N.; Hopkinson, D.A., and Harris, H.: Phosphoglucomutase polymorphism in man. Nature, Lond. *204:* 742–745 (1964).

Stadtman, E.R.: The role of multiple enzymes in the regulatioon of branched metabolic pathways. Ann. N.Y. Acad. Sci. *75:* 516–530 (1968).

Stambaugh, R. and Post, D.: A spectrophotometric method for the assay of lactic dehydrogenase subunits. Analyt. Biochem. *15:* 170–180 (1966a).

Stambaugh, R. and Post, D.: Substrate and product inhibition of rabbit muscle lactic dehydrogenase heart (H_4) and muscle (M_4) isozymes. J. biol. Chem. *241:* 1462–1467 (1966b).

Starkweather, W.H.; Cousineau, L.; Schoch, H.K., and Zarafonetis, C.J.: Alterations of erythrocyte lactate dehydrogenase in man. Blood *26:* 63–73 (1965).

Steinberg, D. and Vaughan, M.: Observations of intracellular protein catabolism studied *in vitro.* Arch. Biochem. *65:* 93–105 (1956).

Stent, G.S.: The operon: on its third anniversary. Science *144:* 816–820 (1964).

Swick, R.W.; Barnstein, P.L., and Stange, J.L.: The metabolism of mitochondrial proteins. 1. Distribution and characterization of the isozymes of alanine-amino transferase in rat liver. J. biol. Chem. *240:* 3334–3340 (1965).

Thorne, C.J.R.: Characterization of two malate dehydrogenases from rat liver. Biochem. biophys. Acta *42:* 175–176 (1960).

Thorne, C.J.R. and Cooper, P.M.: Preparation of pig heart supernatant malate dehydrogenase. Biochem. biophys. Acta *81:* 397–399 (1964).

Tomkins, G.M.; Gelehrter, T.D.; Granner, D.; Martin, D., Jr.; Samuels, H.H., and Thompson, E.B.: Control of specific gene expression in higher organisms. Science *166:* 1474–1480 (1969).

Vesell, E.S.: Lactate dehydrogenase isozyme patterns of human platelets and bovine lens fibers. Science *150:* 1735–1737 (1965).

Vesell, E.S.: Genetic control of isozyme patterns in human tissues. Progr. med. Gen. *5:* 128–175 (1966).

Vesell, E.S.: Introduction. Ann. N.Y. Acad. Sci. *151:* 5–13 (1968).

Vesell, E.S. and Bearn, A.G.: Localization of lactic dehydrogenase activity in serum fractions. Proc. Soc. exp. Biol., N.Y. *94:* 96–99 (1957).

Vesell, E.S. and Bearn, A.G.: Observations on the heterogeneity of malic and lactic dehydrogenase in human serum and red blood cells. J. clin. Invest. *37:* 672–677 (1958).

Vesell, E.S. and Bearn, A.G.: Isozymes of lactic dehydrogenase in human tissues. J. clin. Invest. *40:* 586–591 (1961).

Vesell, E.S. and Bearn, A.G.: Variations of lactic dehydrogenase of vertebrate erythrocytes. J. gen. Physiol. *45:* 553–565 (1962a).

Vesell, E.S. and Bearn, A.G.: Localization of a lactic dehydrogenase isozyme in nuclei of young cells in the erythrocyte series. Proc. Soc. exp. Biol., Wash. *111:* 100–104 (1962b).

Vesell, E.S. and Brody, I.A.: Biological applications of lactic dehydrogenase isozymes: certain methodological considerations. Ann. N.Y. Acad. Sci. *121:* 544–559 (1964).

Vesell, E.S.; Fritz, P.J., and White, E.L.: Effects of buffer, pH, ionic strength and temperature on lactate dehydrogenase isozymes. Biochim. biophys. Acta *159:* 236–243 (1968).

Vesell, E.S.; Philip, J., and Bearn, A.G.: Comparative studies of the isozymes of lactic dehydrogenase in rabbit and man. Observations during development and in tissue culture. J. exp. Med. *116:* 797–806 (1962).

Vesell, E.S. and Pool, P.E.: Lactate and pyruvate concentrations in exercised ischemic canine muscle: relationship of tissue substrate level to lactate dehydrogenase isozyme pattern. Proc. nat. Acad. Sci., Wash. *55:* 756–762 (1966).

Vesell, E.S. and Yielding, K.L.: Effect of pH, ionic strength, and metabolic intermediates on the rates of heat inactivation of lactate dehydrogenase isozymes. Proc. nat. Acad. Sci., Wash. *56:* 1311–1324 (1966).

Whitt, G.S.: Homology of lactate dehydrogenase genes: E gene function in the teleost nervous system. Science *166:* 1156–1158 (1969).

Wieland, T. und Pfleiderer, G.: Nachweis der Heterogenität von Milchsäure-Dehydrogenasen verschiedenen Ursprungs durch Tragerelektrophorese. Biochem. Z. *329:* 112–116 (1957).

Wilson, A.C.; Cahn, R.D., and Kaplan, N.O.: Functions of the two forms of lactic dehydrogenase in the breast muscle of birds. Nature, Lond. *197:* 331–334 (1963).

Wilson, A.C.; Kaplan, N.O.; Levine, L.; Pesce, A.; Reichlin, M., and Allison, W.S.: Evolution of lactic dehydrogenases. Fed. Proc. *23:* 1258–1266 (1964).

Wittenberger, C.L.: Unusual kinetic properties of a DPN-linked lactate dehydrogenase from *Butyribacterium rettgeri.* Biochem. biophys. res. Commun. *22:* 729–736 (1966).

Wittenberger, C.L. and Haaf, A.S.: Lactate-degrading system in *Butyribacterium rettgeri* subject to glucose repression. J. Bact. *88:* 896–903 (1964).

Wuntch, T.; Chen, R.F., and Vesell, E.S.: Lactate dehydrogenase isozymes: kinetic properties at high enzyme concentrations. Science *167:* 63–65 (1970).

Wuntch, T.; Vesell, E.S., and Chen, R.F.: Studies on rates of ternary complex formation of lactate dehydrogenase isozymes. J. biol. Chem. *244:* 6100–6104 (1969).

Yielding, K.L. and Tomkins, G.M.: Studies on the interaction of steroid hormones with glutamic dehydrogenase. Recent Progr. Hormone Res. *18:* 467–485 (1962).

Zinkham, W.H.; Blanco, A., and Kupchyk, L.: Lactate dehydrogenase in pigeon testes: genetic control by three loci. Science *144:* 1353–1354 (1964).

Zondag, H.A.: Lactate dehydrogenase isozymes: lability at low temperature. Science *142:* 965–967 (1963).

Authors' address: Dr. *E.S. Vesell* and Dr. *P.J. Fritz,* Department of Pharmacology, The Milton S. Hershey Medical Center, The Pennsylvania State University, College of Medicine, *Hershey, PA 17033* (USA)

Enzyme Synthesis and Degradation in Mammalian Systems, pp. 375–402
(Karger, Basel 1971)

Synthesis and Degradation
of Proteins in Relation to Cellular Structure

B. Poole

The Rockefeller University, New York, N.Y.

Contents

I.   Introduction

The problem of biochemistry is to study the complex networks of enzyme catalyzed chemical reactions that lead to the formation of the enormous variety of chemical substances found within living cells. The problem of cell biology is to study the way in which these chemical systems are organized within the living cell, giving rise to and maintaining the complex cellular substructure and the complex activities of cells. In recent years remarkable progress has been made in unravelling the chemical events that lead to the formation of specific proteins whose primary sequence is specified by DNA nucleotide sequences. However for the cell biologist the matter becomes much more complex than a series of chemical reactions. DNA occurs not only in the cell nucleus but also in the mitochondria (and in the chloroplasts of plant cells, which are outside the scope of this review). Moreover the protein formed within cells does not simply float around in solution, but is organized into a complex system of membranes and subcellular organelles containing specific proteins. These organelles themselves in many cases show a complex chemical substructure. But the problem of the formation of proteins and their final organization within the cell is not the only one for cell biologists.

In 1916 *Jacques Loeb* wrote:

> The problem is: first, how does it happen that as soon as respiration has ceased only for a few minutes the human body is dead, that is to say, will commence to undergo disintegration, and second, what protects the body against this decay while respiration goes on, although temperature and moisture are such as to favour decay?[1]

At that time biologists tended to regard living cells as more or less stable structures, as chemical systems close to equilibrium. The disintegration observed after death was assumed to be the result of some process prevented while the organism was alive. Studies with radioisotopic tracers, first started by *Schoenheimer* (95), demonstrated that this static concept of living cells was incorrect. Living cells are in a very dynamic steady state where the structure is maintained by a vigorous flow of energy and matter. Once synthesized, all cellular constituents with the exception of nuclear DNA are subject to degradation and replacement by newly synthesized material. When the lack of respiration deprives the synthetic processes of an energy supply, the effects of the degradative processes become evident. We will see later that some degradative processes may themselves actually be slowed down in the absence of respiration.

These processes of synthesis, transport, and degradation are somehow organized and coordinated within the cell in such a way that the structure and integrity of the cell is maintained. The enormous problems associated with this organization rest almost completely unsolved. In this review we will try to summarize the small amount already known and perhaps to hazard some speculations on the basis of the very tenuous data available.

We will consider first the sites within the mammalian cell at which protein is synthesized, and then the process of transfer of proteins from their site of synthesis to the subcellular structure of which they become a part. Then we will

1 The organism as a whole (G.P. Putnam's Sons, New York 1916).

consider the problem of protein turnover within these structures. The data we will present will be limited to mammalian systems, except in a few cases where important discoveries in lower forms appear to throw some light on general properties of cells. In no sense will this be a complete review of the vast literature on intracellular synthesis and degradation. We will try to assemble those observations that throw some light on the relationship of these processes to subcellular structure.

Mammalian cells present an extensive array of subcellular structures, some generally present in cells, and some specific to particular cell types. Unfortunately experimental evidence available is limited at the present time to a few structures, endoplasmic reticulum, mitochondria, peroxisomes, and pancreatic zymogen granules. We have no information whatever about the synthesis and turnover of lysosomes, despite their ubiquity in mammalian cells and their importance in cellular physiology.

II. Synthesis of Subcellular Structures

A. Sites of Protein Synthesis

1. Endoplasmic Reticulum (microsomes)

The first study of protein synthesis in subfractions of mammalian cells was that of *Siekevitz* (98). He found that no fraction alone incorporated amino acids to a significant extent but that a mixture of microsomes, supernatant protein, and mitochondria showed vigorous incorporation. Later it was shown that the mixture of microsomes and supernatant enzyme would incorporate actively provided they were supplied with a source of ATP (114). Many additional studies were undertaken to clarify the chemical reactions and specific enzymes involved in protein synthesis in this system.

This microsomal supernatant protein synthesizing system is inhibited by ribonuclease (60) and cycloheximide (112) but not by chloramphenicol (113). It has been shown (34) that the ribosomes attached to the endoplasmic reticulum membranes of rat liver account for most of the protein synthesis in that tissue. However it has become clear that protein is not synthesized exclusively on endoplasmic reticulum bound ribosomes in mammalian cells.

2. Mitochondria

Simpson and McLean (100) first reported a high rate of amino acid incorporation in mitochondria isolated from rat muscle. This work was later extended to liver mitochondria (60) and it was shown that RNase had no inhibitory effect on the system, in contrast to the system from microsomes. The importance of an energy supply in sustaining synthesis was demonstrated (84) and it was shown that synthesis of protein took place even after mitochondria had been disrupted sonically (46, 49). However the possibility that contaminating bacteria accounted for the RNase resistant protein synthesis in these mitochondrial preparations was not completely ruled out until the work of *Roodyn et al.* (87). There were a number of reports that isolated mitochondria were able to synthesize cytochrome *c* (4, 5, 6), but these were later retracted (101). *Roodyn et al.* (88) showed clearly that neither cytochrome *c* nor malate dehydrogenase is labelled to any significant extent when mitochondria are incubated with labelled amino acids. *Roodyn* (89) and later *Truman* (109) showed that the vast bulk of the label incorporated went into an insoluble lipoprotein complex. At least four electrophoretically separable components have been shown to become labelled in isolated mitochondria (29, 65). *Truman and Korner* (110) claimed that mitochondrial electron transport provided a more efficient energy source for mitochondrial protein synthesis than did an external ATP source, but this was not confirmed by other workers (50, 103, 113). *Wheeldon and Lehninger* (113) have made a very careful study of the process of mitochondrial protein synthesis *in vitro.* Their data in many cases explain the discrepancies among earlier reports. In addition they discovered that a large fraction of the amino acid incorporated into acid insoluble protein was later converted to an acid soluble form. The physiological implications, if any, of this observation are not clear at the present time.

Chloramphenicol treated HeLa cells show deficiencies in their content of a number of cytochromes (19) and the drug reduced the increase in total liver cytochrome oxidase after partial hepatectomy in rats (52). These observations indicate that the choramphenicol-sensitive intramitochondrial protein synthesis is necessary for mitochondrial biogenesis. They should not be taken as proof that the enzymes are synthesized within mitochondria. More recently *Neupert et al.* (64) *and Beattie et al.* (8) have subfractionated mitochondria after *in vitro* amino acid incorporation and found that essentially all the newly synthesized protein is in the inner membrane fraction. The lack of mitochondrial synthesis of the outer membrane is interesting in view of the results of a number of workers (69, 94, 104) showing that the outer membrane has an enzymic composition similar to that of endoplasmic reticulum.

In *Neurospora Luck* (56) has shown that new mitochondria arise from old ones, presumably by fission. The protein synthesizing activity of isolated mitochondria and the demonstration within them of DNA, first by morphological (62) and then by biochemical (63, 76) methods have suggested that mitochondria may be at least partially autonomous (25) within cells and that they may have originated from some symbiotic bacterium (29, 86). The fact that mitochondrial protein synthesis is inhibited by actinomycin D under some conditions (45, 51, 113) suggests very strongly that at least part of the genetic information for the proteins synthesized is contained in the mitochondrial DNA.

However hybridization experiments with mitochondrial RNA from mouse (35) seem to indicate that part, at least, of this RNA is complementary to nuclear DNA. No complementarity was found between nuclear and mitochondrial DNA's.

The idea of the bacterial origin of mitochondria was strengthened by the demonstration that mitochondrial DNA is circular (11, 102). This idea has stimulated efforts to characterize the protein synthesizing system of mitochondria and to demonstrate its similarity to that of bacteria. For some time it has been known that mitochondrial protein synthesis is sensitive to chloramphenicol but not to cycloheximide (51, 113). Differences between nuclear and mitochondrial t-RNA's and acylating systems (55) and DNA polymerase systems (47) have been shown. Mitochondrial ribosomes have proven more difficult to characterize. Such preparations have been reported (67, 77) but their nature is not yet clear.

So it appears that some part at least of the inner mitochondrial membrane is synthesized within the mitochondrion by means of a complete intramitochondrial protein synthesizing system different from that in the microsomes. However the bulk of the mitochondrial protein, and all the specific enzymes so far studied, are synthesized elsewhere. The genetic information for the amino acid sequence of these enzymes is presumably contained in the nucleus. This has been shown in yeast for cytochrome c (97). In any case the loops of DNA found in mitochondria are too small to contain the genetic information necessary for the synthesis of more than a small number of proteins (102).

3.   Cell Sap

Free ribosomes are seen frequently in mammalian cells but their significance for protein synthesis in the cell is unclear. *In vitro* free polysomes from rat liver incorporate amino acids into proteins as actively per unit of RNA as do those bound to endoplasmic reticulum membranes (80). However there seems to be some difference in the nature of the proteins synthesized. The relative rate of synthesis of serum proteins has been shown to be higher in membrane bound polysomes than it is in free polysomes, while the reverse is true for ferritin (31, 81). It would be tempting to hypothesize that ribosomes on the endoplasmic reticulum synthesize proteins for secretion or for transfer into subcellular particles while the ribosomes of the cell sap synthesize the cell sap proteins. The same idea is suggested by the observation of *Redman et al.* (82), that proteins synthesized on guinea pig liver microsomes are not released into the medium but are transferred into microsomal vesicles. However kinetic studies of protein synthesis in rat liver (26) and in tumor cells in culture (29) do not support this hypothesis. We see in figure 1 the results of *Haldar et al.* (29). After a chase of cold amino acid, the specific activity of the cell sap proteins continues to rise while that of the microsomal proteins is decreasing. This suggests very strongly a

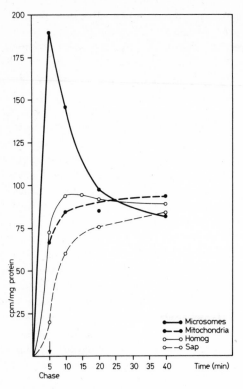

Fig. 1. The kinetics of incorporation of labelled amino acids into subcellular fractions of 'Krebs' ascites tumour cells. A chase of unlabelled amino acids was added to the medium at the time indicated. Redrawn from *Haldar et al.* (29).

transfer of newly synthesized protein from the endoplasmic reticulum into the cell sap. No firm conclusions can be drawn without some information about the nature of the labelled proteins in the cell sap, but it would appear that the bulk of the soluble proteins in these cells is synthesized on the endoplasmic reticulum.

4. Nuclei

There have been a number of reports over the last few years of amino acid incorporation by isolated nuclei from calf thymus (1) and rat liver (85). The physiological significance of this apparent protein synthesis by nuclei was doubtful because of uncertainties about cytoplasmic contamination in the nuclear preparations and the presence of ribosomes on the outer nuclear membrane. Moreover there was no evidence of any very striking properties of this amino acid incorporating activity that would distinguish it from the activity of cytoplasmic contamination. In particular there was no demonstration that particular protein species were synthesized by these nuclear preparations.

Three very recent reports suggest that nuclei may be the sites of some protein synthesis in living cells. *Kuehl* (53) found that the specific activity of hepatic nuclear proteins after the injection of labelled leucine into rats was affected less by puromycin treatment than was that of cytoplasmic proteins. This result suggests strongly that some, at least, of the labelled nuclear proteins may have been synthesized in a cell compartment, presumably the nucleus, different from the size of synthesis of cytoplasmic proteins. *Zimmerman et al.* (115) found that the removal of the outer membrane of HeLa cell nuclei by detergent did not abolish their amino acid incorporating activity, nor could bacterial contamination account for the activity observed. Furthermore on disc electrophoresis only one major and one minor band of protein appeared to be labelled. *Gallwitz and Mueller* (23), also in HeLa cells, compared the electrophoretic pattern of proteins labelled *in vitro* in isolated nuclei and isolated microsomes. These patterns were distinctly different.

These results seem to indicate that protein synthesis in isolated nuclei is something more than evidence of cytoplasmic contamination. However the precise nature of the proteins synthesized remains a mystery. One might expect distinctively nuclear proteins to be synthesized there but some histones, at least, seem to be synthesized in the cytoplasm (22). The highest specific activity of proteins within nuclei occurs in the nucleoli after labelling *in vivo* (33) and *in vitro* (115). This suggests that ribosomal proteins could be involved. However there are many different ribosomal proteins and the number of proteins synthesized by isolated nucleoli appears to be very limited (115).

B.   Transport of Newly Synthesized Protein within Cells

1.   Protein of Secretion Granules
Secretion granules differ from all other subcellular particles in that they do not turn over within the cell, but are transported to the outside. Of the many sorts of secretion granules present in various types of mammalian cells, the only one which has been extensively studied is the pancreatic zymogen granule. *Redman et al.* (83) first showed that amylase in pigeon pancreas is synthesized *in vitro* in microsomal preparations. These authors showed further that the newly synthesized enzyme is not released into the medium but appears inside the microsomal vesicles after incubation. Further studies in this system (79) showed that the transfer of protein into the vesicles is independent of an ATP generating system, that it occurs even at $0°C$, and that it will occur even with unfinished peptide chains released from the ribosomes with puromycin. It appears then that synthesis on membrane bound ribosomes of certain protein species at least in-

volves the transport of nascent protein through the membrane into the cisternae of the rough endoplasmic reticulum.

A series of studies by *Jamieson and Palade* (36–39) on guinea pig pancreas slices has traced the path of these proteins through the rest of their intracellular life. These authors first showed (36) that after a pulse of labelled amino acid, labelled protein appeared first in the rough microsome fraction, then in the smooth microsome fraction, and then in the zymogen granule fraction. Their zymogen granule fraction contained also condensing vacuoles from the Golgi complex and they showed by autoradiography (37) that these vesicles become labelled before the zymogen granules themselves do.

Thus the transport route from cisternae of the rough endoplasmic reticulum, into those of the smooth endoplasmic reticulum, into the condensing vacuoles of the Golgi complex, and finally into the zymogen granules was demonstrated. In further studies (38, 39) these same authors showed that the transport of nascent protein through this system was independent of the synthesis of more protein and of glycolysis. However it depended on respiration, and in the presence of respiratory inhibitors transport was blocked somewhere between the transitional elements of the rough endoplasmic reticulum and the smooth surfaced vesicles at the periphery of the Golgi complex.

2. The Endoplasmic Reticulum and Plasma Membranes

The biogenesis of endoplasmic reticulum membranes has been studied by *Dallner et al.* (15, 16) in the livers of foetal and newborn rats. The system is interesting since the foetal rat liver contains mostly rough endoplasmic reticulum and around the time of birth there is a proliferation of membrane, mostly of the smooth variety. These authors studied the kinetics of incorporation of labelled leucine into membranes of the 'smooth' and the 'rough' microsome fractions separated by specific agglutination in CsC1. Their data are somewhat different for rats of different ages but in general the membranes of the 'rough' fraction had a higher specific activity at early times after labelling than did those of the 'smooth' fraction. However there was no clear indication of a precursor-product relationship. Studies on the time course of the appearance of enzyme activity in the two microsomal fractions of developing rats showed somewhat the same thing. Most enzymic activities were somewhat higher in the 'rough' fraction at early times than they were in the 'smooth', but again there was no clear relationship. *Dallner et al.* (15) concluded that rough endoplasmic reticulum membranes were the precursors for smooth membranes. They suggested that this hypothetical transformation could occur by ribosome detachment or by membrane flow. However the experimental data show very few points at early times after labelling and there is no indication that the activity in the 'smooth' fraction is increasing at a time when that in the 'rough' is decreasing. Consequently it would be dangerous to conclude anything about their biogenetic relationships.

Ray et al. (78) studied the amino acid labelling kinetics of the plasma membrane fraction from rat liver. Their data show that it is labelled over a period of about four hours during the time when the microsomal label is being lost. Whether the plasma membrane forms from some precursor membrane or from soluble precursors, it is clear that the process of transfer of new membrane protein from the ribosomes to the membrane itself is rather slow.

3. Mitochondria

The earliest attempt to study the kinetics of *in vivo* incorporation of amino acids into rat liver mitochondrial proteins (9) gave rather unclear results, but the data suggested that the water soluble mitochondrial proteins were labelled more

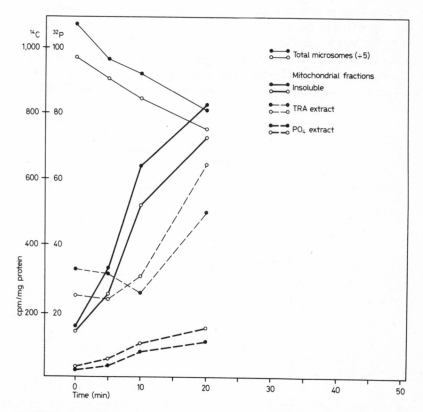

Fig. 2. Kinetics of transfer of label from microsomes (labelled *in vivo* with ^{14}C-leucine and ^{32}P$_i$) to mitochondria (from an unlabelled animal) after the two preparations were mixed. The mitochondria were separated by centrifugation and extracted successively with triethanolamine (TRA) buffer and phosphate buffer, both at pH 7.2. Open symbols: ^{32}P, closed symbols: ^{14}C. Redrawn from *Kadenbach* (43).

slowly than the rest. This finding agrees with the results of experiments discussed above showing that isolated mitochondria are capable of synthesizing only certain insoluble components of the inner membrane. These findings were extended by later workers (26, 71) to the demonstration that cytochrome *c* is labelled first in the microsome fraction of rat liver and then as the specific activity of this protein is decreasing in the microsomes, it is still increasing in mitochondria. Similar results were obtained for the synthesis of cytochrome *c* in mammalian tumor cells in culture (21).

Kadenbach (41) has shown that proteins synthesized in isolated microsomal preparations can be transferred *in vitro* into mitochondria by a process requiring ATP. He then showed the same thing for cytochrome *c* (42). More recently this same author (43) has shown a transfer of phospholipids from microsomes to mitochondria in parallel with the transfer of protein. This is shown in figure 2. He suggests that the mitochondrial proteins are transported as a lipoprotein complex which has the effect of masking the enzymic activity outside the mitochondria and rendering easier passage through the mitochondrial membrane. *Amar-Costesec et al.* (2) have demonstrated the presence in liver microsome fractions of a particulate component with an enzymic composition similar to the

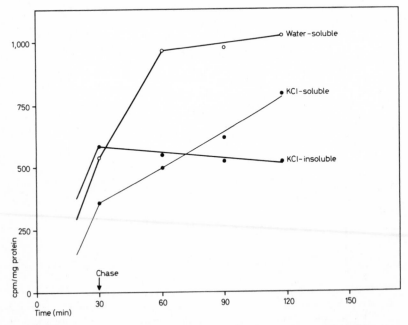

Fig. 3. Kinetics of incorporation of labelled leucine into mitochondrial subfractions from rat liver slices. A chase of unlabelled leucine was added after 30 min as indicated. Taken from *Beattie* (7).

mitochondrial outer membrane. It could be that vesicles of this type are responsible for the transfer observed by *Kadenbach* (41—43).

Work by *Beattie* (7) on rat liver tissue slices has shown a difference in the kinetics of labelling of the water soluble and KCl soluble mitochondrial proteins. These data are illustrated in figure 3. These results suggest that there may be differences in the transport mechanisms of different mitochondrial components.

4. Peroxisomes

Higashi and Peters (32) studied the kinetics of labelling of catalase *in vivo* in various subcellular fractions from rat liver. Their results, illustrated in figure 4, suggest a synthesis of catalase in the endoplasmic reticulum and subsequent transfer to peroxisomes (contained in their mitochondrial fraction). The curious peak of specific radioactivity in the supernatant fraction is hard to explain. Perhaps at some point very early in the transport process the newly synthesized catalase must pass through some fragile structure which is destroyed during homogenization. Nothing is known of the route of transfer of peroxisomal proteins into peroxisomes but apparent sleeve-like attachments of peroxisomes to endoplasmic reticulum have been observed in electron micrographs (66).

De Duve and Baudhuin (17) suggested a possible mechanism of biogenesis and turnover for peroxisomes, which will be discussed in more detail when we come to consider turnover. However one feature of the model was a growth of peroxisomes during a certain period of time, as a consequence of a transfer of

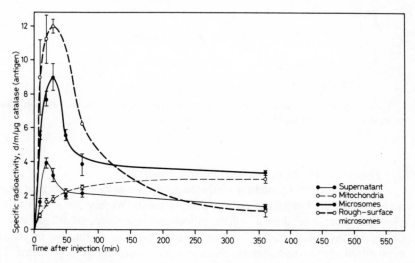

Fig. 4. Kinetics of *in vivo* incorporation of ^{14}C-leucine into various cell fractions. Taken from *Higashi and Peters* (32).

newly synthesized protein from the endoplasmic reticulum to the particles. *Poole et al.* (74) tried to demonstrate this growth by studying the distribution of newly synthesized catalase in rat liver peroxisomes of various size classes at various times after labelling *in vivo*. The results of this study are illustrated in figure 5. In a population of growing particles the ratio of new material to old material (and hence the specific activity after labelling) should be higher in the smaller particles than it is in the larger ones immediately after labelling. Later the specific activity in the smaller particles should decrease relative to that of the larger ones as new unlabelled small particles are formed, *Poole et al.* (74) were unable to demonstrate either of these two effects. Either the time of formation of a peroxisome is short relative to the time scale of the experiments, or the contents of peroxisomes form a single pool as a consequence of the exchange of material between particles.

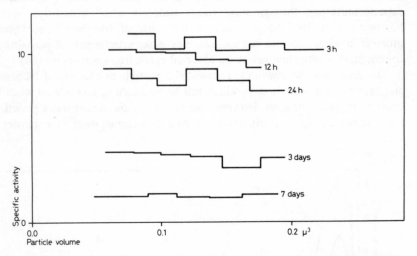

Fig. 5. Specific activity of catalase in rat liver peroxisomes of various sizes at various times after injection of labelled leucine. Redrawn from *Poole et al.* (74).

5. The Mechanism of Intracellular Protein Transport

We have seen that protein is synthesized within mammalian cells in at least 3 and probably in 4 different cellular compartments. We have no reason to suppose that protein synthesized within mitochondria, nucleus, or cell sap is transported to any other cellular location. But the eventual destination of proteins synthesized on the endoplasmic reticulum poses very complex problems. This protein may be packaged into secretion granules. It may be transferred to mitochondria, peroxisomes, and presumably to other subcellular organelles. It may be incorporated into the endoplasmic reticulum membranes. And apparently some of it may eventually come to be free in the cell sap. Some progress has been made in tracing the routes of protein transport but the mechanisms involved are completely unknown. The biggest conceptual problem is the specificity which ensures that each protein species finds its way to its correct subcellular destination. Electron microscopy has increased enormously our knowledge of subcellular structure, but when we consider dynamic problems of cellular function, we may be inhibited by the impression of a static system arising from electron micrographs. There are a number of observations that suggest that the apparent individual existence of discrete subcellular structures may be illusory. *Luck* (56) has shown that in *Neurospora* there is no distinction between newly synthesized and old mitochondria. Mitochondrial substance seems to constitute a single pool. The results of *Poole et al.* (74) on the synthesis of rat liver peroxisomes are consistent with a complete and rapid mixing of peroxisomal proteins. Material entering liver cells by pinocytosis fairly rapidly becomes distributed within the whole lysosomal population (111). While mixing seems to occur within the population of a particular type of subcellular structure, there is no evidence of exchange between different types of particles. The specificity for exchange may reside in the membranes of particles. It could be that particular groups of enzymes are somehow packaged at their site of synthesis (for example the endoplasmic reticulum) within vesicles bounded by a specific type of membrane. The vesicles could then fuse with the corresponding type of subcellular structure and new vesicles could bud off from these structures. As a consequence of a purely random exchange, there would be a net transport of material from the endoplasmic reticulum to subcellular structures and the substance of these structures would form a single pool within which there would be continuous mixing. Oriented vesicle movement would not be required, although it seems to occur in the case of pinocytic vesicles in macrophages (14).

However while the packaging of proteins within the appropriate type of membrane can explain specific transport of proteins from the endoplasmic reticulum to subcellular particles, it raises the question of how this specific packaging is managed. Two possibilities spring immediately to mind. Either there is some mechanism of segregation within the cisternae of the endoplasmic reticulum or particular parts of the endoplasmic reticulum are specialized and synthesize only proteins to be packaged in the same way. We can say nothing about the former possibility except to mention it and to point out that it merely raises again on a smaller scale the problem of specific protein transport. The latter possibility, local endoplasmic reticulum specialization, places the problem of specificity at other levels. The ribosomes on an hypothetical specialized region of the endoplasmic reticulum transcribe RNA messages for a particular class of proteins only. Either they somehow select among messages received, or they receive only messages of a particular type. Both these possibilities involve totally unknown mechanisms of cellular physiology. We can only hope that further research will throw some light on the problem.

III. Turnover

A. Introduction

In the adult animal constant amounts of any protein species are maintained by an adjustment of the rates of synthesis and degradation. It would be beyond the scope of this review to discuss the mechanism of this regulation. We will simply assume the existence of a steady state. The problem in studying protein turnover is to determine the kinetics of protein replacement. This is usually done by injecting some isotopically labelled amino acid into an animal and then isolating at various times thereafter some protein or mixture of proteins in order to measure its loss of radioactivity. However this rate of loss is only a lower limit measure of the actual rate of protein turnover. When a protein is broken down in

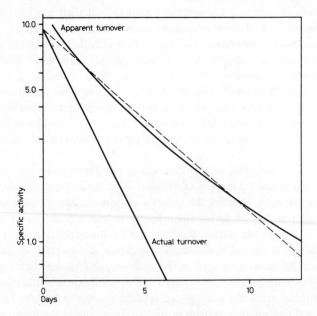

Fig. 6. The effect of reutilization on the decay of labelled amino acid incorporated into protein. The mathematical model for reutilization is described by *Poole* (72). In this case the apparent half life has been increased from one and one half to three and one half days without significant curvature of the semilogarithmic plot of the decay.

a living organism its constituent amino acids go back into the free amino acid pool and may then be reincorporated into protein. The extent to which this will occur is hard to predict and will depend on many factors, including particularly the amino acid composition of the animal's diet.

Ordinarily proteins turn over by a process of random destruction, so that after labelling the specific radioactivity in a particular protein species should decrease exponentially in the absence of appreciable reutilization of label. If such data are plotted against time on a semilogarithmic scale, they should fall on a straight line. In the case of reutilization the mathematical expression for the decay of specific activity becomes the sum of at least two exponential terms and the data should not give a straight line on a semilogarithmic plot. However Poole (72) has calculated that even in the case of extensive reutilization where the apparent decay rate is less than half the true turnover rate, only a very slight curvature appears on the semilogarithmic plot and unless the data were extremely complete and precise, it could never be detected. This is illustrated in figure 6.

We have considered so far the decay of a single protein species. If a protein *fraction* contains protein species with different turnover rates, then the specific activity data plotted on a semilogarithmic scale will show curvature. Thus the decay rates measured for proteins by following the loss of labelled amino acid from the proteins gives only lower limits for the true turnover rates, and data with different precursor amino acids, different tissues, or different animals cannot be compared directly. If significant curvature appears in the semilogarithmic plot of the specific activity of a protein fraction after labelling, it is most likely due to a heterogeneity of turnover rates within the protein species present in that fraction.

B.   Turnover of Proteins in Subcellular Structures

1.   Membrane Proteins of the Endoplasmic Reticulum

One very interesting question that can be asked about the turnover of endoplasmic reticulum membranes is whether the rough and the smooth components of these membranes turn over at the same rate, and whether within the membranes themselves different components have different turnover rates. This matter has been studied by *Omura et al.* (68) but the answer is still not completely clear. These authors found the same half life of about 100 h within experimental error for both rough and smooth membrane proteins with no indication of a complex decay curve. NADPH cytochrome c reductase purified from the membranes appeared to have the same half life as total membrane proteins but cytochrome b_5 appeared to have a somewhat longer half life. These authors made no statistical analysis of their data, but since they had only two experiments each with only four experimental points and a reasonable degree of noise in the data no firm conclusions can be drawn from this work.

Many workers have studied the effect of phenobarbital (a drug which causes an apparent proliferation of smooth endoplasmic reticulum). The general conclusion is that this drug acts both to stimulate synthesis of membrane proteins and to reduce their rate of turnover although the response of different specific proteins is not the same (40, 54).

2. Mitochondria

Fletcher and Sanadi (20) were the first to study the turnover of mitochondria. After injection of labelled methionine into rats they measured the specific activity of three protein fractions from liver mitochondria, water soluble proteins, insoluble proteins, and cytochrome *c*. They found the same apparent half life for all these fractions, about ten days. Many later workers (reviewed by *Gross et al.* [27]) studied the turnover rates of various protein and other fractions from mitochondria of various mammalian tissues. Since these studies were performed with different labelled precursors on different tissues of different animals no useful comparisons can be drawn. The interesting question is not the numerical value for the mitochondrial half life, but whether all components of the particle turn over at the same rate. There are a number of reports in the literature indicating a heterogeneity of turnover rates of mitochondrial protein components. Decay curves for whole mitochondrial protein often show deviation from a simple exponential form (3, 12, 57). Also small differences in turnover rates of mitochondrial protein subfractions have been observed (3, 10, 107). It is difficult to judge to what extent impurities in the mitochondrial fractions were responsible for the apparent heterogeneity observed in these turnover studies. However the rates of change of activity of the mitochondrial enzymes alanine aminotransferase and ornithine aminotransferase are considerably greater than the turnover rate of bulk mitochondrial proteins (107). The activity of δ-aminolevulinic acid synthetase in rat liver mitochondria decays with a half life of about one hour after inhibition of protein synthesis (30, 59). Whether these changes of activity reflect normal turnover rates of these enzymes within mitochondria or whether some special mechanism of inactivation is involved is not known. *Bucher* (12) has shown that the outer membrane of the mitochondrion turns over more rapidly than does the rest of the particle.

It is clear from the work of *Swick et al.* (107) that the true half life of most components, including cytochrome *c*, of rat liver mitochondria is about five days rather than the 8–10 days found by earlier workers. These authors studied the rate of incorporation of C^{14}-carbonate given in the diet into the guanidino group of protein bound arginine. Since they measured simultaneously the specific activity of urea in the urine, they knew the specific activity of the precursor pool and their conclusions are not complicated by the problem of reutilization. For a fuller account of this technique see (106).

Kadenbach (44) used a rather involved method to calculate the half life of rat liver cytochrome *c* to be 13 days. The source of the discrepancy of this result with the more direct measurement of *Swick et al.* (107) is not clear.

3. Peroxisomes

Price et al. (75) first studied the rate of turnover of catalase in rat liver. They used two different methods, both of which avoided the complication of precursor reutilization. They injected 3-amino-1,2,4-triazole into rats and measured the catalase activity in the livers at various times thereafter. This drug causes a rapid irreversible inhibition of catalase activity. Thus the catalase activity measured after the injection represented newly synthesized enzyme. The half time for return to the normal activity level (equal to the half life in the steady state) was about one and one half days. A similar half life value was found when they studied the decrease of hepatic catalase activity after synthesis of the enzyme was blocked by repeated injections of allylisopropylacetamide. These data are compatible with a simple random turnover of catalase with a half life of a day and a half.

However *de Duve and Baudhuin* (17) pointed out that these data are equally compatible with a totally different model of peroxisome biogenesis and turnover, where particles grow linearly for a certain time and then are destroyed. The studies of *Poole et al.* (73) with radioactively labelled precursors ruled out the possibility of this growth and ripening model. Labelled peroxisomal proteins decay in an apparently simple exponential manner over a period of two weeks. According to the growth model of *de Duve and Baudhuin* (17) all label should have been gone from the particles after about 5 days.

Poole et al. (73) fractionated labelled peroxisomal proteins isolated at various times after the injection of label. They were unable to detect any significant heterogeneity in the turnover rates of different protein subfractions. These results suggest that peroxisomes are destroyed as wholes within the liver cell.

The turnover rate of catalase measured by *Poole et al.* (73) using labelled δ-aminolevulinic acid as catalase precursor was less than two days, in agreement with the results of *Price et al.* (75) discussed above. The much longer apparent half lives of catalase and other peroxisomal proteins calculated from experiments with labelled leucine (73) are probably the consequence of reutilization but *Poole* (72) has suggested that the protein part of the molecule may have a slower rate of turnover than the prosthetic groups. The half life of catalase in guinea pig liver appears to be considerably longer than that in rat liver (108).

4. Cell Sap

The proteins of the cell sap show a great diversity of turnover rates. The half life of tyrosine transaminase in rat liver is 3 h or less (28) while that of arginase is 4 to 5 days (91). Much research has been devoted to a few enzymes whose levels of activity in liver can be dramatically modified by hormone treatment or dietary changes. In a number of cases changes in enzyme level are at least partly the consequence of changes in the rate of enzyme turnover. The increase activity of tryptophan pyrrolase in rat liver following the administration of tryptophan is the consequence of a cessation of degradation of the enzyme (92).

There is evidence that the intracellular degradation of tryptophan pyrrolase involves an energy dependent step (93) and that the degradation of tyrosine transaminase requires protein synthesis (48). We will consider these results in more detail in the next section.

C. The Mechanism of Intracellular Turnover

In all cases so far studied intracellular protein turnover appears to be a random process. The probability of destruction is no different for an old molecule than it is for a newly synthesized one. Protein degradation involves the hydrolysis of many peptide bonds in each molecule. As *Cohen* (13) has pointed out, the randomness in the degradation of biologically active polymers cannot be a random hydrolysis of linkages within the entire pool of macromolecules. In such a system the vast bulk of the molecules would very soon sustain some damage rendering them biologically inactive, while only a tiny fraction of the molecules would have been totally hydrolyzed. Once the cellular machinery has selected randomly a molecule for degradation, it must then be hydrolyzed rapidly and completely. *Cohen* (13) has suggested that hydrolytic enzymes may attach themselves to their macromolecular substrates and remain attached until all bonds have been hydrolyzed. This mechanism was suggested for turnover in bacteria, and it probably does not occur in mammalian cells where hydrolytic enzymes have been studied fairly extensively without any indication of such peculiar properties, although the possibility cannot be ruled out completely. One simple way in which the cell could effect complete hydrolysis of a small number of molecules without damaging the others would be to segregate these molecules in a small part of the cell volume along with hydrolytic enzymes. In such a small subcellular compartment optimal conditions for hydrolysis could be maintained without disturbing other cellular processes. It is clear that extracellular material entering cells by pinocytosis is digested within lysosomes (18). At the moment there is no experimental evidence that lysosomes are involved in normal intracellular turnover but this seems likely for a number of reasons. Not only proteins but all other cellular constituents turn over. Lysosomes contain hydrolytic en-

Fig. 7. Electron micrograph of autophagic vacuoles in rat liver. The lower vacuole, containing a mitochondrion, seems to be in the process of fusing with a lysosome, containing ferritin. Taken from *de Duve and Wattiaux* (18).

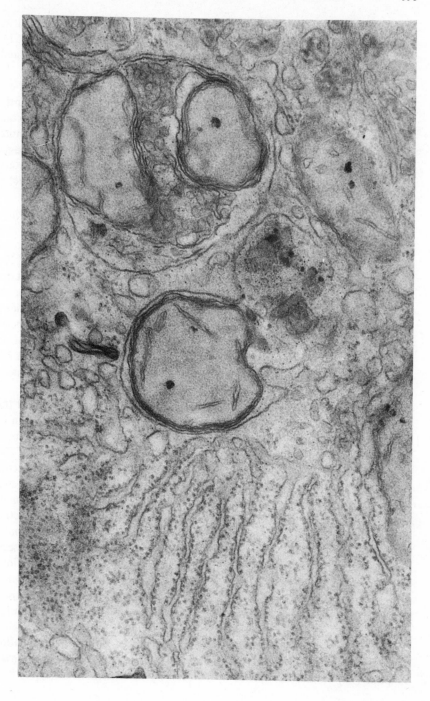

zymes specific for proteins, nucleic acids, carbohydrates, and lipids all together. They are capable of digesting such complex structures as mitochondria (61). There are a few reports of neutral proteases (58, 70) but their nature and significance is not clear.

If digestion of intracellular material does occur within lysosomes, then there must be some mechanism for transporting this material into them. Frequently electron micrographs of mammalian cells show membrane bounded regions containing recognizable cell constituents like mitochondria or peroxisomes in more or less advanced states of disintegration. These structures, called autophagic vacuoles, appear more frequently in certain pathological conditions but they are certainly a feature of normal tissue as well. Figure 7 shows 2 autophagic vacuoles from rat liver.

It appears that bits of cytoplasm are surrounded by membrane to form a vacuole. Then lysosomes fuse with the vacuole releasing their hydrolytic enzymes into it. The contents of the vacuole are then digested and the small molecules resulting from the digestion diffuse across the membrane into the cytoplasm. This sort of random destruction of parts of the cell could account for the turnover of cellular constituents if the turnover rate were the same for all molecular species. However turnover rates vary widely. With subcellular particles we could imagine mechanical factors affecting the inclusion of the particles within the autophagic vacuoles, but even in the cell sap there is a wide variation in turnover rates from one enzyme to another. If autophagy is responsible for turnover, then there must be some selectivity in the formation of autophagic vacuoles.

Presumably the adaptive value of the energetically expensive turnover of the molecular components of cellular structure is to prevent the accumulation of defective molecules. All enzymes studied in the test tube are subject to thermal denaturation at a finite rate. We have no way of knowing at what rate this process occurs within living cells but different protein species would be expected to denature at different rates. If thermal denaturation were the rate limiting step in protein turnover, this would explain different turnover rates for different proteins. This system would also be much more efficient at eliminating defective molecules than would a random replacement of molecules in the whole pool. *Segal et al.* (96) have observed a denaturation rate of alanine aminotransferase *in vitro* more than one hundred times less than its turnover rate in rat liver. However the denaturation rate *in vivo* is not known. It could be that denatured proteins are somehow segregated in autophagic vacuoles. An alternative hypothesis is that neutral proteases in the cell sap degrade specifically denatured proteins.

The few experimental results reported on the process of protein degradation in tissue slices suggest that this process is more complex than simple proteolysis. *Simpson* (99) studied the release of radioactivity into the medium by rat liver

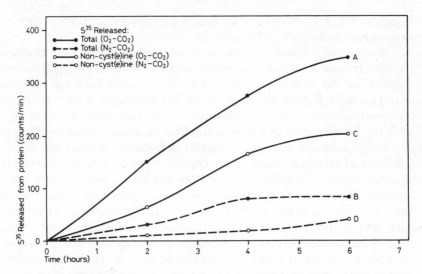

Fig. 8. The effect of anaerobiosis on the kinetics of release of S^{35} labelled amino acids from rat liver slices. Taken from *Simpson* (99).

slices after the proteins had been labelled *in vivo* with S^{35}. As illustrated in figure 8 he found the process to be inhibited by anaerobiosis or mitochondrial inhibitors. These results were confirmed by *Steinberg and Vaughan* (105) who showed in addition that phenylalanine analogues also inhibit intracellular protein breakdown. None of these agents had any effect on the rate of autolysis of homogenates at pH 5. Whatever the mechanism of protein breakdown in cells, it appears that at least one step is energy dependent. If lysosomes are involved in the way we have suggested above, the energy requiring process could be the formation of autophagic vacuoles or perhaps the maintenance of low pH inside the lysosome. If neutral proteases in the cell sap were responsible for turnover it is difficult to see how energy could be involved in the process.

Schimke et al. (93) have made a careful analysis of the turnover of trypto-phan pyrrolase in rat liver. This enzyme can be induced to high levels by the administration of corticoids. The activity then drops precipitously with a half life of 2 or 3 h. These authors studied the loss of enzymic activity and the loss of antigenic activity against an antiserum to purified tryptophan pyrrolase. In the intact animal and in aerobic liver slices enzymic and antigenic activity decreased in parallel. At 37° in whole excised liver, in anaerobic liver slices, or in anaerobic liver homogenates, there was no decrease either in activity or antigen. In aerobic homogenates there was a loss of enzymic activity but no loss of antigen. The physiological significance of this last observation is not clear. Since no significant excess of antigenic over enzymic activity was found *in vivo* or in slices we must conclude that if denaturation is the first step in protein degradation, then proteolysis must follow very rapidly after denaturation.

Kenney (48) has shown that after cycloheximide or puromycin treatment, rat liver tyrosine transaminase is neither synthesized nor destroyed. This result suggests that the normal turnover of the enzyme, which has a half life of less than two hours, requires protein synthesis. The observation of small increases in the hepatic content of tyrosine transaminase after small doses of cycloheximide (90) can be interpreted as a differential effect on actual enzyme synthesis and the protein synthesis required for enzyme degradation. Whether the effect of inhibitors of protein synthesis on the degradation of tyrosine transaminase is a direct or indirect effect, there is no evidence that it is a general effect on protein turnover. The activity of tryptophan pyrrolase decreases after puromycin treatment at a rate comparable to its normal turnover rate measured by more direct methods (24). However if enzyme denaturation is the first step in the degradation of most proteins, then disappearance of activity may not always be a measure of turnover. It would be interesting to know the effect of inhibitors of protein synthesis on the release of amino acids from labelled tissue slices.

This whole problem of the fate of cellular proteins (and other macromolecules) certainly deserves further study. But like all processes that are carried out only in intact cells it will be very difficult to analyze. Once an enzyme has started on the path to proteolysis it may very soon lose its enzymic and antigenic properties, so that there remains no way of recognizing it.

IV. References

1   *Allfrey, V.G.; Littau, V.C., and Mirsky, A.E.:* Methods for the purification of thymus nuclei and their application to studies of nuclear protein synthesis. J. Cell Biol. *21:* 213–231 (1964).

2   *Amar-Costesec, A.; Beaufay, H.; Feytmans, E.; Thines-Sempoux, D., and Berthet, J.:* Subfractionation of rat liver microsomes; in *Gillette, Conney, Cosmides, Estabrook, Fonts and Mannering* Microsomes and drug oxidations, pp. 41–58 (Academic Press, New York 1969).

3   *Bailey, E.; Taylor, C.B., and Bartley, W.:* Turnover of mitochondrial components of normal and essential fatty acid-deficient rats. Biochem. J. *104:* 1026–1032 (1967).

4   *Bates, H.M.; Craddock, V.M., and Simpson, M.V.:* The biosynthesis of cytochrome *c* in cell-free system. I. The incorporation of labelled amino acids into cytochrome *c* by rat liver mitochondria. J. biol. Chem. *235:* 140–148 (1960).

5   *Bates, H.M.; Craddock, V.M., and Simpson, M.V.:* The incorporation of valine-1-C^{14} into cytochrome *c* by rat liver mitochondria. J. amer. chem. Soc. *80:* 1000 (1958).

6   *Bates, H.M. and Simpson, M.V.:* The net synthesis of cytochrome *c* in calf-heart mitochondria. Biochim. biophys. Acta *32:* 597–599 (1959).

7   *Beattie, D.S.:* Studies on the biogenesis of mitochondrial protein components in rat liver slices. J. biol. Chem. *243:* 4027–4033 (1968).

8   *Beattie, D.S.; Basford, R.E., and Koritz, S.B.:* The inner membrane as the site of the *in vitro* incorporation of L-(^{14}C)-leucine into mitochondrial protein. Biochemistry *6:* 3099–3106 (1967).

9   *Beattie, D.S.; Basford, R.E., and Koritz, S.B.:* Studies on the biosynthesis of mitochondrial protein components. Biochemistry *5:* 926–930 (1966).

10   *Beattie, D.S.; Basford, R.E., and Koritz, S.B.:* The turnover of the protein components of mitochondria from rat liver, kidney, and brain. J. biol. Chem. *242:* 4585–4586 (1967).

11   *Bruggen, E.F.J. van; Borst, P.; Ruttenberg, G.J.C.M.; Gruber, M., and Kroon, A.M.:* Circular mitochondrial DNA. Biochim. biophys. Acta *119:* 437–439 (1966).

12   *Bucher, T.:* Biogenesis of mitochondria. 12th int. Congr. Cell Biol., Brussels (1968).

13   *Cohen, D.:* A necessary mechanism for the intracellular degradation of metabolically active macromolecules. J. theor. Biol. *24:* 126–127 (1969).

14   *Cohn, Z.A. and Benson, B.:* The *in vitro* differentiation of mononuclear phagocytes. II. The influence of serum on granule formation, hydrolase production and pinocytosis. J. exp. Med. *121:* 835–848 (1965).

15   *Dallner, G.; Siekevitz, P., and Palade, G.E.:* Biogenesis of endoplasmic reticulum membranes. I. Structural and chemical differentiation in developing rat hepatocyte. J. Cell Biol. *38:* 73–96 (1966).

16   *Dallner, G.; Siekevitz, P., and Palade, G.E.:* Biogenesis of endoplasmic reticulum membranes. II. Synthesis of constitutive microsomal enzymes in developing rat hepatocyte. J. Cell Biol. *30:* 97–117 (1966).

17   *Duve, C. de and Baudhuin, P.:* Peroxisomes (microbodies and related particles). Physiol. Rev. *46:* 323–357 (1966).

18   *Duve, C. de and Wattiaux, R.:* Functions of lysosomes. Ann. Rev. Biochem. *28:* 435–492 (1966).

19   *Firkin, F.C. and Linnane, A.W.:* Differential effects of chloramphenicol on the growth and respiration of mammalian cells. Biochem. biophys. res. Comm. *32:* 398–402 (1968).

20   *Fletcher, M.J. and Sanadi, D.R.:* Turnover of rat-liver mitochondria. Biochim. biophys. Acta *51:* 356–360 (1961).

21   *Freeman, K.B.; Haldar, D., and Work, T.S.:* The morphological site of synthesis of cytochrome *c* in mammalian cells (Krebs cells). Biochem. J. *105:* 947–952 (1967).

22   *Gallwitz, D. and Mueller, G.C.:* Histone synthesis *in vitro* by cytoplasmic microsomes from HeLa cells. Science *163:* 1351–1353 (1969).

23   *Gallwitz, D. and Mueller, G.C.:* Protein synthesis in nuclei isolated from HeLa cells. Europ. J. Biochem. *9:* 431–438 (1969).

24   *Garren, L.D.; Howell, R.R.; Tomkins, G.M., and Crocco, R.M.:* A paradoxical effect of actinomycin D: The mechanism of regulation of enzyme synthesis by hydrocortisone. Proc. nat. Acad. Sci., Wash. *52:* 1121–1129 (1964).

25   *Gibor, A. and Granick, S.:* Plastids and mitochondria: Inheritable systems. Science *145:* 890–897 (1964).

26   *González-Cadavid, N.F. and Campbell, P.B.:* The biosynthesis of cytochrome *c.* Sequence of incorporation *in vivo* of (^{14}C) lysine into cytochrome *c* and total proteins of rat-liver subcellular fractions. Biochem. J. *105:* 443–450 (1967).

27   *Gross, N.J.; Getz, G.S., and Rabinowitz, M.:* Apparent turnover of mitochondrial deoxyribonucleic acid and mitochondrial phospholipids in the tissues of the rat. J. biol. Chem. *244:* 1552–1562 (1969).

28   *Grossman, A. and Mavrides, C.:* Regulation of tyrosine-α-ketoglutarate aminotransferase induction and subsequent inactivation. Fed. Proc. *25:* 285 (1966).

29 *Haldar, D.; Freeman, K., and Work, T.S.:* Biogenesis of mitochondria. Nature, Lond. *211:* 9–12 (1966).

30 *Hayashi, N.; Yoda, B., and Kikuchi, G.:* Mechanism of allylisopropylacetamide-induced increase of δ-aminolevulinate synthetase in liver mitochondria. IV. Accumulation of the enzyme in the soluble fraction of rat liver. Arch. Biochem. *131:* 83–91 (1969).

31 *Hicks, S.J.; Drysdale, J.W., and Munro, H.N.:* Preferential synthesis of ferritin and albumin by different populations of liver polysomes. Science *164:* 584–585 (1969).

32 *Higashi, T. and Peters, T., Jr.:* Studies on rat liver catalase. II. Incorporation of ^{14}C-leucine into catalase of liver cell fractions *in vivo.* J. biol. Chem. *238:* 3952–3954 (1963).

33 *Hnilica, L.S.; Liau, M.C., and Hurlbert, R.B.:* Biosynthesis and composition of histones in Novikoff hepatoma nuclei and nucleoli. Science *152:* 521–523 (1966).

34 *Howell, R.R.; Loeb, J.N.; and Tomkins, G.M.:* Characterization of ribosomal aggregates isolated from liver. Proc. nat. Acad. Sci., Wash. *52:* 1241–1248 (1964).

35 *Humm, D.G. and Humm, J.H.:* Hybridization of mitochondrial RNA with mito-chondrial and nuclear DNA in agar. Proc. nat. Acad. Sci., Wash. *55:* 114–119 (1966).

36 *Jamieson, J.D. and Palade, G.E.:* Intracellular transport of secretory protein in the pancreatic exocrine cell. I. Role of the peripheral elements of the Golgi complex. J. Cell Biol. *34:* 577–596 (1967).

37 *Jamieson, J.D. and Palade, G.E.:* Intracellular transport of secretory protein in the pancreatic exocrine cell. II. Transport to condensing vacuoles and zymogen granules. J. Cell Biol. *34:* 597–615 (1967).

38 *Jamieson, J.D. and Palade, G.E.:* Intracellular transport of secretory proteins in the pancreatic exocrine cell. III. Dissociation of intracellular transport from protein synthesis. J. Cell Biol. *39:* 580–588 (1968).

39 *Jamieson, J.D. and Palade, G.E.:* Intracellular transport of secretory proteins in the pancreatic exocrine cell. IV. Metabolic requirements. J. Cell Biol. *39:* 589–603 (1968).

40 *Jick, H. and Shuster, L.:* The turnover of microsomal reduced nicotinamide adenine dinucleotide phosphate-cytochrome *c* reductase in the livers of mice treated with phenobarbital. J. biol. Chem. *241:* 5366–5369 (1966).

41 *Kadenbach, B.:* Synthesis of mitochondrial proteins: demonstration of a transfer of proteins from microsomes into mitochondria. Biochim. biophys. Acta *134:* 430–442 (1967).

42 *Kadenbach, B.:* Synthesis of mitochondrial proteins. The synthesis of cytochrome *c in vitro.* Biochim. biophys. Acta *138:* 651–654 (1967).

43 *Kadenbach, B.:* Transfer of proteins from microsomes into mitochondria. Biosynthesis of cytochrome *c;* in *Slater, Tager, Papa and Quagliariello* Biochemical aspects of the biogenesis of mitochondria, pp. 415–429 (Adriatica Editrice, Bari 1968).

44 *Kadenbach, B.:* A quantitative study of the biosynthesis of cytochrome *c.* Europ. J. Biochem. *10:* 312–318 (1969).

45 *Kalf, G.F.:* Deoxyribonucleic acid in mitochondria and its role in protein synthesis. Biochemistry *3:* 1702–1706 (1964).

46 *Kalf, G.F. and Simpson, M.V.:* The incorporation of valine-1-C^{14} into the protein of submitochondrial fractions. J. biol. Chem. *234:* 2943–2947 (1959).

47 *Kalf, G.F. and Ch'ih, J.J.:* Purification and properties of deoxyribonucleic acid poly-merase from rat liver mitochondria. J. biol. Chem. *243:* 4904–4916 (1968).

48 *Kenney, F.T.:* Turnover of rat liver tyrosine transaminase: stabilization after inhibition of protein synthesis. Science *156:* 525–528 (1967).

49 *Kroon, A.M.:* Amino acid incorporation into the protein of mitochondria and mito-chondrial fragments from beef heart. Biochim. biophys. Acta *69:* 184–185 (1963).

50 Kroon, A.M.: Protein synthesis in heart mitochondria. I. Amino acid incorporation into the protein of isolated beef heart mitochondria and fractions derived from them by sonic oscillation. Biochim. biophys. Acta 72: 391–402 (1963).

51 Kroon, A.M.: Protein synthesis in mitochondria. III. On the effects of inhibitors on the incorporation of amino acids into protein by intact mitochondria and digitonin fractions. Biochim. biophys. Acta 108: 275–284 (1965).

52 Kroon, A.M. and Vries, H. De: The effect of chloramphenicol on the biogenesis of mitochondria of rat liver in vivo. F.E.B.S. Letters 3: 208–210 (1969).

53 Kuehl, L.: Effects of various inhibitors on nuclear protein synthesis in rat liver. J. Cell Biol. 41: 660–663 (1969).

54 Kuriyama, Y.; Omura, T.; Siekevitz, P., and Palade, G.E.: Effects of phenobarbital on the synthesis ·and degradation of the protein components of rat liver microsomal membranes. J. biol. Chem. 244: 2017–2026 (1969).

55 Lietman, P.S.: Enzymatic acylation of phenylalanyl transfer ribonucleic acids from mitochondria and cytosol of rat liver. J. biol. Chem. 243: 2837–2839 (1968).

56 Luck, D.J.L.: Genesis of mitochondria in Neurospora crassa. Proc. nat. Acad. Sci., Wash. 49: 233–240 (1963).

57 Lusena, C.V. and Depocas, F.: Heterogeneity and differential fragility of rat liver mitochondria. Canad. J. Biochem. 44: 497–508 (1966).

58 Marks, N. and Lajtha, A.: Protein breakdown in the brain. Subcellular distribution and properties of neutral and acid proteinases. Biochem. J. 89: 438–447 (1963).

59 Marver, H.S.; Collins, A.; Tschudy, D.P., and Rechcigl, M., Jr.: δ-Aminolevulinic acid synthetase. II. Induction in rat liver. J. biol. Chem. 241: 4323–4329 (1966).

60 McLean, J.R.; Cohn, G.L.; Brandt, I.K., and Simpson, M.V.: Incorporation of labeled amino acids into the protein of muscle and liver mitochondria. J. biol. Chem. 233: 657–663 (1958).

61 Mellors, A.; Tappel, A.L.; Sawant, P.L., and Desai, I.D.: Mitochondrial swelling and uncoupling of oxidative phosphorylation by lysosomes. Biochim. biophys. Acta 143: 299–309 (1962).

62 Nass, M.M.K. and Nass, S.: Intramitochondrial fibers with DNA characteristics. I. Fixation and electron staining reactions. J. Cell Biol. 19: 593–612 (1963).

63 Nass, S.; Nass, M.M.K., and Hennix, U.: Deoxyribonucleic acid in isolated rat-liver mitochondria. Biochim. biophys. Acta 95: 426–435 (1965).

64 Neupert, W.; Brdiczka, D., and Bücher, T.: Incorporation of amino acids into the outer and inner membrane of isolated rat liver mitochondria. Biochem. biophys. res. Comm. 27: 488–493 (1967).

65 Neupert, W.; Brdiczka, D., and Sebald, W.: Incorporation of amino acids into the outer and inner membrane of isolated rat-liver mitochondria; in Slater, Tager, Papa and Quagliariello Biochemical aspects of the biogenesis of mitochondria, pp. 395–408 (Adriatica Editrice, Bari 1968).

66 Novikoff, A.B. and Shin, W.Y.: The endoplasmic reticulum in the Golgi zone and its relations to microbodies, Golgi apparatus and autophagic vacuoles in rat liver cells. J. Microscopie 3: 187–206 (1964).

67 O'Brien, T.W. and Kalf, G.F.: Ribosomes from rat liver mitochondria. I. Isolation procedure and contamination studies. J. biol. Chem. 242: 2172–2179 (1967).

68 Omura, T.; Siekevitz, P., and Palade, G.E.: Turnover of constituents of the endoplasmic reticulum membranes of rat hepatocytes. J. biol. Chem. 242: 2389–2396 (1967).

69 Parsons, D.F.; Williams, G.R.; Thompson, W.; Wilson, D., and Chance, B.: Improvements in the procedure for purification of mitochondrial outer and inner membrane.

Comparison of outer membrane with smooth endoplasmic reticulum; in *Quagliariello, Papa, Slater and Tager* Mitochondrial structure and compartmentation, pp. 29–70 (Adriatica Editrice, Bari 1967).

70 *Penn, N.W.:* The requirements for serum albumin metabolism in subcellular fractions of liver and brain. Biochim. biophys. Acta *37:* 55–63 (1960).

71 *Penniall, R. and Davidian, N.:* Origin of mitochondrial enzymes. I. Cytochrome *c* synthesis by endoplasmic reticulum. F.E.B.S. Letters *1:* 38–41 (1968).

72 *Poole, B.:* The biogenesis and turnover of rat liver peroxisomes. Ann. NY Acad. Sci. *168:* 229–243 (1969).

73 *Poole, B.; Leighton, F., and Duve, C. de:* The synthesis and turnover of rat liver peroxisomes. II. Turnover of peroxisome proteins. J. Cell Biol. *41:* 536–546 (1969).

74 *Poole, B.; Higashi, T., and Duve, C. de:* The synthesis and turnover of rat liver peroxisomes. III. The size distribution of peroxisomes and the incorporation of new catalase. J. Cell Biol. *45:* 408–415 (1970).

75 *Price, V.E.; Sterling, W.R.; Tarantola, V.A.; Hartley, R.W., Jr., and Rechcigl, M., Jr.:* The kinetics of catalase synthesis and destruction *in vivo.* J. biol. Chem. *237:* 3468–3475 (1962).

76 *Rabinowitz, M.; Sinclair, J.; Salle, L. de; Haselkorn, R., and Swift, H.H.:* Isolation of deoxyribonucleic acid from mitochondria of chick embryo heart and liver. Proc. nat. Acad. Sci., Wash. *53:* 1126–1132 (1965).

77 *Rabinowitz, M.; Salle, L. de; Sinclair, J.; Stirewalt, R., and Swift, H.:* Ribosomes isolated from rat liver mitochondrial preparations. Fed. Proc. *25:* 581 (1966).

78 *Ray, T.K.; Lieberman, I., and Lansing, A.I.:* Synthesis of the plasma membrane of the liver cell. Biochem. biophys. res. Comm. *31:* 54–58 (1968).

79 *Redman, C.M.:* Studies on the transfer of incomplete polypeptide chains across rat liver microsomal membranes *in vitro.* J. biol. Chem. *242:* 761–768 (1967).

80 *Redman, C.M.:* The synthesis of serum proteins on attached rather than free ribosomes of rat liver. Biochem. biophys. res. Comm. *31:* 845–850 (1968).

81 *Redman, C.M.:* Biosynthesis of serum proteins and ferritin by free and attached ribosomes of rat liver. J. biol. Chem. *244:* 4308–4315 (1969).

82 *Redman, C.M. and Sabatini, D.D.:* Vectorial discharge of peptides released by puromycin from attached ribosomes. Proc. nat. Acad. Sci., Wash. *56:* 608–615 (1966).

83 *Redman, C.M.; Siekevitz, P., and Palade, G.E.:* Synthesis and transfer of amylase in pigeon pancreatic microsomes. J. biol. Chem. *241:* 1150–1158 (1966).

84 *Reis, P.J.; Coote, J.L., and Work, T.S.:* Protein biosynthesis and oxidative phosphorylation in isolated rat liver mitochondria. Nature, Lond. *184:* 165–167 (1959).

85 *Rendi, R.: In vitro* incorporation of labelled amino acids into nuclei isolated from rat liver. Exp. Cell Res. *19:* 489–498 (1960).

86 *Roodyn, D.B. and Wilkie, D.:* The biogenesis of mitochondria (Methuen, London 1968).

87 *Roodyn, D.B.; Reis, P.J., and Work, T.S.:* Protein synthesis in mitochondria. Requirement for the incorporation of radioactive amino acids into mitochondrial protein. Biochem. J. *80:* 9–21 (1961).

88 *Roodyn, D.B.; Suttie, J.W., and Work, T.S.:* Protein synthesis in mitochondria. II. Rate of incorporation *in vitro* of radioactive amino acids into soluble proteins in the mitochondrial fraction, including catalase, malic dehydrogenase, and cytochrome *c.* Biochem. J. *83:* 29–40 (1962).

89 *Roodyn, D.B.:* Protein synthesis in mitochondria. III. The controlled disruption and subfractionation of mitochondria labelled *in vitro* with radioactive valine. Biochem. J. *85:* 177–189 (1962).

90 *Rosen, F. and Milholland, R.J.:* Selective effects of cycloheximide on the adaptive enzymes tyrosine-a-ketoglutarate transaminase (TT) and tryptophan pyrrolase (TP). Fed. Proc. *25:* 285 (1966).

91 *Schimke, R.T.:* The importance of both synthesis and degradation in the control of arginase levels in rat liver. J. biol. Chem. *239:* 3808–3817 (1964).

92 *Schimke, R.T.; Sweeney, E.W., and Berlin, C.M.:* The roles of synthesis and degradation in the control of rat liver tryptophan pyrrolase. J. biol. Chem. *240:* 322–331 (1965).

93 *Schimke, R.T.; Sweeney, E.W., and Berlin, C.M.:* Studies of the stability *in vivo* and *in vitro* of rat liver tryptophan pyrrolase. J. biol. Chem. *240:* 4609–4620 (1965).

94 *Schnaitman, C. and Greenawalt, J.W.:* Enzymatic properties of the inner and outer membranes of rat liver mitochondria. J. Cell Biol. *38:* 158–175 (1968).

95 *Schoenheimer, R.:* The dynamic state of body constituents (Harvard University Press, Cambridge 1942).

96 *Segal, H.L.; Matsuzawa, T.; Haider, M., and Abraham, G.J.:* What determines the half-life of proteins *in vivo*? Some experiences with alanine aminotransferase of rat tissues. Biochem. biophys. res. Comm. *36:* 764 (1969).

97 *Sherman, F.; Stewart, J.W.; Margoliash, E.; Parker, J., and Campbell, W.:* The structural gene for yeast cytochrome c. Proc. nat. Acad. Sci., Wash. *55:* 1498–1503 (1966).

98 *Siekevitz, P.:* Uptake of radioactive alanine *in vitro* into the proteins of rat liver fractions. J. biol. Chem. *195:* 549–565 (1952).

99 *Simpson, M.V.:* The release of labelled amino acids from the proteins of rat liver slices. J. biol. Chem. *201:* 143–154 (1953).

100 *Simpson, M.V. and McLean, J.R.:* The incorporation of labelled amino acids into the cytoplasmic particles of rat muscle. Biochim. biophys. Acta *18:* 573–575 (1955).

101 *Simpson, M.V.; Skinner, D.M., and Lucas, J.M.:* On the biosynthesis of cytochrome c. J. biol. Chem. *236:* PC81 (1961).

102 *Sinclair, J.H. and Stevens, B.J.:* Circular DNA filaments from mouse mitochondria. Proc. nat. Acad. Sci., Wash. *56:* 508–514 (1966).

103 *Singh, V.N.; Raghupathy, E., and Chiakoff, I.L.:* Incorporation of amino acid carbon into proteins by sheep thyroid gland mitochondria. Biochem. biophys. res. Comm. *16:* 12–18 (1964).

104 *Sottocasa, G.; Kuylenstierna, B.; Ernster, L., and Bergstrand, A.:* An electron-transport system associated with the outer membrane of liver mitochondria. A biochemical and morphological study. J. Cell Biol. *32:* 415–438 (1967).

105 *Steinberg, D. and Vaughan, M.:* Observations on intracellular protein catabolism studied *in vitro.* Arch. Biochem. *65:* 93–105 (1956).

106 *Swick, R.W.:* Measurement of protein turnover in rat liver. J. biol. Chem. *231:* 751–764 (1958).

107 *Swick, R.W.; Rexroth, A.K., and Stange, J.L.:* The metabolism of mitochondrial proteins. III. The dynamic state of rat liver mitochondria. J. biol. Chem. *243:* 3581–3587 (1968).

108 *Theorell, H.; Béznak, M.; Bonnichsen, R.; Paul, K.G., and Akeson, A.:* On the distribution of injected radioactive iron in guinea pigs and its rate of appearance in some hemoproteins and ferritins. Acta chem. scand. *5:* 445–475 (1951).

109 *Truman, D.E.S.:* The fractionation of proteins from ox-heart mitochondria labelled *in vitro* with radioactive amino acids. Biochem. J. *91:* 59–64 (1964).

110 *Truman, D.E.S. and Korner, A.:* Incorporation of amino acids into the proteins of isolated mitochondria. A search for optimum conditions and a relationship to oxidative phosphorylation. Biochem. J. *83:* 588–596 (1962).

111 *Wattiaux, R.:* Etude expérimentale de la surcharge des lysosomes. Thèse d'agrégation, Louvain (1966).

112 *Wettstein, F.O.; Noll, H., and Penman, S.:* Effect of cycloheximide on ribosomal aggregates engaged in protein synthesis *in vitro.* Biochim. biophys. Acta *87:* 525–528 (1964).

113 *Wheeldon, L.W. and Lehninger, A.L.:* Energy-linked synthesis and decay of membrane proteins in isolated rat liver mitochondria. Biochemistry *5:* 3533–3545 (1966).

114 *Zamecnik, P.C. and Keller, E.B.:* Relation between phosphate energy donors and incorporation of labeled amino acids into proteins. J. biol. Chem. *209:* 337–354 (1954).

115 *Zimmerman, E.F.; Hackney, J.; Nelson, P., and Arias, I.M.:* Protein synthesis in isolated nuclei and nucleoli of HeLa cells. Biochemistry *8:* 2636–2644 (1969).

Author's address: Dr. *Brian Poole,* The Rockefeller University, *New York, NY 10021* (USA)

Glossary

Actinomycin D. An antibiotic, produced by Actinomyces, capable of inhibiting DNA-dependent RNA synthesis, by binding to guanine residues and preventing elongation of the growing RNA chain.

Activator RNA. A (postulated) product of an integrator gene; it activates by binding to a receptor gene of a specific gene set or several gene sets.

Active site. That portion of an enzyme where the substrate molecules combine and are transformed into their reaction products.

Adaptation. Any change in an organism's structure or function that allows it to better cope with conditions in the environment.

Adaptive enzymes. See *Inducible enzymes.*

Adrenalectomy. The surgical removal of the adrenal glands.

Adrenocorticotrophic hormone. The hormone secreted by anterior lobe of pituitary gland and which controls activity of the adrenal cortex; ACTH.

Alleles. Two or more genes that occupy the same locus on homologous chromosomes; when in the same cell, undergo pairing during meiosis, but produce different effects on the same developmental process.

Allosteric effector. An agent that binds to a site other than the catalytic site and

exerts a conformational change in the secondary or tertiary structure of an enzyme resulting in an alteration in its functional activity.

Allylisopropylacetamide. A porphyria-inducing compound which has the property of preventing catalase synthesis and stimulating δ-aminolevulinate synthetase formation.

3-Amino-1, 2, 4-triazole. A potent herbicide producing a rapid and irreversible inactivation of catalase resulting from the binding of the inhibitor with the protein moiety of the enzyme molecule.

Apoenzyme. A specific protein part of an enzyme, requiring coenzyme for action.

Aporepressor. A product of a regulatory gene which can specifically bind to an effector called corepressor.

Architectural gene. A gene that determines the site of an enzyme within a cell.

Autophagy. An intracellular process in which portions of the cell are segregated within a membrane and digested.

8-Azaguanine. An antimetabolite that causes the formation of false or inactive species of RNA.

Basal enzyme. The kind and amount of enzyme normally found in cells in the absence of regulatory effectors.

Blastocyst. In mammals, a fluid-filled vesicle that forms after cleavage of the ovum has produced a ball of cells (morula);

the blastocyst consists of an outer layer of cells, the throphoblast, and an inner cell mass from which the embryo is formed.

Carbohydrate repression. Inhibition of the rate of synthesis of degradative enzymes by carbohydrate.

Catabolite repression. The phenomenon by which compounds that can serve efficiently as a source of intermediary metabolites and of energy have a repressive effect on the biosynthesis of many enzymes in different pathways.

Cell sap. The fluid phase of the cell.

Cellular differentiation. A process brought about by an orderly progression of gene activation and inactivation leading to a structural and functional specialization (see *Differentiation*).

Chloramphenicol. An antibiotic produced by *Streptomyces venezuelae;* it is a potent inhibitor of protein synthesis by attaching to the 70s ribosome and preventing addition of an amino acid to the growing polypeptide chain.

Chromatin proteins. Proteins associated with DNA; some are enzymes and some control the structure and function of chromosomes.

Chromosome. A linkage structure consisting of a linear collection of specific genes controlling the hereditary properties of all genetic systems.

Circadian rhythm. A rhythm of approximately 24-hour duration.

Codon. The smallest combination of bases in a polynucleotide which determines the insertion of a specific amino acid at a specific position into a polypeptide chain.

Coenzyme. A small molecule essential to the catalytic activity of an enzyme.

Cofactor induction. A term applied to increased rate of enzyme synthesis following the administration of the enzyme's cofactor.

Constitutive enzymes. Enzymes that are synthesized in fixed amounts, irrespective of the growth conditions and presence or absence of inducers.

Corepressor. An effector with a great affinity towards a specific aporepressor, presum-

ably causing an allosteric change in its conformation. Effector combined with aporepressor leads to an enhancement (repressible systems) or to a decrease (inducible systems) in affinity of the complex toward the operator region of a given operon.

Cumulative repression. Each final product of a branched pathway acts as a corepressor, producing only partial repression of the common enzymes responsible for its biosynthesis, full repression being achieved only by all the end products acting together.

Cyclic AMP. Cyclic adenosine-3′,5′-monophosphate.

Cycloheximide. An antibiotic that blocks protein synthesis in animal cells by inhibiting peptide bond formation.

Decay curve. A curve showing the relative amount of biological substance remaining during the normal process of degradation in the absence of protein synthesis.

Degradation rate. Rate at which a substance is degraded.

Derepression. A release of a repression leading to the transcription of a structural gene (genes) and the synthesis of a corresponding messenger RNA(s) and protein(s).

Developmental pattern. The consistent orderly progression of a series of events brought about by the process of maturation.

Dexamethasone. 1-Dehydro-16α-methyl-9α-fluorohydrocortisone, a synthetic analogue of the glucocorticoid hormone, hydrocortisone.

Differentiation. The actual appearance in a cell or tissue of new properties, whether defined in biochemical or structural terms. In embryology it is generally distinguished from the prior state of *determination,* in which the tissue is restricted to develop along a specific pathway.

Diurnal. Daily; recurring every day.

Dynamic equilibrium. A state in which the proteins are continually being broken down and replaced by resynthesis.

Effector. A general name for chemical substances having a great affinity for one or several proteins; the biological activity and function of these proteins is altered after binding to such compounds.

Embryonic induction. The process by which a specific course of differentiation is elicited in a tissue, or part of an embryo by contact with or proximity to a second tissue or part; it is generally believed that induction involves transfer of material from the inducing cells to those they affect.

End product inhibition. A biological control mechanism in sequential enzyme systems in which the accumulation of the final product of a sequence of metabolic reactions causes inhibition of its own function.

Endogenous enzyme. See *Basal enzyme.*

Endometrium. The mucosal lining of the uterus, in which the blastocyst implants.

Enzyme degradation. Loss of an enzyme structure due to dynamic or proteolytic degradation *in vivo.*

Enzyme induction. Initiation or increase in the rate of enzyme synthesis following chemical stimulus.

Enzyme realization. The set of processes acting to produce the final phenotype of a protein.

Enzyme repression. A decrease or complete abolition of the transcription of a specific gene or functional genetic unit which is reflected in a decrease of the corresponding protein or proteins of the cell.

Enzyme synthesis. Synthesis of an enzyme from its constituent amino acids.

Enzyme turnover. A continuous synthesis and degradation of an enzyme.

Erythroid. Relating to red blood cells or their precursors.

Euchromatin. Template active, extended parts of genome which can be distinguished, even visually, from the inactive parts of the genome, e.g. as light bands in salivary chromosomes of diptera larvae.

Eukaryotes. Organisms whose cells have, in contrast to prokaryotes, well developed nuclei with nuclear envelopes, chromosomes, as well as undergoing nuclear divisions.

Fallopian tube. The mammalian oviduct; it is complex in structure, and includes an ex-

panded segment, the ampulla, in which the lumen is partially obstructed by labyrinthine folds, and a narrow isthmus, leading to the uterus.

First order reaction. A chemical reaction in which one molecular species disappears at a rate proportional at each instant to the amount of that species.

Follicle cells. Small cells that surround the developing oocyte, presumably contributing to its growth and maintenance; some of the cells remain in contact with the ovum when it is expelled from the ovary.

Gastrin. A hormone secreted by pyloric mucosa which stimulates gastric secretion.

Gastrulation. A series of processes by which the primary germ layers (ectoderm, mesoderm, endoderm) are laid down; in mammals, gastrulation begins immediately after the blastocyst is formed.

Gene amplification. A phenomenon whereby certain genes are replicated without replication of other DNA sequences; the products may remain in the vicinity of the copied original chain, or they may form satellites outside of the chromosome.

Gene battery. A (postulated) group of gene sets the activity of which is controlled by a specific activator RNA.

Gene expression. Only some of the genes in mammalian genome are expressed; i.e., act as templates which are transcribed to messenger RNAs coding for the synthesis of specific proteins. The expression of a gene is controlled by several mechanisms both at the nuclear and cytoplasmic levels; a major portion of the genome in mammalian cells is not expressed.

Gene locus. The position of a gene on the genetic map.

Genome. The complete set of chromosomes in each nucleus of a somatic cell.

Genotype. The sum of the genetic information contained in chromosomes; the gene constitution of an individual organism.

Glucagon. A pancreatic hormone found in *a* cells of islets of Langerhans which stimulates glycogenolysis and gluconeogenesis in the liver, causing increase in blood sugar.

Glucocorticoids. Steroid hormones having numerous physiological effects, including the ability to promote the deposition of glycogen in liver; these may be the natural products of the adrenal cortex, e.g. cortisone, corticosterone and cortisol, or synthetic steroids, e.g. dexamethasone, prednisone and prednisolone.

Gluconeogenesis. Formation of glucose from non-carbohydrate sources.

Glucose effect. Glucose inhibition of the biosynthesis of many unrelated enzymes when fed to animals or added to culture medium. Since similar effect can be obtained by some other compounds, the effect has also been known under a more general term as carbohydrate or catabolite repression.

Glycogenolysis. The ability of a compound to cause the breakdown of glycogen in a specific tissue.

Golgi apparatus. A complex structure within a cell made up of closely packed broad cisternae and small vesicles, containing lipoprotein, and concerned with cellular synthesis and secretion.

Half-life. Applied to protein turnover, the time required for one half of the protein to be degraded in the absence of synthesis.

Hepatoma. A liver neoplasm.

Heterochromatin. Template inactive, tightly packed and condensed parts of genome, which can also be observed visually, e.g. dark bands in salivary chromosomes of diptera larvae.

Heterokaryon. A cell containing two or more nuclei of different genotype.

Holoenzyme. An enzyme consisting of an apoenzyme and coenzyme, neither of which is active by itself.

Homeostasis. The tendency to maintain a relatively stable internal environment in living organism.

Hyperphagia. An abnormally high intake of food.

Hypophysectomy. The surgical removal of the pituitary gland.

Immuno-electrophoresis. A combination of electrophoresis and specific precipitation by double diffusion in 2 directions in a gel.

Inducer. The effector agent of enzyme induction, usually a small molecule, which may or may not be metabolically related to the enzyme induced.

Inducible enzymes. Enzymes that are synthesized only in the presence of their substrates or other inducers.

Induction. See *Enzyme induction.*

Insulin. The anti-diabetic endocrine product of pancreas, found in β-cells of islets of Langerhans.

Integrator gene. Gene which forms an activator RNA following activation by a sensory gene.

Intestinal microvilli. Projections approximately $0.5-1.0$ μ long and $80-90$ mμ in diameter that cover the lumenal surface of intestinal epithelial cells. Each microvillus consists of a covering membrane enclosing a core of fibers.

Isozymes. Multiple molecular forms of an enzyme with similar substrate specificity.

Ketogenesis. The formation of ketone bodies.

L cell. One of a strain of cells grown in tissue culture for 20 years derived from normal fibroblasts in the subcutaneous tissue of C3H mice.

Lifetime. A finite time period measuring the total existence of a specific biological parameter such as messenger RNA.

Ligand. A molecule that will bind to a complementary site on a given structure.

Lipogenesis. The formation of lipids.

Long term repression. A phenomenon whereby certain genes of differentiated mammalian cells are physically in such a state that they are not transcribed under normal physiological circumstances.

Lysosome. Any of a number of morphologically heterogeneous cell organelles present in animal cells and containing hydrolytic enzymes.

Meal-fed. Presented with food for only a short period once a day.

Messenger RNA (mRNA). Polyribonucleotide intermediate conveying the genetic information encoded in DNA to the cellular sites of protein synthesis where it acts as a template for the synthesis of a specific polypeptide.

Microbodies. A separate group of intracytoplasmic particles limited by a single membrane, containing catalase and several oxidative enzymes.

Microsomes. Small vesicles arising from the membranes of the endoplasmic reticulum when cells are homogenized.

Mitochondria. Self-replicating organelles in cytoplasm, surrounded by a double limiting membrane. They are the principal energy source of the cell and contain the cytochrome enzymes of terminal electron transport as well as enzymes of the citric acid cycle, fatty acid oxidation, and oxidative phosphorylation.

Mitomycin. Antibiotic produced by *Streptomyces caespitousus* which prevents DNA replication of crosslinking the complementary strands of DNA double helix.

Modulation. The shift between the alternate states of differentiation, depending on the presence or absence of hormones or other controlling factors.

Multisensitive repression. The final product and pathway intermediates capable of repressing the synthesis of pathway enzymes; all these active effectors are not needed for full repression.

Multivalent repression. Similar to cumulative repression in that all end products of a branched pathway are required for full repression, but differing from cumulative repression in that none of the end products represses at all.

Mutation. Any detectable, heritable change in the genetic material transmitted to succeeding generations and giving rise to mutant cells or constituents.

Myoblast. A mesodermal cell at an early stage of differentiation into muscle.

Myotube. A multinucleate structure formed by the fusion of a number of myoblasts. The myotube is a step in the development of muscle fiber.

Neuro-transmitter. A chemical mediator of nerve transmission, released by one neuron (nerve cell) and affecting an adjoining neuron.

Oocyte. Developing female germ cell, up to the point of fertilization.

Operon. A linear array of structural genes whose expression is coordinately controlled by their associated regulatory genes.

Organelle. Any structure of characteristic morphology and function within the cytoplasm of the cell.

Ouchterlony technique. A gel-diffused antibody-antigen precipitation test depending upon horizontal diffusion from two or more opposite sources.

Overshoot. The phenomenon where after starvation and refeeding the enzyme activity exceeds the pre-starvation values.

Perfusion. Artificial passage of fluid through blood vessels of an organ or animal.

Peroxisomes. A class of subcellular particles containing catalase and oxidative enzymes.

Phenotype. The observable structure and functional properties of an organism, produced by the interaction between the organism's genetic potential and the environment upon which it finds itself.

Plasma membrane. A membrane surrounding the outer layer of the cell's cytoplasm.

Polyribosome. A cluster of ribosomes strung along a strand of messenger RNA which may occur free in the cytoplasm or attached to endoplasmic reticulum ('membrane-bound'); the site of cellular protein synthesis.

Polysome. See *Polyribosome.*

Post-transcriptional regulation. The regulation of protein synthesis at a step beyond gene transcription, beyond mRNA synthesis.

Prednisolone. An analogue of hydrocortisone.

Preimplantation. The period before attachment occurs, in those mammals in which the conceptus becomes embedded in the uterine endometrium. In the mouse, implantation begins on day 5 after conception.

Producer gene. A gene carrying the structural information of a certain protein.

Prokaryotes. Organisms which lack the eukaryote organization of the genetic material, e.g. no membrane bounded nuclei, etc. (see *Eukaryotes*).

Protein hormones. Those hormones having a structure made up of several amino acids arranged as a polypeptide, e.g. growth hormone, insulin, and glucagon.

Protein turnover. Concurrent synthesis and degradation of protein within the cell.

Puromycin. An antibiotic which blocks protein synthesis by forming a peptidyl-puromycin compound which causes the growing polypeptide to be released prematurely from the ribosome.

Rate constant for enzyme degradation. The fraction of enzyme molecules being destroyed per unit time.

Rate constant for enzyme synthesis. The amount of enzyme activity being synthesized per unit of time.

Receptor gene. A gene in a set of genes which controls the function (transcription) of the producer (structural genes) of the same gene set.

Redundancy. A repetitive DNA sequence, the length of which varies; the genome of higher organisms like mammals contains many such sequences, which recur anywhere from a thousand to a million times per cell.

Regulatory genes: A linear sequence of DNA encoding for a gene product, either RNA or protein, which regulates the expression of certain structural genes.

Regulon. A set of structural genes, situated in different genome regions but having a common mechanism for the regulation of their expression.

Repressible enzymes. The biosynthesis of these enzymes decreases or is completely abolished, when an effector concentration in the cell increases high enough. Such effector is often a product of the repressible enzyme or of the whole pathway into which the enzyme belongs.

Repressor. A small molecule that exerts an inhibitory action on the rate of synthesis of an enzyme; it may act either at the translational or the transcriptional level.

Reticulum. Network of intracellular membranes seen in most mammalian cells.

Ribosomes. Particles composed of RNA and proteins involved in protein synthesis.

Sensor gene. This gene controls the activity of one or several integrator genes.

Sequential repression. This type of repression occurs in some pathways in which the structural genes are located in separate functional genetic units. Each intermediary product in this case represses the synthesis of the enzyme responsible for its own biosynthesis.

Somite. A block of mesoderm lying alongside the neural tube in an early stage of vertebrate development. One pair of somites forms in each segment of the body, their appearance occurring in antero-posterior progression. Each somite gives rise to a mass of loose wandering cells that cluster around the notochord and differentiate into the cartilage of the vertebral column; most of the remaining somite tissue becomes skeletal musculature.

Steady state. A state of quasi-equilibrium in which the input into the system equals the output, such as the steady state wherein enzyme synthesis equals enzyme degradation.

Stem cells. A population of undifferentiated cells that can undergo proliferation to supply new cells that differentiate to become part of a continuously renewed tissue (e.g. blood, intestinal epithelium).

Structural gene. Linear sequence of DNA encoding information for the amino acid sequence of a structural or catalytic polypeptide.

Superinduction. A further increase in cellular content of an enzyme beyond the 'maximally' induced amount, brought about, for example, by complete inhibition of RNA synthesis by actinomycin or other inhibitors.

Suppression. If a mutation has caused a loss of a certain protein or an appearance of an inactive or impaired protein, the lost biological function may be restored by another mutation which suppresses the first mutation. This suppression may be due to another mutation in the same gene, which restores the logic of the code, or to an intergenic suppressive mutation, which may be specific or nonspecific in nature. Since the same biological phenomenon may have a great variety of molecular explanations in different cases, we suggest a good textbook of biochemistry or molecular biology for further reference.

Synchronized cells. Cells which are all in the same phase of the cell cycle.

Template. Structure giving pattern to subsequent synthetic products, such as DNA templates, messenger RNA templates, or membrane templates.

Temporal genes. Genes that program the system in time determining the activation of structural, architectural and regulatory genes at different developmental stages in various cell types.

Transcription. The synthesis of a polyribonucleotide complementary in sequence to the DNA template; the transmission of the genetic information encoded in the DNA to an RNA intermediate.

Transduction. The transfer of DNA or a gene from a donor cell to a recipient cell; or the change of a signal from one form to another.

Translation. The synthesis of a polypeptide whose amino acid sequence is determined by its RNA template; the transmission of the genetic information encoded in an RNA intermediate to the amino acid sequence of a polypeptide.

Turnover. The general phenomenon whereby tissue constituents are continually being synthesized and degraded.

Turnover number. The number of molecules of a substrate transformed per minute by a single enzyme molecule under optimum conditions.

Ventromedial hypothalamus. A region of the brain which has as one of its functions the regulation of food intake.

Yolk sac. In mammals, a hollow structure that is homologous with the yolk-filled sac that provides nourishment for avian and reptilian embryos. In primate and rodent embryos the yolk sac is very small, but serves as a source of the earliest red blood cells, as well as the primordial germ cells.

Zero-order reaction. A reaction which has a constant rate.

Zygote. The newly fertilized egg of any animal.

Zymogen. The mother substance of an enzyme; precursor of an enzyme.

Zymogen granules. Enzyme-containing particles elaborated by the cells of the pancreas.

Author Index

Numbers in italics indicate the pages on which the references are listed

Subject Index